INTRODUCTION TO NANOMATERIALS AND DEVICES

INTRODUCTION TO NANOMATERIALS AND DEVICES

OMAR MANASREH

WILEY

A JOHN WILEY & SONS, INC., PUBLICATION

Published by John Wiley & Sons, Inc., Hoboken, New Jersey
Published simultaneously in Canada

Library of Congress Cataloging-in-Publication Data:

Manasreh, Mahmoud Omar.
Introduction to nanomaterials and devices / Omar Manasreh. – 1st ed.
p. cm.
Includes bibliographical references.
ISBN 978-0-470-92707-6 (hardback)
1. Nanostructured materials. 2. Optoelectronic devices. 3. Semiconductor nanocrystals. 4. Quantum electronics. I. Title.
TA418.9.N35M37 2012
620.1′15–dc23

2011022713

Printed in the United States of America

oBook ISBN: 978-1-118-14841-9
ePDF ISBN: 978-1-118-14837-2
ePub ISBN: 978-1-118-14840-2
Mobi ISBN: 978-1-118-14839-6

10 9 8 7 6 5 4 3 2 1

To my wife Taeko who focuses me on the essentials

CONTENTS

PREFACE

Investigating materials and devices at the nanoscale level has become the topic of discussion in our daily life even at the dinner table. The behavior of nanoscale materials is close, to atomic behavior rather than that of bulk materials. This leads to vivid properties and well-defined concepts, even though the description of these properties can be understood in terms of quantum mechanics, which provides only a fuzzy picture. The growth of nanomaterials, such as quantum dots (also known as *atomic designers*), has the tendency to be viewed as an art rather than science. These nanostructures have changed our view of Nature. The invention of the transmission electron microscope, scanning electron microscope, and atomic force microscope provided a method for us to observe materials down to the atomic scale, and yet these microscopes are based on quantum mechanics concepts.

To understand quantum wells, wires, and dots, it is imperative to possess a basic knowledge of quantum mechanics and how one can apply the Schrödinger equation to calculate the quantized electronic energy levels in such a tiny structure. This requirement is due to the fact that classical mechanics is limited in providing an explanation to almost all the properties of the nanomaterials. Quantum mechanics, however, can provide an insight and reasonable predictions of phenomena observed in case of semiconductor nanomaterials. This book is by no means a complete or ideal textbook, but it is one step in a changing field full of limitless possibilities of innovations and inventions.

This book is designed to cover topics on the subject of nanomaterials growth, electrical and optical properties of nanomaterials, and devices based on these nanomaterials. It is designed to provide an introduction to nanomaterials and devices to graduate students in electrical engineering, materials engineering, and applied physics. Advanced undergraduate students as well as researchers in the field of semiconductor heterojunctions and nanostructures may benefit from it. I imagine that graduate students, who use this textbook in their studies, will continue to use it as a reference book after their graduation.

The basic properties of bulk and nanomaterial systems are discussed in order to point out the advantages of nanomaterials. This book is structured such that the discussion starts with bulk crystalline materials, which is the basis for understanding the basic properties of semiconductor. The discussion then evolves to cover quantum structures, such as single and multiple quantum wells. Then, attempts are made to discuss and explain the properties of nanomaterial systems, such as quantum wires and dots. However, since the field is still in its infancy, there are too many unknowns and many of the properties of the nanomaterial systems are yet to be understood or have even reached their full potential. Thus, the discussion regarding quantum wires and dots is limited to the more mature properties of these quantum structures.

Topics covered include an introduction to growth issues, quantum mechanics, quantization of electronic energy levels in periodic potentials, tunneling, distribution functions and density of states, optical and electronic properties, and devices. The chapters are devoted to the growth of bulk materials and nanomaterials; the introduction of quantum mechanics; calculations of the energy levels in periodic potentials, quantum wells, and quantum dots; derivation of the density of states in bulk materials, quantum wells, wires, quantum dots, and the density of states under the influence of electric or magnetic fields; optical properties; electrical and transport properties; electronic devices based on heterojunctions and nanostructures such as ohmic and Schottky contacts, diodes, resonant tunneling diodes, MODFETs, HFETs, Coulomb blockade, and single-electron transistors (known as *SETs*); and optoelectronic devices such as light-emitting diodes, photodetectors based on quantum wells and quantum dots, edge-emitting lasers, VCSELs, quantum cascade lasers, and laser diodes based on quantum dots. End of chapter problems, appendices, tables, and references are included.

Students registering for courses based on this textbook are required to have a basic knowledge of semiconductor materials and devices. Although knowledge in quantum mechanics is not required, it is, however, recommended that the students take undergraduate physics courses, such as university physics and/or modern physics. Chapter 1 of this book covers the growth of various material systems ranging from bulk materials to nanorods. Chapter 2 covers the basic formalism of quantum mechanics needed for a student in electrical engineering to grasp the basic idea of how to calculate the energy levels in a simple quantum well. To understand and appreciate the beauty of quantum transport in quantum structures, the readers must have some knowledge of the classical-type transport. This leads us to focus on both quantum transport, such as tunneling and coherent transport in mesoscopic systems, and classical transport, such as Boltzmann transport equation formalisms. Density of states and distribution functions are discussed in Chapter 3. Chapter 4 focuses on the optical properties, while Chapter 5 is directed toward understanding the electrical properties. In Chapter 6, we discuss several electronic devices, and optoelectronic devices are introduced in Chapter 7.

When an electronic or optoelectronic device is under the influence of an applied electric field and/or photonic excitation, the device is no longer at equilibrium, and its transport properties become more complicated. The limits of various transport regimes, which are classified according to the electron phase coherent length and compared to the de Broglie wavelength, are discussed in Chapter 5. Various scattering mechanisms, which dominate the classical regime, are also discussed. When a nanomaterial possesses a capacitance on the order of *atto Farad*, a new transport phenomenon occurs, which is known as *Coulomb blockade*. This phenomenon, in conjunction with the quantum tunneling effect, forms the basis for single-electron transistors. Discussion regarding this new class of devices is presented in Chapter 6. While these devices have the potential to revolutionize the current technology, it should be pointed out that current technology is still based on carrier-injected and CMOS devices, in which transport is dominated by carrier scattering rather than by ballistic or coherent transports.

In addition to electronic devices, a new generation of optoelectronic devices is under intense research. Recently, we heard of topics such as long-wavelength infrared detectors based on intersubband transitions, edge-emitting quantum well laser diode, vertical cavity surface-emitting lasers, and quantum cascade lasers. All these devices are discussed in Chapte 7.

Excitons play a major role in optoelectronic and photonic devices. Theoretically, the exciton binding energy in a quantum well is larger than that of excitons in the constituent bulk materials by a factor of four as discussed in Chapter 4. Furthermore, it is predicted that the exciton binding energies are even higher in quantum wires and dots. This can be translated to very fast optoelectronic devices that can operate at room temperature. We have presented a detailed discussion and derivation of the exciton binding energies in direct band gap bulk semiconductors, quantum wells, and quantum dots.

OMAR MANASREH
University of Arkansas
Spring, 2011

FUNDAMENTAL CONSTANTS

Name	Symbol	Value	CGS	SI
Velocity of light in vacuum	c	2.997925	10^{10} cm/s	10^{8} m/s
Electron or proton charge	e	1.60219	—	10^{-19} C
	—	4.803259	10^{-10} esu	—
Electron rest mass	m_0	9.110	10^{-28} g	10^{-31} kg
Planck's constant	h	6.62620	10^{-27} erg s	10^{-34} J s
	$h = h/(2\pi)$	1.05459	10^{-27} erg s	10^{-34} J s
	h/e	6.58218×10^{-16} eV	—	—
Avogadro's number	N_a	6.02217×10^{23} mol^{-1}	—	—
Gas constant	R	8.31433	10^{7} erg/mol/K	J/mol/K
Boltzmann constant	$k_B = R/N_a$	1.38062	10^{-16} erg/K	10^{-23} J/K
	k_B/e	8.6171×10^{-5} eV	—	—
Permittivity of vacuum	$\epsilon_0 = 10^7/$	8.85433	10^{-14} F/cm	10^{12} F/m
Permeability of vacuum	$\mu_w = 4\pi \times 10^{-7}$	1.25664	10^{-8} H/cm	10^{-6} H/m
Electron volt	eV	1.60219	10^{-12} erg	10^{-19} J
Frequency	eV/h	2.41797	10^{14} Hz	10^{14} Hz
Wavelength	$hc/(eV)$	1.23985	10^{-4} cm	10^{-6} m
Wavenumber	$eV/(hc)$	8.06546	10^{3} cm^{-1}	10^{5} m^{-1}
Temperature	eV/k_B	1.16048	10^{4} K	10^{4} K
Thermal energy at 300 K	k_BT/e	0.0258512 eV	—	—
Atomic mass unit	amu	1.660531	10^{-24} g	10^{-27} kg

SI UNITS

Quantity	Name	Symbol	Fundamental SI Units	Other
Frequency	Hertz	Hz	s^{-1}	—
Force	Newton	N	$Kg\ m/s^2$	—
Pressure	Pascal	Pa	$Kg/m/s^2$	N/m^2
Energy	Joule	J	$Kg/m^2/s^2$	N m
Power	Watt	W	$Kg\ m^2/s^3$	J/s
Electric charge	Coulomb	C	A s	A s
Electrical potential	Volt	V	$Kg\ m^2/s^3/A$	W/A
Capacitance	Farad	F	$s^4\ A^2/m^2/\ kg$	C/V
Electric resistance	Ohm	Ω	$Kg\ m^2/s^3/A^2$	V/A
Electric conductance	Siemens	S	$Kg^{-1}/m^2\ s^3\ A^2$	A/V
Magnetic flux	Weber	Wb	$Kg\ m^2/s^2/A$	V s
Magnetic induction	Tesla	T	$Kg/s^2/A$	Wb/m^2
Inductance	Henry	H	$Kg\ m^2/s^2/A^2$	Wb/A

PREFIXES AND THEIR SYMBOLS FOR DECIMAL MULTIPLES AND SUBMULTIPLES OF UNITS

Factor	Name	Symbol	Factor	Name	Symbol
10^{24}	yotta	Y	10^{-24}	yocto	y
10^{21}	zetta	Z	10^{-21}	zepto	z
10^{18}	exa	E	10^{-18}	atto	a
10^{15}	peta	P	10^{-15}	femto	f
10^{12}	tera	T	10^{-12}	pico	p
10^{9}	giga	G	10^{-9}	nano	n
10^{6}	mega	M	10^{-6}	micro	μ
10^{3}	kilo	k	10^{-3}	milli	m
10^{2}	hecto	h	10^{-2}	centi	c
10^{1}	deka	da	10^{-1}	deci	d

EFFECTIVE MASSES, IN UNIT OF M_0, IN SELECTED SEMICONDUCTORS

Semiconductor	Electron Effective mass	Heavy Hole Effective Mass	Light Hole Effective Mass
GaAs	0.067	0.45	0.082
AlAs	0.124	0.5	0.22
InP	0.077	0.60	0.012
InAs	0.026	0.41	0.025
InSb	0.014	0.44	0.016
GaSb	0.043	0.33	0.056
AlSb	0.12	—	—
GaP	0.17	0.67	—
GaN	0.20	0.80	0.30
InN	0.11	0.50	0.17
AlN	0.27	—	0.25
Si	$m_l = 0.98$	0.49	0.16
	$m_t = 0.19$	—	—
SiC	0.60	—	—
ZnO	0.28	0.59	—
ZnS	0.28	0.49	—
CdS	0.14	0.51	—
CdTe	0.09	0.40	—

BAND GAP, LATTICE CONSTANT, THERMAL CONDUCTIVITY, AND DENSITY OF SELECTED SEMICONDUCTORS

Semiconductor	Band Gap (eV)	Lattice Constant (nm)	Thermal conductivity (W/cm/C°)	Density (g/cm^3) (300 K)
GaAs (ZB, D)	1.424 (300 K)	0.5653	0.50	5.318
	1.519 (0 K)	—	—	—
AlAs (ZB, I)	2.153 (300 K)	0.5660	—	3.717
	2.229 (2 K)	—	—	—
GaSb (ZB, D)	0.75 (300 K)	0.609	—	5.63
	0.811 (2 K)	—	—	—
GaP (ZB, I)	2.270 (300 K)	0.5451	0.50	4.129
	2.350 (0 K)	—	—	—
InP (ZB, D)	1.344 (300 K)	0.586	—	4.81
	1.424 (2 K)	—	—	—
InAs (ZB,D)	0.36 (300 K)	0.6050	—	5.69
	0.418 (2 K)	—	—	—
InSb (ZB, D)	0.17 (300 K)	0.647	—	5.80
	0.237 (2 K)	—	—	—
AlSb (ZB, I)	1.615 (300 K)	0.61355	—	4.29
	1.686 (30 K)	—	—	—

Semiconductor	Band Gap (eV)	Lattice Constant (nm)	Thermal conductivity (W/cm/C°)	Density (g/cm^3) (300 K)
Si (Di, I)	1.12 (300 K)	0.54311	1.48	2.329
	1.17 (2 K)	—	—	—
Ge (Di, I)	0.66 (300 K)	0.5658	—	5.3234
GaN (W, D)	3.44 (300 K)	$a_0 = 0.3189$	2.20	6.095
	—	$c_0 = 0.5185$	—	—
AlN (W, D)	6.2 (300 K)	$a_0 = 0.3111$	<3 0	3.255
	6.28 (10 K)	$c_0 = 0.4978$	—	—
InN (W, D)	0.69-0.9 (300 K)	$a_0 = 0.3544$	0.80	6.81
	—	$c_0 = 0.5718$	—	—
ZnO (W, D)	3.37 (300 K)	$a_0 = 0.32495$	0.60	5.606
	3.438 (2 K)	$c_0 = 0.52069$	—	—
ZnS (W, D)	3.741 (300 K)	$a_0 = 0.3811$	—	4.11
	3.84 (4 K)	$c_0 = 0.6234$	—	—
SiC (H, I)	4H: 3.26 (300 K)	$a_0 = 0.30730$,	3.7	—
		$c_0 = 1.0053$		—
	6H: 3.03 (300 K)	$a_0 = 0.30806$,	3.8	3.211
		$c_0 = 1.51173$		—

Abbreviations: D, direct, Di, diamond; H, hexagonal; I, indirect; W, wurzite; ZB, zinc blende.

DIELECTRIC CONSTANT OF SELECTED SEMICONDUCTORS

Semiconductor	Dielectric Constant
AlP	9.8
AlAs	10.06
AlSb	12.04
GaP	11.1
GaAs	12.5
GaSb	15.7
InP	12.4
InAs	14.6
InSb	17.7
Si	11.9
Ge	16.0
GaN (W)	Average: 10.0
AlN (W)	Average: 9.14
InN	Average: 15.0
ZnO	7.8
ZnS	9.6
CdTe	9.8
CdS	9.4
CdSe	9.3
SiC (6H)	$9.66 \perp c$-axis
	$10.3 \| c$-axis

1

GROWTH OF BULK, THIN FILMS, AND NANOMATERIALS

1.1 INTRODUCTION

The evolution of the growth of the high quality semiconductor materials from bulk to nanomaterials enables researchers to fabricate devices with a continued enhancement of the properties and performance. In this chapter, the discussion is directed toward the growth of semiconductor single crystals by using various techniques ranging from bulk crystal growth to the epitaxial growth of quantum dots and core/shell nanocrystals. Bulk crystal growth techniques include liquid-encapsulated Czochralski (LEC), horizontal Bridgman (HB), liquid-encapsulated Kyropoulos (LEK), and vertical gradient freezing (VGF) methods. There are many improved methods available for the growth of bulk semiconductor crystals. For example, magnetic LEC, direct synthesis-LEC, pressure-controlled LEC, and thermal baffle LEC methods are all variations of the original LEC technique, but with improved growth conditions. Other bulk growth techniques include dynamic gradient freezing, horizontal gradient freezing, magnetic LEK, and vertical Bridgman methods. The widely used epitaxial growth techniques are the molecular beam epitaxy (MBE), metal–organic chemical vapor deposition (MOCVD) techniques, and liquid-phase epitaxy (LPE). The word *epitaxy* is a Greek origin composed of two parts, *epi* (placed or resting on) and *taxis* (arrangement). Thus, *epitaxy* refers to the formation of single-crystal nanomaterials on top of a substrate. The techniques used to grow bulk materials are usually equilibrium growth techniques, while the epitaxial growth techniques, used for the production of nanomaterials, are considered nonequilibrium techniques.

Introduction to Nanomaterials and Devices, First Edition. Omar Manasreh.

The growth techniques of bulk semiconductor crystals are designed to produce large-volume crystals under equilibrium conditions with almost no flexibility in the production of alloy composition. These techniques, however, lack the ability to produce heterojunctions, ternary or quaternary semiconductor compounds needed for advanced semiconductor devices. Silicon single-crystal boules as large as 12 in. (~300 mm) in diameters and over a meter in length are currently produced by LEC technique. GaAs single-crystal boule diameter is usually smaller than that of Si boules. Epitaxial growth is performed on submillimeter thick substrates cut from these bulk boules.

The process of preparing the boules into wafers that are used as substrates for epitaxial growth is called the *wafering process*. This process includes slicing, lapping, polishing, and cleaning. Since most wafers are used as substrates for epitaxial growth, the wafering process technology and the bulk crystal growth are very important for successful epitaxial growth. For example, the surface orientation accuracy, which is determined during the slicing process, affects the morphology of the epitaxial layer surface. Wafer flatness is another important parameter for high quality epitaxial growth. Single- or double-sided polished wafer flatness is defined by specific parameters, such as total thickness variation, total indicator reading, or focal plane deviation. These parameters are needed for precise photolithography. The surface roughness is also important aspect of the wafering process, since surface roughness in a subnanometer scale is required for many epitaxially grown nanomaterials.

Epitaxial growth, such as MBE and MOCVD growth, requires ready-to-use wafers. For example, thermal oxidation and/or ultraviolet/ozone oxidation processes have been effective in producing thin oxide layers, which protect the wafer surface. These oxide layers can be removed by heating prior to epitaxial growth. Packaging the wafers in nitrogen gas is an effective method against residual oxidation of polished surfaces during storage.

Crystallographic orientation of the wafers is very important for the MBE and MOCVD growth methods. The orientation is determined by Miller indices. These indices are defined as the smallest possible integers with the same rations as the inverse of the intersection of a plane with a set of axes defined by the unit vectors of the crystal. An illustration of this concept is shown in Fig. 1.1 for a cubic crystal where a plane is intersecting x-, y-, and z-axes at a distance $2a$, $3a$, and $4a$, respectively, where a is the interatomic distance. To obtain the miller indices, one may follow the following steps: identify the intersections of the plane with the axes (in the case of Fig. 1.1, these intersections are 2, 3, and 4); take the inverse of these intercepts, which results in 1/2, 1/3, and 1/4; find the smallest multiplier factor of the denominators, which 12; and multiply the factor the inverse of the intercepts to give 6, 4, and 3. These last numbers are called *miller indices*, and they are usually written in the following format (643) to indicate the crystallographic orientation of the wafers. If the intercept is negative, then the negative sign "–" is usually place on the top of the index. A group of Miller indices, such as (100), (010), (001), $(\bar{1}00)$, $(0\bar{1}0)$, and $(00\bar{1})$,

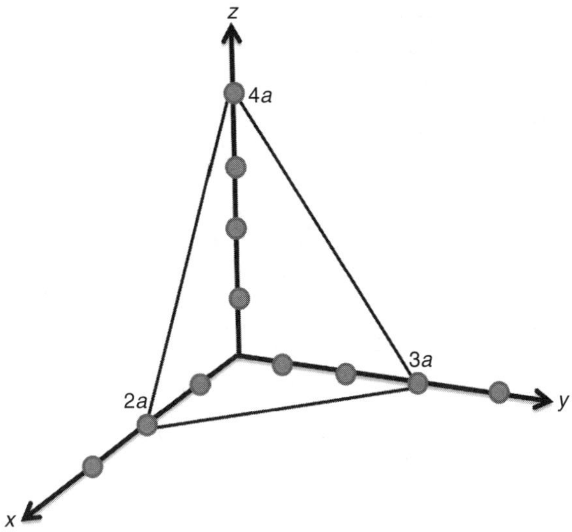

FIGURE 1.1 A plane intersecting the x-, y-, and z-axes at $2a$, $3a$, and $4a$, respectively, to determine the Miller indices. The interatomic distance is designated as "a."

is given the following notation $\{100\}$. For hexagonal crystal structure, the Miller indices are $(a_1a_2a_3c)$.

Most of the semiconductor materials are produced by artificial methods. Semiconductor binary, ternary, quaternary alloys (Fig. 1.2), heterojunctions, and other quantum structures such as superlattices and quantum dots are currently grown by two main epitaxial growth techniques, namely MBE and MOCVD. These growth techniques enable the synthesis of high quality single-crystal nanomaterials deposited layer by layer on suitable substrates.

Both equilibrium and nonequilibrium growth of semiconductors are based on chemical reactions. Furthermore, the thermodynamic analysis provides information about the feasibility of chemical reactions that can lead to the production of compound semiconductors. For example, the free energy function, G, can be written as

$$G = H - TS, \tag{1.1}$$

where H is the enthalpy, S is the entropy, and T is the temperature. For a system that is undergoing a chemical reaction, the change in the free energy can be written as

$$\Delta G = G_\mathrm{f} - G_\mathrm{i} = \Delta H - T\Delta S, \tag{1.2}$$

where G_i and G_f are the free energy of the initial and final states of the reaction, respectively. According to the second law of thermodynamics, "*in all energy exchanges, if no energy enters or leaves the system, the potential energy of the final state will always be less than that of the initial state*," or $G_\mathrm{f} < G_\mathrm{i}$. Thus, the

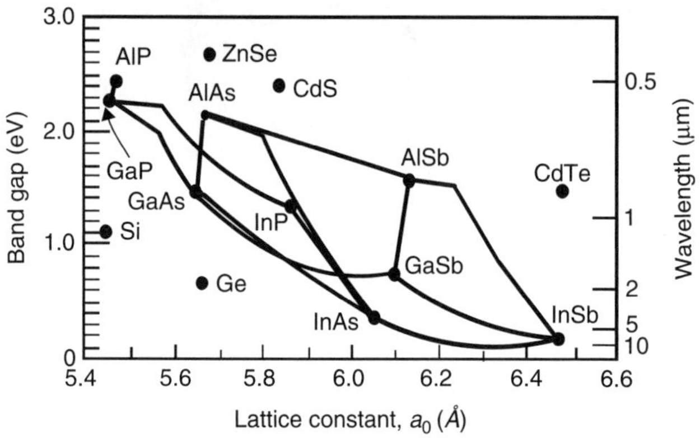

FIGURE 1.2 The band gap as a function of the lattice constant plotted for several binary semiconductors. Silicon and Germanium are also shown. The solid lines represent the ternary compounds.

system tends to minimize its free energy to a lower value than the initial state. For a forbidden process, we have $\Delta G > 0$, and for a system at equilibrium, the change in the free energy is zero ($\Delta G = 0$).

Let us consider the following chemical reaction between materials X and Y, which yields material Z,

$$xX + yY \rightarrow zZ, \tag{1.3}$$

where x, y, and z are called the *stoichiometric coefficients*. If one assumes that the materials X, Y, and Z are at equilibrium, then the change in the free energy is given by

$$\Delta G = zG_z - xG_x - yG_y. \tag{1.4}$$

The free energy of individual reactants is usually given as

$$G_j = G_j^{\mathrm{o}} + RT \ln a_j, \tag{1.5}$$

where j is the reactant (X, Y, or Z), a_j is called the *activity*, which reflects the change in the free energy when the material is not in its standard state, G_j^{o} is the free energy of the jth reactant in its standard state, and R is the gas constant [8.3143 J/(K/mol), 1.9872 cal mol^{-1}.K^{-1}, or 62.363 L mol^{-1}K^{-1}]. The standard state is defined as one atmospheric pressure for a gas at a temperature of 25°C. Substituting Equations (1.5) into (1.4) and solve for $\Delta G = 0$ to obtain

$$-\Delta G^{\mathrm{o}} = RT \ln k, \quad \text{where} \quad k = \frac{(a_Z)^z}{(a_X)^x (a_Y)^y}. \tag{1.6}$$

The values of a_j (j = X, Y, Z) are usually taken while the system is at the equilibrium.

In addition to the above simple thermodynamic description, the inspection of changes of composition of a material from one phase to another is usually accomplished by visualizing the phase diagram. The phase diagram helps understanding the chemical and physical properties of the material and how one can produce nanomaterials. For example, when a material fails to perform, one can refer to the phase diagram and deduce what might have happened to cause the failure. Then, one can revisit the thermodynamic laws that govern the phase diagram and extrapolate information.

1.2 GROWTH OF BULK SEMICONDUCTORS

Single-crystal growth of various bulk semiconductor materials has been performed by several methods, such as the LEC method, modified LEC methods, VGF methods, and HB methods. In this section, the most commonly used methods are discussed.

1.2.1 Liquid-Encapsulated Czochralski (LEC) Method

This growth method of bulk semiconductors was first developed by Czochralski in 1916. It uses what is called a *crystal puller* as illustrated in Fig. 1.3. The crystal puller consists of a high purity quartz crucible filled with polycrystalline materials, which are heated above their meting point by induction using radio frequency (RF) energy. The crucible holder is usually made of graphite. What is called a *seed*, or a small single crystal with a specific orientation is lowered into the molten material and then drawn upward using a pulling–rotation mechanism. The material in the melt makes a transition into a solid-phase crystal at the solid–liquid interface. The new solid-phase crystallographic structure is a replica of that of the seed crystal. During the growth process, the crucible rotates in one direction (12–14 rotations/min) while the seed holder rotates in the opposite direction (6–8 rotation per minute). At the same time, the boule is slowly pulled upward. The crystal diameter is usually monitored by an optical pyrometer, which is focused at the interface between the crystal (boule) and the melt. An automatic diameter control system maintains the desired crystal diameter through a feedback loop control. An inert gas such as argon is usually used as the ambient gas during the crystal-pulling process.

In the LEC crystal growth, boric oxide (B_2O_3) is used as an encapsulant to prevent the decomposition of the melt. Boric oxide is extracted as $Na_2B_4O_7$ solution from minerals including boron and then precipitated as boric acid (H_3BO_3). Boric acid is refined by repeated recrystallization and dehydrated by heating. The purity of boric oxide is very important since the impurities in boric oxide could contaminate the melt.

The molten semiconductor and solid are usually kept at the same pressure and have approximately the same composition. The crystallization resulted from a

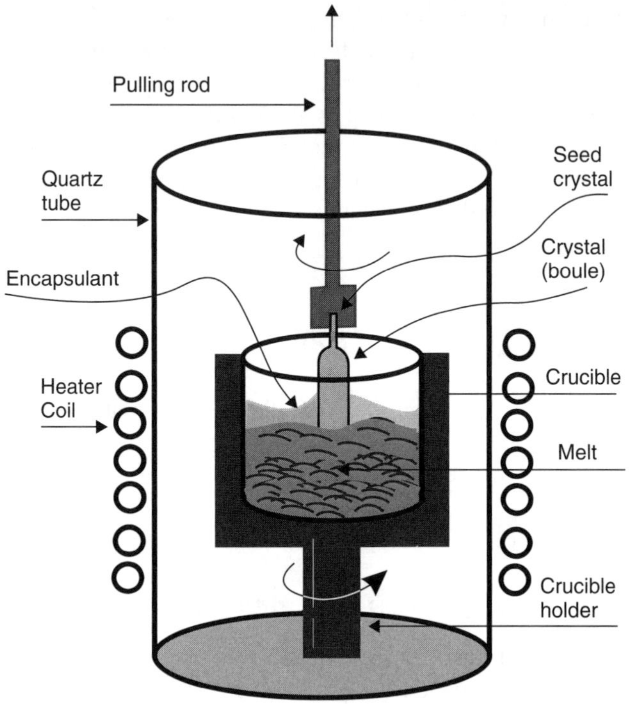

FIGURE 1.3 A illustration of a furnace (crystal puller) typically used in LEC single-crystal growth.

reduction in temperature; as the melt is pulled up, it loses heat by radiation and convection to the inert gas (for example, argon). The heat loss to the inert gas causes a substantial thermal gradient across the liquid–solid interface. Additional energy loss is due to solidification (latent heat of fusion). For a fixed volume, a one dimensional energy balance for the interface can be expressed as

$$\left[-k_l A \left.\frac{\mathrm{d}T}{\mathrm{d}x}\right|_l\right] - \left[-k_s A \left.\frac{\mathrm{d}T}{\mathrm{d}x}\right|_s\right] = L\frac{\mathrm{d}m}{\mathrm{d}t}, \tag{1.7}$$

where k_l and k_s are the thermal conductivities at the melting point of the liquid and solid, respectively, A is the boule's cross-sectional area, T is the temperature, L is the latent heat of fusion, m is the mass of the growing solid, and t is the time. Generally speaking, the heat diffusion from the liquid is small as compared to that from the solid. Thus, Equation (1.7) can be approximated by neglecting the first term on the left-hand side. With this approximation, one can express the maximum velocity, $\upsilon_{\max}$, at which the solid can be pulled as

$$\upsilon_{\max} \approx \frac{k_s A}{L}\frac{\mathrm{d}T}{\mathrm{d}m} = \frac{k_s}{M_\upsilon L}\left.\frac{\mathrm{d}T}{\mathrm{d}x}\right|_s, \tag{1.8}$$

where M_v is the solid density of the crystal being grown. If the crystal is pulled with a velocity larger than this maximum velocity, it will not conduct heat fast enough and the formation of a single crystal becomes difficult to achieve. In general, the pull rate of the *seed* crystal varies during the growth cycle. It is faster when growing the narrow neck so that the generation of dislocations is minimized and slower during the growth of the boule. Figure 1.4 shows a picture of 75-mm diameter and 240-mm long InP single-crystal grown by pressure-controlled LEC method (Oda *et al.*). The seed, neck, and shoulder of the crystal are indicated.

Crystals grown by LEC technique are susceptible to the incorporation of unwanted impurities. For example, quartz is used as a crucible when growing silicon crystals and the growth temperature is on the order of 1500°C. Thus, a small amount of oxygen will be incorporated into the boule. For extremely low concentration of oxygen impurities in silicon, the boule can be grown under the influence of a magnetic field. The magnetic field is usually directed perpendicular to the pull direction, where Lorentz force will change the motion of the ionized impurities in the melt in such a way as to keep them away from the liquid–solid interface. This configuration leads to a substantial decrease in the impurity incorporation in the crystal.

The LEC growth of compound semiconductors, such as GaAs and InP, is more difficult when compared to that of silicon crystals. For example, pyrolytic boron nitride (PBN) crucibles are used for compound semiconductors instead of quartz crucibles and B_2O_3 is used as encapsulant. The thermal conductivity of GaAs ($\sim$0.5 W.cm^{-1}.K^{-1}) is about one-third that of silicon ($\sim$1.4 W/cm/K). Thus, GaAs cannot dissipate the latent heat of fusion as fast as silicon. Additionally, the sheer stress required to generate a dislocation in GaAs at the melting point is about one-fourth that in silicon. These thermal and mechanical properties only permit the growth of 125-mm diameter GaAs boules as compared to that of silicon of 300 mm diameter. Furthermore, the GaAs boules contain defect densities a few orders of magnitude higher than those defect densities found in silicon boules.

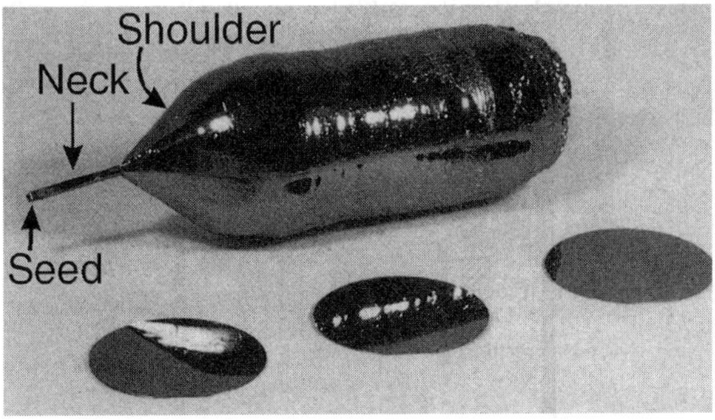

FIGURE 1.4 A 75-mm diameter and 240-mm long InP single-crystal grown by pressure-controlled LEC method (after Oda *et al.*).

In addition to the above difficulties of LEC semiconductor growth, there are other issues worth mentioning. One of them is called *stacking fault energy*. Small stacking fault energy promotes the formation of twins. This is the main reason why many of the compound crystals are difficult to grow as single crystals. An example is InP where twining, which is more common in InP as compared to GaAs, is the main growth obstacle. The stacking fault energy in InP is smaller than those of GaAs or GaP. Another issue worth mentioning here is the sheer stress, which is a major factor in predicting the generation of dislocations. Most compound semiconductors have lower shear stress as compared to that of silicon. Thus, the reduction of dislocation densities becomes a key issue for realizing high quality materials.

Doping single crystals during LEC growth is very important since it is desired to produce n-type, p-type, or semi-insulating substrates. For example, introducing boron or phosphorus into silicon melt produces p-type or n-type silicon, respectively. In the case of GaAs, the undoped or Cr-doped material is usually semi-insulating. Adding silicon as a dopant to GaAs produces n-type materials, while the addition of carbon or beryllium produces p-type GaAs. During LEC growth, the dopant concentration in the boule is usually different from the dopant concentration in the melt. The ratio between the two concentrations is known as the *equilibrium segregation coefficient*, k_0, which expressed as

$$k_0 \equiv \frac{C_s}{C_l}, \tag{1.9}$$

where C_s and C_l are the equilibrium dopant concentrations in the solid and liquid in the vicinity of the interface. Usually k_0 is less than unity. For example, k_0 values for B and P dopants in silicon are 0.80 and 0.35, respectively, while the values of k_0 for Si and C dopants in GaAs are 0.185 and 0.8, respectively.

It is desired to obtain an expression for the dopant concentration in the solid crystal as it is pulled out of the melt during the LEC growth. Assume that the initial crystal weight and dopant concentration in the crystal are m_0, C_s, respectively. Also, assume that the amount of dopants by weight remaining in the melt is σ when the crystal weight increases to m during growth. When the crystal weight increases by the amount dm, the corresponding reduction of the dopant weight from the melt is

$$-\mathrm{d}\sigma = C_s \mathrm{d}m. \tag{1.10}$$

On the other hand, the remaining weight of the melt is $(m_0 - m)$. Hence, the doping concentration in the liquid, C_l, is given by

$$C_l = \frac{\sigma}{m_0 - m}. \tag{1.11}$$

By combining Equations 1.9–1.11, the reduction of the dopant (by weight) in the melt can be written as

$$\frac{d\sigma}{\sigma} = -k_0 \frac{dm}{m_0 - m}. \tag{1.12}$$

Integrate Equation (1.12) using $C_0 m_0$ and σ for the initial and final weights of the dopant in the melt, respectively, and the initial and final weights of the crystal are 0 and m, respectively. The final result is given as

$$C_s = k_0 C_0 \left[1 - \frac{m}{m_0}\right]^{k_0 - 1}. \tag{1.13}$$

A plot of C_s versus m for different values of k_0 is left as an exercise. Notice that C_0 is the initial dopant concentration in the melt. For $k_0 = 1$, we have a constant concentration profile. On the other hand, C_s increases as m is increased for $k_0 < 1$, while C_s decreases as a function of m for $k_0 > 1$. Equation (1.13) tells us that there is a concentration gradient along the length of the crystal. In other words, the dopant concentration near the seed is different than that near the tail of the crystal. There is also a radial gradient dopant concentration. In other words, the dopant concentration near the center of the boule is different than that near the rim. Usually, the mapping of dopant concentration across the wafer using techniques, such as photoluminescence or absorption at a fixed wavelength, is very helpful in determining the carrier concentration in wafers.

The segregation coefficient may not be constant for dopants in LEC-grown semiconductor materials. The segregation coefficient discussed above is derived for the system near the liquid–solid interface. Away from this interface, the segregation coefficient can be different. To derive an expression for the effective segregation coefficient, let us assume that the dopant distributions in the solid and liquid phases are given by the profiles shown in Fig. 1.5 (see Ohring). As mentioned earlier, the segregation coefficient at equilibrium is defined near $x = 0$ as $k_0 = C_s/C_l(0)$. However, one can define the effective segregation coefficient, k_e, as the ratio between C_s and C_l, where C_l is the dopant concentration away from the solid–liquid interface. Now, let us consider small layer of the melt with width ϵ, in which the only flow is that required to replace the crystal being withdrawn from the melt, as shown in Fig. 1.5. Outside this layer, the dopant concentration is almost constant with a value of C_l, while inside the layer, the dopant concentration, $C(x)$, can be described by the steady-state continuity equation as

$$\upsilon \frac{dC(x)}{dx} + D_d \frac{d^2C(x)}{dx^2} = 0, \tag{1.14}$$

where υ is the velocity at which the crystal is being pulled out the melt (Eq. (1.8)), and D_d is the dopant diffusion coefficient. A possible solution of this equation is

$$C(x) = Ae^{-\upsilon x/D_d} + B, \tag{1.15}$$

where A and B are constants that need to be determined from the boundary conditions. The first boundary condition is $C(x) = C_l(0)$ at $x = 0$, which gives

$$C_l(0) = A + B. \tag{1.16}$$

The second boundary condition is that the sum of the dopant fluxes at the interfaces must be zero, which yields

$$D_d \left. \frac{dC(x)}{dx} \right|_{x=0} + (C_l(0) - C_s)\upsilon = 0.. \tag{1.17}$$

Substituting Equations (1.15) into (1.17) gives

$$A = (C_l(0) - C_s). \tag{1.18}$$

Combining Equations (1.15), (1.16), and (1.18) yields

$$C(x) = (C_l(0) - C_s)(e^{-\upsilon x/D_d} - 1) + C_l(0). \tag{1.19}$$

From Fig. 1.5, one can see that $C(x) \approx C_l$ at $x = \epsilon$. Thus, Equation (1.19) becomes

$$e^{-\upsilon\epsilon/D_d} = \frac{(C_l - C_s)}{(C_l(0) - C_s)}, \tag{1.20}$$

which yields

$$k_e \equiv \frac{C_s}{C_l} = \frac{k_0}{k_0 + (1 - k_0)e^{-\upsilon\epsilon/D_d}}. \tag{1.21}$$

A plot of Equation (1.21) is shown in Fig. 1.6 for different values of the growth parameters ($\upsilon\epsilon/D_d$) ranging from 0 to 10. For low values of $\upsilon\epsilon/D_d$ and $k_0 < 1$,

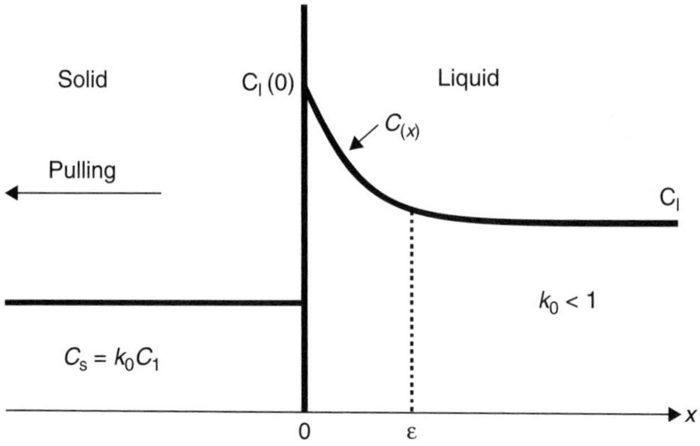

FIGURE 1.5 The distribution of a dopant near the solid–liquid interface in LEC crystal.

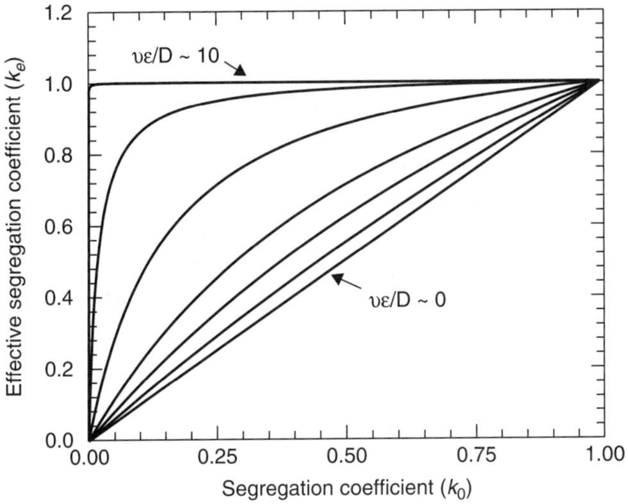

FIGURE 1.6 The effective segregation coefficient, k_e, is plotted as a function of k_0 for different values of $\upsilon\epsilon/D_d$.

the effective segregation coefficient is approximately the same as k_0, but it is always larger than k_0 and approaches unity for large values of $\upsilon\epsilon/D_{\mathrm{d}}$. Thus, a uniform doping distribution, where k_{e} approaches unity, in the crystalline solid can be achieved by increasing the pull maximum velocity and a low rotation speed. Owing to the centripetal force, the rotation speed is inversely proportional to ϵ.

1.2.2 Horizontal Bridgman Method

As the case of LEC growth method, the HB growth method needs a crystal seed. This growth technique consists of melt, crystal, and seed, which all kept inside a crucible during the entire heating and cooling processes. This technique is illustrated in Fig. 1.7, where two variations of the same method are presented. In the case of silicon growth using Bridgman method, a quartz crucible filled with polycrystalline material is placed inside a furnace tube and the heater is pulled. As the heater is drawn slowly away from the seed (Fig. 1.7a), the polycrystalline material is melted near the seed. The heater continues to move away from the seed's region, the molten material solidifies into a single crystal with a crystallographic structure similar to that of the seed. The shape of the resulting crystal is determined by the shape of the crucible. Another variation of this growth technique is shown in Fig. 1.7b, where the crucible is pulled slowly from the heater region into a colder region. The seed crystal induces single-crystal growth.

The drawback of Bridgman growth method is that the material is constantly in contact with the crucible, which produces two effects. First, the silicon crystals tend to adhere to the crucible, and second, the crucible wall introduces stress in

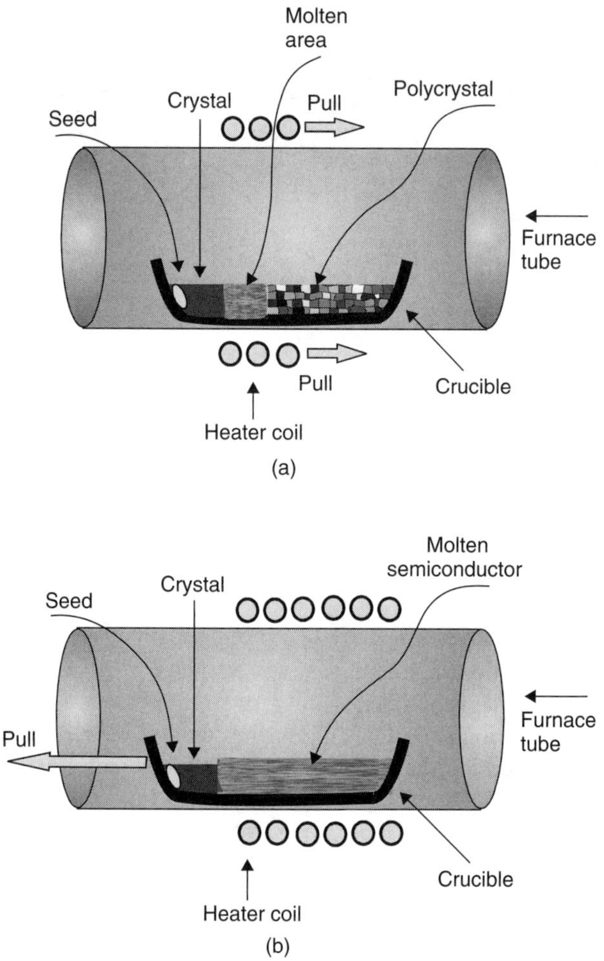

FIGURE 1.7 Schematic diagrams of the horizontal Bridgman growth method. (a) The heater is pulled over the polycrystalline material causing it to melt. As the heater moved away form the seed's region, the molten material solidifies into a single crystal. (b) A variation of Bridgman method, where the crucible is pulled away from the heater region.

the solidifying crystal. The presence of stress causes deviations from the ideal crystal structure.

The growth of compound semiconductors using Bridgman method is somewhat different than the growth of silicon crystals. For example, the growth of GaAs crystals is illustrated in Fig. 1.8a. In this growth method, both gallium and arsenic are loaded onto a fused silica ampoule, which is then sealed. The addition of the solid arsenic in the chamber provides the overpressure necessary to maintain stoichiometry. The furnace tube is slowly pulled past the charge (the charge is a term used to describe the semiconductor components that are placed in the

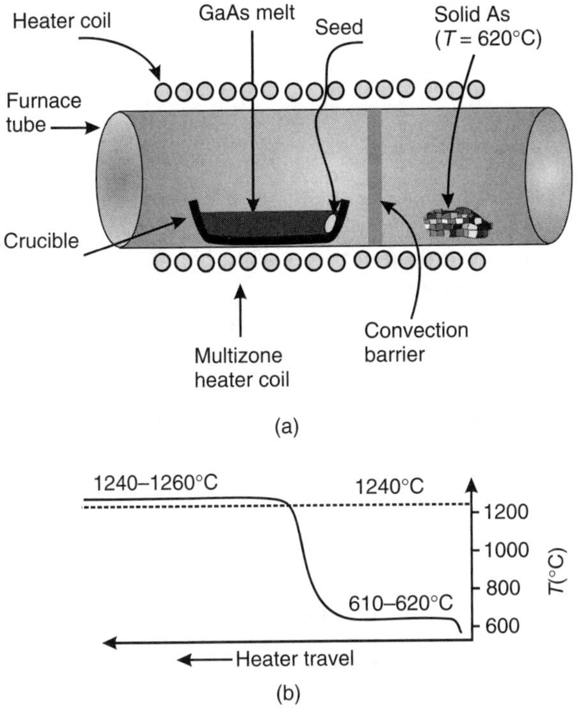

FIGURE 1.8 (a) A schematic diagram of the horizontal Bridgman growth method for GaAs and other compound semiconductors. (b) A sketch of the temperature for two-stage furnace.

crucible). The temperature of the furnace is set to melt the charge when it is completely inside the furnace. As the furnace is pulled past the ampoule, the molten GaAs charge recrystallizes with a structure similar to that of the seed crystal.

The heater coil in the Bridgman growth method is actually a multizone furnace. Figure 1.8b shows the temperature as a function of the direction of the heater travel for a two-stage heater. The first stage (on the right) is kept at about 610–620°C to maintain the required overpressure of arsenic. The second stage (on the left) is held at about 1240–1260°C, which is just above the melting point of GaAs (~1240°C). It is possible to grow GaAs with this method using GaAs polycrystalline as the charge (the starting material) instead of gallium and arsenic components.

Compound semiconductor boules grown by HB method are usually ~50 mm in diameter, which is small as compared to those boules grown by LEC method. However, with precise control of the stoichiometry and the radial and axial temperature gradients, a large boule size can be grown. The advantage of the HB method is that the dislocation densities in materials, such as GaAs, is of the order of 10^3 cm^{-2}, which is about an order of magnitude smaller than the dislocation densities found in LEC-grown material.

1.2.3 Float-Zone Growth Method

The float-zone growth method is mostly used to grow high purity silicon boules directly from a high purity rod of polycrystalline material obtained from other methods such as purification processes. With this growth method, a background carrier concentration lower than 10^{11} cm^{-3} can be easily achieved. Compound semiconductor materials are not generally grown by this technique. A schematic of the float-zone growth apparatus is shown in Fig. 1.9. A seed crystal is attached at the bottom of the polycrystalline rod in a vertical position. The rotating polycrystalline rod is enclosed in a quartz tube. An inert gas (argon) flows in the tube such that a one atmospheric pressure is maintained during growth. A small

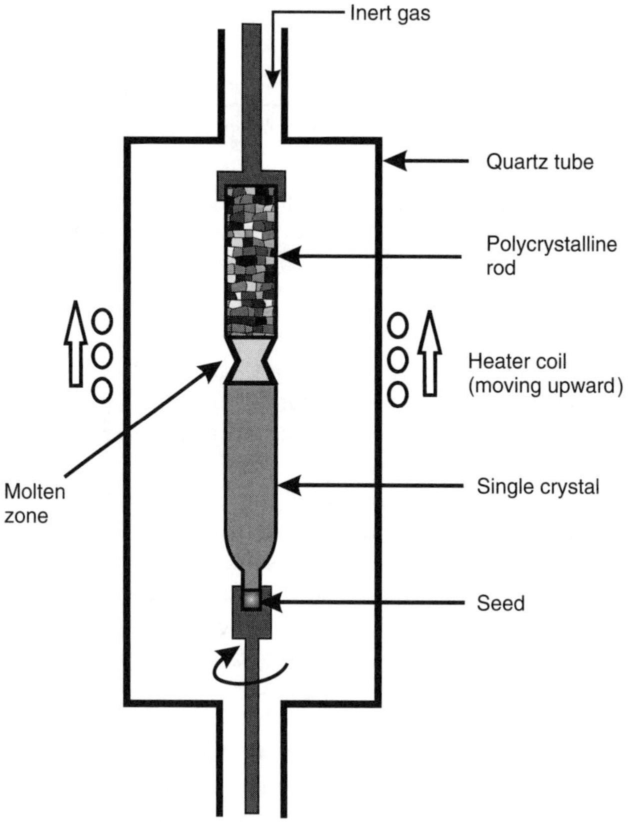

FIGURE 1.9 A cross-section of the float-zone apparatus used to grow silicon single crystals.

region of the polycrystalline rod is melted by passing an RF heater, which is moved upward from the seed. A float zone (a few centimeters in length) of melt is formed between the seed crystal and the polysilicon rod and retained by the surface tension between the melting and the growing solid phases. The molten zone that solidifies first remains in contact with the seed crystal retaining the same crystallographic structure of the seed. As the molten region is moved along the length of the polycrystalline rod, the rod melts and then solidifies throughout its entire length, it becomes a single crystal. The motion of the heater controls the diameter of the crystal.

Difficulties in preventing the molten zone from collapse have limited the float-zone method to growing small diameter crystals. The maximum crystal diameter is on the order of 70 mm. However, one of the advantages of this technique is that there is no crucible involved, so oxygen contamination is eliminated. Another advantage of this growth method is that the background impurities can be substantially reduced by passing the heater coil over the crystal several times. The background impurities can be reduced by a few orders of magnitude when the heater coil is passed over the crystal seven or eight times (see Pfann).

The introduction of doping in the growth method is more difficult when compared to the LEC and Bridgman growth methods. There are four techniques used to introduce dopants in float-zone growth method. First, core doping is based on the introduction of doped polysilicon as the starting material. Second, gas doping is based on the injection of gases, such as $AsCl_3$, BCl_3, or PH_3, into the polysilicon rod as it is being deposited or into the molten zone region during refining. Third, pill doping is based on the insertion of a small pill of dopant, such as gallium or indium, into a hole from at the top of the polysilicon rod. Dopants with small segregation coefficients will diffuse into the rod as the melt passes over the polysilicon rod. Fourth, neutron transmutation doping is based on irradiating the silicon single crystal by thermal neutrons. This process produces a fractional transmutation of silicon into phosphorus, which leads to n-type silicon. The neutron transmutation equation is given by

$$\mathrm{Si}_{14}^{30} + \text{thermal neutron} \rightarrow \mathrm{Si}_{14}^{31} + \gamma\text{ray} \rightarrow \mathrm{P}_{15}^{31} + \beta\text{ray}. \tag{1.22}$$

The life time of the intermediate Si_{14}^{31} is 2.62 h. Neutron transmutation doping is very uniform since thermal neutrons penetration length in silicon is $\sim$100 cm.

The doping distribution in the float-zone process can be understood by considering the model illustrated in Fig. 1.10. Assume that the initial uniform doping concentration in the rod is C_0, L is the length of the molten zone at a distance x along the rod, A is the cross-sectional area of the rod, ρ is the specific density of silicon, and σ is the dopant concentration in the molten zone. When the molten zone is moved a distance dx, the amount of dopant added to it at its advancing end can be expressed as $C_0 \rho A \mathrm{d}x$, while the amount of dopant removed at the retreating end is $k_e \sigma \mathrm{d}x/L$. The differential amount of dopant, $d\sigma$, remaining in

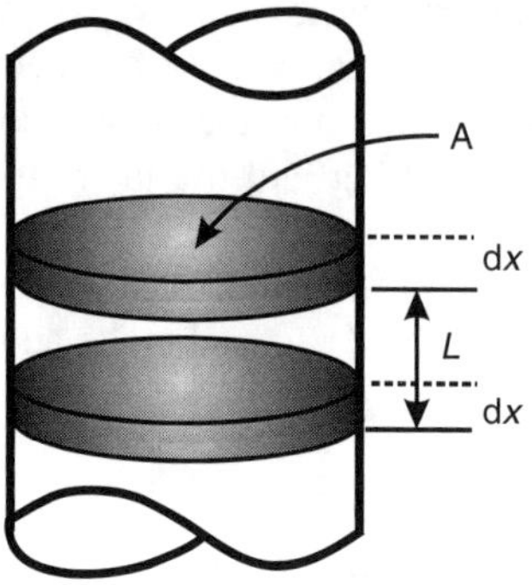

FIGURE 1.10 A sketch of a portion of a semiconductor boule used to illustrate the calculation of the doping profile.

the molten zone as it moves a distance *dx* can be expressed as

$$d\sigma = C_0 \rho A dx - \frac{k_e \sigma}{L} dx, \tag{1.23}$$

which can be rewritten as

$$\int_0^x dx = \int_{\sigma_0}^{\sigma} \frac{d\sigma}{C_0 \rho A - k_e \sigma / L}, \tag{1.24}$$

where σ_0 is the amount of dopant in the molten zone when it was first formed at the front end of the rod and is given by $C_0 \rho A L$. An expression for the dopant concentration in the crystal at the retreating end is given by

$$C_s = \frac{k_e \sigma}{A \rho L}. \tag{1.25}$$

An expression can be obtained for σ by integrating Equation (1.24), which can be substituted into Equation (1.25) to yield the following relation

$$C_s = C_0 \{1 - (1 - k_e) e^{-k_e x / L}\}. \tag{1.26}$$

A plot of this equation is left as an exercise. For small values of $k_e x / L$, C_s is nearly constant.

1.2.4 Lely Growth Method

The growth techniques discussed earlier could not be employed to grow wide band gap materials such as GaN and SiC. This is due to the fact that these materials do not have a liquid phase under reasonable thermodynamic conditions. The growth conditions of the materials require high temperature and high pressure environment. For example, SiC melt exists only at pressures in the excess of 10^5 atmosphere and temperatures higher than 3200°C. Under these extreme conditions, the stoichiometry and stability of the melt are difficult to maintain. Silicon carbide material is grown by Lely method, which is schematically shown in Fig. 1.11a. The growth process is driven by a temperature gradient, which is maintained between the outer and inner areas of the crucible. The system is kept near equilibrium with lower partial pressures of the SiC precursor in the inner and colder zone. The two areas are separated by a porous graphite material, which provides nucleation centers. Since the inner region is colder, the chemical gradient causes a mass transport from the outer region to the inner region. Single crystals of SiC start to nucleate on the inner side of the porous graphite. As

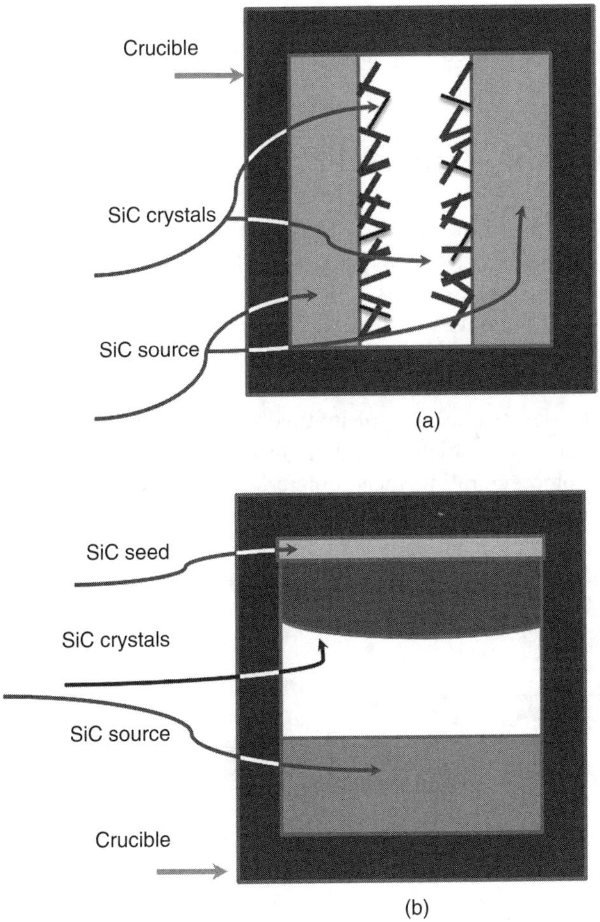

FIGURE 1.11 (a) A cross-sectional diagram of a cylindrical crucible for the Lely growth of SiC. (b) A cross-sectional diagram of the modified Lely growth technique.

illustrate in Fig. 1.11a, the crystals are limited in size with random dimensions, but nonetheless, they are of a high quality in terms of low defect densities. Atypical size of these crystals is ~1 cm, but they are used as seed crystals for other bulk SiC growth techniques including the modified Lely method described below.

The modified Lely method is based on a seeded sublimation growth or physical vapor transport technique. It is basically similar to the Lely method except that a SiC seed crystal is used as shown in Fig. 1.11b to achieve a controlled nucleation. According to Fig. 1.11, the cooler seed is placed at the top to minimize falling contaminations. Polycrystalline SiC source is heated to ~2600°C at the bottom of the crucible and sublimes at low pressure. Mass transport of SiC occurs and recrystallizes through supersaturation at the seed. The disadvantages of this technique include the poor control of the polytype and shape of the crystals, nonuniform doping, and high density of defects. Furthermore, the screw dislocations have been one of the long-standing problems of commercial bulk SiC wafers. This class of dislocations, which are also known as *nano- or micropipes*, can be closed or hollow. These micropipes have detrimental effects on SiC-based devices.

1.3 GROWTH OF SEMICONDUCTOR THIN FILMS

The growth of thin films requires finely polished substrates (wafers) cut from single-crystal boules grown by bulk crystal growth methods described in the previous sections. The growth of thin films, quantum wells, superlattices, quantum wires, and quantum dots requires a precise knowledge of the crystallographic structure of the substrates. Silicon substrates are used in the vast majority of silicon-based devices and technology. Silicon also has been used as a substrate for GaN-based devices because of its favorable physical properties, high quality, large sizes, and above all low cost. It has a diamond structure, which can be seen as two interpenetrating face-centered cubic sublattices with one sublattice displaced from the other by one-quarter of the distance along the diagonal of the cube. Each atom in the lattice is surrounded by four equivalent nearest neighbors that lie at the corners of a tetrahedron. The three commonly used structural orientations of Si are shown in Fig. 1.12.

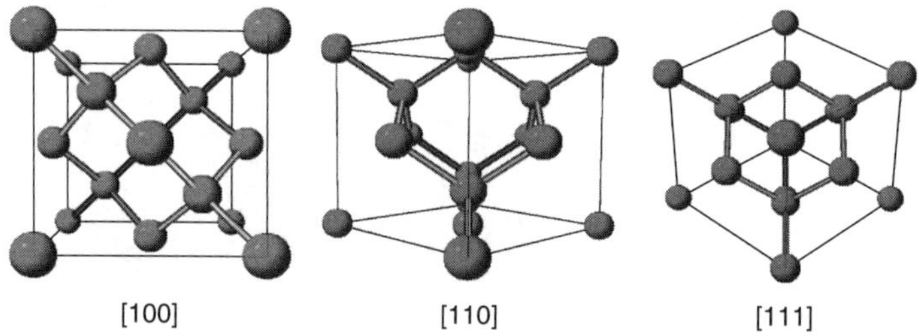

FIGURE 1.12 Views of the three commonly used crystallographic directions of Si wafers.

Gallium arsenide-based technology, which includes both electronic and optoelectronic devices, has been advancing rapidly since the epitaxial growth is mature for many III–V semiconductor nanomaterials. The reason for this advancement is the availability of GaAs substrates with many structural orientations. Gallium arsenide single crystal has a zincblende structure. A view of the main three structural orientations is shown in Fig. 1.13. Recently, GaAs substrates have been used for the growth of III-nitride materials. However, because the relatively low melting point of GaAs, it is less stable when compared to SiC and sapphire substrates.

1.3.1 Liquid-Phase Epitaxy Method

The LPE growth method is basically a precipitation of materials from supercooled solution onto a substrate. The LPE reactor is shown in Fig. 1.14, which consists of a horizontal furnace system and a sliding graphite boat. An enlarged illustration

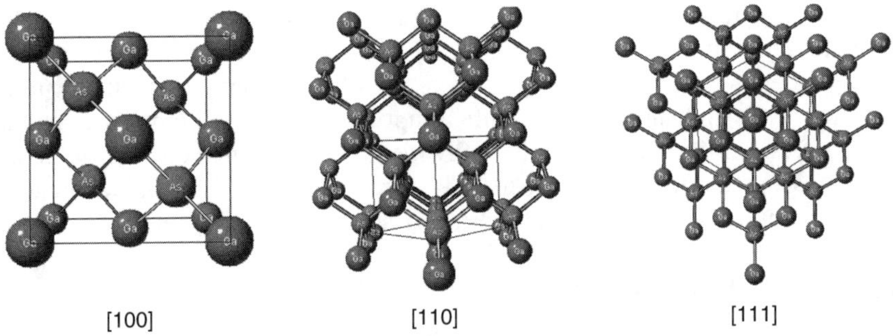

FIGURE 1.13 A view of three different structural directions of GaAs, which has a zincblende crystal structure.

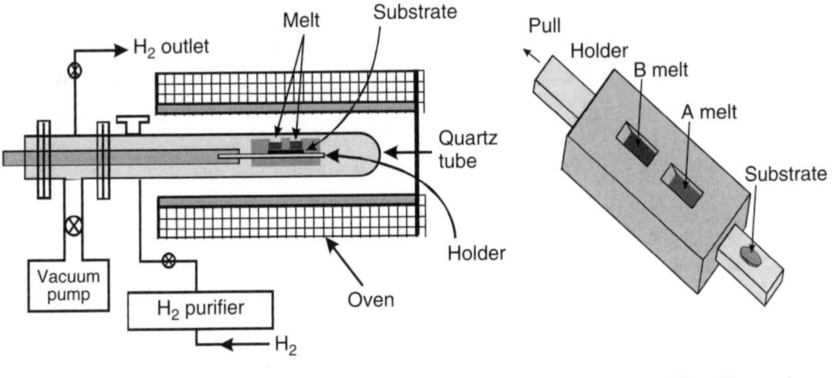

FIGURE 1.14 A cross-section of the liquid-phase epitaxy reactor. The illustration on the right is an enlarged crucible showing the melts and the substrate on the holder, which slides under the melt.

of the sliding graphite boat is shown on the right-hand side of Fig. 1.14, where two different melts A and B are used as an example of which a heterojunction can be made. If the melt A is GaAs, the melt B is AlGaAs, and the substrate is semi-insulating GaAs, then a GaAs/AlGasAs heterojunction can be grown with this technique. This simple reactor usually produces high purity thin films. The epitaxial growth processes in this technique are usually maintained at thermodynamic equilibrium. The composition of the thin films depends mainly on the equilibrium phase diagram of the material and to a lesser extent on the orientation of the substrate. The molten material is placed in a graphite boat and is slid inside the heated furnace of a suitable atmosphere. A subsequent cooling causes the solute to come out and deposit on the underlying substrate forming, an epitaxially grown layer. Growth using the LPE method is affected by the melt composition, growth temperature, and growth duration.

The advantages of the LPE method are the simplicity of the equipment used, higher deposition rates, low defect concentrations, excellent control of stoichiometry, and high purity materials. Background impurities are minimized by using high purity melt materials and by the inherent purification process that occurs during the liquid-to-solid-phase transition. Disadvantages, on the other hand, include poor thickness uniformity, high surface roughness, melt back effect, and high growth rates, which prevent the growth of multilayer structures, such as multiple quantum wells and superlattices, with abrupt interfaces. Additionally, only small size wafers can be used with the LPE method, which makes it a small-scale process. Contrary to the growth materials from the melt, LPE grown materials are temperature independent and thermal gradients are usually neglected.

1.3.2 Vapor-Phase Epitaxy Method

The name vapor-phase epitaxy (VPE) implies that the growth of thin films is based on reactive compounds in their gaseous form. The recent development of III-nitride materials renewed the interest in the VPE growth method. This growth technique is performed at thermodynamic equilibrium. A generic sketch of the VPE reactor is shown in Fig. 1.15 (see Razeghi). As shown in the figure, the reactor consists of a quartz tube (chamber), gas inlets, exhaust, and a furnace with different temperature zones. Three zones are shown in the temperature trace, which are called *synthesis, pyrolysis*, and *growth zones*. The growth of InP and GaAs samples is taken as an example. The gaseous species for the group III source materials are synthesized by reacting hydrogen chloride gas with a melted pure metal placed in crucibles. This process occurs in the first zone, the synthesis zone, which is maintained at temperature, T_s, of 750 and 850°C for GaAs and InP, respectively. The reaction between the metals and hydrogen chloride results in group III-chloride vapor compounds as follows

$$\mathrm{In_{liquid}} + \mathrm{HCl_{gas}} \rightarrow \mathrm{InCl_{gas}} + \frac{1}{2}\mathrm{H_{2gas}}, \tag{1.27a}$$

$$\mathrm{Ga_{liquid}} + \mathrm{HCl_{gas}} \rightarrow \mathrm{GaCl_{gas}} + \frac{1}{2}\mathrm{H_{2gas}}. \tag{1.27b}$$

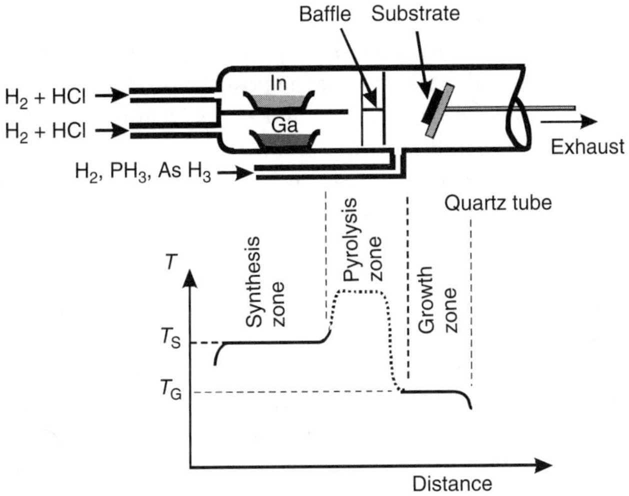

FIGURE 1.15 A cross-section of a vapor-phase epitaxy reactor with a multitemperature zones heater.

The group V source materials are provided in the form of hydride gases, such as arsine ($As\,H_3$) and phosphine (PH_3). The hydride gas is pyrolized in the second zone, which is maintained at temperatures $T > T_S$. The decomposition of group V can be described as follows

$$AsH_3 \rightarrow \frac{u}{4}As_4 + \frac{1-u}{2}As_2 + \frac{3}{2}H_2, \tag{1.28a}$$

$$PH_3 \rightarrow \frac{v}{4}P_4 + \frac{1-v}{2}P_2 + \frac{3}{2}H_2, \tag{1.28b}$$

where u and v are the mole fractions of AsH_3 and PH_3 that are decomposed into As_4 and P_4, respectively.

The gas flow is cooled by a temperature gradient between the second and the third zones. The cooling of reactants results in the growth of semiconductor materials, such as GaAs and InP, on the substrate in the growth zone. The growth zone is maintained at temperatures, T_G, of 680 and 750°C for GaAs and InP, respectively. It is clear from the above equations that there are several chemical reactions taking place in the VPE reactor. These reactions can be classified as heterogeneous reactions, which occur between solids and liquids, solids and gases, and liquids and gases and homogeneous reactions that occur in the gas phase. During the steady-state film growth, the overall growth process is limited by the heterogeneous reactions, whereas the change in the composition of the grown semiconductor in the process (for example, switching the growth from InP to InGaAs) is limited by the mass transport in the gas phase.

The advantages of the VPE method include high deposition rate, multi-wafer growth, flexibility in introducing dopants into the materials, and good control of the composition gradients by accurate control of the gas flows. The disadvantages include difficulties in growing multilayer quantum structures, potential formation of hillocks and haze, and interfacial decomposition during the preheat stage.

The renewed interest in the VPE method stems from its ability of high deposition rates under reasonable growth conditions. This advantage has been used for the growth of thick GaN films, on the order of 100 μm or thicker, where native bulk substrates are not available. The idea here is to replace the current substrates, which are mainly sapphire and SiC, by producing thick GaN films that can be used as compliant (quasi) substrates. The thick GaN films can be grown on other substrates, such as sapphire, then lifted and used as substrates in other growth techniques, such as the MBE and MOCVD techniques. The VPE growth of GaN utilizes hydrogen chloride gas that passes over a crucible containing metallic gallium at a temperature of $\sim 850^{\circ}C$ to form gaseous GaCl. Ammonia (NH_3) and HCl are then injected into the hydride pyrolysis zone using N_2 as a carrier gas. Gallium chloride is injected through a showerhead into the growth zone, which is kept at temperatures in the range of $950-1050^{\circ}C$. Gallium chloride then reacts with NH_3 on the substrate surface to produce GaN according to the following reaction

$$GaCl + NH_3 \rightarrow GaN + HCl + H_2. \tag{1.29}$$

The $NH_3 : HCl$ ratio is typically 30 : 1 with a growth rate of ~0.3 μm/min. There are, however, problems associated with the VPE growth of GaN. First, it is quite possible for NH_3 to dissociate and react with *HCl* to produce NCl_3, which is highly explosive. Second, *HCl* can potentially causes leaks in the reactor. Third, undesired by-products such as NH_3Cl and $GaCl_3$ can clog the exhaust system unless heated to temperatures higher than $150^{\circ}C$. Fourth, because of the exchange reactions with the quartz chamber walls of the reactor, AlGaN growth and p-type-doped GaN are difficult to realize.

1.3.3 Hydride Vapor-Phase Epitaxial Growth of Thick GaN Layers

The hydride vapor-phase epitaxy (HVPE) is essentially a variation of the VPE method, but it has been developed to grow thick GaN films (for further details on the subject, see Paskova *et al.*). The increasing interest in III-nitride materials and devices has led to the long-standing demand for GaN substrates for homoepitaxy of GaN, which has yet to be satisfied. There are substantial difficulties in growing large-volume GaN single crystals at the high equilibrium vapor pressure of N_2 and the high growth temperature needed in bulk growth from Ga solution. Currently, there are three promising techniques that can be used to obtain GaN bulk crystals—high pressure crystal growth from Ga solution, the sublimation technique, and HVPE growth.

HVPE growth of thick GaN layers was developed by several research groups to provide quasi-bulk thick GaN layers. GaN layers with a thickness of several hundred micrometers have been achieved in HVPE atmospheric pressure reactors at temperatures of about 1050°C and at a reasonable cost. The deposition process of GaN for substrate application requires a high growth rate and the ability to produce low defect material, since the threading defects are likely to extend into the subsequently grown epilayers. The growth rates in HVPE have been reported to be as high as 100 μm/h with a crystalline quality comparable to the best quality reported for MOCVD-grown GaN films. Several substrate pretreatments such as a GaCl sapphire pretreatment, a sapphire nitridation pretreatment, different buffer layers such as ZnO, reactive sputtered AlN, MOCVD-grown GaN, and epitaxial lateral overgrowth technique have greatly improved the quality of thick HVPE-GaN films. Despite rapid progress in the HVPE technique, a number of basic issues remain to be solved. One of them is the presence of a high density of extended defects such as dislocations, domain boundaries, and cracks. Efforts to further develop the HVPE-GaN thick layers for substrate use are concentrated on two main issues. The first is focused on the reduction of defect density and the control of the initial stage of the growth, which is the source for most defects. Secondly, an optimal procedure for subsequent removal of the foreign substrate from the GaN layer is far from complete, although very intense investigations of chemical, reactive ion etching, laser-induced liftoff, and polishing separation have been reported in the literature.

The basic HVPE reactions that describe the GaN deposition process can be written as follows:

$$x\mathrm{Ga(l)} + \mathrm{HCl(g)} \rightarrow x\mathrm{GaCl(g)} + (1-x)\mathrm{HCl(g)} + x/2\mathrm{H_2(g)}, \tag{1.30a}$$

$$\mathrm{GaCl(g)} + \mathrm{NH_3(g)} \rightarrow \mathrm{GaN(s)} + \mathrm{HCl(g)} + \mathrm{H_2(g)}, \tag{1.30b}$$

where l is liquid, g is gas, s is solid, and x is the mole fraction of HCl reacting in the process. Notice that Equation (1.30bb) is the same as Equation (1.29). The GaN deposition is determined by the efficiency of both chemical reactions. Values of x in reaction (Eq. 1.30a) were found to be in the range from 0.70 to 0.86 depending on the temperature, the position of the HCl inlet, the carrier gas ambient, and the liquid Ga surface exposed to the HCl gas. The chemical reaction (Eq. 1.30b) depends on the fraction of ammonia not decomposed into nitrogen and hydrogen, since GaN cannot be formed by direct reaction between GaCl and N_2. It is known that ammonia is a thermodynamically unstable gas at the temperatures employed in the GaN growth. Fortunately, the thermal decomposition of NH_3 is a very slow reaction, and when no catalyst is present, no more than about 4% of the NH_3 is typically decomposed at temperatures higher than 950°C. Equation 1.30a, 1.30b is accompanied by GaN decomposition via the following two reactions:

$$\mathrm{GaN(s)} + \mathrm{HCl(g)} \rightarrow \mathrm{GaCl(g)} + \frac{1}{2}\mathrm{N_2(g)} + \frac{1}{2}\mathrm{H_2(g)} \tag{1.31a}$$

$$\mathrm{GaN(s)} \rightarrow \mathrm{Ga(l)} + \frac{1}{2}\mathrm{N_2(g)} \tag{1.31b}$$

These decomposition reactions are unlikely to occur in the growth temperature range of 950–1150°C.

The basic design of the HVPE reactor is similar to the VPE reactor with some modifications as shown in Fig. 1.16. These modifications can be summarized into two groups: horizontal and vertical reactor design. The horizontal reactor shown in Fig. 1.16a typically has five main temperature zones. In the first upstream zone, HCl reacts with metallic Ga forming GaCl and H_2. The area of the liquid Ga source is increased as much as allowed (typically 10–100 cm^2) to achieve a large reactive Ga surface area for efficient GaCl production. The optimum temperature in the first zone is about 850°C. The second zone may be used for other metallic sources, such as In or Al when needed, or for dopants. The temperature of the third zone is kept in the range of about 1000–1060°C where GaCl and NH_3 are introduced and mixed. The substrate holder is placed in the fourth region of the reactor where the temperature is kept at ~1080°C. The most

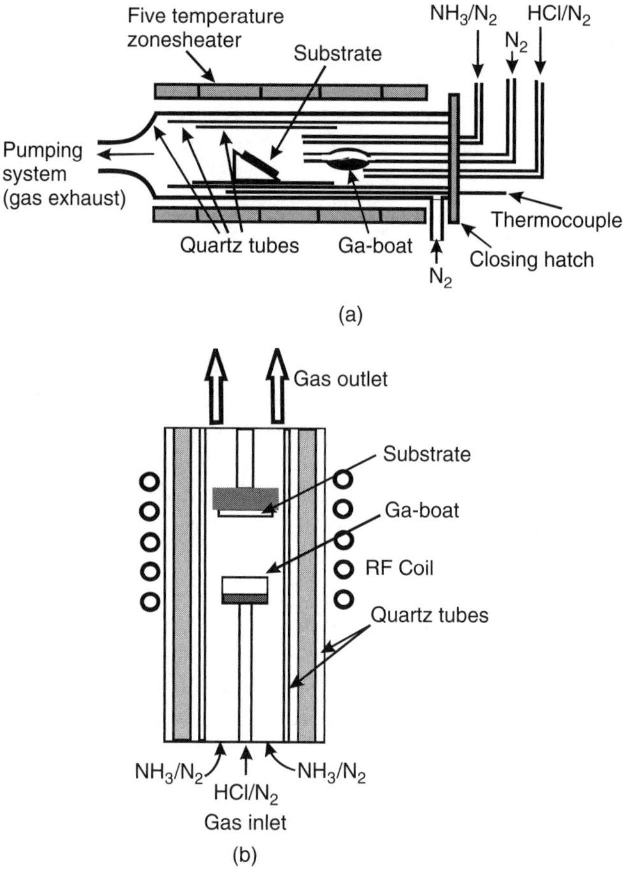

FIGURE 1.16 A schematic diagrams of (a) horizontal and (b) vertical HVPE reactors.

common method of heating in this design is resistive heating. The horizontal reactors utilize a susceptor that is situated approximately parallel to the gas flow direction. Uniform growth can be improved by tilting the substrate holder to eliminate reactant depletion along the flow direction. Another approach used in some horizontal reactors is the rotation of the substrate holder.

In the vertical design, the reactants are typically introduced through the top. The substrate is held flat on a susceptor that is perpendicular to the gas flow direction. The vertical reactor design facilitates substrate rotation during the growth to improve film uniformity. Heating is accomplished by resistance or RF induction, and temperature monitoring is accomplished by an infrared pyrometer or a thermocouple. An alternative modification is an inverted vertical reactor as shown in Fig. 1.16b, where the process gases are supplied through the bottom inlet flange, while the top flange can be lifted for loading and unloading. The substrates are placed in the upper part where the gases are mixed. The inverted reactor keeps all advantages of the vertical design and also provides the possibility for raising the substrate holder. An additional advantage of the inverted vertical reactor is the significant reduction of solid particle contamination.

1.3.4 Pulsed Laser Deposition Technique

The pulsed laser deposition (PLD) is a relatively new technique widely used for the growth of oxide thin films, such as ferroelectrics and superconductors (for detailed discussion, see Huang and Harris). There are, however, several advantages of the PLD for depositing high quality thin films and make it worthy of study as a method of growing III-nitride materials. One of these advantages is the simplicity of the technique. PLD is typically accomplished with a high power pulsed laser beam irradiating a bulk stoichiometric target. Through the interaction of the laser beam with the target, a forward-directed flux of material is ejected. A plasma is formed which is then transported toward a heated substrate placed directly in the line of the plume. This is illustrated in Fig. 1.17. The congruent ablation achieved with short laser pulses enables stoichiometric composition transfer between targets and films and allows deposition of multicomponent materials by employing a single target. This feature makes the PLD the best initial investigation tool for complex materials because the stoichiometry control is vastly easier.

A useful feature of the PLD method is that multiple targets can be loaded inside the chamber on a rotating holder, which can then be used to sequentially expose different targets to the laser beam, thereby enabling *in situ* growth of heterostructures and superlattices with relatively clean interfaces. Virtually, any material can be laser evaporated, leading to the possibility of multilayers of a variety of materials. Therefore, PLD is suitable for rapid exploration of new materials-integration strategies for developing heterostructures and performing basic studies at the laboratory scale. The growth rate achieved by PLD can be varied from subangstrom per second to a few micrometers per hour by adjusting the repetition rate and the laser fluence, which is useful for both atomic

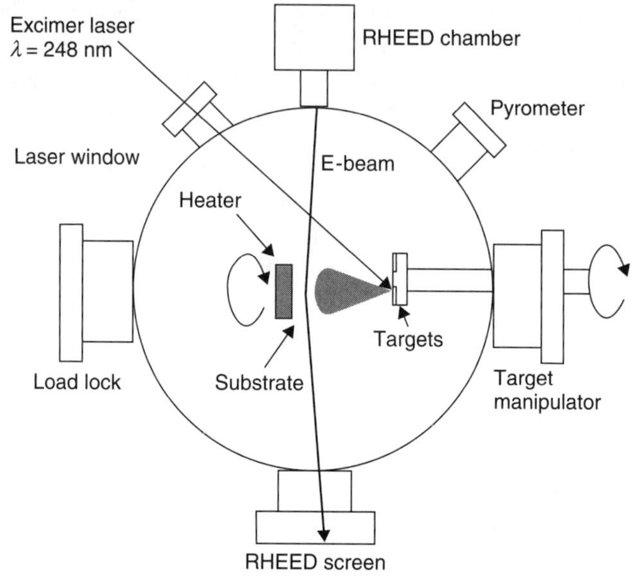

FIGURE 1.17 A sketch of the basic configuration of the pulsed laser deposition chamber (after Huang and Harris).

level investigations and thick layer growth. Moreover, the strong nonequilibrium growth conditions of PLD may allow a much broader range of metastable materials to be grown, including the introduction of higher dopant concentrations and alloy compositions that in equilibrium phase normally segregate.

There are three main stages of the PLD process: laser–target interaction, laser–plume interaction, and subsequent deposition of the thin film. In the beginning of the laser pulse, the optical energy is largely absorbed by the surface of the target. Since the laser energy is supplied to a small volume of 10^{-13} m^3 in a short time (typically 30 ns), the local temperature of the target, frequently on the order of 10^4 K. Therefore, all the species of the target evaporate simultaneously, that is, congruent evaporation. This condition ensures that the ejected materials have the same stoichiometry as the target which makes the PLD process particularly suitable for exploring binary, ternary, and more complicated systems without having to adjust fluxes from multiple sources. Continuous interaction of the laser beam with the plume results in the photodissociation and photoionization of the evaporated material. This interaction breaks molecular species and clusters and ionizes the evaporated material by nonresonant multiphoton process, leading to the formation of expanding plasma above the target surface, and transport toward the substrate.

The evaporation of the materials from targets by laser irradiation depends on the laser parameters (such as laser fluence, pulse duration, and wavelength) and material properties (such as reflectivity, absorption coefficient, and thermal conductivity). According to Fig. 1.17a KrF excimer laser operating at a wavelength of 248 nm and pulse duration of 20 ns are used to ablate materials from the targets. The laser is incident at 45° from the target normal, and the substrate is centered along the target normal. The system is capable of holding six targets for multilayer growth. Each target is rotated about its axis to ensure uniform wear on the targets, and individual targets can be successively clocked into the position for the ablation of multitargets. A load lock chamber with a magnetically coupled transfer rod is equipped to facilitate the transfer of both targets and substrates without breaking the vacuum of main chamber. The base pressure, on the order of 10^{-8} torr, is achieved by pumping the chamber with turbo and mechanical pumps. The target to substrate distance can be varied over 15 cm to operate in different pressure regions. The substrate is rotated to enhance the temperature and thickness uniformity during deposition. The substrate heater is capable of reaching 800°C in either an oxygen ambient for oxide growth or a nitrogen ambient for nitride growth. Another attractive feature of PLD apparatus is the capability of *in situ* monitoring of the growth process, such as the reflection high energy electron diffraction (RHEED). The RHEED patterns offer abundant information on the crystal structure and the quality of the growing film. It also provides means to study surface structure and growth kinetics.

1.3.5 Molecular Beam Epitaxy Growth Technique

Despite the high price tag on the MBE reactor, it is one of the most versatile and widely used nonequilibrium growth techniques. While the MBE growth processes are under continuous development ranging from effusion cell shape to the addition of many *in situ* diagnostic tools, it has been used to grow almost any kind of doped and undoped semiconductor materials ranging from thin films and quantum wells to quantum dots and nanomaterials. It is capable of controlling the deposition of a submonolayer on substrates with various crystallographic structures. A schematic of an MBE growth chamber is shown in Fig. 1.18, which is showing the sources for the growth of GaAs and GaN with two different dopants (Si and Mg).

A thin film deposition process is performed inside the MBE chamber in which thermal beams of atoms or molecules react on the clean surface of a single-crystalline substrate that is held at high temperatures under ultrahigh vacuum conditions ($\sim 10^{-10}-10^{-11}$ Torr) to form an epitaxial film. It turns out that this ultrahigh vacuum is a major advantage to the MBE growth. This is primarily due to this very low impurity environment and the fact that many *in situ* tools can be added to the vacuum chamber. The most common way to create molecular beam for MBE growth is through the use of Knudsen effusion cells. The crucibles employed in Knudsen cells are mostly made of PBN. The temperatures of different crucibles are usually independently controlled to within ±1°C.

The material sources could be solid, gas, or metal–organic materials. Solid precursor sources are generally solids that are heated above their melting point in effusion cells, Knudsen cells. The atoms of source material escape the cell

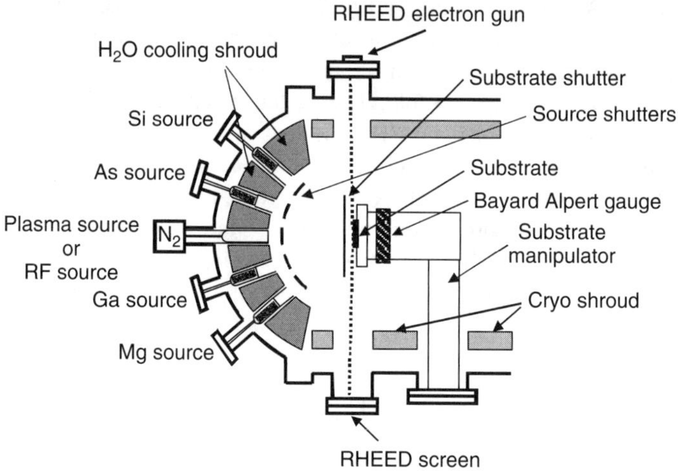

FIGURE 1.18 A sketch of a molecular beam epitaxy growth chamber showing the configuration of the sources and RHEED electron guns.

into the vacuum chamber by thermoionic emission. The beam flux is a function of its vapor pressure, which can be controlled by the source temperature. In the gas source MBE, group V of III–V semiconductors are connected through an injector or cracker. The gas source beam flux is controlled by using a mass flow controller. The metal–organic sources are either liquids or powders. An inert carrier gas is usually used to control the beam flux. The thickness and compositions of the epitaxial layers are controlled by the interruption of the unwanted atomic beam using shutters, which are usually remotely controlled by a computer. The beam of atoms and molecules will attach to the substrate forming the epitaxial layers. The growth rate is generally about a monolayer per second. The layers crystallize through the reaction between the atomic or molecular beams of the source materials and the substrate surface that is maintained at a certain temperature.

Another major difference between the MBE growth and other growth techniques is that it is far from thermodynamic equilibrium conditions. It is mainly governed by the kinetics of the surface processes. The five major kinetic processes are illustrated in Fig. 1.19 where the blocks represent the atoms and molecules that reach the surface of the substrate. Process (a) is the adsorption of the atoms or molecules impinging on the substrate surface; process (b) is the thermal desorption of the atoms or molecules that are not incorporated in the epitaxial layer; process (c) is the surface migration and dissociation of the absorbed atoms and molecules; process (d) is the incorporation of the atoms and molecules into the epitaxial layer or the surface of the substrate; and process (e) is the interdiffusion between the substrate and the epitaxial layer.

In order to grow smooth surfaces, the atoms impinging on the substrate surface should be given enough time to reach their proper position at the edge before

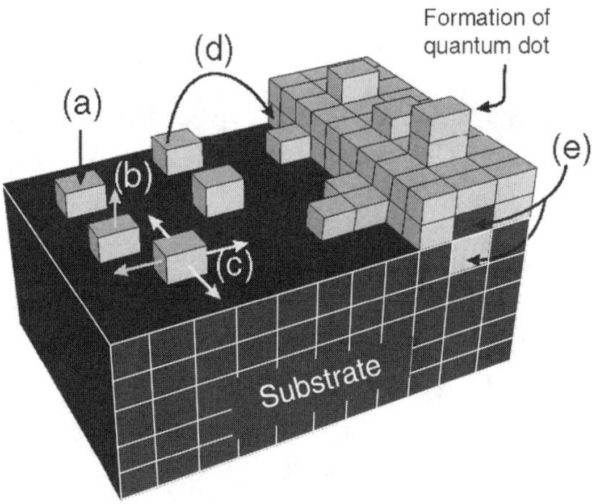

FIGURE 1.19 A schematic illustration of the kinetic processes that occur at the surface of the substrate during the MBE growth.

the formation of the next entire new layer. This also reduces the formation of defects and dislocations. The atoms in the MBE growth chamber have a long mean-free path where collisions and scattering with other atoms are infrequent before reaching the surface of the substrate. This mean-free path, $\mathcal{L}$, can be written in terms of the atoms or molecules concentration, $\mathcal{N}$, according to the following relation

$$\mathcal{L} = \frac{1}{\sqrt{2}\pi \mathcal{N} d^2}, \tag{1.32}$$

where d is the diameter of the species. The concentration, $\mathcal{N}$, is determined by the vapor pressure, P, and temperature, T, inside the MBE chamber according to the following relation

$$\mathcal{N} = \frac{P}{k_B T}, \tag{1.33}$$

where k_B is the Boltzmann constant.

As mentioned earlier, one of the advantages of having an ultrahigh vacuum in the MBE growth chamber is that *in situ* tools can be added to monitor the epitaxial layer during growth. One of these tools is the reflection high energy electron diffraction abbreviated or known as *RHEED*. The electron energy in the RHEED gun is typically 5–50 keV. The electrons are directed toward the substrate at a grazing angle $\theta \leq 1$. A schematic showing the RHEED configuration is shown in Fig. 1.20a. The electrons are then diffracted by the epitaxial layer formed at the substrate surface. This leads to the appearance of intensity-modulated streaks on a fluorescence screen. The results obtained by RHEEDS are generally characterized as static or dynamic mode. In the static mode, the atomic construction of the surface can be determined from RHEED diffraction patterns. These patterns (Fig. 1.20b) usually provide information on the atomic

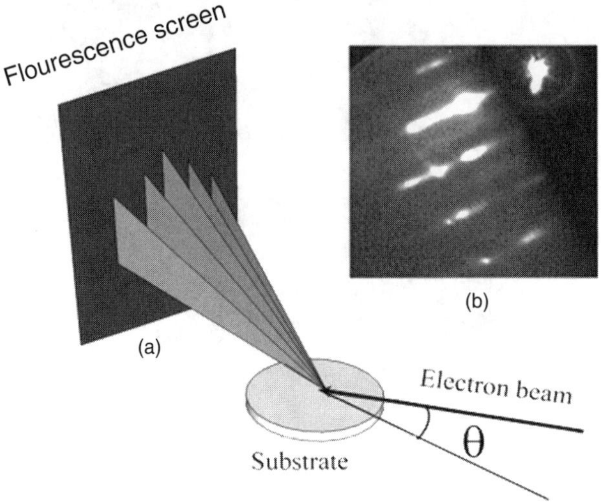

FIGURE 1.20 (a) The configuration of the RHEED inside the MBE chamber, and (b) RHEED patterns formed on a fluorescence screen.

surface construction, which is a function of the incoming electron beam flux, the substrate temperature, and the strain of the epitaxial layer. On the other hand, the dynamic mode is based on the change of the intensity of the central diffraction streak as the wafer roughness changes over time. This process is illustrated in Fig. 1.21, where the formation of a single complete monolayer is shown. The fractional layer coverage is represented by the factor S. During the epitaxial growth process, the roughness of the epitaxial layer increases as a new atomic layer forms. When the surface coverage reaches 50% or S = 0.5, the roughness is at maximum and begins to decrease as the growing layer is complete, which corresponds to minimum roughness (S = 1.0). The intensity of the main RHEED streak follows the period oscillation of the layers roughness with the maximum intensity corresponding to the minimum roughness. This is illustrated in the RHEED oscillation signal depicted in the right-hand side panel. The time separation between the two adjacent peaks in the RHEED oscillations provides the time needed for the growth of a single layer of a crystal.

Another *in situ* tool that is often used in MBE growth chamber is Auger electron spectroscopy. This technique is based on the Auger effect of measuring the elemental composition surface. This technique uses an electron beam with energy ranging between 3 and 25 keV which when hits the substrate surface, it excites the atoms at the surface of the substrate by knocking a core level electron to a higher energy level. When the excited electrons relax, the atoms release the extra energy by emitting Auger electrons with characteristic energies. These energies are measured, and the quantity of Auger electrons is proportional to the concentration of the atoms on the substrate surface. Thus, Auger electron spectroscopy technique measures the planar distribution of elements on a surface.

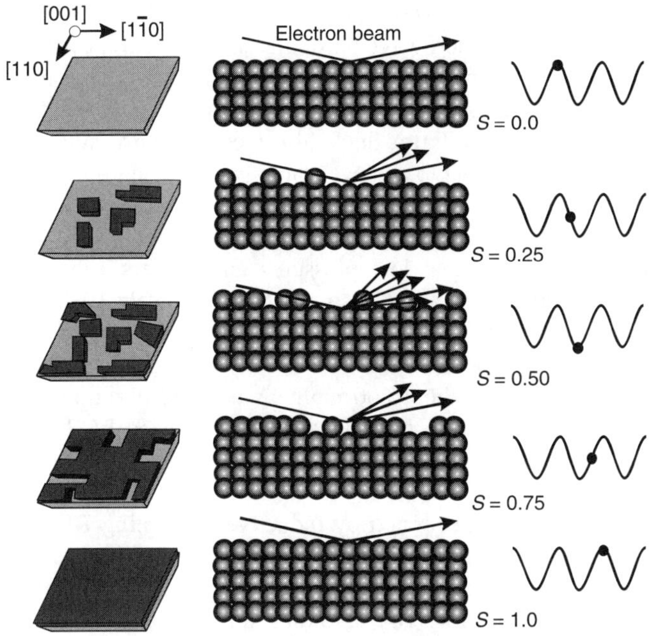

FIGURE 1.21 An illustration of the formation of a single monolayer as seen by an RHEED *in situ* instrument. The corresponding RHEED oscillation signal is shown.

This technique can also be used to measure the depth profile when used in conjunction with an ion sputtering method.

1.3.5.1 Example of a Molecular Beam Epitaxy Growth: III-Nitrides The growth of many semiconductor material thin films and quantum structures by MBE technique, in particular III–V semiconductors, is approaching maturity in almost every aspect. Recently, this technique has been employed in growing GaN and related compounds. The MBE sketch shown in Fig. 1.17 is in fact configured to grow GaN structures. One of the major issues in the deposition of GaN and group III-nitrides by MBE is the incorporation of an appropriate nitrogen source, since molecular nitrogen (N_2) does not chemisorb on Ga because of its large binding energy of 9.5 eV. To solve this problem, different approaches are currently being reported for the growth of cubic and hexagonal group III-nitrides. The first approach is the use of gaseous sources such as ammonia (NH_3) or dimethylhydrazine (DMH), this kind of MBE is also called *chemical beam epitaxy* (CBE) or *reactive ion molecular beam epitaxy* (RMBE). This compound is quite thermally stable and as a result limits the growth temperature significantly. Therefore, lower growth temperatures, such as those needed for low temperature nucleation buffers or for layers containing indium, cannot be grown as easily with NH_3. DMH has higher reactivity than NH_3 and is expected to produce better quality crystals.

The second approach utilizes plasma-activated molecular nitrogen supplied via DC plasma sources, microwave plasma-assisted electron cyclotron resonance (ECR) plasma sources, or RF plasma sources. However, because of the low growth rate of 10–30 nm/h, imposed by the limited nitrogen flux of the DC source, the synthesis of a 1 μm thick film would require approximately 50 h, making it almost impossible to achieve stable growth conditions throughout such a run. ECR sources rely on coupling microwave energy at 2.45 GHz with the resonance frequency of electrons in a static magnetic field. Such coupling allows for ignition of the plasma at low pressures and powers and produces a high concentration of radicals. In an ECR source, approximately 10% of the molecular nitrogen is converted into atomic nitrogen. Because these sources operate very efficiently at fairly low powers, they are usually cooled by air. A typical growth rate of an ECR source is about 200 nm/h. A detailed description of the design and principle of operation of microwave plasma-assisted ECR sources is given by Moustaka. The physical properties of binary nitrides, namely GaN, AlN, and InN, are tabulated in Table 1.1.

Nitrogen plasmas are generated by inductively coupling RF energy at a frequency of 13.56 MHz into a discharge chamber filled with nitrogen at pressures of $>10^{-6}$ mbar. The discharge tube and the beam exit plate can be fabricated from pyrolytic boron nitride (BN) avoiding quartz, which may be a source of residual Si or O doping of GaN. The plasma sheath effect confines ions and electrons within the plasma discharge regions allowing only low energy (<10 eV) neutral species to escape. Therefore, these sources are believed to produce significant concentration of atomic nitrogen. Owing to the very high powers used

TABLE 1.1 A List of Physical Parameters of GaN, AlN, and InN

Parameter	Notation	Unit	GaN	AlN	InN
Lattice constant	a	Å	3.189	3.112	3.548
Lattice constant	c	Å	5.185	4.982	5.760
Thermal coefficient	$\Delta a/a$	10^{-6} K^{-1}	5.59	4.2	—
Thermal coefficient	$\Delta c/c$	10^{-6} K^{-1}	3.17	5.3	—
Band gap, 300 K	E_g	eV	3.42	6.2	0.68
Band gap, 4 K	E_g	eV	3.505	6.28	—
Electron effective mass	m_e	m_0	0.22	—	—
Hole effective mass	m_h	m_0	>0.8	—	—
Elastic constant	C_{13}	GPa	94	127	100
Elastic constant	C_{33}	GPa	390	382	392
Static dielectric Constant	ϵ_r	ϵ_0	10.4	8.5	15.3
Spontaneous polarization	P_{spon}	C/m^2	−0.029	−0.081	−0.032
Piezoelectric coefficient	e_{31}	C/m^2	−0.49	−0.60	−0.57
Piezoelectric coefficient	e_{33}	C/m^2	0.73	1.46	0.97
Binding energy, exciton A	E_{xb}	meV	21	—	—
Thermal conductivity	κ	W/cm.K	2.1	2.85	—
Refractive index	n_r	—	2.2	2.15	—
Melting point	T_m	K	>2573	>3000	—

in these sources, up to 600 W, the plasma chambers are usually water cooled. RF sources permit growth rates up to about 1 μm/h.

All epitaxial growth requires bulk substrates. Bulk GaN is difficult to grow since high pressure and high temperature growth conditions are required (see Manasreh and Ferguson). Thus, substrates other than GaN are currently being used for the growth of epitaxial GaN thin films, heterojunctions, and quantum wells. The most common used substrates are sapphire (Al_2O_3) and SiC. The properties of these substrates are listed in Table 1.2. Although sapphire has a rhombohedral structure, it can be described by a hexagonal material. The hexagonal structure of sapphire provides the same orientation of the grown GaN layer of wurtzite

TABLE 1.2 A List of Physical Parameters of Sapphire and SiC

Parameter	Sapphire		SiC	
Lattice constant (nm)				
	a = 0.4765 @20°C	3C	a = 0.43596	
	c = 1.2982 @20°C	2H	a = 0.30753, c = 0.50480	
		4H	a = 0.30730, c = 1.0053	
		6H	a = 0.30806, c = 1.51173	(Harris)
Melting point (°C)	2030	3C	2793 (Weast *et al.*)	
Density (g/cm^3)	3.98 @20°C	3C	3.166	
		2H	3.214	
		6H	3.211	(Harris)
Thermal expansion coefficients ($10^{-6}K^{-1}$)				
	6.66$\parallel$ to c-axis @20–50°C	3C	3.9	
	9.03$\parallel$ to c-axis @20–10^{3}°C	4H	4.46 a-axis	
	5.0 ⊥ to c-axis @20–10^{3}°C		4.16 c-axis	(Ambacher)
% change in lattice constants between 293–1300 K for Al_2O_3 and 300–1400 K for SiC				
	a/a_0 = 0.83	6H	$\Delta a/a_0$:0.4781	$\Delta c/c_0$:0.4976
	c/c_0 = 0.892	3C	$\Delta a/a_0$:0.5140	(Reeber *et al.*)
Thermal conductivity (W/cm·K)				
	0.23$\parallel$ to c-axis @296 K	3C	3.2	
	0.25$\parallel$ to a-axis @299 K	4H	3.7	
		6H	3.8	(Harris)
Heat capacity (J/K·mol)				
	77.9 @298 K	6H	0.71	(Harris)
Dielectric constants	8.6$\parallel$ to c-axis @10^2–10^8 Hz	3C	$\epsilon(0)$:9.75 $\epsilon(\infty)$:6.52	(Harris)
	10.55$\parallel$ to a-axis @10^2–10^8 Hz	6H	$\epsilon(0)$:9.66, $\epsilon(\infty)$:6.52 ⊥ c-axis	
			$\epsilon(0)$:10.3, $\epsilon(\infty)$:6.70 $\parallel$ c-axis	
Refractive index	1.77 @ λ = 577 nm	3C	2.6916 @ λ = 498 nm	
	1.73 @ λ = 2.33μm	2H	2.6686 @ λ = 500 nm	
		4H	2.6980 @ λ = 498 nm	
		6H	2.6894 @ λ = 498 nm	(Harris)
Resistivity (Ω·cm)	> 10^{11}@300 K	4H	10^2–10^3	(Siergiej)
Young's modulus (GPa)				
	452–460 in [0001] direction	3C	~440	(Harris)
	352–484 in the [$11\bar{2}0$] direction			

The parameters of sapphire were obtained from Belyaev.

symmetry. The growth of GaN on sapphire suffers from the lattice mismatch of interatomic separation in the (0001) interface and from the mismatch of thermal expansion coefficients. The large lattice constant mismatch between GaN and sapphire causes the film to be completely relaxed (not strained). This large lattice constant mismatch must be improved by introducing various processing scheme, such as surface preparation, substrate nitridation, and the growth of buffer layers.

Surface Preparation: Wet and *in situ* methods of etching sapphire include phosphoric acid (H_3PO_4), sulfuric–phosphoric acid mixtures, H_2SO_4–H_3PO_4 fluorinated and chlorofluorinated hydrocarbons, tetrafluoro sulfur (SF_4), and sulfur hexafluoride (SF_6). However, the most common substrate preparation procedure before the growth of GaN is to simply heat the sapphire under flowing hydrogen at high temperatures.

Sapphire Nitridation: Sapphire is nitridated by exposure to nitrogen plasmas or thermally cracked ammonia in MBE systems. Under the conditions of temperature used for MBE growth, AlO_xN_{1-x} should be unstable and nitridation of sapphire results in the formation of AlN. The AlN layer promotes GaN nucleation and increases the wetting of the GaN over layer from 550 to 820°C in MBE growth. The benefits of the nitridation layer are due to a change in the surface energy in the low temperature GaN or AlN buffer layer. The nitridation of sapphire before the growth of a low temperature buffer AlN or GaN is an important step for reducing the defect density, enhancing the electron mobility, and reducing the yellow luminescence in subsequently deposited films. The chemical alternation of surfaces of sapphire substrates using particle beams can be used as an alternative process to nitridation. The advantage of this method over nitridation is its simplicity and room temperature operation. The reactive ion (N_2^+) beam has also been used for pretreatment of sapphire substrates.

Buffer Layer: A low temperature (LT) GaN or AlN buffer (the growth temperature is usually about 400°C for MBE) is an important technique for III-nitride growth, since it can dramatically improve the surface morphology and crystalline quality of GaN sequentially deposited at high temperature (700–850°C for MBE growth).

Polarity: Control of polarity of GaN film is critical in epitaxial growth. This is because it will change surface morphology, doping characteristics, and most importantly, it will determine the direction of piezoelectric field, which is crucial to the device performance.

Other Substrates: In addition to sapphire substrate, III-nitride materials have been grown on other substrates such as SiC, GaAs, and Si (for more details, see Liu and Edgar). The SiC substrates are second to sapphire for epitaxial growth of GaN thin films and quantum structures. The most common polytype SiC structures are illustrated in Fig. 1.22. It has several advantages over sapphire including a smaller lattice constant mismatch (3.1%) for [0001] oriented films, a much higher thermal conductivity (3.8

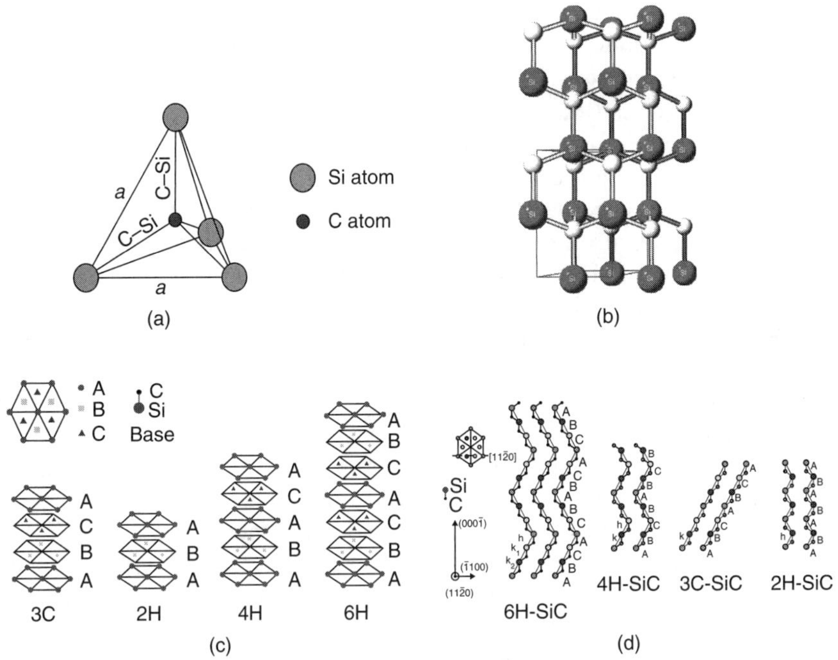

FIGURE 1.22 (a) The tetragonal bonding of a carbon atom with the four nearest silicon neighbors. The distance a and C–Si bond are approximately 3.08 and 1.89 Å, respectively, (b) the three-dimensional structure of 2H-SiC structure, (c) the stacking sequence of double layers of the four most common SiC polytypes, and (d) the $[11\bar{2}0]$plane of the 6H-, 4H-, 3C-, and 2H-SiC polytypes (after Gith and Pentusky).

$W cm^{-1} K^{-1}$), and low resistivity, so electrical contacts to the backside of the substrate are possible.

The lattice constant mismatch for SiC is smaller than that of sapphire, but it is still sufficiently large to cause a large density of defects to form in the GaN layers. The crystal planes in epitaxial GaN parallel those of the SiC substrate, making facets for lasers easier to form by cleaving. It is available with both carbon and silicon polarities, potentially making control of the GaN film polarity easier. High gain heterojunction bipolar transistors taking advantage of the discontinuity created at the GaN–SiC interface are possible. However, SiC does have its disadvantages. Gallium nitride epitaxy directly on SiC is problematic, because of poor wetting between these materials. This can be remedied by using an AlN or $Al_xGa_{1-x}N$ buffer layer but at the cost of increasing the device resistance. This roughness and also remnant subsurface polishing damage are sources of defects in the GaN epitaxial layer. The screw dislocation density in SiC is 10^3-10^4 cm^{-2}, and these defects may also propagate into the GaN epitaxial layer and/or degrade device performance. The thermal expansion coefficient of SiC is less than that of AlN or GaN; thus, the films are typically under biaxial tension at room temperature. Finally, the cost of silicon carbide substrates is high.

Silicon Substrate: Silicon substrates possess physical properties, including high quality and low cost, that are very attractive to GaN-based devices. The physical parameters of bulk silicon are listed along those of GaAs in Table 1.3. The crystallographic structure of silicon is shown in Fig. 1.12. Silicon wafers are very low priced and are available in very large sizes because of its mature development and large-scale production. Silicon has good thermal stability under GaN epitaxial growth conditions. The crystal perfection of silicon is better than any other substrate material used for GaN epitaxy, and its surfaces can be prepared with extremely smooth finishes. The possibility of integrating optoelectronic GaN devices with Si electronic devices is another advantage. To date, the quality of GaN epitaxial layers on silicon has been much poorer than that on sapphire or silicon carbide, because of a large lattice constant and thermal expansion coefficient mismatch, and the tendency of silicon to form an amorphous silicon nitride layer when exposed to reactive nitrogen sources. Gallium nitride and AlN grown on Si(111) are highly defective and nonradiative carrier recombination channels severely limit the luminescence efficiency for device application.

GaAs Substrate: The crystallographic structure of GaAs is zincblende, which is shown in Fig. 1.13. Gallium nitride materials have been grown on zincblende GaAs (for more details on the subject, see As), which is most widely used III–V compound semiconductor as a substrate for zincblende GaN epitaxy since it is well developed and large area substrates are commercially available. The properties of GaAs are listed in Table 1.3. In principle, zincblende structures of GaN possess superior electronic properties for device applications, such as a higher mobility, isotropic properties

TABLE 1.3 Physical Parameters of Silicon and GaAs

Properties	Silicon	GaAs
Lattice constant (nm)	0.543102	0.56536
Density (g/cm^3)	2.3290	5.32
Melting point (°C)	1410	1240
Heat capacity (J/g·K)	0.70	0.327
Thermal conductivity (Wcm·K)	1.56	0.45
Thermal diffusivity (cm^2/s)	0.86	0.26
Thermal expansion(linear) $\times 10^{-6}$ K^{-1}	2.616	6.03
% change in lattice (298K$\sim$ 1311K)	$\Delta a/a_0$: 0.3995	$\Delta a/a_0$: 0.5876
Bulk modulus (GPa)	97.74	75.0
Young's modulus (GPa)	165.6	85.5
Refractive index	3.42	3.66
Dielectric constant	11.8	13.1

because of the cubic symmetry and high optical gain. These advantages may not have been realized due to the difficulty in producing low defect content material. The growth of zincblende GaN requires (001)-oriented substrates having fourfold symmetry. Several substrates, such as GaAs, Si, 3C-SiC, GaP, and MgO, can be used to grow zincblende GaN. The isoelectronic structure (i.e., both GaAs and GaN are III–V compounds), shared element (Ga), potential to convert the surface of GaAs to GaN, and cleavage planes parallel to the epitaxial GaN cleavage planes are the major material advantages of GaAs as a substrate for GaN epitaxy. Technological advantages include a well-established process technology, several readily available substrate orientations of both polar and nonpolar varieties, and low resistance ohmic contacts. There are several disadvantages to GaAs substrates, including a large lattice constant and thermal expansion coefficient mismatch, a poor thermal conductivity, and perhaps most problematic, low thermal stability.

GaAs is much more readily wet etched than sapphire, making GaN films easier to separate from GaAs than sapphire. Thus, GaAs(111) substrates are considered a better template for creating free-standing thick GaN films for subsequent epitaxy and device fabrication, with the ultimate goal of eliminating the problems associated with heteroepitaxy. Since the decomposition rate of GaAs in $N H_3$ or an ultrahigh vacuum rapidly increases at temperatures above 700°C, this imposes limits on the epitaxial growth temperature of GaN and hence its maximum growth rate. Even a small amount of GaAs decomposition is detrimental to zincblende GaN epitaxy, as surface roughening or faceting enhances the onset of wurtzite growth. Since MBE is capable of depositing epitaxial GaN films at a lower temperature, it has been more commonly employed than MOCVD or HVPE when GaAs is the substrate.

1.3.5.2 Growth Rate The gas impingement flux, Φ, on the surface of a substrate is a measure of the frequency with which atoms and molecules impinge on, or collide with, the surface. This flux can be defined in one dimension as the number of molecules or atoms striking a surface per unit area and unit time, assume that the surface is perpendicular to the direction of the atoms or molecules motion, which can be expressed as (see Ohring)

$$\Phi = \int_0^\infty \upsilon_x \mathrm{d}\mathcal{N}_x, \tag{1.34}$$

where

$$\mathrm{d}\mathcal{N}_x = \mathcal{N} f(\upsilon_x)\mathrm{d}\upsilon_x, \tag{1.35}$$

$f(\upsilon_x)$ is the velocity distribution function described by a Maxwell–Boltzmann form as

$$f(\upsilon_x) = \sqrt{\frac{M}{2\pi RT}} e^{-\frac{M\upsilon_x^2}{2RT}}, \tag{1.36}$$

M is the atomic or molecular weight, R is the gas constant, T is temperature, and υ_x is the velocity of the atoms or molecules. By combining Equations 1.34–1.36, the flux is obtained as

$$\Phi = \mathcal{N}\sqrt{\frac{M}{2\pi RT}}\int_0^\infty \upsilon_x e^{-\frac{M\upsilon_x^2}{2RT}}\,\mathrm{d}\upsilon_x = \mathcal{N}\sqrt{\frac{M}{2\pi RT}}\frac{RT}{M} = \mathcal{N}\sqrt{\frac{RT}{2\pi M}}. \qquad (1.37)$$

Substituting the ideal gas equation, $P = \mathcal{N}RT/N_A$, into Equation (1.37) yields

$$\Phi = \frac{PN_A}{\sqrt{2\pi MRT}} \approx 3.513 \times 10^{22}\frac{P}{\sqrt{MT}} \text{ molecules/(cm}^2\text{/s)}, \qquad (1.38)$$

where P is the gas vapor pressure in Torr. Consider a gas escaping a container through an opening of an area $\mathcal{A}$ into a region where the gas concentration is zero. Thus, the rate at which the molecules leave the container is $\Phi\mathcal{A}$ and the corresponding volume flow per second is $\dot{V} = \Phi\mathcal{A}/\mathcal{N}(\text{cm}^3\text{sec}^{-1})$, which can be rewritten as $\dot{V} = 3.64 \times 10^3\sqrt{T/M}\mathcal{A}cm^3 \cdot \text{sec}^{-1}$.

Another aspect of the gas impingement flux is to calculate the time required for a surface to be coated with one monolayer of gas molecules. The characteristic deposition time, $\mathcal{T}$, is inversely proportional to the impingement flux. It is the ratio between the surface density and the impingement flux. The surface density of most semiconductor crystals is on the order of $\sim 7 \times 10^{14}$ atoms/cm^2. Thus, $\mathcal{T}$ can be obtained as

$$\mathcal{T} = \frac{7 \times 10^{14}\text{atoms/cm}^2}{\Phi} = \frac{7 \times 10^{14}\sqrt{MT}}{3.513 \times 10^{22}P} \approx 2.0 \times 10^{-8}\frac{\sqrt{MT}}{P}\text{sec}. \qquad (1.39)$$

The pressure is measured in Torr. For a gas with an atomic weight of 30 g/mole, the deposition time at T = 300 K in 1 Torr pressure is $\sim 1.9 \times 10^{-6}$ sec. On the other hand, if the pressure is 10^{-10} Torr, then the deposition time is ~5.3 h.

Now, let us consider a substrate positioned at a distance l from an aperture of an area $\mathcal{A}$ of a source in an MBE growth chamber. The number of molecules, $\mathcal{M}$, striking the substrate per unit area per second can be expresses as (see Cho)

$$\mathcal{M} = 3.513 \times 10^{22}\frac{P\mathcal{A}}{\pi l^2\sqrt{MT}} \text{ molecules/(cm}^2\text{s)}. \qquad (1.40)$$

For a Ga source in a BN crucible with an opening of $\mathcal{A} = 10$ cm^2 and $l = 20$ cm, the deposition rate can be calculated as follows: assume the source temperature is 970°C or 1243 K and the vapor pressure is 1×10^{-4} Torr. With an atomic weight of 70 g/mole, the arrival rate of Ga atoms at the substrate can be calculated from (1.40) as $\sim 9.5 \times 10^{13}$ atoms/(cm^2.sec). The average GaAs monolayer thickness is 2.83 Å and contains $\sim 6.3 \times 10^{14}$ Ga atoms/cm^2. Hence, the growth rate is $[(9.5 \times 10^{13})/(6.3 \times 10^{14})] \times 2.83 \times 60 \simeq 25.5$ Å/min.

The layer thickness can also be measured using the optical interference method. Consider Fig. 1.23 where the incident light reaches the thin film at an angle. Part of the light will be transmitted through the thin film and the substrate, but a portion of the light will be reflected back and forth between the two interfaces of the thin film. As the photons are bounced between the interfaces, the intensity of the light decreases and an interference pattern is formed due to the difference in the phase of the electromagnetic wave. If the mth order maximum occurs at wavelength λ_1 and the (m+1)th order occurs at λ_2, we have

$$2n_r d \cos\theta = m\lambda_1 = (m+1)\lambda_2, \tag{1.41}$$

where d is the thickness of the layer, n_r is the refractive index of the layer material, and θ is the diffraction angle inside the layer. Equation (1.41) can be

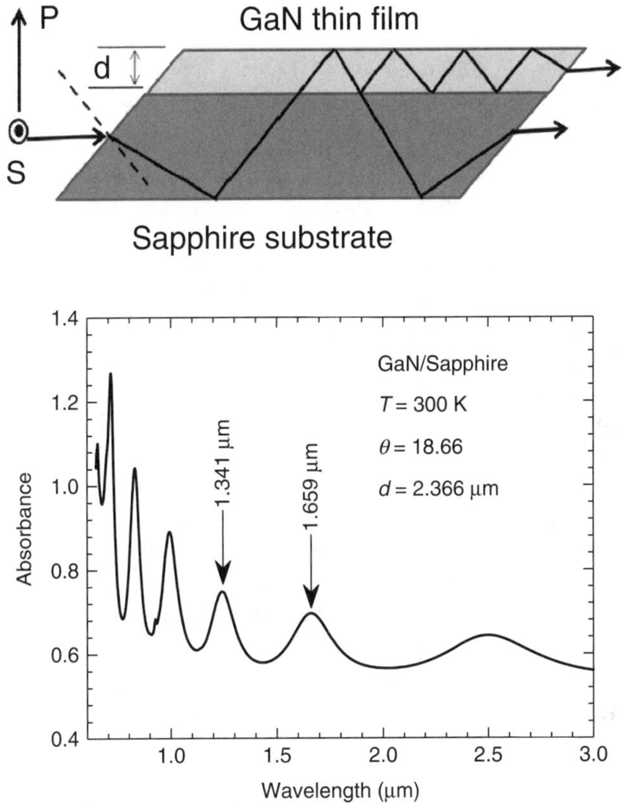

FIGURE 1.23 (a)Waveguide configuration of GaN thin film on a sapphire substrate with an internal diffraction angle of 18.66°. (b) An optical interference pattern in a thin layer is illustrated as the electromagnetic wave bounce back and forth between the two layer interfaces.

rewritten in a more general form as

$$d = \frac{N_f}{2n_r\left(\frac{1}{\lambda_2} - \frac{1}{\lambda_1}\right)\cos\theta}, \tag{1.42}$$

where N_f is the number of fringes between λ_1 and λ_2. The separation between the fringes decreases as the layer thickness is increased. The lower panel in Fig. 1.23 is an actual interference pattern observed in a GaN thin film grown by MOCVD on sapphire. By choosing any two adjacent peaks, the thickness can be calculated by using Equation (1.42). For normal incident light, θ is zero and cos θ is 1. The diffraction internal angle is calculated to be 18.66° for the waveguide configuration shown in Fig. 1.23a.

Experimentally, the actual growth rate of thin films is determined by the measured layer thickness divided by the growth time. Finally, the mole fraction, x, of a ternary compound $A_xB_{1-x}C$ can be determined from the growth rates as

$$x = \frac{\mathcal{G}(A_xB_{1-x}C) - \mathcal{G}(BC)}{\mathcal{G}(A_xB_{1-x}C)}. \tag{1.43}$$

For example, if the growth rates of $Al_xG_{1-x}As$ and GaAs are 36 and 25 Å, respectively, then $x = (36-25)/36 \sim 0.30$.

1.3.5.3 Metal–organic Chemical Vapor Deposition Growth Technique Metalorganic chemical vapor deposition (MOCVD) also known as *metal–organic vapor-phase epitaxy* (MOVPE) is becoming one of the most widely used techniques for the growth of various semiconductor films and structures. It is capable of mass production, where several wafers can be used at the same time for a single run. Thus, most industrial applications rely on this technique for mass production. For the growth of III–V semiconductor compounds, this technique relies on the pyrolysis of metal–organic compounds containing group III elements in an atmosphere of hydrides containing group V elements. Both the metal–organic compounds and the hydride gases are introduced in the reactor chamber in which a bulk semiconductor substrate is placed on a heated susceptor. The substrate has a catalyst effect on the decomposition of the gaseous products. The substrate temperature is usually higher than the temperature of the precursor sources. A sketch of the MOCVD reaction is shown in Fig. 1.24. The gas handling system includes the metal–organic sources, hydride sources, valves, pumps, and any other instruments needed to control the gas flows. The most common carrier gases in MOCVD reactors are hydrogen, nitrogen, argon, and helium.

The metalorganic compounds are either liquids or powders contained in stainless steel cylinder known as *bubblers*. The partial pressure of the source is regulated by controlling the temperature and total pressure inside the bubbler. Mass flow controllers are used to control the mass flow rate of hydride and carrier gases. By sending a controlled flow of carrier gas through the bubbler,

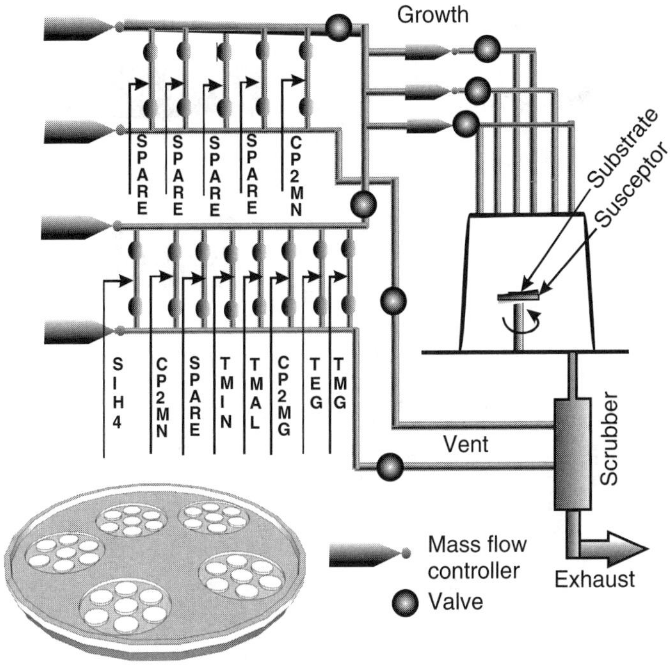

FIGURE 1.24 A sketch of an MOCVD reactor showing gas inputs with valves and mass flow controllers. The sketch in lower left corner is a sample holder with several wafers to illustrate the mass production capability of this growth technique.

the mass flow in a form of dilute vapors of the metal–organic compounds is obtained. The purity of the sources is of paramount importance in the growth of layered structures such as quantum wells and quantum dots. Thus, efforts are devoted to constantly purify source materials.

The reactor chamber is made of stainless steel containing the susceptor, which can hold one or several substrates as illustrated in Fig. 1.24. A commercial MOCVD reactor can hold many 2 in. wafers. The susceptor can be heated by different methods including RF induction heating, radiative heating (lamp), and resistance heating. Knowing and controlling the temperature of the substrate is extremely important for the growth of thin films and quantum structures. One of the recent schemes used to heat the substrate and control its temperature in MOCVD reactors is shown in Fig. 1.25.

In the MBE world, the substrate temperature is controlled by monitoring its band gap absorption edge as a function of temperature. In case of MOCVD, the temperature is basically controlled by measuring the wafer holder (Pocket) temperature as shown in Fig. 1.25. The schematic in Fig. 1.25a consists of LED, filter, photodiode, electronics, and software. An enlarged portion of the substrate is shown in Fig. 1.25b where the substrate temperature is actually the Pocket temperature. Pyrometers are used to measure the Pocket temperature. This technique

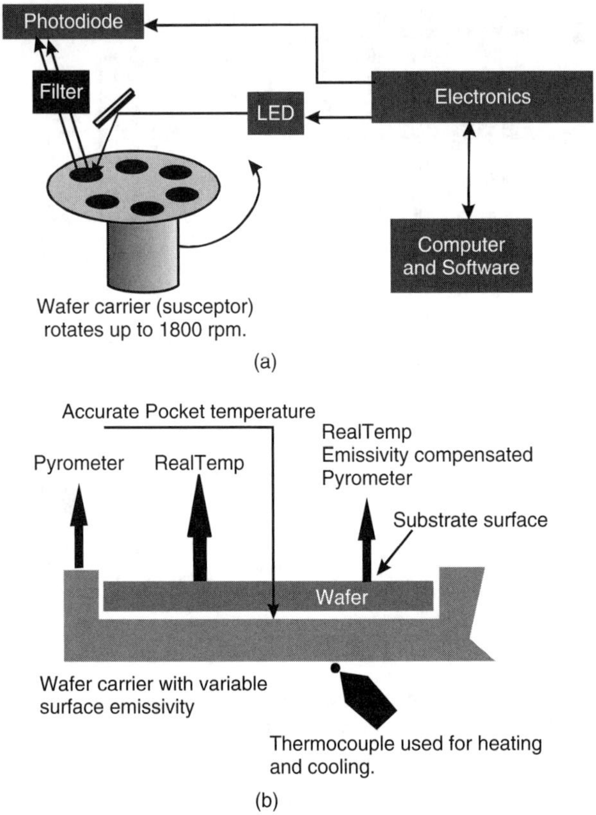

FIGURE 1.25 (a) A schematic of an *in situ* substrate temperature measurement is shown. (b) The substrate in the Pocket is enlarged as shown for clarity.

is actually the only method that one can use to measure the substrate temperature very accurately in case of transparent substrates such as sapphire, which is the commonly used substrate for the III-nitride heterostructures and nanostructures.

The band-gap of sapphire is over 7.3 eV (~170 nm) and the band gap measurement as a function of temperature is difficult to implement, since light sources in this spectral region are difficult to find and thermal radiance from the wafer surface is below the detection level. As the nitride materials are grown on the substrate, the emissivity from the deposited material can be measured. The basic idea is that the temperature of the substrate can be determined accurately and repeatedly by accurate wafer carrier Pocket temperature measurements. The thermocouple in Fig. 1.25b is used to give feedback to cooling and heating only, while the substrate temperature is measured by pyrometers as shown in the figure.

Two fundamental processes occur during the epitaxial growth. First, the thermodynamic process which determines the overall epitaxial growth. Second, the kinetic process which defines the growth rates. The thermodynamic calculations provide information about the solid composition of multicomponent materials when vapor-phase compositions are known. The MOCVD growth is a nonequilibrium process, which cannot provide any information about the time required to reach equilibrium. It also cannot provide information about the transition from the initial input gases to the final semiconductor solid.

While the MOCVD growth technique has been used extensively in the epitaxial growth of materials, such as III-nitride compounds which require growth temperature over 1000°C, it has its own limitations. In particular, many *in situ* characterization tools, such as RHEED, scanning tunneling microscopy (STM), and Auger electron microscopy, cannot be used in the MOCVD chamber due to the fact that the MOCVD growth occurs at atmospheric pressure. However, other techniques, such photoreflectance, ellipsometry, and optical transmissions, have been used recently during the MOCVD growth to monitor growth rates and thin film uniformity.

1.3.5.4 Chemical Vapor Deposition Technique The chemical vapor deposition (CVD) technique has many variations and is used to grow nanorods and nanotubes. A typical configuration is shown in Fig. 1.26. The general configuration of the growth apparatus is composed of a furnace, two quartz tubes, at least one mass flow controlled (MFC), gas cylinders, and a vacuum pump. In the case of ZnO nanorods growth, the zinc powder was place on a wafer in the center of the furnace while several silicon or sapphire substrates with gold nanoparticles were lined up in the inner quartz tube as shown in the figure. This growth technique does not require specific substrates or crystallographic orientations. The gold nanoparticles are used as catalyst for the growth of the nanorods. The chamber is usually evacuated to about one milli-Torr and then flushed several times with nitrogen gas prior to the introduction of oxygen. The Zn powder temperature should be raised above the melting point of the Zn. A typical temperature ranges between 575 and 800°C depending on the desired nanorod size. The oxygen is introduced through the MFC with a rate of 10 sccm. Higher rates produced different nanorod sizes. The growth time ranges between 30 min and several hours. The nanomaterials grown on the substrate are usually removed from the furnace after cooling the system to room temperature. This technique produces all kind of nanomaterials with different shapes and orientations. An example of the material grown by this technique is shown in Fig. 1.27 for ZnO nanorods.

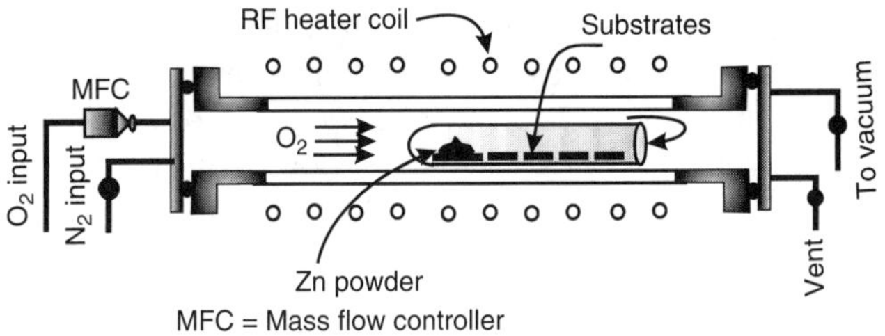

FIGURE 1.26 A sketch of the chemical vapor deposition furnace configuration. The growth of ZnO nanorods is taken as an example.

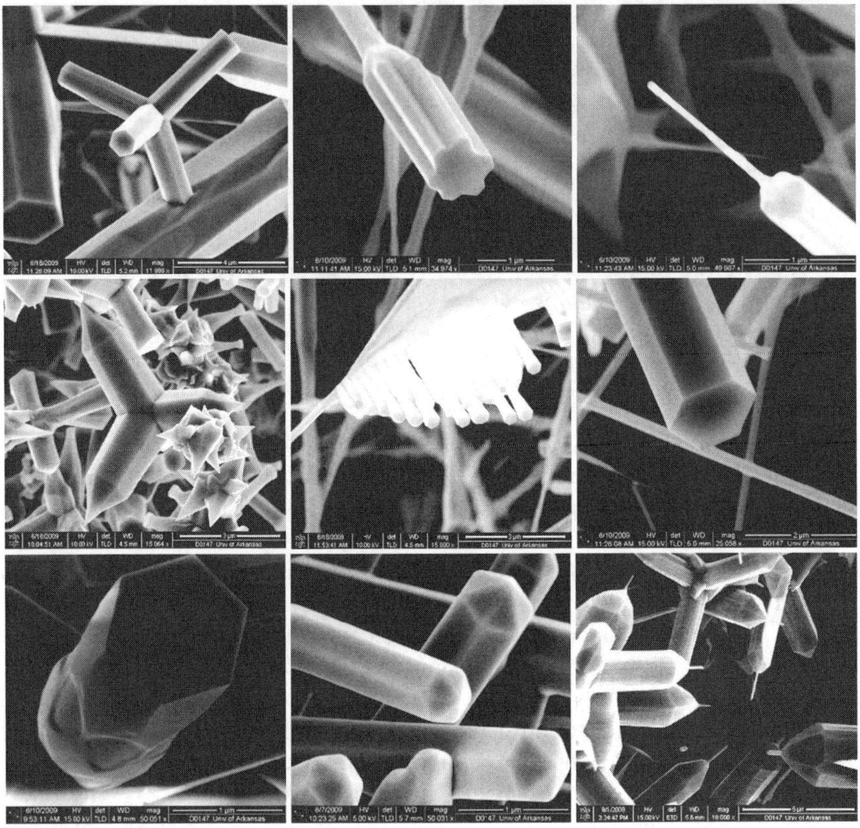

FIGURE 1.27 Several scanning electron microscope images of ZnO nanorods that were grown by the chemical vapor deposition technique. The shape and the size are due to the variation of the growth parameters.

The growth kinetics of CVD nanomaterials depends on a few factors associated with the heterogeneous reaction (gas–substrate interface), such as the transport of reactants through the boundary layer to the substrate, adsorption of reactant at the substrate, atomic and molecular surface diffusion, and transport of by-products away from the substrate through the boundary layer. While the microscopic details of these factors are difficult to model, the growth kinetics is often modeled in macroscopic terms and is capable of predicting the growth rate and uniformity of the grown materials. Following the discussion by Ohring, the reactor configuration is shown in Fig. 1.28. By assuming that the gas has a constant velocity component along the axis of the furnace tube, a constant temperature, and the reactor extends a large distance in the z-direction, the mass flux ($\mathcal{J}$) can be written as

$$\mathcal{J} = C(x, y)\overline{v} - D\nabla C(x, y), \tag{1.44}$$

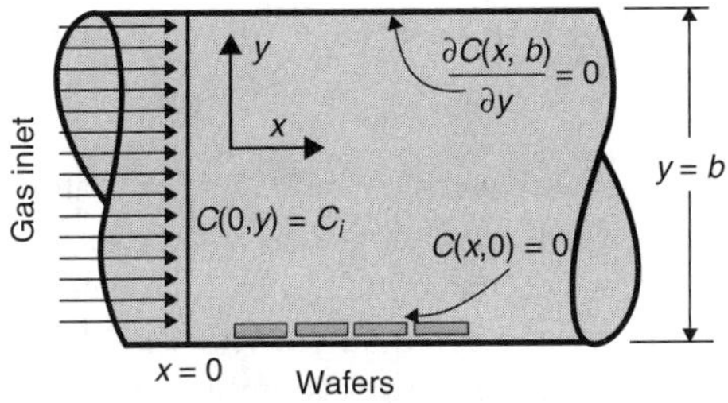

FIGURE 1.28 Horizontal reactor geometry used to obtain the growth rate.

where the first term represents a bulk viscous flow where the source of concentration $C(x,y)$ moving as a whole at a drift velocity $\overline{\upsilon}$. The second term represents the diffusion of individual gas molecules, with a diffusion coefficient D, along the concentration gradients. The flux source at the substrate surface is given by

$$\mathcal{J}(x) = D\left.\frac{\partial C(x, y)}{\partial y}\right|_{y=0} \quad \text{g/(cm}^2\text{s)} \tag{1.45}$$

where $C(x,y)$ is the solution of the steady-state continuity equation

$$D\left(\frac{\partial^2 C(x, y)}{\partial x^2} + \frac{\partial^2 C(x, y)}{\partial y^2}\right) - \overline{\upsilon}\frac{\partial C(x, y)}{\partial x} = 0, \tag{1.46}$$

and is given by

$$C(x, y) = \frac{4C_i}{\pi}\sin\left(\frac{\pi y}{2b}\right)\mathrm{e}^{-\frac{\pi^2 Dx}{4\overline{\upsilon}b^2}}. \tag{1.47}$$

where C_i and b are defined in Fig. 1.28. Equation (1.46) is subject to three conditions that are shown in figure, which are $C = 0$ for $y = 0$ and $x > 0$; $C = C_i$ for $x = 0$ and $b \geq y \geq 0$; and $\partial C/\partial y = 0$ for $y = b$ and $x \geq 0$. For an elemental semiconductor system, such as silicon (Si), the resultant deposition growth rate, $\mathcal{G}$, is related to $\mathcal{J}(x)$ according to the following relation

$$\mathcal{G} = \frac{m_{\mathrm{Si}}}{\rho m_{\mathrm{s}}}\mathcal{J}(x) \text{ cm/sec} \tag{1.48}$$

where m_{Si} and m_{s} are molecular weight of the silicon and the source gas, respectively. Combing Equations (1.45), (1.47), and (1.48) yields

$$\mathcal{G} = \frac{2C_i m_{\mathrm{Si}}}{b\rho m_{\mathrm{s}}}D\mathrm{e}^{-\frac{\pi^2 Dx}{4\overline{\upsilon}b^2}} \text{ cm/sec.} \tag{1.49}$$

This equation predicts an exponential decay of the growth rate as function of the distance along the reactor length, which is quite reasonable, since the input gases are progressively depleted of reactants. The expression of the growth rate provides design guidelines even though these guidelines are not always simple to implement.

1.4 FABRICATION AND GROWTH OF SEMICONDUCTOR NANOMATERIALS

Semiconductor nanomaterials have received significant attention in recent years. While the early techniques of producing these materials, such as quantum dots, relied on optical lithography, X-ray lithography, and electron beam lithography, the preferred method today is the epitaxial growth. The MBE, CVD, and MOCVD growth of nanomaterials is basically self-assembled growth. It is also possible to epitaxially grow nanomaterials on prepatterned substrates. Earlier production of quantum dots, for example, is made by lithography techniques with the multiple quantum wells as the starting materials. This production technique usually yields regular and uniform arrays of quantum dots where the charge carriers are confined in the three dimensions inside the dots.

During the early stages of epitaxial thin film formation, a small number of vapor atoms or molecules condense on the surface of the substrate. This stage is called *nucleation*. Modern *in situ* techniques such as STM and RHEED imaging provide useful information between the end of nucleation and the onset of nucleus growth. A nucleus here refers to as a collection of atoms that form the building block of the quantum dots. When the substrate is exposed to the incident vapor (atomic or molecular beams), a uniform distribution of small and highly mobile clusters or islands (three-dimensional structures or 3D) is observed. In this early growth stage, the prior nuclei incorporate impinging atoms to grow in size. As the growth continues, the islands merge together to form liquidlike materials especially at high substrate temperatures. Coalescence decreases the island density. Further deposition and coalescence causes the islands to connect forming unfilled channels. Additional deposition fills the channel and finally thin films are formed (two dimensional structures or 2D). The idea of the quantum dot growth is to form islands (3D structures) and discontinue the vapor depositions before a thin film (2D structure) is formed. The degree of confinement and the degree of freedom are illustrated in Table 1.4. One should distinguish between the degree of freedom and the degree of confinement. For example, 3D growth means three degree of confinement while the 0D degree of freedom means the charge carriers are confined by three directions, which indicates a quantum dot.

There are three well-known modes of heteroepitaxial growth, which are illustrated in Fig. 1.29. First, the island or Volmer-Weber mode, Fig. 1.29a, which is characterized by the island growth when the smallest stable clusters nucleate

TABLE 1.4 The Confinement and the Degree of Freedom for Bulk and Quantum Wells, Wires, and Dots

	Confinement	Degree of Freedom
Bulk	0D	3D
Quantum well	1D	2D
Quantum wire	2D	1D
Quantum dot	3D	0D

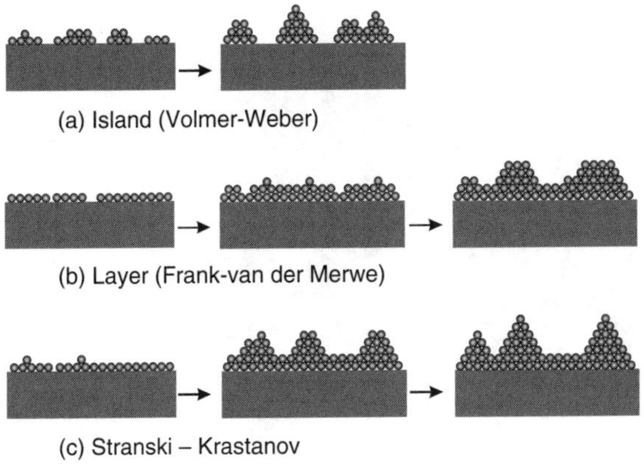

FIGURE 1.29 The three common growth modes of heteroepitaxy. (a) island or Volmer-Weber mode, (b) layer or Frank-van de Merwe mode, and (c) layer-island or Stranski–Krastanov mode.

on the substrate and grow in 3D to form islands (quantum dots). This mode occurs when the atoms or molecules in the deposit are more strongly bound to each other than to the substrate. An example of this growth mode is the deposition of metals on insulators. Second, the layer or Frank-van de Merwe mode as demonstrated in Fig. 1.29b, which is opposite to the island mode during the layer growth. The extension of the smallest stable clusters occurs primarily in two dimensions, resulting in the formation of planar film. The atoms in this mode are more strongly bound to the substrate than to each other. The first complete layer is then covered with a less tightly bound second layer. An example of this mode is the growth of single-crystal semiconductor thin films. Third, the Stranski–Krastanov (SK) mode shown in Fig. 1.29c, which is known as *layer plus island mode*, is an intermediate combination of both the island and layer modes. After forming one or two monolayers, subsequent layer growth becomes unfavorable and islands form. The layer composed of the first two or three monolayers formed at the surface of the substrate, or even the buffer layer, is often called "*wetting layer*."

The transition from 2D to 3D growth is still not well understood. However, any effect that disturbs the monotonic decrease in binding energy characteristic of the layer growth mode may cause the 2D to 3D growth transformation. As an example, the film–substrate lattice mismatch causes strain energy to accumulate in the growing film. The release of this energy from the deposit–intermediate layer interface may trigger the formation of the islands. This process is illustrated in Fig. 1.30, where the formation of InAs (lattice constant = 6.0564 Å) quantum dot on GaAs (lattice constant = 5.65321 Å) is illustrated.

1.4.1 Nucleation

There are a few theories dealing with nucleation. One is the capillarity theory, which is a simple qualitative model that describes the film nucleation. It does

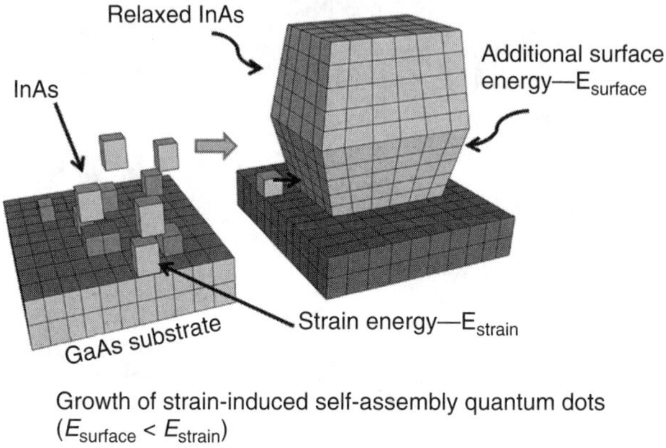

FIGURE 1.30 A schematic illustrates the formation of InAs quantum dot (island) on GaAs substrate.

not provide quantitative information since it lacks the detailed atomistic assumption. However, it provides attractive broad generality, where useful connections between variables, such as substrate temperature, deposition rate, and critical film nucleus size, can be deduced. Atomic nucleation processes theory, introduced by Walton *et al*., is based on the atomistic approach to nucleation. It treats clusters as macromolecules and applies concepts of statistical mechanics in describing them. Another useful model is based on cluster coalescence and depletion. Brief descriptions of these three models are presented in this section.

1.4.1.1 Capillarity Theory Island formation is assumed when the atoms and molecules are impinging on the substrate. The change of the free energy accompanying the formation of islands of mean dimension r can be written as

$$\Delta G = \alpha_3 r^3 \Delta G_V + \alpha_1 r^2 \gamma_{vf} + \alpha_2 r^2 \gamma_{fs} - \alpha_2 r^2 \gamma_{sv}, \quad (1.50)$$

where $\Delta G_{\rm V}$ is the chemical free energy change per unit volume which drives the condensation reaction, $\gamma_{\rm vf}$ is the interfacial tension between the vapor and the film, $\gamma_{\rm fs}$ is the interfacial tension between the film and the substrate, and $\gamma_{\rm sv}$ is the interfacial tension between the substrate and the vapor. The parameters, α_1, α_2, and α_3, are geometric constants given by $2\pi(1 - \cos\theta), \pi \sin^2\theta$, and $\pi(2 - 3\cos\theta + \cos^3\theta)/3$, respectively, for the nucleus shape shown in Fig. 1.31. The curved surface area is $\alpha_1 r^2$, and the volume is $\alpha_3 r^3$. The projected circular area on the substrate is $\alpha_2 r^2$. Young's equation between the interfacial tensions at equilibrium yields

$$\gamma_{\rm sv} = \gamma_{\rm fs} + \gamma_{\rm vf} \cos\theta, \quad (1.51)$$

where the angle θ depends on the surface properties of the involved materials. The three growth modes described in Fig. 1.29 can now be distinguished according to

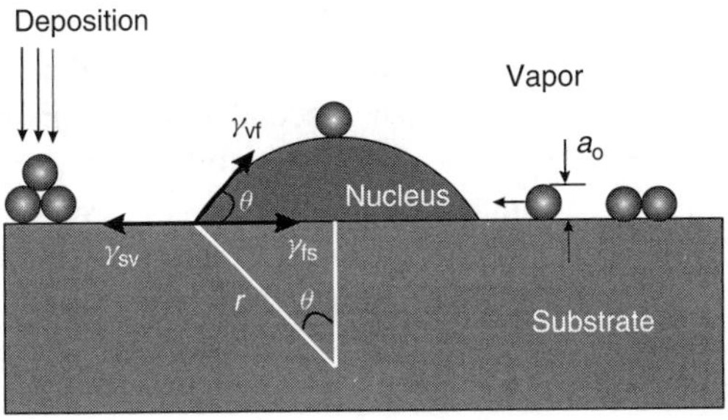

FIGURE 1.31 An illustration of the basic processes of vapor deposition on a surface of a substrate.

the following relationships between the interfacial tensions. For island (Volmer-Weber) growth mode, $\theta > 0$ which yields

$$\gamma_{\text{sv}} < \gamma_{\text{fs}} + \gamma_{\text{vf}}. \tag{1.52}$$

For layer growth (Frank-van der Merwe), the deposit wets the substrate and $\theta = 0$, hence

$$\gamma_{\text{sv}} = \gamma_{\text{fs}} + \gamma_{\text{vf}}. \tag{1.53}$$

An ideal homoepitaxy implies that $\gamma_{fs} = 0$, which yield $\gamma_{sv} = \gamma_{vf}$. The SK growth mode fulfills the inequality

$$\gamma_{\text{sv}} > \gamma_{\text{fs}} + \gamma_{\text{vf}}, \tag{1.54}$$

where the strain energy per unit area of film growth is larger than the interfacial tension between the vapor and film. This condition triggers the formation of quantum dots on the top of the wetting layer.

Figure 1.30 indicates that when a new interface appears there is an increase in the surface free energy. This implies that the second and third terms in Equation (1.50) are positive. The loss of the circular substrate–vapor interface under the film nucleus indicates, as shown in Fig. 1.31 a reduction of the system energy. Thus, the fourth term in Equation (1.50) is negative. The energy barrier to a nucleation process, ΔG^*, can be obtained by first finding the critical radius of the film nucleus. This critical radius, r^*, is obtained by evaluating $\partial \Delta G/\partial r = 0$. Second, substitute r^* backs into Equation (1.50) to obtain

$$\Delta G^* = \frac{4(\alpha_1\gamma_{\text{vf}} + \alpha_2\gamma_{\text{fs}} - \alpha_2\gamma_{\text{sv}})^3}{27\alpha_3^2\Delta G_{\text{V}}^2}. \tag{1.55}$$

By substituting the geometrical factors, α_1, α_2, and α_3 into Equation (1.55), the energy barrier, ΔG^*, takes the following form

$$\Delta G^* = \left(\frac{16\pi\gamma_{vf}^3}{3\Delta G_V^2}\right)\left(\frac{2 - 3\cos\theta + \cos^3\theta}{4}\right). \tag{1.56}$$

An island or aggregate smaller in size than r^* disappears by shrinking, thus lowering ΔG in the process. Equation (1.56) indicates that the heterogeneous nucleation depends on the angle θ. The second term in this equation is called the *wetting factor*. For $\theta = 0$, the wetting factor is zero, and for $\theta = \pi$, the wetting factor is unity. When the deposited film wets the substrate ($\theta = 0$), ΔG^* is zero and there is no barrier to nucleation. On the other hand, when the wetting factor is unity ($\theta = \pi$), ΔG^* is maximum and the growth is identical to that for homogeneous growth.

In the case where the strain energy per unit volume, ΔG_s, is considered in the analysis, the denominator of Equation (1.56) is modified to $3(\Delta G_V + \Delta G_s)^2$. The chemical free energy per unit volume, ΔG_V, is usually a negative quantity, while ΔG_s is a positive quantity. Thus, the overall energy barrier to nucleation is increased. If the substrate is initially strained, then release of this stress during nucleation would be indicated by a reduction in ΔG^*.

The nucleation rate, $\dot{N}$, is one of the parameters that has to be considered during quantum dot growth. According to the capillarity model, the nucleation rate can be written as

$$\dot{N} = N^* A^* \Phi \text{ nuclei/(cm}^2\text{.s)} \tag{1.57}$$

where N^* is the equilibrium concentration per square centimeter of stable nuclei, A^* is the nucleus critical area, and Φ is the overall impingement flux. The equilibrium number of nuclei of critical size per unit area on the substrate is given by

$$N^* = n_s e^{-\frac{\Delta G^*}{k_B T}}, \tag{1.58}$$

where n_s is the total nucleation site density, k_B is the Boltzmann constant, and T is the temperature. A certain number of these sites are occupied by monomers (adatoms) whose surface density, n_a, is the product of the vapor impingement flux and the adatom lifetime, τ_s, which is given by

$$n_a = \frac{\tau_s P N_A}{\sqrt{2\pi MRT}}, \tag{1.59}$$

and τ_s is the given by

$$\tau_s = \frac{1}{\nu} e^{\frac{E_{des}}{k_B T}} \tag{1.60}$$

where E_{des} is the energy required to desorb the adatom back into vapor, and ν is vibrational frequency of the atom ($\sim 10^{12}$ s^{-1}). The area of the nucleus, depicted in Fig. 1.31, can be expressed as

$$A^* = 2\pi r^* a_0 \sin\theta, \tag{1.61}$$

where a_0 and θ are defined in Fig. 1.32. The overall impingement flux is the product of the jump frequency and n_a, where the jump frequency is defined

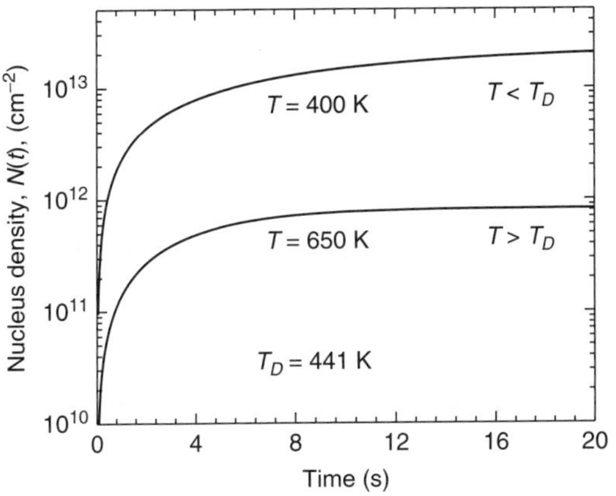

FIGURE 1.32 The nucleus density, $N(t)$, is obtained as a function of deposition time for both $T > T_D$ and $T < T_D$ using Equation (8.74).

as the adatom diffuse jumps on the substrate with a frequency given by $\nu \exp(-E_s/k_B T)$ and E_s is the activation energy of the surface diffusion. Thus, Φ can be expressed as

$$\Phi = \frac{\tau_s P N_A \nu e^{\frac{-E_s}{k_B T}}}{\sqrt{2\pi M R T}} (cm^{-2}.\text{sec}). \tag{1.62}$$

Combine Equations (1.57), (1.58), (1.61), and (1.62) to obtain the following expression for the nucleation rate

$$\dot{N} = 2\pi r^* a_0 \sin\theta \frac{P N_A}{\sqrt{2\pi M R T}} n_s e^{\frac{(E_{des} - E_s - \Delta G^*)}{k_B T}} \text{nuclei/(cm}^2\text{.sec)}. \tag{1.63}$$

The nucleation rate is a strong function of the desorption energy, surface diffusion energy, and nucleation energy.

1.4.1.2 Atomistic Nucleation Processes (Walton et al. Theory) The Walton *et al*. model of nucleation describes the role of individual atoms and the association of small numbers of atoms during the earliest stage of film formation. The model introduces the critical dissociation energy E_{i^*}, which is defined as the energy required to dissociate a critical cluster containing i atoms into i separate monomers. The critical concentration of clusters per unit area, N_{i^*}, is given by

$$\frac{N_{i^*}}{n_0} = \left| \frac{N_1}{n_0} \right|^{i^*} e^{\frac{E_{i^*}}{k_B T}}, \tag{1.64}$$

where n_0 is the total number of adsorption sites and N_1 is the monomer density, which can be written as

$$N_1 = \Phi \tau_s = \Phi \nu^{-1} e^{\frac{E_{des}}{k_B T}}, \tag{1.65}$$

where Φ and τ_s are defined earlier. The critical monomer supply rate is given by the impingement rate and the area over which the monomers are capable of diffusing before desorbing. This area, $\mathcal{L}^2$, can be defined as

$$\mathcal{L}^2 = 2D_s\tau_s, \quad (1.66)$$

where $\mathcal{L}$ is the diffusing mean distance and D_s is the surface diffusion coefficient given by

$$D_s = \frac{1}{2}a_0^2\nu e^{-\frac{E_S}{k_B T}}. \quad (1.67)$$

Thus, the critical monomer supply rate, $\Phi\mathcal{L}^2$, is given by

$$\Phi\mathcal{L}^2 = \Phi a_0^2 \mathbf{e}^{\frac{(E_{des}-E_S)}{k_B T}}. \quad (1.68)$$

The critical nucleation rate, $\dot{N}_{i^*}$, can now be obtained by combining Equations (1.64)-(1.68) such as

$$\dot{N}_{i^*} = N_{i^*}\Phi\mathcal{L}^2 = \Phi a_0^2 n_0 \left|\frac{\Phi}{\nu n_0}\right|^{i^*} e^{\frac{\{(i^*+1)E_{des}-E_S+E_{i^*}\}}{k_B T}} (\text{cm}^2\ \text{sec}^{-1}) \quad (1.69)$$

This expression has been used extensively in determining the nucleation rates in many materials including metals and semiconductors. One of the advantage of this model over the capillarity model is that the uncertainties are in i^* and E_{i^*}, while the uncertainties in the capillarity model are in more parameters (ΔG^*, γ, and θ).

Equation (1.69) can be used to predict the thermally activated nucleation rate whose energy depends on the size of the critical nucleus. This means that there are critical temperatures where the nucleus size and orientation undergo change. As an example, let us consider the temperature $T_{1\to2}$ at which there is a transition from one- to two-atom nucleus. This temperature can be obtained by equating the rates $\dot{N}_{i^*=1} = \dot{N}_{i^*=2}$, which leads to

$$T_{1\to2} = -\frac{E_{des}+E_{21}}{k_B \ln\left(\frac{\Phi}{\nu n_0}\right)} \quad \text{or} \quad \Phi = \nu n_0 e^{-\frac{E_{des}+E_{21}}{k_B T_{1\to2}}} \quad (1.70)$$

where $E_{21} = E_{i^*=2} - E_{i^*=1}$. According to Walton *et al*. when the cluster is composed of one atom, then $E_{i^*=1} = 0$. Thus, E_{21} is simply E_2 and E_{31} is E_3 and so on. As an example, let us assume that a deposition growth rate of a compound semiconductor on a substrate was obtained as a function of temperature. An Arrhenius plot of the data yields an activation energy of $(E_{des} + E_{21}) = 2.1$ eV. The deposition rate was estimated to be 1.0×10^{14} atoms/(cm/s) and $\nu n_0 = 7 \times 10^{27}$ atoms/(cm.sec). The critical transition temperature can now be calculated from Equation (1.70) to be ~764 K. The derivations of expressions for transitions from $i^* = 1$ to $i^* = 3$, $i^* = 1$ to $i^* = 4$, and $i^* = 2$ to $i^* = 4$ were left as an exercise.

Kinetic modeling of nucleation has been the subject of many complex mathematical and physical theoretical models in recent years. Detailed discussions of these models are outside the scope of this textbook. However, the general form of the rate equation for clusters with size i is

$$\frac{\mathrm{d}N_i}{\mathrm{d}t} = K_{i-1}N_1N_{i-1} - K_iN_1N_i, \tag{1.71}$$

where N_i are the densities of clusters and K_i are the rate constants. The first term on the right-hand side represents the increase in the clusters size by attaching monomers to smaller (i-1) sized clusters. The second term expresses the decrease in the cluster density when it reacts with monomers to produce larger (i+1) sized clusters. There are i coupled rate equations to work with, each one of which depends directly on the impingement from the vapor as well as desorption through Equation (1.65). The addition of diffusion terms ($\partial^2 N_i/\partial x^2$) to these coupled equations allows one to account for the change in the cluster shape. A more complete nucleation event can be obtained by including the cluster mobility and coalescence.

Robinson and Robins presented a model for the nucleation and growth kinetics for a one-atom critical nucleus ($i^* = 1$). They considered two temperature regimes separated by a characteristic temperature, T_{D}, given by

$$T_{\mathrm{D}} = \left| \frac{2E_{\mathrm{s}} - 3E_{\mathrm{des}}}{k_{\mathrm{B}} \ln[(C\alpha^2/\beta)(\Phi/\nu n_0)]} \right|, \tag{1.72}$$

where C is a number of pair formation sites ($C = 4$ for a square lattice) and α and β are dimensionless constants with typical values of 0.3 and 4, respectively. For temperatures higher than T_{D}, the reevaporation rate from the surface will control the adatom density and exceed the rate of diffusive capture into growing nuclei. In this regime, the adsorption–desorption equilibrium is rapidly established where $N_1 = \Phi\tau_{\mathrm{s}}$ and incomplete condensation is said to occur. The second regime, where the temperatures are lower than T_{D}, is characterized by high desorption energy (E_{des}) and reevaporation is insignificant. Thus, the condensation is complete and the monomer capture rate by growing nuclei exceeds the rate at which they are lost due to desorption (evaporation).

The analytical expressions for the time-dependent transient density of stable nuclei, $N(t)$, and the saturation value of $N(t = \infty) = N_{\mathrm{s}}$ are given as follows

$$N(t) = \begin{cases} N_{\mathrm{s}} \tan h\left(\dfrac{\dot{N}(0)t}{N_{\mathrm{s}}}\right) & \text{for } T > T_{\mathrm{D}} \\ N_{\mathrm{s}}\left\{1 - \mathrm{e}^{-3\eta^2\frac{\dot{N}(0)t}{N_{\mathrm{s}}^3}}\right\} & \text{for } T < T_{\mathrm{D}} \end{cases} \tag{1.73}$$

$$N_{\mathrm{s}} = \begin{cases} \sqrt{Cn_0/(\beta\nu)}\sqrt{\Phi}\mathrm{e}^{\frac{E_{\mathrm{des}}}{2k_{\mathrm{B}}T}} & \text{for } T > T_{\mathrm{D}} \\ (Cn_0^2/(\alpha\beta\nu))^{1/3}\Phi^{1/3}\mathrm{e}^{\frac{E_{\mathrm{s}}}{3k_{\mathrm{B}}T}} & \text{for } T < T_{\mathrm{D}} \end{cases} \tag{1.74}$$

where

$$\dot{N}(0) = \left.\frac{\partial N(t)}{\partial t}\right|_{t=0} = \frac{C\Phi^2}{\nu n_0}\mathrm{e}^{\frac{2E_{\mathrm{des}} - E_{\mathrm{s}}}{k_{\mathrm{B}}T}}, \tag{1.75}$$

TABLE 1.5 Nucleation Parameters P and E in Equation (1.77) as Reported by Venables *et al*. for Different Regimes

Regime	Three Dimensional	Two Dimensional
Extreme incomplete	$P = 2i^*/3$	$P = i^*$
	$E = (2/3)\{E_{i^*} + (i^* + 1)E_{\text{des}} - E_{\text{s}}\}$	$E = E_{i^*} + (i^* + 1)E_{\text{des}} - E_{\text{s}}$
Initially incomplete	$P = 2i^*/5$	$P = i^*/2$
	$E = (2/5)\{E_{i^*} + i^* E_{\text{des}}\}$	$E = (1/2)\{E_{i^*} + i^* E_{\text{des}}\}$
Complete	$P = i^*/(i^* + 5/2)$	$P = i^*/(i^* + 2)$
	$E = \{E_{i^*} + i^* E_{\text{s}}\}/(i^* + 5/2)$	$E = \{E_{i^*} + i^* E_{\text{s}}\}/(i^* + 2)$

and

$$\eta = \frac{n_0}{\alpha} e^{\frac{E_{\text{S}} - E_{\text{des}}}{k_{\text{B}}T}}. \tag{1.76}$$

These equations indicate that $N(t)$ increases with time and reaches saturation at the value N_{S}. Let us consider the deposition of clusters of a material where $E_{\text{s}} = 0.29$ eV and $E_{\text{des}} = 0.7$ eV. The rest of the parameters are $n_0 = 5 \times 10^{15}$ cm^{-2}, $\nu = 1.65 \times 10^{12}$ sec^{-1}, $C = 4$, $\alpha = 0.3$, $\beta = 4$, $\Phi = 8.5 \times 10^{14}$ nucleus/(cm/s). The characteristic temperature is calculated from Equation (1.72) to be $T_{\text{D}} = 441.0$ K. The cluster (nucleus) density $N(t)$ is calculated from Equation (1.73) for the following temperatures $T = 700$ K (larger than T_{D}) and $T = 430$ K (smaller than T_{D}). The results are shown in Fig. 1.32. It is clear from this figure that the number of nuclei (clusters) is larger when the deposition temperature, T, is smaller than the characteristic temperature T_{D}. When $T > T_{\text{D}}$, the desorption process (reevaporation rate) dominates over the condensation process. This leads to a lower cluster density as compared to the case when $T < T_{\text{D}}$.

For the case where i^* is any integer, the analysis becomes complicated. However, a review article on the subject was presented by Venables *et al*. in which the nucleation parameters for 3D and 2D were summarized. The stable cluster density is given by Venables *et al*. as

$$N_{\text{s}} = An_0 |\frac{\Phi}{\nu n_0} e^{\frac{E}{k_{\text{B}}T}}, \tag{1.77}$$

where A is a dimensionless constant that depends on the substrate coverage. The parameters P and E depend on the condensation regimes, which are summarized in Table 1.5.

The complete and extreme incomplete regimes are similar to those discussed earlier. The extreme incomplete regime occurs when the reevaporation process (desorption) is dominant, and the complete regime occurs when the monomers capture rate exceeds the desorption rate. The initially incomplete regime can be thought of as an intermediate regime which is applicable to SK growth mode.

1.4.1.3 Cluster Coalescence, Sintering, and Migration Models According to the kinetic models, the initial stages of growth are characterized by an increase in the density of the stable clusters as a function of growth time until it reaches a maximum level before starts to decrease (saturation effect). The process that describes the cluster behavior beyond saturation is called *coalescence*. This coalescence process is usually characterized by a decrease in the total number of clusters and an increase in the height of surviving clusters. Other features that describe the coalescence process include the following: clusters with well-defined crystallographic facets tend to become rounded; the clusters take a crystallographic shape with time; the process appears to be liquidlike in nature with clusters merging and changing shapes, where the crystallographic structure of the larger clusters dominates during the merger of smaller clusters; and the clusters are observed to migrate prior to their merger into one another.

There is a size variation when clusters are deposited on the surface of a substrate. The larger clusters tend to grow in size with time at the expense of the smaller ones. This is called *ripening effect*. The time evolution of cluster distribution was investigated by Vook using both statistical models involving single atom process and macroscopic surface diffusion-interface transfer models.

The coalescence process occurs by several methods. One method is called *Ostwald ripening* where the diffusion of adatoms will proceed from the smaller to larger cluster until the smaller cluster disappears completely. The diffusion of atoms occurs without having the clusters in direct contacts. A second method is called *sintering*, where the clusters are in contact. A neck is formed between the clusters and then thickens as the atoms are transported in the contact region. Cluster migration is another mechanism for coalescence where the clusters on the surface of the substrate actually migrate. Coalescence occurs in this mechanism as a result of collisions between separate clusters (droplet) as they randomly move around.

1.4.2 Fabrications of Quantum Dots

The production of low dimensional semiconductor systems, where the charge carriers are confined in two directions (quantum wires) and/or three dimensions (quantum dots), is of interest to those who are involved in the basic understanding of the nature of these systems, as well as those who are interested in producing devices based on the novelty of these quantum structures. Since the early 1980s, many research groups throughout the world have been focused on the production of quantum wires and dots using lithograph techniques. All the early efforts were focused on processing quantum wells into quantum wires or quantum dots by patterning. While, in many cases, the patterned techniques have proven to be difficult or expensive to perform, they offer several advantages, such as good control on the lateral shape, size, and arrangement.

Optical lithography techniques using lasers and ultraviolet optics in conjunction with photoresists are used to produce quantum dots with a resolution as high as 100 nm. This technique may not be able to reach dimensions on the order of

20 nm or less. However, X-ray has the potential to mass produce nanostructures, since it has a shorter wavelength.

Electron beam lithography has been used to produce quantum dots and wires. The electron beam is usually emitted from a high brightness cathode or a field emission gun. Since electrons are charged, it is very easy to focus them with magnetic lens system. A resolution of 10–20 nm has been achieved by this technique. The electron beam is computer controlled, and images can be defined on the substrate with the help of deflection fields system. The final resolution of the pattern is limited by the resists due to finite length of the organic molecules and the grain size. Periodic nanostructures can be produced by using electron beam interference technique.

In addition to electron beam lithography, focused ion beam lithography has been used in production of patterned quantum dots. This technique has been used for maskless etching, maskless implantation of dopants, deposition of metallic structures, and patterning of resists. However, the resolution of the focused ion beam lithography is not as high as the electron beam lithography. There are many other techniques used to pattern quantum dots or grow quantum dots on patterned substrate. For a detailed review of the subject, see Bimberg *et al*.

1.4.3 Epitaxial Growth of Self-Assembly Quantum Dots

The growth of self-assembly quantum dots has been widely made by MBE and MOCVD techniques. Our knowledge of the quantum dots structural characteristics has been obtained by tools, such as scanning electron microscopy (SEM), STM, atomic force microscopy (AFM), transmission electron microscopy (TEM), RHEED images, and X-ray diffraction. AFM has an atomic resolution, and the limiting factor of AFM is the size and shape of the microscope tip. The STM, on the other hand, has the advantage of revealing the morphology of a surface on an atomic scale. It also allows the manipulation of the surface atoms by lining them up and forming shapes and figures.

One of the most important aspects of quantum dot growth is the starting surface of the substrate. The surface construction, surface strain, and crystallographic orientations of the substrate play a major role in the growth of self-assembly quantum dots. As an example, let us consider the formation of InAs quantum dots on a GaAs (001) surface. In this case, the RHEED image exhibits a streaky pattern before the deposition of InAs atoms indicating a flat surface. RHEED pattern remains streaky before the 2D–3D transition occurs, which corresponds to 1–1.5 monolayer deposition of InAs. When more InAs atoms are deposited, the RHEED pattern changes from streaky to spotty, which is an indication of 3D islands being formed. Generally speaking, a monolayer of InAs remains stable for a short growth interruption time on the order of 10 s or so. As the growth interruption continues (more than 1000 s), the InAs material transitions into 3D islands with anisotropic shapes and sizes.

A simple model was presented by Leonard *et al*., where the 3D islands start to develop on top of a 2D wetting layer above a critical thickness, Θ_c. The planar density of the MBE-grown InAs 3D islands, ρ_{SAD}, was determined from AFM images versus the amount of InAs deposited, Θ. The relation between ρ_{SAD} and Θ was described as being similar to that of a first-order phase transition

$$\rho_{SAD} = \rho_0(\Theta - \Theta_c)^{\alpha}, \tag{1.78}$$

where α is the exponent and ρ_0 is the normalization density. A fit of the results is shown in Fig. 1.33 using Equation (1.78), which yields $\alpha = 1.76$ and $\rho_0 = 2 \times 10^{11}$ cm^{-2}.

Since Leonard *et al*. reported their finding, several research groups have investigated the 2D to 3D transition in many quantum dot systems and their reports provided complex structures and behaviors depending on the growth temperature, substrate starting surface, and crystallographic orientations. For example, AFM images of various self-assembled quantum dots grown by the MBE technique are shown in Fig. 1.34. The quantum dot size and density depend strongly on the growth temperature and post-growth annealing. An illustration of these effects on InAs quantum dots is shown in Fig. 1.35, where the growth temperature (substrate temperature) varied from 400°C to 500°C at a fixed deposition time of 20 s. The quantum dots were MBE grown on ~0.20 μm thick AlAs (lattice constant = 5.660 Å) layer, which was grown on a GaAs substrate. The InAs growth rate was ~0.30 Å/s, which yields a total deposited material of ~2 monolayers. The quantum dot density decreases as the growth temperature increases. Accompanying the decrease in density is an increase in the quantum dot size, as shown in Fig. 1.35a. The post-growth annealing has a similar effect. Figure 1.35b shows STM images of the same sample that was grown at 500°C but annealed as a function of time. In this case, the density of the quantum dots decreases and their size increases as

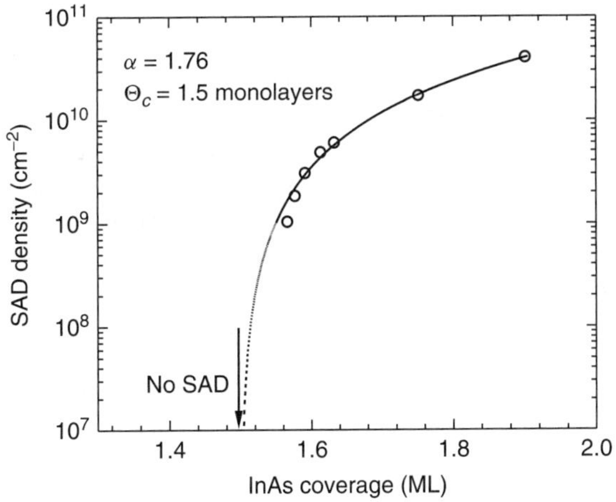

FIGURE 1.33 Density of self-assembled dots (SAD) versus InAs coverage. Treating these data as a first-order phase transition gives a critical thickness of 1.50 monolayers (ML) (after Leonard *et al*.).

FIGURE 1.34 Atomic force microscopy images of different types of quantum dots grown by the molecular beam epitaxy.

the annealing time is progresses from $t = 0$ to $t = 10$ min. The behavior of the quantum dots as a function growth temperature and post-growth annealing time can be understood in terms of the kinetics model discussed in Section 1.4.1.

The growth of quantum dots is highly influenced by the growth conditions. The growth temperature and post-growth annealing discussed above are examples of how the density and size of the quantum dots are drastically changed. There are many other parameters that affect the structural and physical properties of quantum dots. The growth modes such as the simultaneous deposition mode and alternate deposition modes can yield different results. In the simultaneous deposition mode, the constituent atoms are deposited at the same time. This is accomplished by opening the sources shutters at the same time. On the other hand, the alternative deposition mode relies on the alternative deposition of the constituents of the quantum dots. Quantum dots grown by the simultaneous deposition mode were found to be affected by the quantum dot arrangement on the surface of the substrate, while the alternative deposition mode produces higher densities of quantum dots grown on vicinal substrates.

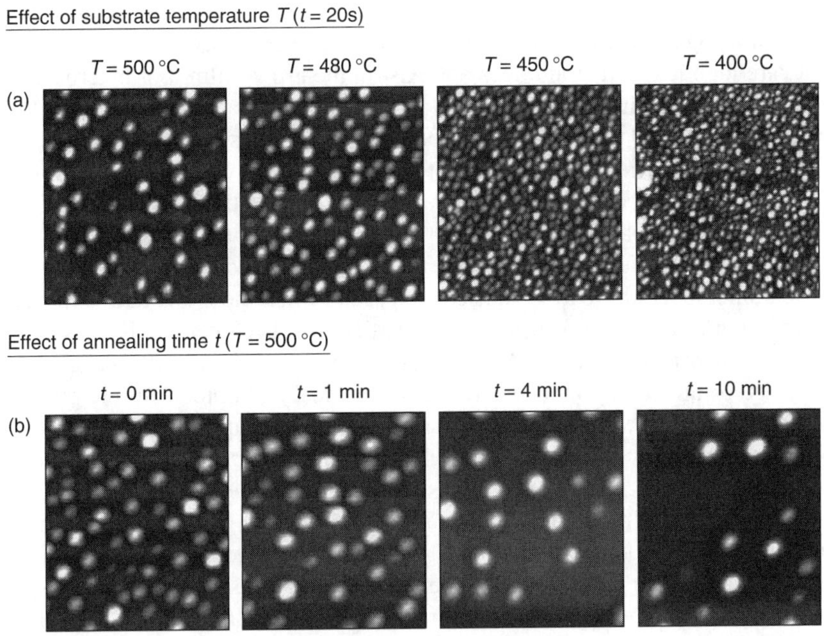

FIGURE 1.35 Illustration of the growth temperature and post-growth annealing effects on InAs quantum dots grown by MBE on AlAs layer, which was grown on GaAs (001) substrate. (a) An STM images obtained for different samples that were grown at different temperatures. (b) The STM image illustrates the post-growth annealing on the quantum dots size and density as the annealing time is increased.

Arsenic pressure in the MBE growth of InAs affects the stability of the quantum dots. An increase of the arsenic pressure in the MBE chamber can dramatically affect the morphology of the quantum dots. The size of the quantum dots is reduced, and significant dislocation densities appear in larger InAs clusters. The arsenic pressure can also affect the transitions from 2D to 3D growth. The stoichiometry of (001) surfaces of III–V compound semiconductors that are in equilibrium with the gas is known to depend on the partial pressure of group V elements. Thus, a change in the arsenic pressure leads to a change in the surface reconstructions and to a change in the surface energy of the GaAs (001) surface. At a lower arsenic pressure, indium is known to segregate to the surface, which can be regarded as a quasi-liquid phase that shows no strain-induced renormalization of surface energy. Thus, an increase in the surface area due to the arsenic pressure reduction may lead to a large surface energy making the formation of 3D islands unfavorable.

Substrate orientation plays a vital role in the growth and formation of the quantum dot systems. It affects all parameters governing the quantum dot formation, such as strain energy and surface energy. The strain caused by the difference of the lattice constants of the quantum dots and the substrate is the primary reason for the formation of self-assembled quantum dots. Additionally, this strain difference affects the electrical and optical properties of the quantum dot systems. For homoepitaxial growth, the lattice constants of the epitaxial film and the substrate are identical and strain does not exist in the grown film. For heteroepitaxial growth, the film being deposited on a substrate does not necessary have a lattice constant similar to that of the substrate. Owing to this lattice constant difference, one can envision three distinct epitaxial cases, which are illustrated in Fig. 1.36. The first case is when there is a lattice match or a very small lattice mismatch between the deposited film and the substrate as shown in Fig. 1.36a. The strain in this case is almost zero, and the heterojunction growth is identical to the homoepitaxial growth. The second case is when there is a lattice mismatch between the film and the substrate. This lattice mismatch causes the strain in the film as shown in Fig. 1.36b. A small lattice mismatch is actually beneficial for many heterojunction systems where the mechanical and optoelectronic properties are enhanced by the strain. Strain usually removes the degeneracy, which leads to an improvement in the electronic and optical properties of the film. In fact, the presence of this strain in the deposited material is the driving mechanism for the formation of self-assembled quantum dots. When the lattice mismatch is substantially large, the thin film is relaxed by the formation of the dislocations at the interfaces as shown in Fig. 1.36c. The relaxed epitaxy is usually reached during later film formation stage (thicker films) regardless of the crystal structures or lattice constant difference. For all quantum dot systems, the dots are strained.

In addition to planar self-organization of quantum dots, the quantum dots can be vertically stacked. The vertical stacking is very important for devices, such as detectors and emitters, since this stacking increases the filling factor of quantum dots. Depending on the growth temperature and the spacer or barrier, the physical, structural, and morphological properties of the vertically stacked quantum dots can be different than those of the planar self-organized dot systems. An illustration of the vertical stacked quantum dots is shown in Fig. 1.37 for InGaAs/GaAs system. The images were obtained using TEM for both the planar and the cross-section (X-TEM) configurations. A remarkable property of the vertically stacked quantum dots is that the dots are self-aligned vertically as it is clearly shown in the cross-sectional TEM images. The vertical correlation has been observed in many quantum dot systems, and it can be lost if the barrier thickness is too large. For vertical self-alignment to occur, it is very important that the quantum dot layer is successfully grown.

The planar TEM image shown in Fig. 1.37 clearly demonstrates the formation of the InGaAs quantum dot chains grown on GaAs (001) surface. The quantum dots shape in the first layer is not well defined. By inspecting the X-TEM image of the quantum dot stack, the quantum dot shape is not well defined in the first layer, but then subsequent layers show that the shape of the quantum dots are well defined and pyramidal in shape.

Additionally, the size of the quantum dots becomes uniform as the multiple layers of quantum dots are stacked. The production of quantum dots uniform in size is a very important and necessary aspect of nanotechnology, since the uniformity of the dots affects the device performance. The size fluctuation impacts the

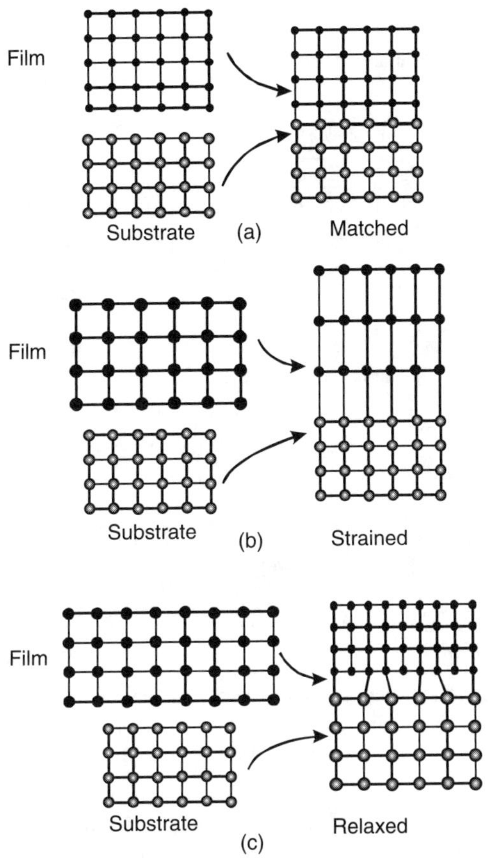

FIGURE 1.36 A sketch of heterojunction growth showing (a) lattice matched, (b) strained, and (c) relaxed structures.

quantized energy levels in the dots causing inhomogeneous energetic broadening. This broadening should be minimized by producing quantum dots with highly uniform size. For example, infrared detectors fabricated from quantum dots usually require structures that are composed of multiple layers of dots and barriers. This structure is required to obtain a minimum number of dots. The detection of light relies on the confined energy levels inside the quantum dots. The positions of these energy levels are very sensitive to the physical dimensions of the dots.

1.5 COLLOIDAL GROWTH OF NANOCRYSTALS

Colloidal growth technique has many variations depending on the type and properties of the nanocrystal. The most common techniques are sol–gel, hydrothermal synthesis, solvothermal synthesis, emulsion, high temperature organic solvent,

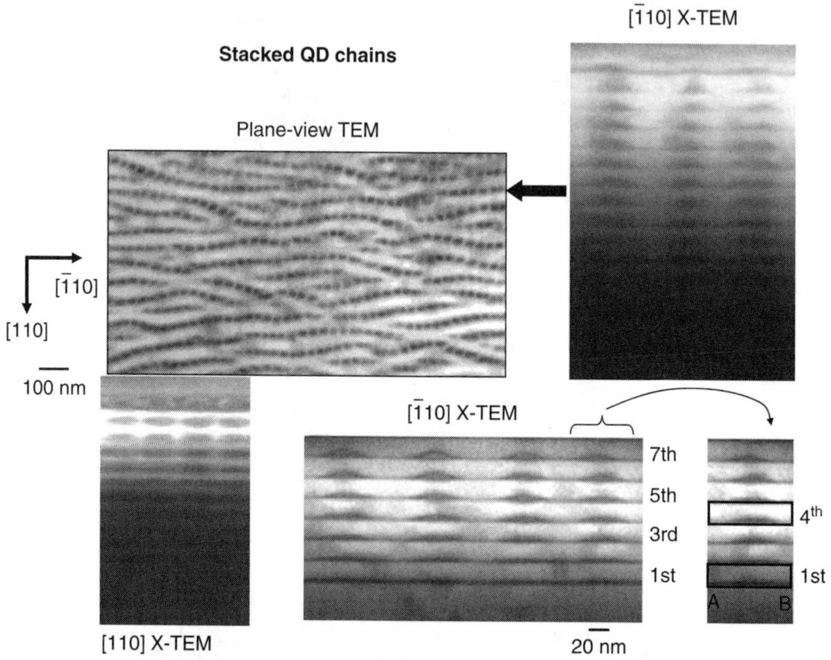

FIGURE 1.37 Tunneling electron microscopy images of InGaAs/GaAs multiple quantum dot layers. The planar view shows the formation of quantum dot chains, while the cross-sectional images show the vertical correlation of the quantum dots.

and co-precipitations. Colloidal growth produces all kind of nanomaterials including metals, semiconductors, and oxide nanoparticles with different shapes and sizes. These colloidal nanocrystals are usually uniform in sizes and properties make them ideal nanomaterials for many applications, such as light emitting diodes and detectors. The percentage of the surface atoms of the nanocrystals is very large depending on the number of the complete shells in the crystal. This is demonstrated in Table 1.6. When the nanocrystals are not stable, a shell from a different stable material is grown leading to what is called core/shell nanocrystals, such as CdSe/ZnS, InP/ZnS, and CdSe/CdS core/shell nanocrystals. The diameters of these nanocrystals are usually ranging between 2 and 30 nm or so.

The optical and electrical properties of the nanocrystal strongly depend on their size. For example, the effective band gap energy (E_g^*) of the nanoparticles is red shifted as the size of the nanocrystal is increased according to the following relationship:

$$E_g^* = E_g + \chi \left(\frac{\pi^2 \hbar^2}{2\mu^* d^2} \right), \tag{1.79}$$

where E_g is the band gap energy of the bulk materials, $\hbar$ is Planck's constant, μ^* is the reduce effective mass of the materials, which depends on the electron (m_e^*) and hole (m_h^*) effective masses ($1/\mu^* = 1/m_e^* + 1/m_h^*$), d is the diameter

TABLE 1.6 The Relationship between the Total Number of Atoms in Full-Shell Clusters and the Percentage of Surface Atoms

Crystal	Number of Shells	Diameter Å	Number of Atoms	Surface Atoms (%)
	One shell	6.0	13	92
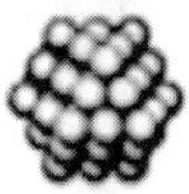	Two shells	10	55	76
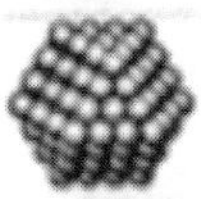	Three shells	14	147	63
	Four shells	18	309	52
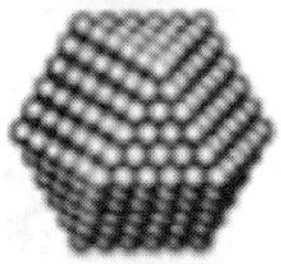	Five shells	22	561	45
	Six shells	26	1415	35

Courtesy of A. Wang.

of the nanocrystal, and χ is a numerical factor which is close to unity for ideal nanocrystals. A plot of the effective band gap as a function of the nanocrystal diameter for different materials is left as an exercise (Problem 1.20).

The cost of the nanocrystals' growth is usually very low as compared to the growth cost by using MBE and MOCVD techniques. Since the nanocrystal growth depends on many chemical reactions, these nanocrystal are usually covered with ligands called *trioctylphosphineoxide* (TOPO). Ligands removal or exchange is very necessary to functionalize the nanocrystals for device applications, such as light emitting device, detectors, sensors, and photovoltaic device.

1.6 SUMMARY

The basic principles of single-crystal growth ranging from bulk semiconductor materials to quantum dots and nanocrystals are discussed. The introduction of this chapter is focused on the importance of bulk materials and the wafering process.

The growth of any materials, elemental or compound, depends on a set of thermodynamic conditions. The phase diagram is a critical aspect in understanding the thermodynamic conditions needed to grow single-crystal materials.

In addition to many bulk semiconductor applications, single-crystal wafers are vital to the growth of thin films and the epitaxial growth of all nanomaterial structures including heterojunctions, quantum wells, quantum wires, and quantum dots. Thus, understanding the growth of bulk semiconductors is essential to this chapter. The most widely used bulk growth methods were discussed, including LEC, Bridgman, float zone, and Lely methods. The segregation coefficient of dopants in bulk material is discussed. The elimination of background impurities or the introduction of a well-controlled dopant in bulk materials are very important. The dopant distribution in bulk semiconductor crystals is usually depends on the radius and the length of the boules.

There are several methods used to grow semiconductor thin films on substrates and wafers. The most common methods used in the growth of thin films are LPE, VPE, HVPE, PLD technique, MBE, and MOCVD technique. Each one of these techniques has advantages and disadvantages, which are briefly discussed in this chapter.

Epitaxial growth rate, nucleation, and growth kinetics of highly nonequilibrium growth were discussed. Several nucleation models were reported to explain the three island nucleation. The CVD growth technique is discussed. This technique is usually used to grow nanorods and nanotubes. The colloidal growth of nanocrystals was briefly discussed. These nanomaterials have myriad applications, ranging from optoelectronic devices to medical diagnosis.

PROBLEMS

1.1. Find the miller indices for a plane intercepting the following crystal coordinates:

$1a, -2a, 3a$

$2a, \infty, \infty$

$\infty, 3a, 5a$

$\infty, \infty,$ &minus1a

$1a, 1a, 1a$

where a is the interatomic distance also known as the *lattice constant*.

1.2. Draw the crystal planes for the following orientations: (200), (222), (311), (133), and (123).

1.3. What are the total number of planes of the following groups: {001}, {011}, and {111}. Show all the possible orientations for each group

1.4. Derive Equation (1.6). Calculate ΔG° for a chemical reaction at $T = 300$ K with $k = 2.3 \times 10^{-9}$.

1.5. Derive Equation (1.8). Calculate the maximum velocity in inches per hour needed to grow silicon single crystal. Assume the following parameters: $L = 340$ cal/g, $M_v = 2.33$ g/cm^3, $dT/dx = 6$ K/cm, and $k_s = 0.21$ W/cm/K.

1.6. Derive Equation (1.13) and plot C_s/C_0 versus m/m_0 for the following values of $k_0 = 0.01$, 0.05, 0.3, 0.5, 0.9, 1, 2, and 3.

1.7. Derive Equation (1.21).

1.8. Derive Equation (1.26), then plot C_s/C_0 as a function of x/L for the following values of $k_e = 0.01$, 0.1, 0.5, 2, and 5.

1.9. Calculate the mean-free path of an atom of a diameter of 2.5 Å in an MBE chamber with a vapor pressure of 7.5×10^{-4} Torr. The substrate temperature is kept at 550°C. Compare your result to the typical source–substrate distance of 30 cm.

1.10. Show that the volume of flow of a gas escaping a container through an opening of an area $\mathcal{A}$ into a region where the gas concentration is zero is given by

$$\dot{V} = 3.64 \times 10^3 \sqrt{T/M}\mathcal{A}cm^3 \cdot \sec^{-1}$$

where T is the temperature and M is the atomic weight of the gas. Calculate the volume flow rate of a gas at 300 K and an atomic mass of 30 g/mole.

1.11. Show that the gas impingement flux, Φ, can be written as $\Phi = \frac{1}{4}\mathcal{N}\overline{\upsilon}$, where $\mathcal{N}$ is molecules concentration and $\overline{\upsilon}$ is the average velocity of the molecules. Consider the following form for the velocity distribution function: $f(\upsilon) = \frac{4}{\sqrt{\pi}}\left(\frac{M}{2RT}\right)^{3/2}\upsilon^2 e^{-\frac{M\upsilon^2}{2RT}}$.

1.12. Derive Equation (1.49). A plot of this equation as a function of distance, x, for Si layer is shown in Fig. P1.12 .The following conditions: $\overline{\upsilon} = 10$ cm/sec, $C_i = 3 \times 10^{-6}$ g/cm^3, $b = 1.5$ cm, $\rho = 2.3$ g/cm^3, and $m_{Si}/m_s = 0.0205$. What would be the value of D needed to generate Fig. P1.12.

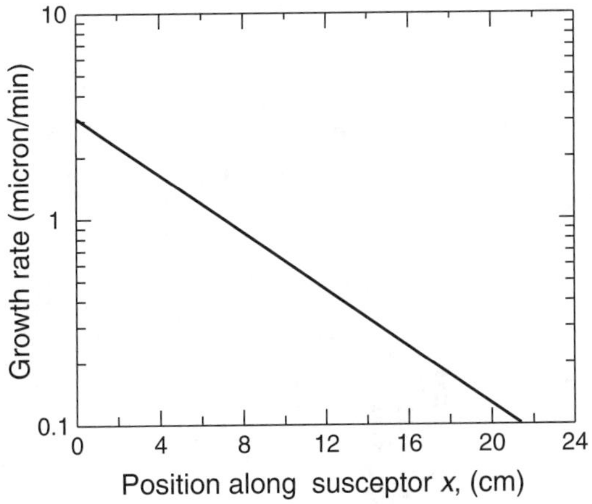

Fig. P1.12

1.13. Derive the expression of the barrier energy, ΔG^*, given by Equation (1.55). Show that this expression can be reduced to the form shown in Equation (1.56). Plot ΔG^* in Equation (1.56) as a function of the angle, θ, assuming that the first term in the right-hand side of the equation is unity.

1.14. The free energy change per unit thickness for a cluster of radius r is given by $\Delta G = \pi r^2 \Delta G_V + 2\pi\gamma r + A - B\ln(r)$, where $A - B\ln(r)$ is the energy contributed from dislocations within the cluster. Determine the critical radius r* and the nucleation barrier energy ΔG^*. Sketch ΔG versus r for the following parameters (assume unitless): $\pi \Delta G_V = -10, 2\pi\gamma = 50$, $A = 10$, and $B = 0.1$. From the sketch shows the locations of r* and ΔG*.

1.15. Derive Equation (1.69). Then obtain expressions for the critical temperatures at which the following transitions occur for one- to three-atom nucleus, one- to four-atom nucleus, and two- to three-atom nucleus.

1.16. Consider the kinetics of one-atom critical nucleus where $E_s = 0.2$ eV and $E_{des} = 0.8$ eV, $n_0 = 5 \times 10^{15}$ cm^{-2}, $\nu = 0.10 \times 10^{12}$ sec^{-1}, $C = 4$, $\alpha = 0.3$, $\beta = 4$, and $\Phi = 1.0 \times 10^{14}$ nucleus/cm^2.sec^{-1}. Calculate the characteristic temperature T_D using Equation (1.72). Plot the clusters (nucleus) density $N(t)$ as a function of time for $T = 730$ K and $T = 530$K.

1.17. Consider the following relations between the free energy, $\Delta G(i^*)$ and the number of atoms, i^*, in 3D deposited clusters on a surface of a substrate. $\Delta G(i^*) = -i^*\Delta\mu + (i^*)^{2/3}\chi$, where $\Delta\mu$ is the chemical potential energy and $\chi = 3.9$ eV is the surface free energy. Derive expressions for r^* and $\Delta G^*(i^* = r^*)$ from $\Delta G(i^*)$, where r^* is the critical number of atoms in the 3D clusters. Plot $\Delta G(i^*)$ vs. i^* for $\Delta\mu = -1$, 0,1, and 2 eV. From the graph, find the values of $\Delta G^*(i^* = r^*)$ and the critical number of atoms, r^*, in the clusters for each value of $\Delta\mu$. When you plot $\Delta G(i^*)$ vs. i^*, use the limits of 0–100 atoms for i^*.

1.18. The image shown here is an SEM image of InAs quantum dots grown on GaAs. The dimension of the image is 1 μm × 1 μm. Determine the quantum dots density in units of dots/cm^2.

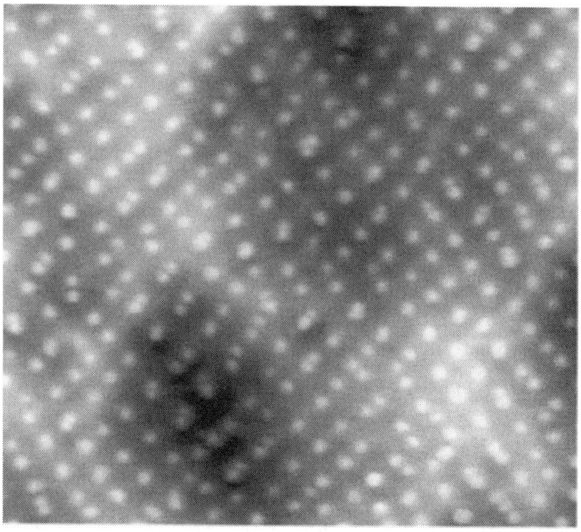

1.19. Assume that the dimensions of the STM images of InAs quantum dots in Fig. 1.35 are 0.5μm × 0.5 μm. Estimate the quantum dots density (ρ) in

cm^{-2} in all images. Plot ρ as a function of the growth temperature and as a function of the annealing time.

1.20. Plot the effective band gap energy as a function of the nanocrystal diameters in nanometers for the nanocrystals fabricated from the following materials.

Materials	E_g	m_e^*/m_0	m_h^*/m_0	χ
CdSe	1.75	0.11	0.45	0.75
CdS	2.48	0.14	0.51	0.80
ZnS	3.84	0.34	0.49	0.60
GaAs	1.52	0.067	0.45	0.70
InP	1.42	0.077	0.60	0.90
GaN	3.45	0.20	0.80	0.80
ZnO	3.43	0.28	0.59	0.70

BIBLIOGRAPHY

Belyaev LM. Rubby and sapphire. New Delhi: Amerind Publishing Co.; 1980, translated from Russian, RUBIN I SAPFIR. Moscow: Nauka Publishers; 1974.

Bimberg D, Grundmann M, Ledentsov NN. Quantum dot heterostructures. New York: John Wiley & Sons, Inc; 1999.

Cho AY. Thin Solid Films 1983;100:291.

Gith J, Petusky WT. J Phys Chem Solids 1987;48:541 (1987).

Huang T-F, Harris SJ Jr. In: Manasreh MO, Ferguson IT, editors. Volume 19, III-nitride semiconductor growth. New York: Taylor & Francis; 2003. Chapter 10.

Liu L, Edgar JH. Substrates for gallium nitride epitaxy. Mater Sci Eng 2002;R37:61.

Manasreh MO, Ferguson IT. III-nitride semiconductors growth. New York: Taylor & Francis; 2003.

Morkoç H. Handbook of nitride semiconductors and devices. Amsterdam: Springer-Verlag; 1999.

Moustakas TD. Volume 57, Semiconductors and semimetals. New York: Academic; 1999. p. 33.

Oda O, Fukui T, Hirano R, Muchida M, Kohiro K, Kurita H, Kainosho K, Asahi S, Suzuki K, Manasreh MO. InP and related compounds. New York: Taylor & Francis; 2000. Chapter 2.

Ohring M. The materials science of thin films. New York: Academic; 1992. Chapter 4.

Paskova T, Monemar B, Manasreh MO, Ferguson IT. Volume 19, III-nitride semiconductor growth. New York: Taylor & Francis; 2003. Chapter 6.

Pfann WG. Zone melting. 2nd ed. New York: John Wiley & Sons, Inc; 1966.

Razeghi M. Volume 1, The MOCVD challenge. Bristol: IOP Publishing Ltd.; 1989.

Venables JA, Spiller GDT, Hanbücken M. Rep Prog Phys 1984;47:399.

Vook RW. Int Met Rev 1982;27:209.

Walton D, Rhidin TN, Rollins RW. J Chem Phys 1963;38:2698.

2

APPLICATION OF QUANTUM MECHANICS TO NANOMATERIAL STRUCTURES

2.1 INTRODUCTION

Quantum mechanics can provide precise answers to many properties of nanomaterials, but it provides only the average value of individual measurements made on a given dynamical system in a certain initial state. One of the fundamental differences between classical mechanics and quantum theory is that quantum mechanics does not measure all variables with specific accuracy at the same time, while classical mechanics can do so. Another difference is that, classically, the effects of the disturbances due to the measurements can be exactly allowed in predicting the future behavior of the system, whereas quantum mechanically, the exact effects of the disturbances accompanying any measurements are inherently unknown. An example is that the measurement of the position of a particle introduces an unpredictable uncertainty regarding its momentum.

The development of quantum mechanics is made possible due to the failure of classical mechanics in explaining many phenomena. For example, the blackbody radiation from incandescent hot bodies was actually the first step in developing quantum mechanics. In 1901, Planck described the spectral intensity of blackbody radiation by assuming that oscillators at equilibrium with radiation can have certain discrete energy E_n, given by

$$E = E_n = n\hbar\omega_0, \quad \text{for } n = 0, 1, 2, 3, \ldots \tag{2.1}$$

where ω_0 is the oscillator frequency and $\hbar$ is the Planck's constant. The basic assumption of Planck's work is that for a cavity radiator, the number of internal

Introduction to Nanomaterials and Devices, First Edition. Omar Manasreh.

degrees of freedom (standing waves) can be calculated at a given frequency range per unit volume of the cavity to be $2 \times 4\pi\nu^2/c^3$, where ν is the frequency, c is the speed of light, and the factor 2 is added to account for the fact that each electromagnetic wave can have two orthogonal polarizations. However, each of these standing waves in the cavity cannot take on all possible energies as Maxwell's equations imply, but can take on only certain integrally related discrete energies, 0, $\hbar\omega_0$, $2\hbar\omega_0$, $3\hbar\omega_0$, ... as shown in Equation 2.1.

The photoelectric effect is another example that illustrates the failure of classical mechanics. In 1887, Hertz, while conducting experiments on the generation of electromagnetic waves, discovered that electrons could be ejected from solids by letting radiation fall onto the solid. He and others found that the maximum energy of these photoejected electrons depended only on the frequency of the light falling on the surface and not on its intensity. In 1905, Einstein explained the photoelectric effect in a satisfactory way by making use of Planck's ideas. He assumed that radiation exists in the form of quanta of definite size, that is, light consists of packets of energy of size $\hbar\omega$. He also assumed that when light falls on a surface, the individual electrons in solid can absorb these energy quanta. Thus, the energy received by an electron depends only on the frequency of light and is independent of its intensity. The intensity merely determines how many photoelectrons will leave the surface per second. Thus, the maximum kinetic energy of an electron excited by such light can be expressed as

$$E_{\mathrm{m}} = \hbar(\omega - \omega_0) = \hbar\omega - e\varphi_0, \tag{2.2}$$

where $\hbar\omega_0 = e\varphi_0$ or $\omega_0 = e\varphi_0/\hbar$, which is known as the threshold frequency, and e is the charge of the electron. Above this threshold frequency, light quanta have more-than-enough energy to excite the electrons into vacuum. The quantity φ_0 is a characteristic property of the metal called the work function. The electron must obtain energy $e\varphi_0$ from the incident light to be emitted as a photoelectron. Einstein's analysis of the photoemission phenomenon assumes that it can be considered as a two-body collision in which the light is giving up all its energy to a single electron.

Many other early experiments that were found difficult to explain in terms of classical mechanics are Compton scattering, electron diffraction from solid crystals, and emission and absorption spectra of atoms and molecules. The failure of classical mechanics was associated with two general types of effects. The first one is that physical quantities such as the energies of electromagnetic waves and of lattice vibrations of a given frequency, or the energies and angular momenta associated with electronic orbits in the hydrogen atoms, which in classical theory can take on a continuous range of values, were found to take on discrete values instead of a continuous quantity. The second type of effect is called wave–particle duality, where both the wave nature and particle nature exist. The wave nature of light is demonstrated by diffraction and interference effects, and the particle nature of light is demonstrated by the photoelectric and Compton effects. Particle parameters (the energy E and momentum $\boldsymbol{p}$ of a photon) and wave parameters (angular frequency $\omega = 2\pi v$ and wave vector $\mathbf{k}$, where $\mathbf{k} = \pi/\lambda$, frequency v and wavelength λ) are linked by the fundamental relations $E = \hbar\omega$ and $\boldsymbol{p} = \hbar\boldsymbol{k}$ known as Planck–Einstein relation.

The Bohr model of the atom is one of the early examples presented in support of the newly established field of study called quantum mechanics. Classical mechanics failed to explain the sharply defined spectral lines observed in the optical emission spectra of elements. In 1913, Neils Bohr found a way of quantizing the hydrogen atom that described the spectrum of the element with impressive accuracy. This delay was partly due to the fact that the atomic nucleus was not discovered until 1910 when Rutherford's scattering experiments were performed. It was only then that the concept of an atom as a point nucleus surrounded by a swarm of electrons emerged. In Bohr's model, a single electron of mass m and charge e is assumed to move in a circular orbit around the nucleus with a positive charge of ez, where z is an integer. In classical electrodynamics, accelerated charges such as the orbiting electrons always radiate energy in the form of electromagnetic waves. Classically, one would expect the electron to continually lose energy, spiraling inward toward the nucleus as its energy is depleted by radiation. Bohr suggested that stable nonradiative states of the atom can exist, corresponding to circular electron orbits whose angular momentum L is quantized in integral multiples of $\hbar$ so that

$$L_n = mr_n^2\omega_n = mr_n \upsilon = n\hbar \ \ (n = 0, 1, 2, 3, \ldots), \tag{2.3}$$

where υ is the electron velocity in its orbit. This quantization of the angular momentum also quantizes the orbit radii and angular velocities as indicated in Equation 2.3. The allowed total energies, kinetic plus potential, can be written as

$$E_n = \frac{1}{2}mr_n^2\omega_n^2 - \frac{ze^2}{4\pi\epsilon_0 r_n}, \tag{2.4}$$

where ϵ_0 is the permittivity of free space. By equating Coulomb force to the centripetal force, one can write

$$mr_n\omega_n^2 = \frac{ze^2}{4\pi\epsilon_0 r_n^2}, \tag{2.5}$$

where $r_n\omega_n^2$ is the centripetal acceleration. From Equations 2.3–2.5, one can obtain r_n, ω_n, and E_n according to the following relationships:

$$r_n = \frac{4\pi\epsilon_0 n^2\hbar^2}{mze^2} \tag{2.6}$$

$$\omega_n = \frac{mz^2e^4}{(4\pi\epsilon_0)^2 n^3\hbar^3} \tag{2.7}$$

$$E_n = -\frac{mz^2e^4}{2(4\pi\epsilon_0)^2 n^2\hbar^2}. \tag{2.8}$$

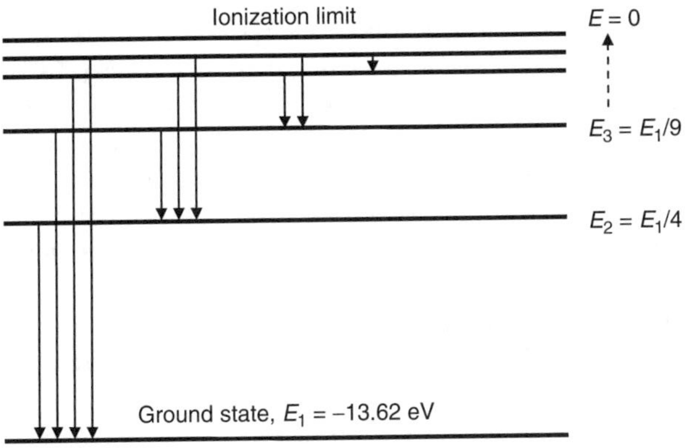

FIGURE 2.1 Energy level diagram for hydrogen atom derived from Bohr model.

By introducing the dimensionless unit α, known as the "fine structure constant"

$$\alpha = \frac{e^2}{(4\pi\epsilon_0)\hbar c} = \frac{1}{137.036} \approx 1/137, \tag{2.9}$$

the quantities in Equations 2.6–2.8 can be rewritten in much simpler expressions such as

$$r_n = \frac{n^2\hbar}{mzc\alpha}, \omega_n = \frac{m(zc\alpha)^2}{n^3\hbar}, \text{ and } E_n = -\frac{m(z\alpha c)^2}{2n^2}. \tag{2.10}$$

The allowed energies are negative, corresponding to stable bound states of the electron. For $z = 1$ (hydrogen atom) and $n = 1$, one can find that $r_1 = 0.5293$ Å and $E_1 = -13.62$ eV. Thus, one can rewrite the orbit radii and the energy as $r_n = n^2 r_1$ and $E_n = -E_1/n^2$. The Energy levels of the hydrogen atom are illustrated in Fig. 2.1.

Bohr's model was significant in the development of quantum theory because it shows the potential usefulness of its concepts in describing the structure of the atoms and molecules. However, attempts to extend Bohr's model to helium and more complex atoms were not very successful. These problems were not fully solved until after 1930, and in order to work them, a completely new and much more general theory of quantum mechanics had to be developed. Still, Bohr's atomic system provided a simple picture of the structure of a one-electron atomic system.

2.2 THE de BROGLIE RELATION

Classical mechanics failed to explain the narrow lines that composed the atomic emission and absorption spectra. In other words, a given atom emits or absorbs only photons having well-defined frequencies. This can be easily understood if one accepts the fact that the energy of the atom is quantized. The emission or absorption of a photon is then accompanied by a jump in the energy of the atom from one permitted value (E_i) to another (E_f). Conservation of energy implies that the photon has a frequency ν_{if} such that $h\nu_{if} = |E_i - E_f|$. In 1923, de Broglie presented the following hypothesis: *Material particles, just like photons, can have a wavelike aspect*. He then derived the Bohr–Sommerfeld quantization rules as a consequence of this hypothesis. The various permitted energy levels appear as analogs of the normal modes of a vibrating string. Electron diffraction experiment by Davisson and Germer in 1927 strikingly confirmed the existence of the wave-like aspect of matter by showing that interference patterns could be obtained with material particles such as electrons. One therefore associates with a material particle of energy E and momentum $\boldsymbol{p}$, a wave whose angular frequency is $\omega = 2\pi\nu$ and a wave vector $\boldsymbol{k}$ given by the same relations presented by Planck–Einstein relations ($E = \hbar\omega$ and $\boldsymbol{p} = \hbar\boldsymbol{k}$). The corresponding wavelength is

$$\lambda = \frac{2\pi}{|k|} = \frac{h}{|p|}. \tag{2.11}$$

The small value of h explains why the wavelike nature of matter is very difficult to demonstrate on a macroscopic scale.

Example 1: Consider a dust particle of diameter $r = 1\ \mu\text{m}$ and mass m $= 10^{-15}$ kg. For such a particle of small mass and a speed of $\upsilon = 10^{-3}$m/sec, the de Broglie wavelength is $\lambda = 6.6 \times 10^{-34}/\ (10^{-15} \times 10^{-3}) = 6.6 \times 10^{-16}$ m $= 6.6 \times 10^{-6}$ Å. This wave length is negligible on the scale of the dust particle. Let us now consider a thermal neutron ($m_n \sim 1.67 \times 10^{-27}$ kg) of energy 1.5 kT. Hence, $1/2m_n\upsilon^2 = \text{p}^2/(2m_n)$, where $k = 1.38 \times 10^{-23}$ J/K. This gives $\lambda = h/(3m_nkT)^{0.5}$, which is ~1.4 Å at T = 300 K. This wavelength is in the order of the lattice constant in crystalline solid. A beam of thermal neutrons falling on a crystal therefore gives rise to diffraction phenomena analogous to those observed with X-rays.

Example 2: Let us now examine the de Broglie wavelengths associated with electrons ($m_e \sim 0.9 \times 10^{-30}$ kg). If the electron is accelerated by a potential difference V, then the electron kinetic energy is $E = qV = 1.6 \times 10^{-19} \times$ V Joules. Since $E = p^2/(2m_e)$, then $\lambda = h/(2m_eE)^{0.5} = 12.3/(V)^{0.5}$ Å. With a potential difference of several hundred volts, one can obtain a wavelength comparable to those of X-rays. Thus, electron diffraction phenomena can be observed in crystals or crystalline powders.

2.3 WAVE FUNCTIONS AND SCHRÖDINGER EQUATION

By considering de Broglie hypothesis, one can apply the wave properties for the case of photon to all material particles. Thus, for the classical concept of a trajectory, the time-varying state is substituted by quantum state characterized by a *wave function*, $\psi(\mathbf{r}, t)$, which contains all the information that is possible to obtain about the particle. ψ ($\mathbf{r}$,t) can be thought of as a *probability amplitude* of the particle's presence. The measurements of an arbitrary physical quantity must belong to a set of eigenvalues. Each eigenvalue is associated with an eigenstate. The equation describing the evolution of the wave function $\psi(\mathbf{r},t)$ remains to be written. The wave equation can be introduced by using Planck and de Broglie relations to yield the fundamental equation known as Schrödinger equation. The form of this equation for a particle of mass m which is subject to the influence of a potential $V(\mathbf{r},t)$ takes the following form:

$$i\hbar\frac{\partial}{\partial t}\psi(\mathbf{r},t) = -\frac{\hbar^2}{2m}\Delta\psi(\mathbf{r},t) + V(\mathbf{r},t)\psi(\mathbf{r},t), \tag{2.12}$$

where Δ is the Laplacian operator given by $\Delta = \partial^2/\partial x^2 + \partial^2/\partial y^2 + \partial^2/\partial z^2$. This equation is linear and homogeneous in $\psi(\mathbf{r},t)$. Consequently, for material particles, there exists a superposition principle.

When $V(\boldsymbol{r},t) = 0$, Schrödinger equation is reduced to

$$i\hbar\frac{\partial}{\partial t}\psi(\mathbf{r},t) = -\frac{\hbar^2}{2m}\Delta\psi(\mathbf{r},t), \tag{2.13}$$

which is the wave equation for a free particle. A solution of this equation has the form

$$\psi(\mathbf{r},t) = Ae^{i(\mathbf{k}\cdot\mathbf{r}-\omega t)}, \tag{2.14}$$

where A is a constant, and the dispersion relation obtained by substituting Equation 2.14 into Equation 2.13 is

$$\omega = \frac{\hbar\mathbf{k}^2}{2m}. \tag{2.15}$$

Equations 2.11 and 2.15 give the relation between energy (E) and momentum ($\boldsymbol{p}$) as

$$E = \frac{\mathbf{p}^2}{2m}, \text{ where } \mathbf{p} = \hbar\mathbf{k}.$$

The constant A in Equation 2.14 can be obtained by normalization. Using the wave function form given by Equation 2.14, one can write

$$|\psi(\mathbf{r},t)\psi^*(\mathbf{r},t)| = |A|^2, \text{ where } \psi^*(\mathbf{r},t) = A^*\mathrm{e}^{-i(\mathbf{k}\cdot\mathbf{r}-\omega t)}. \tag{2.16}$$

Equation 2.16 tells us that a plane wave of this type represents a particle whose probability of presence is uniform throughout the space.

The principle of superposition tells us that every linear combination of plane waves satisfying the dispersion relation given by Equation 2.15 will also be a solution for Equation 2.13. This superposition can be written as

$$\psi(\boldsymbol{r},t) = \frac{1}{(2\pi)^{3/2}}\int g(\boldsymbol{k})\,\mathrm{e}^{i(\mathbf{kr}-\omega t)}\,\mathrm{d}^3\boldsymbol{k}, \tag{2.17}$$

where $d^3\boldsymbol{k}$ is the infinitesimal volume element in **k**-space and $g(\boldsymbol{k})$ can be complex but must be sufficiently regular to allow differentiation inside the integral. A wave function such as Equation 2.17 is called a three-dimensional "*wave packet*."

2.4 DIRAC NOTATION

Each quantum state of a particle is characterized by a state vector belonging to an abstract space, $\mathcal{S}$ called the space state of a particle. Any element or vector of $\mathcal{S}$-space is called a *ket* vector or simply a *ket*, which is represented by the symbol $|\ \rangle$. Inside this *ket* we can place a quantity, which distinguishes it from all others, for example, $|\psi\rangle$. Also, we can define a *bra* vector with every *ket* $|\psi\rangle \in \mathcal{S}$, which is denoted $\langle\psi| \in \mathcal{S}^*$, where $\mathcal{S}^*$ is the complex conjugate of $\mathcal{S}$. The origin of this terminology is based on the word "*bracket*" used to denote the symbol $\langle\ |\ \rangle$, hence the name *bra* for the left-hand side and *ket* for the right-hand side of this symbol. Thus, the notation $\langle\varphi|\psi\rangle$ is identical to the familiar wave mechanics expression

$$\langle\varphi|\psi\rangle = \int_{-\infty}^{+\infty} \varphi^*(x)\psi(x)\,\mathrm{d}x. \tag{2.18}$$

The *bra* and *ket* vectors satisfy the scalar products defined as

$$\langle\varphi|\psi\rangle = (|\varphi\rangle, |\psi\rangle) \tag{2.19a}$$

$$\langle\varphi|\psi\rangle^* = \langle\psi|\varphi\rangle. \tag{2.19b}$$

The product of two linear operators A and B is defined as $(\mathrm{AB})|\psi\rangle = \mathrm{A}(\mathrm{B}|\psi\rangle)$. In general, $\mathrm{AB} \neq \mathrm{BA}$. The commutator [A,B] is by definition given as [A,B] $= \mathrm{AB} - \mathrm{BA}$. Now, let $|\varphi\rangle$ and $|\psi\rangle$ be two *kets*, we define $\langle\varphi|A|\psi\rangle$ as the matrix element of A. Now assume that $|\varphi\rangle$ and $|\psi\rangle$ are written in opposite order $|\varphi\rangle\langle\psi|$. This is actually an operator since applying it to an arbitrary *ket* $|\chi\rangle$ yields $|\varphi\rangle\langle\psi|\chi\rangle = \alpha|\varphi\rangle$, where α is a real constant. Thus, applying $|\varphi\rangle\langle\psi|$ to an arbitrary *ket* gives another *ket*, which is the definition of an operator.

The order of symbols is very important in Dirac notation. The following discussion focuses on the properties of *bras* and *kets* functions. If λ is a complex number and $|\psi\rangle$ is a *ket*, then $\lambda|\psi\rangle$ is *ket*, which can be presented as

$$\lambda|\psi\rangle = |\lambda\psi\rangle. \tag{2.20}$$

One must then remember that $\langle\lambda\psi| = \lambda^*\langle\psi|$ is the *bra* associated with the *ket* $|\lambda\psi\rangle$. In addition,

$$\langle\varphi|\lambda_1\psi_1 + \lambda_2\psi_2\rangle = \lambda_1\langle\varphi|\psi_1\rangle + \lambda_2\langle\varphi|\psi_2\rangle \tag{2.21}$$

$$\langle\lambda_1\varphi_1 + \lambda_2\varphi_2|\psi\rangle = \lambda_1^*\langle\varphi_1|\psi\rangle + \lambda_2^*\langle\varphi_2|\psi\rangle \tag{2.22}$$

$$\langle\psi|\psi\rangle = \begin{cases} \text{A if } |\psi\rangle \neq 0 \\ 0 \text{ if } |\psi\rangle = 0, \end{cases}$$

$$\text{A} = \text{ real positive number.} \tag{2.23}$$

In Dirac notation, the wave function can be written as

$$|\psi\rangle = \sum_i C_i |u_i\rangle, \tag{2.24}$$

where $\{|u_i\rangle\}$ is a discrete set for the basis of the *ket* $|\psi\rangle$.

There are several quantum mechanics postulates that are worth mentioning.

First postulate: At a fixed time, t_0, the state of a physical system is defined by specifying a ket $|\psi(t_0)\rangle$ belonging to the state space $\mathcal{S}$. This postulate implies that a linear combination of state vectors is a state vector. It should be emphasized here that the *ket* is not a statistical mixture of states.

Second postulate: Every measurable physical quantity is described by an operator in $\mathcal{S}$-space. This operator is an observable. Unlike classical mechanics, quantum mechanics describes, in a fundamentally different manner, a system and the associated physical quantities as a state is represented by a vector and a physical quantity by an operator.

Third postulate: The only possible result of the measurement of a physical quantity is one of the eigenvalues (true values) of the corresponding observable.

Fourth postulate: When a physical quantity is measured for a system in the normalized state $|\psi\rangle$, the probability, P, of obtaining a nondegenerate eigenvalue of the corresponding observable is

$$P = \left|\langle u_n|\psi\rangle\right|^2, \tag{2.25}$$

where $|u_n\rangle$ is the normalized eigenvector of the observable associated with the eigenvalue. If the eigenvalues are degenerate, several orthonormal eigenvectors $|u_n\rangle$ correspond to them. The probability can then be rewritten as

$$P = \sum_{i=1}^{g_n} |\langle u_n|\psi\rangle|^2, \tag{2.26}$$

where g_n is the degree of degeneracy. However, for continuous nondegenerate systems, the probability of obtaining a result between α and $\alpha + \mathrm{d}\alpha$ is equal to

$$\mathrm{d}P = \left|\langle u_n|\psi\rangle\right|^2 \mathrm{d}\alpha, \tag{2.27}$$

where $|u_n\rangle$ is the eigenvector corresponding to the eigenvalue α of the observable associated with the physical quantity.

Fifth postulate: If the measurement of a physical quantity of a system in the state $|\psi\rangle$ gives the result a_n, the state of the system immediately after the measurement is the normalized projection, $\frac{P_n|\psi\rangle}{\sqrt{\langle\psi|P_n|\psi\rangle}}$, of $|\psi\rangle$ on to the eigensubspace associated with a_n.

Sixth postulate: The time evolution of the state vector $|\psi(t)\rangle$ is governed by the Schrödinger equation

$$i\hbar\frac{\partial}{\partial t}|\psi(t)\rangle = \boldsymbol{H}(t)|\psi(t)\rangle, \tag{2.28}$$

where $\boldsymbol{H}(t)$ is the observable associated with the total energy of the system. $\boldsymbol{H}(t)$ is called the Hamiltonian operator of the system.

There are also important mathematical tools needed for working with quantum mechanics. Let us first introduce the terms "*wave function space*, $\mathcal{F}$" and "*state space*, $\mathcal{F}$". The wave function introduced earlier belongs to $\mathcal{F}$ and the state vector belongs to $\boldsymbol{\mathcal{E}}$. $\mathcal{F}$ satisfies all criteria of a vector space.

The scalar product: For each pair of elements of $\mathcal{F}$, $|\varphi\rangle$ and $|\psi\rangle$, we associate a complex number denoted $(|\varphi\rangle, |\psi\rangle)$, which by definition is equal to $\langle\varphi|\psi\rangle = (|\varphi\rangle, |\psi\rangle)$ (Eq. 2.19a). The quantity $\langle\varphi|\psi\rangle$ always converges so long as both wave functions belong to $\mathcal{F}$. Based on this definition, we have

$$\begin{aligned}
&(\langle\varphi|\psi\rangle) = (\langle\psi|\varphi\rangle)^* \\
&\langle\varphi|\lambda_1\psi_1 + \lambda_2\psi_1\rangle = \lambda_1(\langle\varphi|\psi_1\rangle) + \lambda_2(\langle\varphi|\psi_2\rangle) \text{ This is called linear.} \\
&\langle\lambda_1\varphi_1 + \lambda_2\varphi_2|\psi\rangle = \lambda_1^*\langle\varphi|\psi_1\rangle + \lambda_2^*\langle\varphi|\psi\rangle \text{ This is called antilinear.}
\end{aligned} \tag{2.29}$$

If $\langle\varphi|\psi\rangle = 0$, then $|\varphi\rangle$ and $|\psi\rangle$ are said to be orthogonal. Furthermore, $|\varphi\rangle$and $|\psi\rangle$ must satisfy Equation 2.23.

Linear operators: Equation 2.20 is a simple definition of a linear operator. Let **A** and **B** be two linear operators. Their product is defined as:

$$(\mathrm{AB})|\psi\rangle = \mathrm{A}(\mathrm{B}|\psi\rangle) \tag{2.30}$$

B is the first to operate on $|\psi\rangle$; then A operates on the new product. In general, AB $\neq$ BA. The commutator of A and B is the operator [A,B] defined as [A,B] = AB − BA. Examples are the operators X and $\frac{\partial}{\partial x}$ that are operating on an arbitrary function $|\psi\rangle$.

$$\begin{aligned}
\left[X, \frac{\partial}{\partial x}\right]|\psi\rangle &= \left(X\frac{\partial}{\partial x} - \frac{\partial}{\partial x}X\right)|\psi\rangle \\
&= X\frac{\partial}{\partial x}|\psi\rangle - \frac{\partial}{\partial x}X|\psi\rangle = X\frac{\partial}{\partial x}|\psi\rangle - \frac{\partial}{\partial x}(X|\psi\rangle)
\end{aligned}$$

$$= X\frac{\partial}{\partial x}|\psi\rangle - |\psi\rangle - X\frac{\partial}{\partial x}|\psi\rangle$$

$$= -|\psi\rangle$$

$$\text{Thus, } \left[X, \frac{\partial}{\partial x}\right] = -1. \tag{2.31}$$

Examples:

a) Show that [A, B] = −[B, A].
Solution: [A, B] = AB − BA = −(BA − AB) = −[B,A].

b) Expand [A,(B + C)] Solution: [A, (B + C)] = A(B + C) − (B + C)A = AB + AC − BA − CA = AB − BA + AC − CA = [A,B] + [A,C]

c) Expand [A,BC]
Solution: [A,BC] = ABC − BCA. Adding BAC and subtracting BAC, we get = ABC − BCA + BAC − BAC = ABC − BAC + BAC − BCA = [A,B]C + B[A,C]

d) Show that $[X,P_x] = i\hbar$, where $P_x = \frac{\hbar}{i}\frac{\partial}{\partial x}$.
Solution:

$$\langle\varphi|[X,P_x]|\psi\rangle = \langle\varphi|[XP_x - P_xX]|\psi\rangle = x\langle\varphi|P_x|\psi\rangle - \langle\varphi|\frac{\hbar}{i}\frac{\partial}{\partial x}X|\psi\rangle$$

$$= x\langle\varphi|P_x|\psi\rangle - \frac{\hbar}{i}(x\langle\varphi|\frac{\partial}{\partial x}|\psi\rangle)$$

$$= x\langle\varphi|P_x|\psi\rangle - x\frac{\hbar}{i}\langle\varphi|\frac{\partial}{\partial x}|\psi\rangle - \frac{\hbar}{i}\langle\varphi|\psi\rangle$$

$$= x\langle\varphi|P_x|\psi\rangle - x\langle\varphi|P_x|\psi\rangle - \frac{\hbar}{i}\langle\varphi|\psi\rangle = i\hbar\langle\varphi|\psi\rangle$$

then $[X,P_x] = i\hbar$

Similarly, $[R_i, P_j] = i\hbar\delta_{ij}$, where $\delta_{ij} = \begin{cases} 1 \text{ for } i = j \\ 0 \text{ for } i \neq j \end{cases}$.

2.4.1 Action of a Linear Operator on a *Bra*

Let $\langle\varphi|$ be a well-defined *bra*, and consider the set of all *kets* $|\psi\rangle$. With each of these *kets* can be associated the complex number $\langle\varphi|\mathbf{A}|\psi\rangle$, which is defined as the matrix element of **A** between $\langle\varphi|$ and $|\psi\rangle$. Since **A** is linear and the scalar product depends linearly on the ket, the matrix element depends linearly on $|\psi\rangle$. Thus, for fixed $\langle\varphi|$ and **A**, we can associate with every ket $|\psi\rangle$ a number that depends on $|\psi\rangle$. The specification of $\langle\varphi|$ and **A** therefore defines a new linear function, that is a new *bra* belonging to the conjugate state space ϵ^*. The new *bra* is denoted $\langle\varphi|\mathbf{A}$. The relation that can define $\langle\varphi|\mathbf{A}$ can be written as

$$(\langle\varphi|\mathbf{A})|\psi\rangle = \langle\varphi|(\mathbf{A}|\psi\rangle). \tag{2.32}$$

This equation defines the linear operation on *bras*.

2.4.2 Eigenvalues and Eigenfunctions of an Operator

The *ket* $|\psi\rangle$ is said to be an eigenvector or eigenket of the linear operator A if

$$\mathrm{A}|\psi\rangle = \lambda|\psi\rangle, \tag{2.33}$$

where λ is a complex number. This equation is called the eigenvalue equation of the linear operator A. In general, this equation has a solution only when λ takes on certain values called eigenvalues. The set of the eigenvalues are called the spectrum of **A**. If $|\psi\rangle$ is an eigenvector for A, then $\alpha|\psi\rangle$ is also an eigenvector, where α is an arbitrary complex number.

$$\boldsymbol{A}(\alpha|\psi\rangle) = \alpha\boldsymbol{A}|\psi\rangle = \alpha\lambda|\psi\rangle = \lambda(\alpha|\psi\rangle). \tag{2.34}$$

To get rid of α, the eigenvectors are usually normalized to 1:

$$\langle\psi|\psi\rangle = 1. \tag{2.35}$$

The eigenvalue λ is called nondegenerate when its corresponding eigenvector is unique within a constant factor, that is, when all its associated eigenkets are collinear. On the other hand, if there exists at least two linearly independent kets that are eigenvectors of $\boldsymbol{A}$ with the same eigenvalue, this eigenvalue is said to be degenerate.

2.4.3 The Dirac δ-Function

The δ-function is a distribution, but it is usually treated as an ordinary function. Consider a δ-function as shown in Fig. 2.2 with a width of ϵ and a height of $1/\epsilon$. By definition, we have

$$\int_{-\infty}^{+\infty} \delta(x)\,\mathrm{d}x = 1. \tag{2.36}$$

Let us evaluate the following integral, where $f(x)$ is an arbitrary function. $\int_{-\infty}^{+\infty} \delta^{(\epsilon)}(x) f(x)\,\mathrm{d}x$. If ϵ is sufficiently small, the variation of $f(x)$ over the effective interval $[-\epsilon/2, +\epsilon/2]$ is negligible and $f(x)$ remains practically equal to $f(0)$. Therefore,

$$\int_{-\infty}^{+\infty} \delta^{(\epsilon)}(x) f(x)\,\mathrm{d}x \cong f(0)\int_{-\infty}^{+\infty} \delta^{(\epsilon)}(x)\,\mathrm{d}x = f(0). \tag{2.37}$$

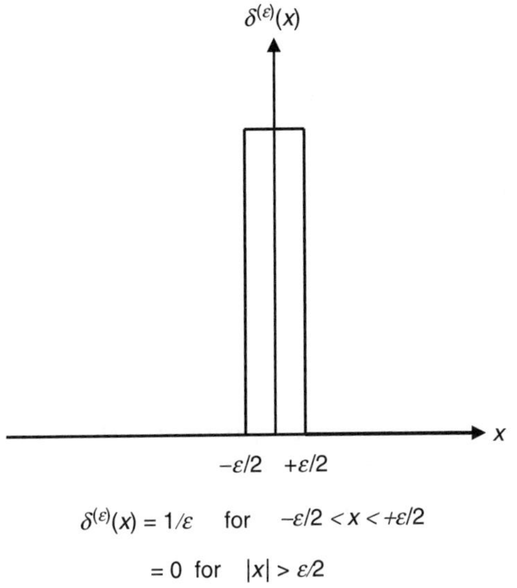

FIGURE 2.2 The δ-function: a square function of width ϵ and height $1/\epsilon$ centered at $x = 0$.

The smaller the ϵ, the better the approximation, and for the limit $\epsilon = 0$, we define the δ-function as

$$\int_a^b \delta(x) f(x)\, dx = f(0), \quad \text{for } 0 \in [a,b] \tag{2.38}$$

$$= 0, \qquad \text{for } 0 \notin [a,b].$$

For a more general form of $\delta(x - x_0)$, we have

$$\int_{-\infty}^{+\infty} \delta(x - x_0) f(x)\, dx = f(x_0). \tag{2.39}$$

Other properties of the δ-function are:

$$\delta(-x) = \delta(x) \tag{2.40a}$$

$$\delta(cx) = \frac{1}{|c|}\delta(x) \tag{2.40b}$$

$$\delta[g(x)] = \sum_j \frac{1}{\left|\dfrac{\partial g_j(x)}{\partial x}\right|}\delta(x - x_j) \tag{2.40c}$$

$$x\delta(x - x_0) = x_0\delta(x - x_0) \tag{2.40d}$$

$$g(x)\delta(x - x_0) = g(x_0)\delta(x - x_0) \tag{2.40e}$$

$$\int_{-\infty}^{+\infty} \delta(x - y)\delta(x - z)\ \mathrm{d}x = \delta(y - z). \tag{2.40f}$$

The Fourier transform $\overline{\delta}_{x_0}(p)$ of $\delta(x - x_0)$ is

$$\overline{\delta}_{x_0}(p) = \frac{1}{\sqrt{2\pi\hbar}} \int_{-\infty}^{+\infty} \mathrm{d}x\ \mathrm{e}^{ipx/\hbar}\delta(x - x_0) = \frac{1}{\sqrt{2\pi\hbar}} \mathrm{e}^{ipx_0/\hbar} \tag{2.41}$$

and

$$\overline{\delta}_{x_0}(p) = \frac{1}{\sqrt{2\pi\hbar}} \text{ is the Fourier transform of } \delta(x).$$

The inverse Fourier transform is:

$$\delta(x - x_0) = \frac{1}{2\pi\hbar} \int_{-\infty}^{+\infty} \mathrm{d}p\ \mathrm{e}^{ip(x-x_0)/\hbar} = \frac{1}{2\pi} \int_{-\infty}^{+\infty} \mathrm{d}k\ \mathrm{e}^{ip(x-x_0)}. \tag{2.42}$$

In addition, $\delta(x)$ is a derivative of a unit-step function $\theta(x)$, that is,

$$\frac{\mathrm{d}}{\mathrm{d}x}\theta(x) = \delta(x) \tag{2.43a}$$

and

$$\theta(x) = \frac{1}{2\pi\hbar} \int_{-\infty}^{x} \delta^{(\epsilon)}(y)dy. \tag{2.43b}$$

Other properties of $\delta(x)$ include the following:

$$\int_{-\infty}^{+\infty} \delta'(x) f(x)\ \mathrm{d}x = -\int_{-\infty}^{+\infty} \delta(x) f'(x)\ \mathrm{d}x = -f'(0) \tag{2.44}$$

and

$$\delta'(-x) = -\delta'(x) \tag{2.45a}$$

$$\delta'(x) = -\delta(x), \tag{2.45b}$$

where the prime indicates the first derivative and Equation 2.42 allows us to write

$$\delta'(x) = \frac{i}{2\pi} \int_{-\infty}^{+\infty} k\ \mathrm{d}k \mathrm{e}^{ip(x-x_0)}. \tag{2.46}$$

The nth-order derivative (n) can be defined in the same way as

$$\int_{-\infty}^{+\infty} \delta^{(n)}(x) f(x)\ \mathrm{d}x = (-1)^n f^{(n)}(0). \tag{2.47}$$

Equation 2.45 can then be generalized to the following:

$$\delta^{(n)}(-x) = (-1)^n \delta^{(n)}(x) \tag{2.48a}$$

$$\delta^{(n)}(x) = -n\delta^{(n-1)}(x) \tag{2.48b}$$

The δ-function is very useful in quantum mechanics as we will see in the following chapters.

2.4.4 Fourier Series and Fourier Transform in Quantum Mechanics

In this section, we review a few definitions that are important in quantum mechanics. A function $f(x)$ is said to be periodic if there exists a real nonzero number L such that for all x: $f(x+L) = f(x)$, where L is called *the period of the function*. If $f(x)$ is periodic with a period L, then all numbers nL, where n is an integer, are also periods of $f(x)$. Another important example of periodic functions is the periodic exponential. For an exponential $e^{\alpha L}$ to have a period L, it is necessary to have $e^{\alpha L} = 1$, that is, $\alpha L = i2n\pi$, where n is an integer. Thus, if $f(x)$ is a periodic function with a fundamental period of L, one can expand this function in the following form, known as *Fourier series*:

$$f(x) = \sum_{n=-\infty}^{\infty} c_n e^{ik_n x} \quad \text{with} \quad k_n = n\frac{2\pi}{L} \tag{2.49}$$

The coefficients c_n are given by the following formula:

$$c_n = \frac{1}{L}\int_{x_0}^{x_0+L} dx\; e^{-ik_n x} f(x). \tag{2.50}$$

where x_0 is an arbitrary number. The coefficients c_n are called Fourier spectrum of $f(x)$. Another useful relation, known as *Bessel–Parseval relation* is

$$\frac{1}{L}\int_{x_0}^{x_0+L} dx\left|f(x)\right|^2 = \sum_{n=-\infty}^{\infty} \left|c_n\right|^2. \tag{2.51}$$

Now assume that we have two functions, $g(x)$ and $f(x)$ with the same period and having Fourier coefficients, d_n and c_n, respectively. We can generalize Equation 2.51 according to the following relation:

$$\frac{1}{L}\int_{x_0}^{x_0+L} dx f(x)g(x) = \sum_{n=-\infty}^{\infty} c_n d_n. \tag{2.52}$$

If $\psi(x)$ is a one-dimensional wave function, its Fourier transform $\overline{\psi}(p)$ is defined as

$$\overline{\psi}(p) = \frac{1}{\sqrt{2\pi\hbar}} \int_{-\infty}^{\infty} \mathrm{d}\, x e^{-ipx/\hbar} \psi(x) \tag{2.53}$$

and the inverse formula is

$$\psi(x) = \frac{1}{\sqrt{2\pi\hbar}} \int_{-\infty}^{\infty} \mathrm{d}p e^{ipx/\hbar} \overline{\psi}(p). \tag{2.54}$$

Another useful relationship in quantum mechanics is the Parseval–Placherel formula, which has the following general form:

$$\int_{-\infty}^{\infty} \varphi^*(x)\psi(x)\, \mathrm{d}x = \int_{-\infty}^{\infty} \overline{\varphi}^*(p)\overline{\psi}(p)\, \mathrm{d}p. \tag{2.55}$$

2.5 VARIATIONAL METHOD

There are several well known approximations used to solve quantum mechanical problems. One approximation is called the variational method, which has numerous applications in solid-state physics, in which the exact solution requires extensive computational analysis. Let us consider a nondegenerate arbitrary physical system with a time-independent Hamiltonian $\boldsymbol{H}$. The general solution of Schrödinger equation can be written as

$$\boldsymbol{H}\left|\varphi_n\right\rangle = E_n\left|\varphi_n\right\rangle, \tag{2.56}$$

where E_n is the discrete eigenvalue and $n = 0, 1, 2 \ldots$ The Hamiltonian is known, whereas the eigenvalues are not necessary to be known. In this case, the variational method can be used to obtain an approximate expression for the eigenvalues. This method is very useful for cases in which $\boldsymbol{H}$ cannot be exactly diagonalized. To proceed, let us choose an arbitrary *ket* where the mean value of the Hamiltonian can be expressed as

$$\langle \boldsymbol{H} \rangle = \frac{\langle\psi\left|H\right|\psi\rangle}{\langle\psi\left|\psi\right\rangle} \geq E_0, \tag{2.57}$$

where E_0 is the ground-state eigenvalue, and the inequality is valid if $|\psi\rangle$ is the eigenvector of $\boldsymbol{H}$ with an eigenvalue of E_0. Without going through the derivation, we simply state the final result as

$$H\left|\psi\right\rangle = \langle H\rangle\left|\psi\right\rangle. \tag{2.58}$$

This method can be generalized and applied to the approximate determination of the eigenvalues of the Hamiltonian. Equation 2.58 tells us that if the function

$\langle H\rangle(\alpha)$ obtained from the trial *kets* $|\psi(\alpha)\rangle$ has several extrema, they give the approximate values of its energies E_n.

As an example, let us find the first energy level of the simple harmonic oscillator with the following Hamiltonian:

$$H = -\frac{\hbar^2}{2m}\frac{\mathrm{d}^2}{\mathrm{d}x^2} + \frac{1}{2}m\omega^2 x^2 \tag{2.59}$$

and the following trial function:

$$\psi(x) = e^{-\alpha x^2}, \quad \text{where } \alpha > 0. \tag{2.60}$$

The objective here is to evaluate $\langle\psi|H|\psi\rangle$ and $\langle\psi|\psi\rangle$, which are

$$\langle\psi|H|\psi\rangle = \sqrt{\pi/2}\left[\frac{\hbar^2}{2m}\alpha^{1/2} + \frac{1}{8}m\omega^2\alpha^{-3/2}\right] \tag{2.61}$$

and

$$\langle\psi|\psi\rangle = \sqrt{\frac{\pi}{2\alpha}}.$$

Dividing the two expressions, we obtain

$$\langle H\rangle(\alpha) = \frac{\hbar^2}{2m}\alpha + \frac{1}{8}m\omega^2\alpha^{-1}. \tag{2.62}$$

We assume at the beginning that the wave function has extrema such that the derivative of Equation 2.62 is zero. This yields $\alpha = m\omega/(2h)$, which can be substituted back into Equation 2.62 to give $\langle H\rangle(\alpha) = \frac{1}{2}\hbar\omega$. This is exactly the ground-state energy obtained from the exact solution (Appendix E).

2.6 STATIONARY STATES OF A PARTICLE IN A POTENTIAL STEP

Consider the potential step shown in Fig. 2.3, and consider that a particle with mass m is traveling from left to right with an energy $E > V_0$, where V_0 is the

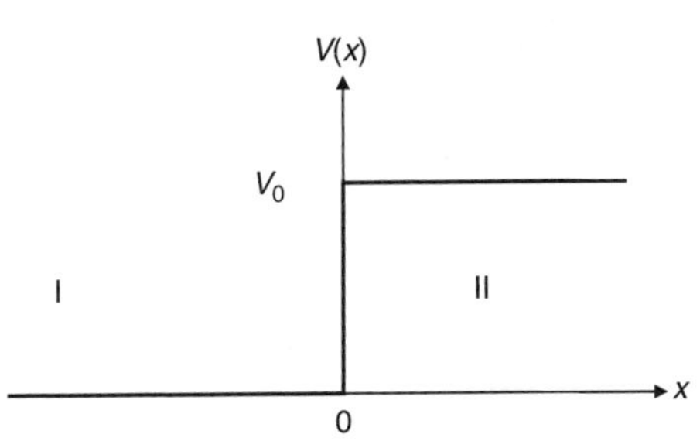

FIGURE 2.3 Step potential where $V(x) = 0$ for $x < 0$ (region I) and $V(x) = V_0$ for $X > 0$ (region II).

height of the potential step. The Schrödinger equation for this potential can be written as

$$-\frac{\hbar^2}{2m}\frac{\mathrm{d}^2}{\mathrm{d}x^2}\varphi(x) + V(x)\varphi(x) = E\varphi(x), \tag{2.63}$$

which can be rearranged in the following form:

$$\frac{\mathrm{d}^2}{\mathrm{d}x^2}\varphi(x) + \frac{2m}{\hbar^2}(E - V_0)\varphi(x) = E\varphi(x). \tag{2.64}$$

The solution for Equation 2.63 has the following form:

$$\varphi(\boldsymbol{r}, t) = \sum_n C_n\varphi_n(r)\mathrm{e}^{-iE_nt/\hbar} \tag{2.65}$$

in both region I ($x < 0$) and region II ($x > 0$). Let us introduce the following positive numbers, known as *propagation vectors*, k_1 *and* k_2 such that

$$k_1^2 = \frac{2m(E)}{\hbar^2} \text{ for region I and } k_2^2 = \frac{2m(E - V_0)}{\hbar^2} \text{ for region II.} \tag{2.66}$$

Thus, the solutions of Equation 2.64 for both regions can be written as

$$\varphi_{\mathrm{I}}(x) = A\mathrm{e}^{ik_1x} + B\mathrm{e}^{-ik_1x} \tag{2.67a}$$

$$\varphi_{\mathrm{II}}(x) = C\mathrm{e}^{ik_2x} + D\mathrm{e}^{-ik_2x}, \tag{2.67b}$$

where A, B, C, and D are complex constants and are equivalent to the constants C_n shown in Equation 2.65.

There are four coefficients and only two equations. In quantum mechanics, the wave functions in both regions must be matched. This requires us to introduce boundary conditions. These boundary conditions are as follows:

(i) The wave functions at the boundary must be continuous. Thus $\varphi_{\mathrm{I}}(x = 0) = \varphi_{\mathrm{II}}(x = 0)$, and

(ii) The first derivative of both wave functions must also be continuous, that is, $\frac{\mathrm{d}}{\mathrm{d}x}\varphi_{\mathrm{I}}(x = 0) = \frac{\mathrm{d}}{\mathrm{d}x}\varphi_{\mathrm{II}}(x = 0)$, or $\varphi_{\mathrm{I}}'(x = 0) = \varphi_{\mathrm{II}}'(x = 0)$, where the prime stands for the first derivative.

By applying the boundary conditions mentioned above to Equation 2.67, we obtain

$$A + B = C + D$$

$$ik_1A - ik_1B = ik_2C - ik_2\mathrm{D}.$$

We still have two equations with four unknowns. If we assume that the particle is coming from $x = -\infty$, then we can choose $D = 0$ or $A + B = D$ and $k_1A -$

$k_1 B = k_2 C$. Even with this simplification, we can only determine the ratios B/A and C/A, which are shown as

$$\frac{B}{A} = \frac{k_1 - k_2}{k_1 + k_2} \quad \text{and} \quad \frac{C}{A} = \frac{2k_1}{k_1 + k_2}. \tag{2.68}$$

Thus far, we have $\varphi_I(x)$ composed of two waves, one represents the particle coming from $x = -\infty$ and the other represents the particle as being reflected form the potential step. As we have chosen $D = 0$, $\varphi_{\text{II}}(x)$ is composed of only one wave representing the particle as being transmitted above the potential step with an energy $E > V_0$.

The concepts of transmissions and reflections of particles based on the ratios shown in Equation 2.68 can be understood in terms of a *probability current*, which can be discussed as follows. Let us consider a system composed of only a single spinless particle with a normalized wave function of $\psi(\mathbf{r}, t)$. A quantity known as a *probability density* is the probability, $\mathrm{d}p(\mathbf{r},\mathrm{t})$, of finding the particle at time t in an infinitesimal volume $\mathrm{d}^3\mathbf{r}$ located at the point $\mathbf{r}$ in the system, and is defined as

$$\mathrm{d}p(r, t) = \rho(\mathbf{r}, t)\ \mathrm{d}^3 r. \tag{2.69}$$

where

$$\rho(r, t) = \left|\psi(\mathbf{r}, t)\right|^2. \tag{2.70}$$

The *probability density* is analogous to an isolated physical system with a volume density charge distribution in space of $\rho(\mathbf{r},\mathrm{t})$. The total charge in this case is conserved over time. But the spatial charge distribution may vary within the system, giving rise to electric currents. More precisely, the variation of the charge, $\mathrm{d}Q$, during a time interval, $\mathrm{d}t$, contained within the volume V is given by $-I\ \mathrm{d}t$, where I is the current. Thus, the current density $\mathbf{J}(\mathbf{r},\mathrm{t})$ according to the classical vector analysis can be written as

$$\frac{\partial}{\partial t}\rho(\mathbf{r},\ t) + \mathrm{div}\mathbf{J}(\mathbf{r},\ t) = 0. \tag{2.71}$$

The objective here is to show that it is possible to find $\mathbf{J}(\mathbf{r},\mathrm{t})$, known as the *probability current* that satisfies an equation identical to Equation 2.71. Let us first assume that the particle under study is subject to a potential, $V(\mathbf{r},\mathrm{t})$, and thus, the Hamiltonian of the particle is

$$H = \frac{p^2}{2m} + V(\mathbf{r}, t). \tag{2.72}$$

The corresponding Schrödinger equation is

$$i\hbar\frac{\partial}{\partial \mathrm{t}}\psi(\mathbf{r}, t) = -\frac{\hbar^2}{2m}\Delta\psi(\mathbf{r}, t) + V(\mathbf{r}, t)\psi(\mathbf{r}, t) \tag{2.73}$$

and the complex conjugate equation is

$$i\hbar\frac{\partial}{\partial t}\psi^*(\mathbf{r},t) = -\frac{\hbar^2}{2m}\Delta\psi^*(\mathbf{r},t) + V(\mathbf{r},t)\psi^*(\mathbf{r},t), \tag{2.74}$$

where $V(\mathbf{r}, \mathrm{t})$ is real and **H** is Hermitian.

Multiplying both sides of Equation 2.73 by $\psi^*(\mathbf{r},t)$ and both sides of Equation 2.74 by $-\psi(\mathbf{r},t)$, and then adding both equations, we obtain the following:

$$i\hbar\frac{\partial}{\partial t}\psi^*(\mathbf{r},t)\psi(\mathbf{r},t) = -\frac{\hbar^2}{2m}[\psi^*(\mathbf{r},t)\Delta\psi(\mathbf{r},t) - \psi(\mathbf{r},t)\Delta\psi^*(\mathbf{r},t)]. \tag{2.75}$$

That is,

$$\frac{\partial}{\partial t}\rho(\mathbf{r},t) + \frac{\hbar}{2mi}[\psi^*(\mathbf{r},t)\Delta\psi(\mathbf{r},t) - \psi(\mathbf{r},t)\Delta\psi^*(\mathbf{r},t)] = 0. \tag{2.76}$$

If we set

$$\boldsymbol{J}(\mathbf{r},t) = -\frac{\hbar}{2mi}[\psi^*(\mathbf{r},t)\nabla\psi(\mathbf{r},t) - \psi(\mathbf{r},t)\nabla\psi^*(\mathbf{r},t)], \tag{2.77}$$

Equation 2.76 can be written in the form of Equation 2.71 since

$$\begin{aligned}
\operatorname{div}\mathbf{J}(\mathbf{r},t) &= \nabla\cdot\mathbf{J}(\mathbf{r},t)\\
&= \frac{\hbar}{2mi}\begin{bmatrix}\nabla\psi^*(\mathbf{r},t).\nabla\psi(\mathbf{r},t) + \psi^*(\mathbf{r},t)\nabla^2\psi(\mathbf{r},t)\\ -\nabla\psi(\mathbf{r},t).\nabla\psi^*(\mathbf{r},t) - \psi(\mathbf{r},t)\nabla^2\psi^*(\mathbf{r},t)\end{bmatrix}\\
&= \frac{\hbar}{2mi}\left[\psi^*(\mathbf{r},t)\nabla^2\psi(\mathbf{r},t) - \psi(\mathbf{r},t)\nabla^2\psi^*(\mathbf{r},t)\right]\\
&\equiv \text{second term of Equation 2.76.}
\end{aligned} \tag{2.78}$$

This proved the equation of local conservation of probabilities, and we have found the expression for the *probability current* in terms of the normalized wave function $\psi(\mathbf{r},\mathrm{t})$. Hence, if we have plane wave $\psi(\mathbf{r},t) = Ae^{ik\cdot x}$, we can calculate the probability density, $\rho(\mathbf{r},\mathrm{t})$ and probability current, $\boldsymbol{J}(\mathbf{r},t)$ such that

$$\rho(\mathbf{r},t) = \left|\psi(\mathbf{r},t)\right|^2 = \left|A\right|^2 \tag{2.79a}$$

and

$$\boldsymbol{J}(\mathbf{r},t) = \left|A\right|^2\frac{\hbar k}{m} = \rho(\mathbf{r},t)V_g, \tag{2.79a}$$

where $\boldsymbol{V}_{\mathrm{g}}$ is the group velocity obtained with the help of $\hbar\omega = \hbar^2k^2/2m$.

The objective from the above derivation is to show that the *probability current* is proportional to $|A|^2$ and thus the definition of transmission (T) and reflection

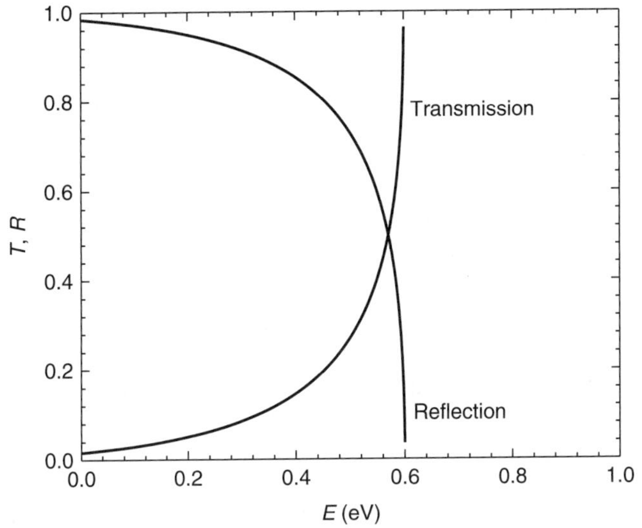

FIGURE 2.4 The transmission and reflection coefficients plotted as a function of the particle energy (E) for a barrier height of $V0 = 0.6$ eV.

(R) coefficients depend on the squares of the ratios of Equation 2.68. In other words, the reflection coefficient can be shown to be

$$R = \left|\frac{C}{A}\right|^2 = \left|\frac{k_1 - k_2}{k_1 + k_2}\right|^2 = 1 - \frac{4k_1k_2}{(k_1 + k_2)^2}. \tag{2.80}$$

The transmission coefficient can thus be obtained from $T + R = 1$ or from $T = (k_2/k_1)|C/A|^2$.

The reflection and transmission coefficients are plotted in Fig. 2.4 as a function of the particle energy with a fixed potential height of $V_0 = 0.6$ eV. It is clear from this figure that the particle is completely transmitted when $E \geq V_0$ and completely reflected when $E \ll V_0$.

For the case when $E < V_0$, the problem is quite different. Let us assume that the propagation vectors are

$$k_1^2 = \frac{2m(E)}{\hbar^2} \text{ for region I and } k_2^2 = \frac{2m(E - V_0)}{\hbar^2} \text{ for region II,} \tag{2.81}$$

The parameter k_2 is a complex quantity since $E < V_0$ and can be replaced by $k_2 = i\rho_2 = i\sqrt{\frac{2m(V_0 - E)}{\hbar^2}}$. The wave function in region II becomes

$$\varphi_{\text{II}}(x) = C\mathrm{e}^{\rho_2 x} + D\mathrm{e}^{-\rho_2 x}. \tag{2.82}$$

For the solution to remain bounded when $x \to \infty$, it is necessary to have the coefficient $C = 0$, reducing the wave function to $\varphi_{\text{II}}(x) = D\mathrm{e}^{-\rho_2 x}$, while $\varphi_I(x)$

remains the same as Equation 2.67a. The boundary conditions at $x = 0$ give

$$\frac{B}{A} = \frac{k_1 - i\rho_2}{k_1 + i\rho_2} \quad \text{and} \quad \frac{D}{A} = \frac{2k_1}{k_1 + i\rho_2}. \tag{2.83}$$

The reflection coefficient can then be given as

$$R = \left|\frac{B}{A}.\frac{B^*}{A^*}\right| = \frac{k_1 - i\rho_2}{k_1 + i\rho_2}.\frac{k_1 + i\rho_2}{k_1 - i\rho_2} = 1 \text{ and } T = 0. \tag{2.84}$$

Equation 2.84 shows that we have a total reflection. This effect demonstrated in Fig. 2.4 for R → 1 and T → 0 when $E \ll V_0$. This is similar to classical mechanics in which the particle is always reflected. In quantum mechanics, however, the wave function in region II of Fig. 2.3 is an evanescent wave with the form $e^{-\rho_2 x}$. Thus the particle has a nonzero probability of being in region II, which is decreased as x is increased.

The ratio B/A is complex, and a certain phase shift appears upon reflection, which is due to the fact that the particle is delayed when it penetrates the region $x > 0$. This effect is analogous to the phase shift that appears when light is reflected from a metallic material.

2.7 POTENTIAL BARRIER WITH A FINITE HEIGHT

Let us now derive the transmission and reflection coefficient for a particle with an energy (E) larger than the potential barrier height (V_0) and a width L as shown in Fig. 2.5. First we assume that the particle is coming from $x = -\infty$, thus the particle cannot be reflect back in region III and there is only one wave vector associated with the particle in this region. The wave functions for the three regions can be written as

$$\phi_{\text{I}} = Ae^{ik_1x} + Be^{-ik_1x} \tag{2.85a}$$

$$\phi_{\text{II}} = Ce^{ik_2x} + De^{-ik_2x} \tag{2.85b}$$

$$\phi_{\text{III}} = Fe^{ik_1x}, \tag{2.85c}$$

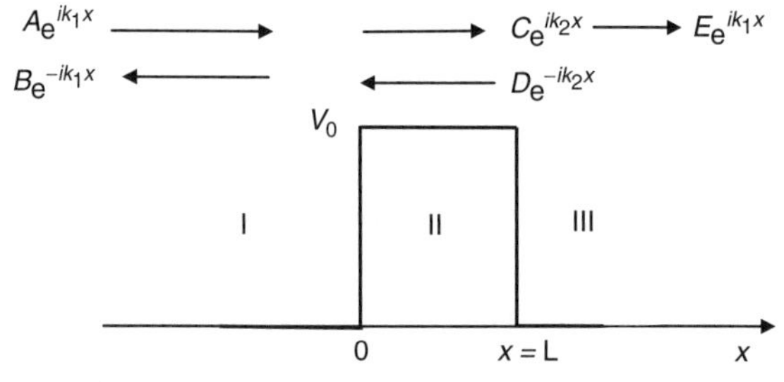

FIGURE 2.5 The wave function of a particle with energy $E > V_0$ is sketched with a potential barrier width of L and a height of V_0.

where A, B, C, D, and F are complex numbers. By applying the boundary conditions at $x = 0$ and $x = L$, one can obtain the ratios A/F and B/A. The best approach to solve this problem is by finding the coefficient A and B in terms of C and D using the boundary conditions at $x = 0$, C and D in terms of E using the boundary conditions at $x = \mathrm{L}$, and then relating A and B to F.

Let us start with the boundary conditions at $x = L$:

$$Ce^{ik_2L} + De^{-ik_2L} = Fe^{ik_1L} \tag{2.86a}$$

$$Ck_2e^{ik_2L} - Dk_2e^{-ik_2L} = Fk_1e^{ik_1L}. \tag{2.86b}$$

Multiplying Equation 2.26a by k_2 and adding (a) to (b) and then subtracting the same two equations, we obtain:

$$\frac{C}{F} = \frac{k_1 + k_2}{2k_2}e^{i(k_1-k_2)L} \tag{2.87a}$$

$$\frac{D}{F} = \frac{k_1 - k_2}{2k_2}e^{i(k_1+k_2)L}. \tag{2.87b}$$

Similarly, the boundary conditions at $x = 0$ gives:

$$A = \frac{k_1 + k_2}{2k_1}C + \frac{k_1 - k_2}{2k_1}D \tag{2.88a}$$

$$B = \frac{k_1 - k_2}{2k_1}C + \frac{k_1 + k_2}{2k_1}D. \tag{2.88b}$$

By substituting Equation 2.87 into Equation 2.88 and with rearrangement, we obtain:

$$\frac{A}{F} = e^{ik_1L}\left[\cos(k_2L) - i\frac{k_1^2 + k_2^2}{2k_1k_2}\sin(k_2L)\right] \tag{2.89a}$$

$$\frac{B}{F} = ie^{ik_1L}\left(\frac{k_1^2 - k_2^2}{2k_1k_2}\right)\sin(k_2L). \tag{2.89b}$$

The transmission coefficient, T, can be obtained from Equation 2.89a as

$$T = \left|\left(\frac{F}{A}\right)\left(\frac{F^*}{A^*}\right)\right| = \frac{4k_1^2k_2^2}{4k_1^2k_2^2 + (k_1^2 - k_2^2)^2\sin^2(k_2L)}. \tag{2.90}$$

Substitute the expressions

$$k_1^2 = \frac{2mE}{\hbar^2} \text{ for regions I and III, and } k_2^2 = \frac{2m(E - V_0)}{\hbar^2} \text{ for region II} \tag{2.91}$$

and insert in Equation 2.90 to obtain

$$T = \left|\frac{F}{A}\frac{F^*}{A^*}\right| = \frac{4E(E - V_0)}{4E(E - V_0) + V_0^2\sin^2\left(\sqrt{\frac{2m(E - V_0)}{\hbar^2}}L\right)}. \tag{2.92}$$

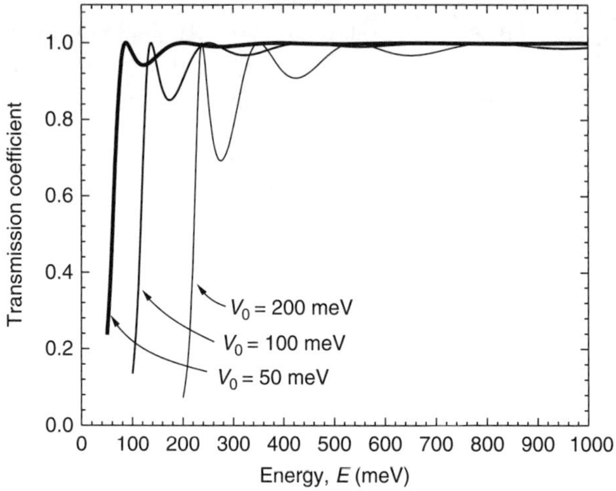

FIGURE 2.6 The transmission coefficient expressed in Equation 2.92 is plotted as a function of particle energy for three different potential heights (50, 100, and 200 meV). The width of the potential barriers is kept constant at 100 Å.

The reflection coefficient, R, can be obtained from the relation $T + R = 1$. The transmission coefficient is plotted as a function of the particle energy and shown in Fig. 2.6 for three different barrier heights. The barrier width, however, was the same for the three cases. Notice the oscillatory behavior of the transmission coefficient, even for a particle with energy higher than the potential barrier.

For the case of $E < V_0$, one can go through the same analysis shown above to obtain an expression for the transmission coefficient. The results are identical to Equation 2.92, except for the fact that the sine argument is $(-2m(V_0 - E)/h^2)^{1/2}L$. This quantity is a complex number. By using the trigonometry relation $\sin(ix) = i \sin h(x)$ in Equation 2.92, we can write the transmission coefficient as follows:

$$T = \left|\left(\frac{F}{A}\right)\left(\frac{F^*}{A^*}\right)\right| = \frac{4E(V_0 - E)}{4E(V_0 - E) + V_0^2 \sinh^2\left(\sqrt{\dfrac{2m(V_0 - E)}{\hbar^2}}L\right)}. \quad (2.93)$$

A plot of T and R as a function of the barrier width is shown in Fig. 2.6. For $\rho_2 L \gg 1$, where $\rho_2 = (2m(V_0 - E)/\hbar^2)^{1/2}$, we have $4E(V_0 - E) \ll V_0^2 \sin h^2(\sqrt{\frac{2m(V_0-E)}{\hbar^2}}L)$ and $\sinh^2(\sqrt{\frac{2m(V_0-E)}{\hbar^2}}L) \approx \frac{1}{4}e^{2\rho_2 L}$. This leads to the following expression for the transmission coefficient:

$$T \approx \frac{16E(V_0 - E)}{V_0^2} e^{-2\rho_2 L}. \quad (2.94)$$

Figure 2.7 shows an additional plot of $T \sim \exp(-2\rho_2 L)$. It is clear from this figure that the particle penetrates the barrier, and the probability of finding the particle at $x > 0$ does exist. This behavior cannot be explained in terms of classical mechanics. The particle has considerable probability of crossing the barrier by the "*tunneling effect*." The evanescent wave has a range of $1/\rho_2$. For a free electron of mass m, this range is $\left(\frac{1}{\rho_2}\right) \approx \frac{1.95}{\sqrt{V_0 - E}}$ Å and for the conduction electron in GaAs with an effective mass of $m* = 0.067m_0$ is $\left(\frac{1}{\rho_2}\right) \approx \frac{7.55}{\sqrt{V_0 - E}}$ Å.

The tunneling of the particle through the barrier is shown schematically in Fig. 2.8 in which the wave functions for the regions are shown assuming the

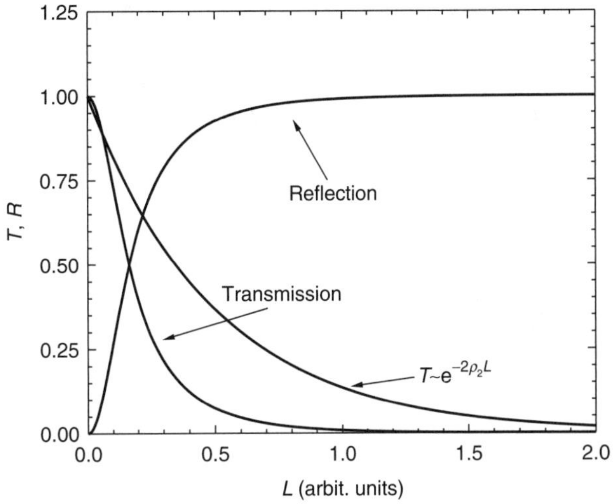

FIGURE 2.7 The reflection $(1 - T)$ and transmission (T) coefficients obtained in Equation 2.93 are plotted as a function of the barrier width. The form of Equation 2.93 is also plotted as $T \approx e^{-2\rho_2 L}$.

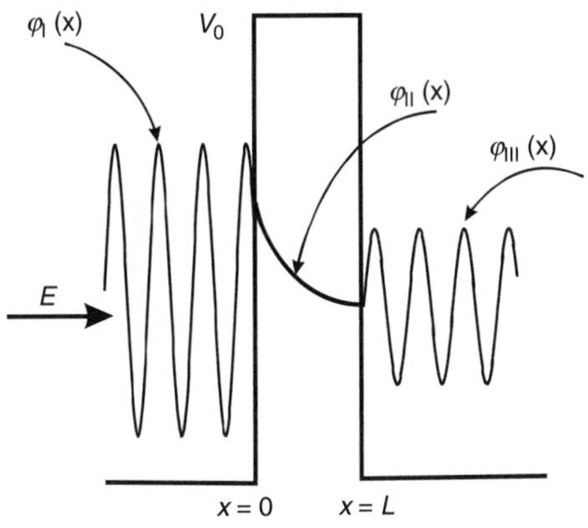

FIGURE 2.8 A schematic diagram showing the tunneling of a particle through a potential barrier is plotted for the three regions. The wave function inside the barrier is a decaying (evanescent) function.

particle is traveling form the left to right. The reflected wave functions in regions I and II are not shown. The wave function inside the barrier is a decaying function with the width of the barrier. This effect is due to the fact that $E < V_0$ and the propagation vector, k_2, is a complex quantity. Thus the wave function takes the following form: $\phi_{\text{II}} \sim e^{ik_2x} = e^{-\rho_2 x}$ since $k_2 = i\rho_2$. As F/A (Eq. 2.89a) is a complex number, a certain phase shift appears on reflection, which is physically due to the fact that the particle is delayed when it penetrates the $x > 0$ region.

2.8 POTENTIAL WELL WITH AN INFINITE DEPTH

Let us consider a potential well with an infinite depth as shown in Fig. 2.9 where $V(x)$ is zero for $0 < x < L$ and infinite everywhere else. Inside the quantum well, the Schrödinger equation is

$$\frac{\mathrm{d}^2}{\mathrm{d}x^2}\varphi(x) + \frac{2mE}{\hbar^2}\varphi(x) = 0. \tag{2.95}$$

By setting the propagation vector as $k = \sqrt{\frac{2mE}{\hbar^2}}$, where $E > 0$, we obtain the following Schrödinger equation

$$\frac{\mathrm{d}^2}{\mathrm{d}x^2}\varphi(x) + k^2\varphi(x) = 0. \tag{2.96}$$

FIGURE 2.9 A schematic of an infinite potential well.

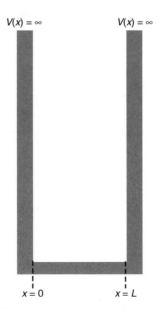

The general solution of this equation is

$$\varphi(x) = A\sin(kx) + B\cos(kx). \tag{2.97}$$

The boundary conditions in this case is $\varphi(0) = \varphi(L) = 0$. Thus, B in Equation 2.97 must be zero, reducing the wave function to $\varphi(x) = A\sin(kx)$. To have a nontrivial solution at $x = L$, $A \neq 0$, which implies that $\sin(kL) = 0$. Hence,

$$kL = n\pi n = 1, 2, 3, \ldots . \tag{2.98}$$

By substituting the expression of k into Equation 2.98, one can obtain the eigenvalues as

$$E_n = \frac{\hbar^2 k^2}{2m} = \frac{\hbar^2 \pi^2 n^2}{2mL^2} \qquad n = 1, 2, 3, \ldots \tag{2.99}$$

By normalizing the wave function and substituting for $k = 2\pi/L$, we can finally write the wave function as

$$\varphi(x) = \sqrt{\frac{2}{L}}\sin\left(\frac{n\pi}{L}x\right). \tag{2.100}$$

The example of the potential well when it is defined such that $V_0 = 0$ for $-L/2 \leq x \leq +L/2$ can be solved by performing the following transformation $x \rightarrow x - L/2$. The wave function becomes

$$\varphi(x) = \sqrt{\frac{2}{L}}\sin\left[\frac{n\pi}{L}\left(x - \frac{L}{2}\right)\right] = \sqrt{\frac{2}{L}}\sin\left(\frac{n\pi}{L}x - \frac{n\pi}{2}\right). \tag{2.101}$$

By expanding the sine function, we obtain the following:

$$\sin\left(\frac{n\pi}{L}x - \frac{n\pi}{2}\right) = \sin\left(\frac{n\pi}{L}x\right)\cos\left(\frac{n\pi}{2}\right) - \cos\left(\frac{n\pi}{L}x\right)\sin\left(\frac{n\pi}{2}\right). \tag{2.102}$$

For $n = 1, 3, 5, \ldots$, the wave function is proportional to $\cos(\frac{n\pi}{L}x)$, which is an even function. For $n = 2, 4, 6, \ldots$, the wave function is proportional to $\sin(\frac{n\pi}{L}x)$, which is an odd function. Thus, for $-L/2 < x < +L/2$, the wave function is given as

$$\varphi(x) = \begin{cases} \sqrt{\frac{2}{L}}\sin\left(\frac{n\pi}{L}x\right) & \text{for } n = 2, 4, 6, \ldots \\ \sqrt{\frac{2}{L}}\cos\left(\frac{n\pi}{L}x\right) & \text{for } n = 1, 3, 5, \ldots \end{cases} . \tag{2.103}$$

The energy levels and the probability functions of a particle in an infinite potential well are shown in Fig. 2.10. The energy levels are proportional to n^2 as indicated in Equation 2.99.

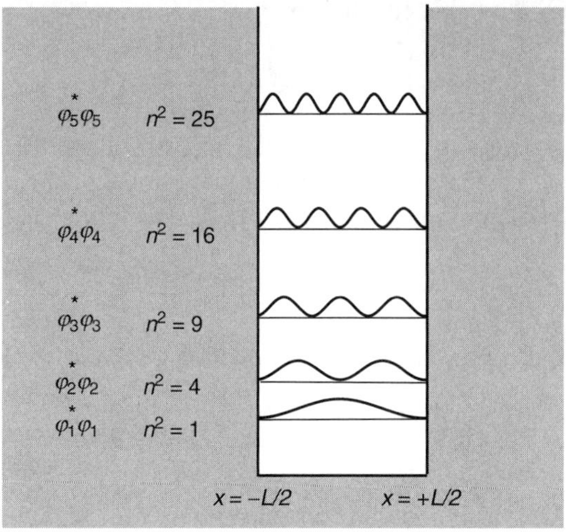

FIGURE 2.10 Energy levels $E_n = \frac{\hbar^2\pi^2 n^2}{2mL^2}$ of a particle in an infinite potential well plotted along with the probability $\varphi^*(x)\varphi(x)$ using the expressions shown in Equation (2.103).

2.9 FINITE DEPTH POTENTIAL WELL

A one-dimensional finite potential well is illustrated in Fig. 2.11, where the potential is defined as

$$\begin{aligned} V(x) &= 0 \quad \text{for } -\frac{L}{2} < x < +\frac{L}{2} \\ &= V_0 \quad \text{for } x < -\frac{L}{2} \text{ and } x > +\frac{L}{2}. \end{aligned} \tag{2.104}$$

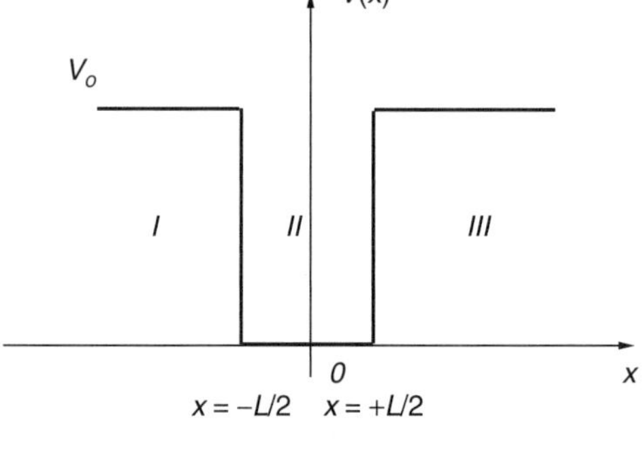

FIGURE 2.11 Finite potential well plotted for a particle with $0 < E < V_0$.

The wave functions for the case where $-V_0 < E < 0$ can be chosen as

$$\phi_{\mathrm{I}} = A\mathrm{e}^{\rho x} + B\mathrm{e}^{-\rho x} \tag{2.105a}$$

$$\phi_{\mathrm{II}} = C\mathrm{e}^{ikx} + D\mathrm{e}^{-ikx} \tag{2.105b}$$

$$\phi_{\mathrm{I}} = F\mathrm{e}^{\rho x} + \mathrm{G}\mathrm{e}^{-\rho x}, \tag{2.105c}$$

where $k = \sqrt{\frac{2mE}{\hbar^2}}$ and $\rho = \sqrt{\frac{2m(V_0 - E)}{\hbar^2}}$. To simplify the analysis, we can assume that the particle is bound in the well so that $\mathrm{B} = 0$.

By imposing the boundary conditions at $x = -L/2$, the relations between A, C, and D are obtained as follows:

$$\begin{aligned} C &= \mathrm{e}^{(-\rho+ik)L/2}\frac{\rho + ik}{2ik}A \\ D &= \mathrm{e}^{-(\rho+ik)L/2}\frac{\rho - ik}{2ik}A. \end{aligned} \tag{2.106}$$

With the help of Equation 2.106 and the matching conditions at $x = +L/2$, we can obtain the relations between A, F, and G according to the following:

$$\begin{aligned} \frac{F}{A} &= \frac{\mathrm{e}^{-\rho L}}{4ik\rho}\left[(\rho + ik)^2\mathrm{e}^{ikL} - (\rho - ik)^2\mathrm{e}^{-ikL}\right] \\ \frac{G}{A} &= \frac{\rho^2 + k^2}{2k\rho}\sin(kL). \end{aligned} \tag{2.107}$$

We still cannot obtain meaningful results without additional assumption. As the particle is bound in the well, one would find it necessary to set $F = 0$. Thus, the first equality in Equation 2.107 leads to the following relation:

$$\left(\frac{\rho - ik}{\rho + ik}\right)^2 = \mathrm{e}^{2ikL}. \tag{2.108}$$

Since ρ and k depend on E, Equation 2.108 can only be satisfied for certain values of E. In solving this problem, we consider the two possible cases for the following relation:

$$\left(\frac{\rho - ik}{\rho + ik}\right) = \pm\mathrm{e}^{ikL}. \tag{2.109}$$

The first case is when $\frac{\rho - ik}{\rho + ik} = -\mathrm{e}^{ikL}$, which yields

$$\frac{\rho}{k} = \tan\left(\frac{kL}{2}\right). \tag{2.110}$$

Let us define k_0 as $k_0^2 = \rho^2 + k^2 = \frac{2mV_0}{\hbar^2}$, which leads to

$$\frac{1}{\cos^2(kL/2)} = 1 + \tan^2(kL/2) = \frac{\rho^2 + k^2}{k^2} = \frac{k_0^2}{k^2}. \tag{2.111}$$

Equation 2.111 is equivalent to the following set of solutions:

$$\left|\cos^2(kL/2)\right| = \frac{k}{k_0}. \tag{2.112}$$

The second case is when $\frac{\rho-ik}{\rho+ik} = +e^{ikL}$. By following the above-mentioned procedure, one can find the following equation:

$$\left|\sin^2(kL/2)\right| = \frac{k}{k_0}. \tag{2.113}$$

It is difficult to solve Equations 2.112 and 2.113; however, graphical solution is possible, and the results are shown in Fig. 2.12. The values of k_0 are taken in units of k. From Fig. 2.12, one can calculate the values of the energy levels from the intersections of k_0 and the sine and cos curves. The intersections give the values of k, from which the energy level values are calculated. The number of confined states in the well can be obtained from the number of intersections that the straight line makes with the curves. For example, when k_0 is $5k$ we have four even states and three odd states. For the line marked $k_0 = 2k$, we have three states (two even and one odd). This figure indicates that there will be at least one bound state in the potential well.

The finite well potential problems can be solved in a different way as discussed in many textbooks. Let us assume that the wave functions for the three regions

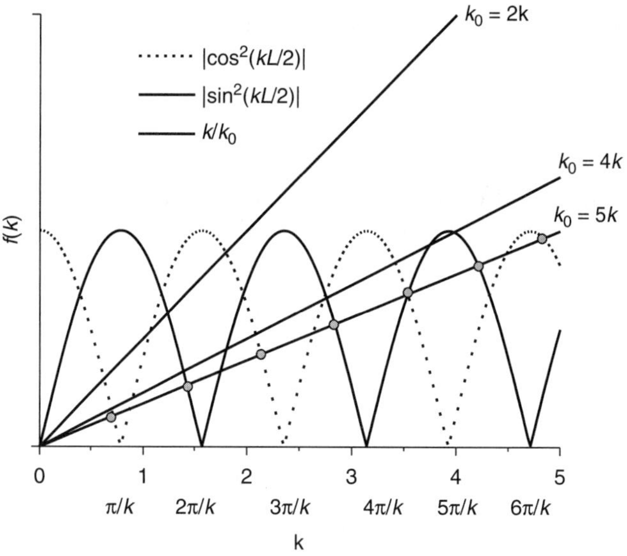

FIGURE 2.12 Graphic solutions of Equations 2.112 and 2.113 giving the bound states of a particle in a finite potential well.

in Fig. 2.11 have the following forms for a particle with energy, E, such that $0 < E < V_0$:

$$\phi_I = Ae^{\rho x} \tag{2.114a}$$

$$\phi_{II} = B\sin(kx) + C\cos(kx) \tag{2.114b}$$

$$\phi_I = De^{-\rho x}, \tag{2.114c}$$

where $k = \sqrt{\frac{2me}{\hbar^2}} and \rho = \sqrt{\frac{2m(V_0-E)}{\hbar^2}}$.

From the boundary conditions at $x = -L/2$ and $x = +L/2$, we have the following set of equations:

$$Ae^{-\rho L/2} + B\sin(kL/2) - C\cos(kL/2) = 0 \tag{2.115a}$$

$$A\rho e^{-\rho L/2} - Bk\cos(kL/2) - Ck\sin(kL/2) = 0 \tag{2.115b}$$

$$B\sin(kL/2) + C\cos(kL/2) - De^{-\rho L/2} = 0 \tag{2.115c}$$

$$Bk\cos(kL/2) - Ck\sin(kL/2) + D\rho e^{-\rho L/2} = 0. \tag{2.115d}$$

This is a system of four homogeneous linear equations for the coefficients A, B, C, and D. For the nontrivial solution, we must set the determinant of the coefficients to zero. While the values of coefficients are arbitrary, one can solve for their ratios. The determinant of the coefficient can be written as

$$\begin{vmatrix} 1 & \sin(kL/2) & -\cos(kL/2) & 0 \\ \rho & -k\cos(kL/2) & -k\sin(kL/2) & 0 \\ 0 & \sin(kL/2) & \cos(kL/2) & -1 \\ 0 & k\cos(kL/2) & -k\sin(kL/2) & \rho \end{vmatrix} e^{-\rho L/2} = 0. \tag{2.116}$$

This determinant can be expanded in minors to give

$$(k\sin(kL/2) - \rho\cos(kL/2))(k\cos(kL/2) + \rho\sin(kL/2)) = 0. \tag{2.117}$$

Dividing by $\cos^2(kL/2)$, we obtain

$$(k\tan(kL/2) - \rho)(\rho\tan(kL/2) + k) = 0. \tag{2.118}$$

By introducing $k_0^2 = \rho^2 + k^2 = \frac{2mV_0}{\hbar^2}$ and by knowing that Equation 2.118 can vanish by setting either of the quantities in the parentheses to zero we get

$$k\tan(kL/2) = \sqrt{k_0^2 - k^2} \quad \text{and} \quad -k\cot(kL/2) = \sqrt{k_0^2 - k^2}. \tag{2.119}$$

Once again, these equations can be solved graphically to obtain k values, from which the energy eigenvalues can be determined. To find the values of k, we plot Equation 2.119 in Fig. 2.13 for both expressions along with the propagation vector $\rho = \sqrt{k_0^2 - k^2}$.

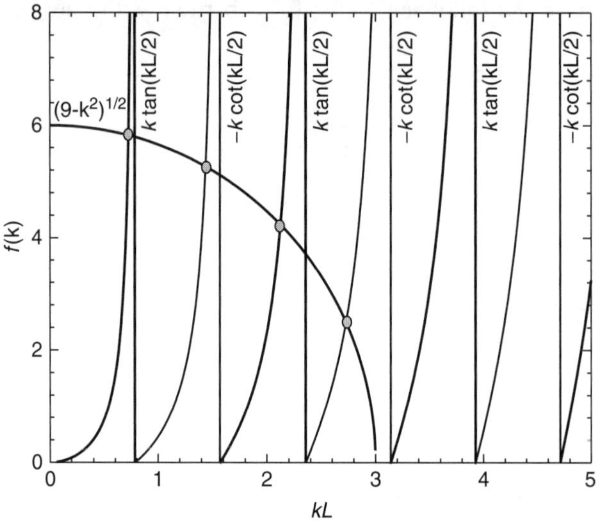

FIGURE 2.13 A plot of the functions shown in (2.119) as a function of kL. The intersections of $(ko - k^{1/2})$ with the tan and cot functions determine the values of k and the number of bound states.

The intersections of ρ with the functions tan(kL/2) and cot(kL/2) are shown as crosses for three values of ρ. The solutions of the potential well problem demonstrate the quantization of the eigenvalues. But, care must be taken when constructing the potential well since the eigenvalues, eigenfunctions, and the number of states depend on the physical structure of the potential well.

2.10 UNBOUND MOTION OF A PARTICLE ($E > V_0$) IN A POTENTIAL WELL WITH A FINITE DEPTH

Consider the potential well shown in Fig. 2.14 and that a particle is traveling from the left ($x = -\infty$) to the right ($x = +\infty$) with energy $E > V_0$.

The propagation vectors are given by

$$\rho = \sqrt{\frac{2mE}{\hbar^2}} \text{ and } k = \sqrt{\frac{2m(E - V_0)}{\hbar^2}}, \tag{2.120}$$

and the wave functions are constructed for the three regions according to Bastard and given as

$$\phi_\text{I} = \mathrm{e}^{i\rho(x+L/2)} + \mathrm{re}^{-i\rho(x+L/2)} \tag{2.121a}$$

$$\phi_\text{II} = \alpha \mathrm{e}^{ikx} + \beta \mathrm{e}^{-ikx} \tag{2.121b}$$

$$\phi_\text{I} = t\mathrm{e}^{i\rho(x-L/2)}. \tag{2.121c}$$

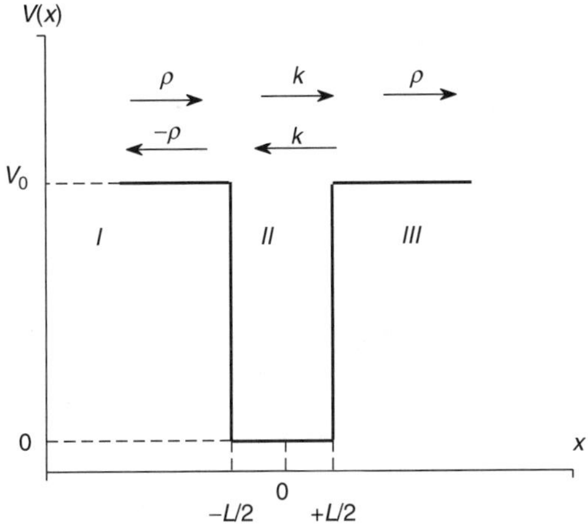

FIGURE 2.14 A schematic of a potential well with a particle of energy $E > V_0$ plotted along the x-axis. The potential well $V(x)$ is zero for $-L/2 < x < +L/2$ and V_0 for $x < -L/2$ and $x > L/2$.

From the boundary conditions at $x = -L/2$, we have

$$1 + r = \alpha e^{-ikL/2} + \beta e^{ikL/2} \tag{2.122a}$$

$$\rho - \rho r = \alpha k e^{-ikL/2} - \beta k e^{ikL/2}, \tag{2.122b}$$

and the boundary conditions at $x = +L/2$ give

$$t = \alpha e^{ikL/2} + \beta e^{-ikL/2} \tag{2.123a}$$

$$\rho t = \alpha k e^{ikL/2} - \beta k e^{-ikL/2}. \tag{2.123b}$$

By multiplying Equation 2.122a by ρ and adding to Equation 2.122b, we obtain

$$2\rho = \alpha(k + \rho)e^{-ikL/2} - \beta(k - \rho)e^{ikL/2}. \tag{2.124}$$

and by multiplying Equation 2.123a by ρ and then subtracting it from Equation 2.123b, we obtain

$$0 = \alpha(k - \rho)e^{ikL/2} - \beta(k + \rho)e^{-ikL/2} \quad \text{or} \quad \beta = \alpha\frac{(k - \rho)}{(k + \rho)}e^{ikL}. \tag{2.125}$$

By dividing Equation 2.124 by $(k - \rho)e^{ikL/2}$ and utilizing Equation 2.125, we can obtain the following:

$$\alpha = \frac{\rho(k + \rho)e^{-ikL/2}}{2k\rho\cos(kL) - i(k^2 + \rho^2)\sin(kL)}$$

$$\beta = \frac{\rho(k - \rho)e^{ikL/2}}{2k\rho\cos(\mathrm{kL}) - i(k^2 + \rho^2)\sin(kL)}. \tag{2.126}$$

Substituting Equation 2.126 into Equations 2.122a and 2.123a, we obtain

$$t = \frac{1}{\cos(kL) - \frac{1}{2}i\left(\frac{k}{\rho} + \frac{\rho}{k}\right)\sin(kL)} \tag{2.127}$$

and

$$r = \frac{\frac{i}{2}\left(\frac{k}{\rho} - \frac{\rho}{k}\right)\sin(kL)}{\cos(kL) - \frac{1}{2}i\left(\frac{k}{\rho} + \frac{\rho}{k}\right)\sin(kL)}.$$

Let the transmission coefficient be $T(E) = |t(E)|^2$ and the reflection coefficient be $R(E) = |r(E)|^2$, where $T(E) + R(E) = 1$. Then,

$$T(E) = \frac{1}{1 + \frac{1}{4}\left(\frac{k}{\rho} - \frac{\rho}{k}\right)^2 \sin^2(kL)} \tag{2.128}$$

and

$$R(E) = \frac{\left(\frac{\mathrm{k}}{\rho} - \frac{\rho}{k}\right)^2 \sin^2(kL)}{4 + \left(\frac{\mathrm{k}}{\rho} - \frac{\rho}{k}\right)^2 \sin^2(kL)}.$$

The transmission coefficient is plotted in Fig. 2.15 for two different potential heights. It reaches unity as indicated in Equation 2.73 for $\sin(kL) = 0$ or $kL = n\pi$, where n is an integer. The form of $T(E)$ corresponds to constructive interference inside the potential well. The discrete energies that fulfill the condition $kL = n\pi$ are called transmission resonances. They correspond to an enhanced probability of finding the particle inside the quantum well. For detailed discussion see Bastard.

2.11 TRIANGULAR POTENTIAL WELL

Another important potential well is the triangular quantum well. This type of well is common at the semiconductor interfaces such as GaAs/AlGaAs heterojunction. In particular, the high electron mobility transistor (HEMT) is based on the energy quantization in the triangular well formed at the heterojunction interface. A schematic representation of the conduction band edge of GaAs/AlGaAs HEMT structure is shown in Fig. 2.16. The space (W) is the undoped barrier region. If N is the number of electrons transferred to the well per unit area

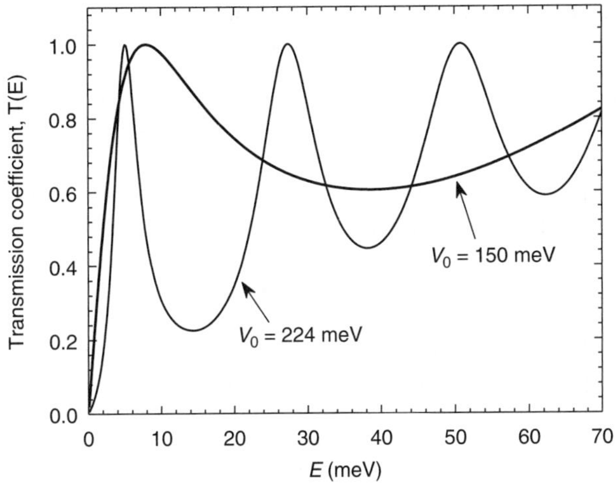

FIGURE 2.15 The transmission coefficient, $T(E)$, of unbound particle ($E > 0$) plotted as a function of energy for an optional well with a height of 224 meV and thickness of 250 Å. $T(E)$ is also plotted for a similar potential well of height 150 meV.

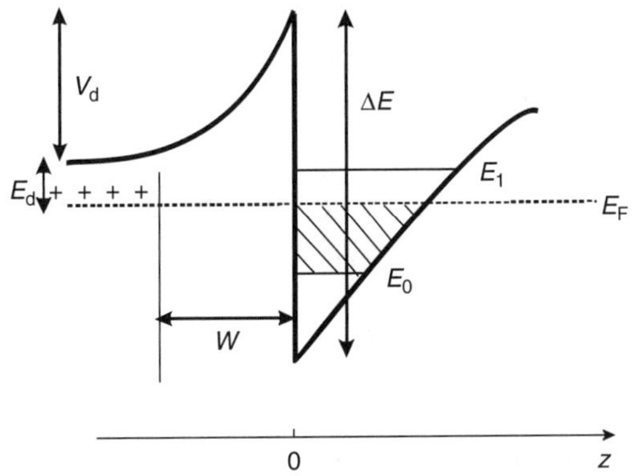

FIGURE 2.16 A schematic plot of the conduction band of a HEMT structure.

(known as the two-dimensional electron gas), the electric field $\mathcal{E}_s$ is given by Gauss's law as

$$\mathcal{E}_s = \frac{eN}{\epsilon_0 \epsilon_r}, \tag{2.129}$$

where ϵ_r is the dielectric constant of the well material and ϵ_0 is the permittivity of free space. For a triangular well as shown in Fig. 2.16, the electrostatic potential $\varphi(z)$ is linear in $z > 0$ region and is given by

$$\varphi(z) = \mathcal{E}_s \ {}_s z. \tag{2.130}$$

The Hamiltonian for an electron in the triangular well assuming that the potential barrier is infinite at $z = 0$ can be written as

$$H = -\frac{\hbar^2}{2m}\frac{d^2}{dz^2} + V_p(z) + e\varphi(z) \tag{2.131}$$

where $V_p(z)$ is the periodic potential energy. Using the envelope function approximation for one-dimensional system, the wave function can be written as

$$\psi(z) = F(z)U(z), \tag{2.132}$$

where $U(z)$ is the conduction band Bloch function for zero wave vector and $F(z)$ is the envelop function that satisfies the effective mass equation

$$\left[-\frac{\hbar^2}{2m^*}\frac{d^2}{dz^2} + e\varphi(z)\right]F(z) = E_n F(z). \tag{2.133}$$

The index n identifies the eigenvalues, and m^* is the conduction electron effective mass of the well material. The wave function $F(z)$ can be further written as

$$F(z) = e^{ik_z z}\chi_n(z), \tag{2.134}$$

where k_z is the two-dimensional wave vector perpendicular to the surface normal, and $\chi(z)$ satisfies the equation

$$\left[-\frac{\hbar^2}{2m^*}\frac{d^2}{dz^2} + e\varphi(z)\right]\chi(z) = E_n\chi(z) \tag{2.135}$$

and

$$E_n = E - \frac{\hbar^2 k_z^2}{2m^*}, \tag{2.136}$$

where E is the total energy eigenvalue of the carriers. The boundary conditions to be satisfied by $\chi(z)$ are $\chi_n(0) = \chi_n(\infty) = 0$. A solution that satisfies the boundary condition at infinity is the Airy function (see Stern, Balanski *et al*., and Ferry) given by $Ai\left[\left(\frac{2m^*}{\hbar^2 e^2 E_s^2}\right)^{1/3}(eE_s z - E_n)\right]$. The boundary condition at $z = 0$ determines the allowed values of E_n as

$$E_n = -\left(\frac{\hbar^2 e^2 \mathcal{E}_s^2}{2m^*}\right)^{1/3} a_n.. \tag{2.137}$$

The quantity a_n is the zero of the Airy function and is approximated as (Stern and Balanski *et al*.)

$$a_n \approx -\left(\frac{3\pi}{2}\left(n + \frac{3}{4}\right)\right)^{2/3}, \tag{2.138}$$

where $n = 0, 1, 2, \ldots$ The values of E_n are then

$$E_n = \left(\frac{\hbar^2}{2m^*}\right)^{1/3}\left(\frac{3\pi e\mathcal{E}_s}{2}\left(n+\frac{3}{4}\right)\right)^{2/3} \quad \text{with } E_0 \approx \left(\frac{\hbar^2}{2m^*}\right)^{1/3}\left(\frac{9\pi e^2 N}{8\epsilon_0\epsilon_r}\right)^{2/3}. \tag{2.139}$$

The triangular potential is a very good approximation for potential distribution near the semiconductor interfaces. The E_n shown in Equation 2.139 is a function of the quantum number n, which represents the energy levels in an infinite triangular quantum well.

2.12 DELTA FUNCTION POTENTIALS

The δ-function problem is discussed in this section for one particular reason. The current technology in optoelectronics is gravitated toward the semiconductor nanostructures. The recent research is focused on the use of quantum dots for lasers and detectors. Quantum dots are a small collection of semiconductor atoms such as InAs sandwiched between GaAs barrier materials. The quantum dots are sometimes called the designer atoms. In other words, quantum dots could be represented or approximated by δ-function wells. A few of δ-function characteristics were discussed earlier in this chapter. Let us define a potential $V(x)$ such as

$$V(x) = -\frac{\hbar^2\lambda}{2ma}\delta(x), \tag{2.140}$$

where a is a quantity with a dimension of length and λ is a dimensionless quantity introduced to characterize the strength of the δ-function (see for example Gasiorowicz). The δ-function well is shown in Fig. 2.17. The Schrödinger equation can be written as

$$-\frac{\hbar^2}{2m}\frac{\mathrm{d}^2u(x)}{\mathrm{d}x^2} - \frac{\hbar^2\lambda}{2ma}\delta(x)u(x) = -\left|E_n\right|u(x). \tag{2.141}$$

By integrating Equation 2.141, we can obtain the condition at $x = 0$

$$-\int_{-\epsilon}^{+\epsilon}\frac{\mathrm{d}^2u(x)}{\mathrm{d}x^2}\mathrm{d}x - \frac{\lambda}{a}\int_{-\epsilon}^{+\epsilon}\delta(x)u(x)\mathrm{d}x = -\frac{2m}{\hbar^2}\left|E_n\right|\int_{-\epsilon}^{+\epsilon}u(x)\mathrm{d}x \tag{2.142}$$

The right-hand side of this equation is zero, since if we choose as an example $u(x) = Ae^{kx}$, the right-hand side can then be proportional to sin $h(\epsilon)$ and $\lim_{\epsilon\to 0}\sin h(\epsilon) \to 0$. With the help of Equation 2.37, we can write Equation 2.142 as

$$\left.\frac{\mathrm{d}u(x)}{\mathrm{d}x}\right|_{-\epsilon}^{+\epsilon} = -\frac{\lambda}{a}u(o) \tag{2.143}$$

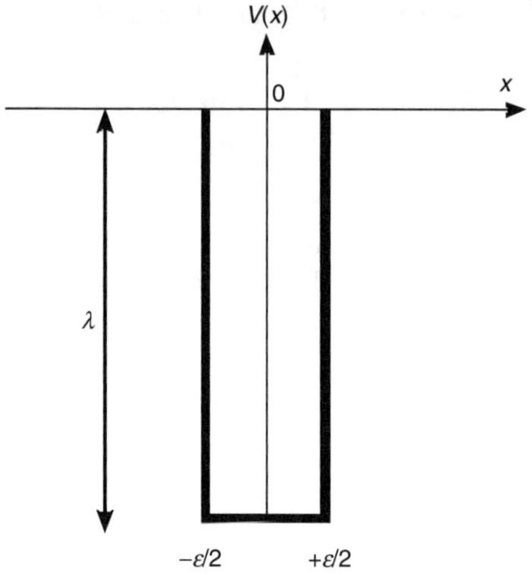

FIGURE 2.17 A sketch of a δ-function with a width of a small quantity ϵ and strength of λ is shown.

or

$$\left.\frac{\mathrm{d}u(x)}{\mathrm{d}x}\right|_{x=+\epsilon} - \left.\frac{\mathrm{d}u(x)}{\mathrm{d}x}\right|_{x=-\epsilon} = -\frac{\lambda}{a}u(o).$$

For the solution of Equation 2.141 at $x \neq 0$, we have

$$\frac{\mathrm{d}^2u(x)}{\mathrm{d}x^2} - \frac{2m}{\hbar^2}|E_n|u(x) = 0 \tag{2.144}$$

or

$$\frac{\mathrm{d}^2u(x)}{\mathrm{d}x^2} - k^2u(x) = 0,$$

where $k = \sqrt{\frac{2m|E_n|}{\hbar^2}}$. The solution that satisfies Equation 2.144 at all x values except for $x = 0$ and vanishes at $x = \pm\infty$ is

$$u(x) = \begin{cases} e^{-kx} & \text{for } x > 0 \\ e^{kx} & \text{for } x < 0 \end{cases}. \tag{2.145}$$

The amplitude of $u(x)$ is the same by symmetry for $x > 0$ and $x < 0$, and for simplicity, it was chosen as unity. From Equations 2.143 and 2.145 we can find that

$$-k - k = -\frac{\lambda}{a} \quad \text{or} \quad 2k = \frac{\lambda}{a}. \tag{2.146}$$

Substituting the value of k into Equation 2.146, we find the energy to be

$$E = \frac{\hbar^2\lambda^2}{8ma^2}, \tag{2.147}$$

which means that there is only one bound state in the δ-function potential well. This is in many ways similar to small size semiconductor quantum dots, where each quantum dot has only one bound state. The situation is different as the quantum dot size is increased beyond approximately five monolayers.

A more interesting problem is the double-narrow deep δ-function potential well shown in Fig. 2.18. The potential can be written as

$$V(x) = -\frac{\hbar^2\lambda}{2ma}[\delta(x-a) + \delta(x+a)]. \tag{2.148}$$

The potential is symmetric under the interchange $x \to -x$, therefore the solutions have definite parity. Let us consider both the solutions.

(i) *Even solution:* Let us consider the following wave function that satisfies the even parity:

$$\begin{aligned} u(x) &= \mathrm{e}^{-kx} \quad \text{for } x > a \\ &= A\cos h(kx) \quad \text{for } a < x < -a \\ &= \mathrm{e}^{kx} \quad \text{for } x < -a. \end{aligned} \tag{2.149}$$

By applying the boundary conditions at $x = a$ and with the help of Equation 2.143, we obtain the relations:

$$\mathrm{e}^{-ka} = A\cos h(ka) \tag{2.150}$$

and

$$-k\mathrm{e}^{-ka} - kA\sin h(ka) = -\frac{\lambda}{a}\mathrm{e}^{-ka}. \tag{2.151}$$

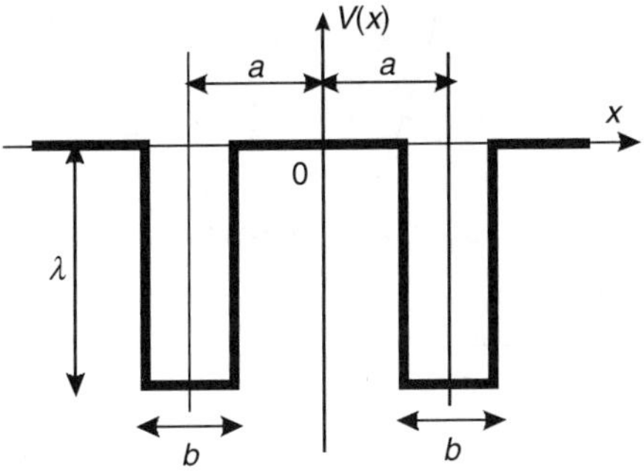

FIGURE 2.18 Double narrow deep δ-function potential.

The constant A can be eliminated by combining Equations 2.149 and 2.150 to yield

$$\tan h(ka) = \frac{\lambda}{ka} - 1 \tag{2.152}$$

Equation 2.152 can be rewritten as

$$\frac{e^{ka} - e^{-ka}}{e^{ka} + e^{-ka}} = \frac{\lambda}{ka} - 1 \tag{2.153a}$$

or

$$e^{-2ka} = \frac{2ka}{\lambda} - 1. \tag{2.153b}$$

Equation 2.151 can be used to obtain the eigenvalues graphically in a manner similar to the finite depth potential well. The result is shown in Fig. 2.19. It is clear from this figure that there is only one solution corresponding to an eigenvalue. In addition, since tan $h(ka) < 1$, it follows form Equations 2.151 that

$$k > \frac{\lambda}{2a} \text{ or } E > \frac{\hbar^2\lambda^2}{8ma^2}. \tag{2.154}$$

By comparing the above equation to Equation 2.147, it implies that the energy level in the double δ-function potential well is lower than the energy level in a single δ-function. The wave function of the double δ-function potential well [Eq. 2.149] is plotted in Fig. 2.20 to show the singularities at $+a$ and $-a$. The reduction of the energy level in the double

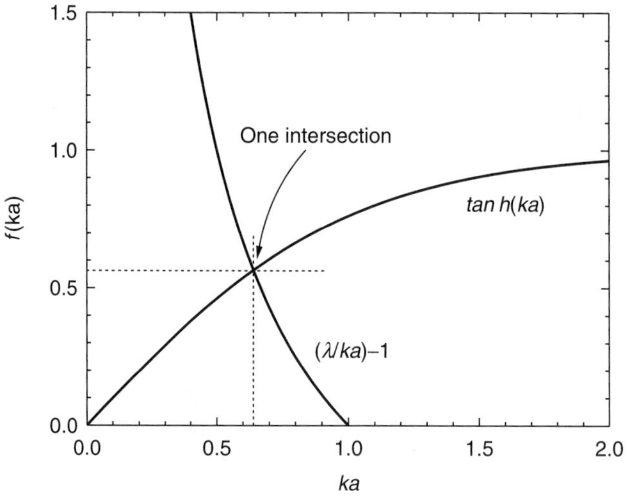

FIGURE 2.19 Graphical solution for the eigenvalue of a double δ-function potential well.

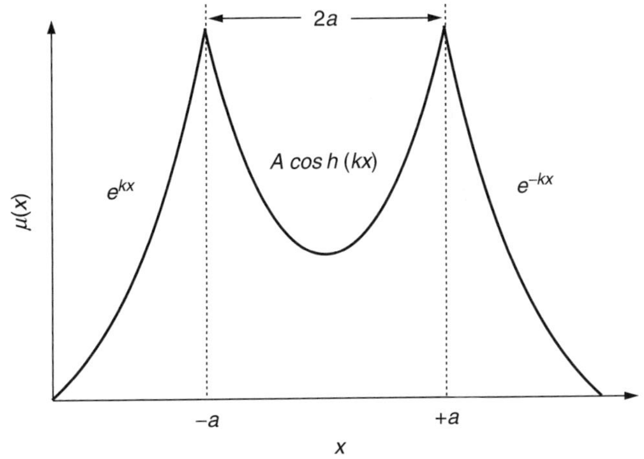

FIGURE 2.20 A plot of the wave function (Eq. 2.149) of double δ-function potential well as a function of x.

δ-function as compared to that of the single δ-function barrier is difficult to explain, but such an effect is observed experimentally in multiple quantum wells. In multiple quantum well system, it was observed that the confined energy levels are reduced as the number of quantum wells in the structure is increased. This, however, could be a coincident and the latter example may not provide a clear explanation of the above effect.

(ii) *Odd solution:* for the odd solution we consider the following wave function:

$$\begin{aligned} u(x) &= \mathrm{e}^{-kx} \text{ for } x > a \\ &= A \sin h(kx) \quad \text{for } a < x < -a \\ &= -\mathrm{e}^{kx} \quad \text{for } x < -a. \end{aligned} \tag{2.155}$$

The analysis here is similar to the analysis followed in the even solution case. By taking the boundary conditions at $x = +a$ and by the using Equation 2.143, we have

$$A \sin h(ka) = \mathrm{e}^{-ka} \tag{2.156a}$$

and

$$-k\mathrm{e}^{-ka} - kA \cos h(ka) = -\frac{\lambda}{a}\mathrm{e}^{-ka}. \tag{2.156b}$$

Substituting Equation 2.156a into Equation 2.156b, we obtain

$$\cot h(ka) = \frac{\lambda}{ka} - 1 \quad \text{or} \quad \tan h(ka) = \left(\frac{\lambda}{ka} - 1\right)^{-1}. \tag{2.157}$$

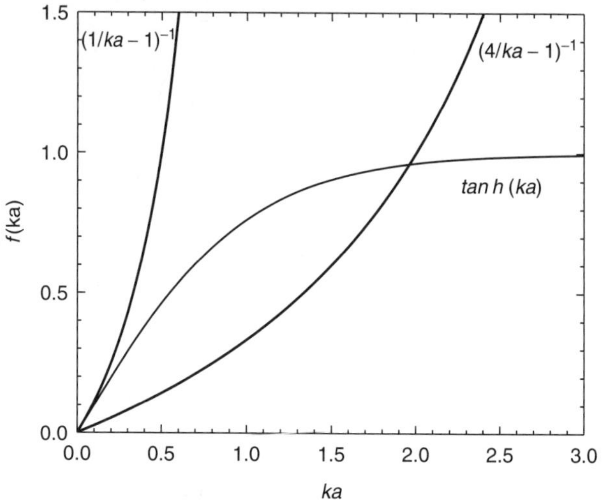

FIGURE 2.21 The graphical representation of the odd solution of the double δ-function potential well.

This equation can be solved graphically, as was the case with the even solution discussed above. The results are shown in Fig. 2.21. Equation 2.116 indicates that there is a singularity when $\lambda = ka$. The vertical line in Fig. 2.21 is due to the singularity when λ is chosen to be one. When λ is larger than one, for example $\lambda = 4$, the singularity occurs at 4, which is not shown in the figure. As indicated in this figure, there is only one bound state when $\lambda \geq 1$; when λ is less than one, we may or may not have a bound state because of the odd parity solution as shown in the figure.

2.13 TRANSMISSION IN FINITE DOUBLE BARRIER POTENTIAL WELLS

This is a more complicated problem, and we will follow the propagation matrix method. This structure, however, is very important to illustrate the basic concept of tunneling resonant diode. Consider the double barrier potential shown in Fig. 2.22. Also, consider the general case where $L_1 \neq L_2 \neq L_3$ and the barrier height is the same (V_0). The problem could be simplified by assuming $L_1 = L_2 = L_3 = L_0$. The aim here is to obtain an expression for the transmission coefficient of a particle traveling from $Z = -\infty$ to $Z = +\infty$ assuming that the particle mass does not change as it travels through the barriers. The structure consists of five regions, and the solutions to Schrödinger equation within each region for $E < V_0$ are:

$$\psi_1(z) = Ae^{ikz} + \mathrm{B}e^{-ikz} \quad \text{for region 1}$$

$$\psi_2(z) = Ce^{\rho z} + \mathrm{D}e^{-\rho z} \quad \text{for region 2}$$

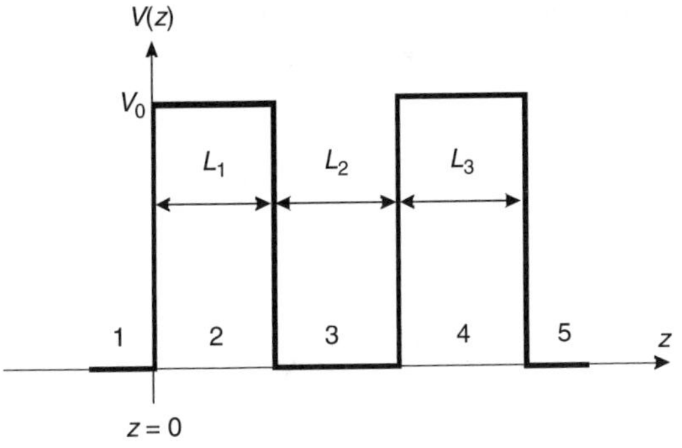

FIGURE 2.22 A sketch of a double barrier potential well.

$$\psi_3(z) = Fe^{ikz} + \mathrm{G}e^{-ikz} \quad \text{for region 3}$$
$$\psi_4(z) = He^{\rho z} + \mathrm{J}e^{-\rho z} \quad \text{for region 4}$$
$$\psi_5(z) = Ke^{ikz} + \mathrm{L}e^{-ikz}, \quad \text{for region 5} \tag{2.158}$$

where $k = \sqrt{\frac{2mE}{\hbar^2}}$ and $\rho = \sqrt{\frac{2m(V_0-E)}{\hbar^2}}$. The boundary conditions at $Z = 0, L_1$, $L_1 + L_2$, and $L_1 + L_2 + L_3$ gives the following relations assuming the mass of the particle is constant throughout the structure. For $Z = I_1 = 0$:

$$A + B = C + D \tag{2.159}$$

$$ikA - ikB = \rho C - \rho D \tag{2.160}$$

For $Z = I_2 = L_1$:

$$Ce^{\rho I_2} + De^{-\rho I_2} = Fe^{ikI_2} + Ge^{-ikI_2} \tag{2.161}$$

$$C\rho e^{\rho I_2} - D\rho e^{-\rho I_2} = Fike^{ikI_2} - Gike^{-ikI_2} \tag{2.162}$$

For $Z = I_3 = L_1 + L_2$:

$$Fe^{ikI_3} + Ge^{-ikI_3} = He^{\rho I_3} + Je^{-\rho I_3} \tag{2.163}$$

$$Fike^{ikI_3} - Gike^{-ikI_3} = H\rho e^{\rho I_3} - J\rho e^{-\rho I_3} \tag{2.164}$$

For $Z = I_4 = L_1 + L_2 + L_3$:

$$He^{\rho I_4} + Je^{-\rho I_4} = Ke^{ikI_4} + Le^{-ikI_4} \tag{2.165}$$

$$H\rho e^{\rho I_4} - J\rho e^{-\rho I_4} = Kike^{ikI_4} - Like^{-ikI_4}. \tag{2.166}$$

The best method to proceed from here is to put the results of Equations 2.159–2.166 in matrix form known as *propagation matrix technique*, which yields

$$M_1 \begin{pmatrix} A \\ B \end{pmatrix} = M_2 \begin{pmatrix} C \\ D \end{pmatrix} \tag{2.167a}$$

$$M_3 \begin{pmatrix} C \\ D \end{pmatrix} = M_4 \begin{pmatrix} F \\ G \end{pmatrix} \tag{2.167b}$$

$$M_5 \begin{pmatrix} F \\ G \end{pmatrix} = M_6 \begin{pmatrix} H \\ J \end{pmatrix} \tag{2.167c}$$

$$M_7 \begin{pmatrix} H \\ J \end{pmatrix} = M_8 \begin{pmatrix} K \\ L \end{pmatrix}. \tag{2.167d}$$

The coefficient of the outer regions can be linked by forming the *transfer matrix* such as

$$\begin{pmatrix} A \\ B \end{pmatrix} = M_1^{-1} M_2 M_3^{-1} M_4 M_5^{-1} M_6 M_7^{-1} M_8 \begin{pmatrix} K \\ L \end{pmatrix}. \tag{2.168}$$

Since we assumed that the particle is traveling from $Z = -\infty$ to $Z = +\infty$, the coefficient L can be set to zero, and if the 2×2 matrix is written as M, we obtain

$$\begin{pmatrix} A \\ B \end{pmatrix} = M \begin{pmatrix} K \\ 0 \end{pmatrix}. \tag{2.169}$$

Thus, we have $A = M_{11} K$, and the transmission coefficient can be written as

$$T(E) = \left| \frac{K \cdot K^*}{A \cdot A^*} \right| = \frac{1}{|M_{11} \cdot M_{11}^*|}. \tag{2.170}$$

The matrix multiplication needed to obtain M_{11} is tedious and time consuming.

The tunneling through a double barrier structure is the basis for the tunneling resonant diode. This structure has already been briefly discussed, but this problem is discussed in more detail in this section. For the simplest case, we have chosen the barriers and well widths to be identical. By following the propagation matrix procedure described above, the transmission coefficient is presented in Equation 2.170. To simplify the analysis for the transmission coefficient of an electron traveling with an energy $E < V_0$ from $x = -\infty$ to $x = +\infty$, one can assume that the widths of the well and the two barriers are the same. Typical materials for this structure are GaAs for the well and AlGaAs for the barriers. Since the electron's effective mass does not change considerably in the well ($m^* = 0.067 m_0$) and in the barrier ($m^* = 0.094\ m_0$), we assumed that it is the same for both materials.

With these assumptions in mind, an expression for the transmission coefficient is derived as

$$T(E) = \frac{64\,[E(V_0 - E)]^2}{D_1 + D_2}, \tag{2.171}$$

where

$$D_1 = \Big[\{(V_0 - 2E)^2 - 4E(V_0 - E)\} \cos h(2\rho L_b) \\ - V_0^2 \{2 \sin h^2(\rho L_b) \cos(2kL_w) + 1\}\Big]^2$$

$$D_2 = \Big[4(V_0 - 2E)\sqrt{E(V_0 - E)} \sin h(2\rho L_b) + 2V_0^2 \sin h^2(\rho L_b) \sin(2kL_w)\Big]^2.$$

L_w is the well width, L_b is the barrier width, $k = \sqrt{\frac{2m^*E}{\hbar^2}}$, and $\rho = \sqrt{\frac{2m^*(V_0-E)}{\hbar^2}}$.

The transmission coefficient is plotted as a function of the electron energy for a barrier of height $V_0 = 0.10$ eV and three different widths ($L = 50$, 100, and 150 Å) as shown in Fig. 2.23. The peaks in the transmission coefficient correspond to the electron energy as being resonant with the confined energy levels in the well. As the well width increases, the number of confined energies is increased for a fixed potential barrier. When the electron energy is larger than the barrier, transmission resonance peaks can be observed, which correspond to virtual states in the continuum. Notice that the number of the resonance peaks and their energy positions change as the well width is varied. For example, there is only one state for a well thickness of 50 Å and barrier height of 0.10 eV.

When the well width is fixed and the barrier height is increased, one would expect to observe additional bound states as illustrated in Fig. 2.24. In this figure, the transmission coefficient is plotted for three different barrier heights. The ground state is expected to slightly shift as the barrier height increases. Furthermore, one can observe the resonance peaks to shift as a function of the barrier height.

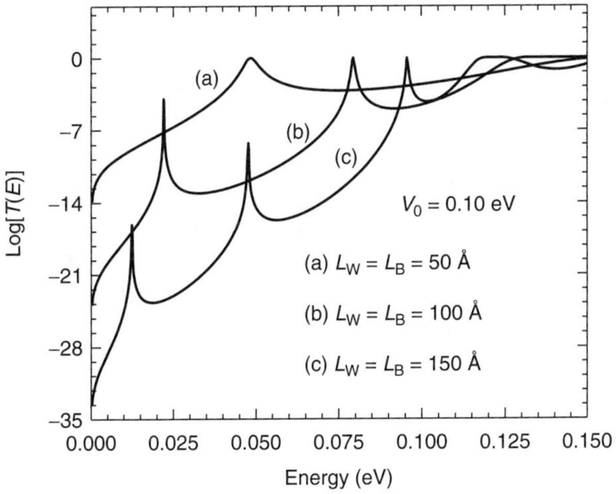

FIGURE 2.23 Transmission coefficient is plotted as a function of electron energy in a double barrier structure for three different well (LW) and barrier (LB) widths and a barrier height of 0.10 eV.

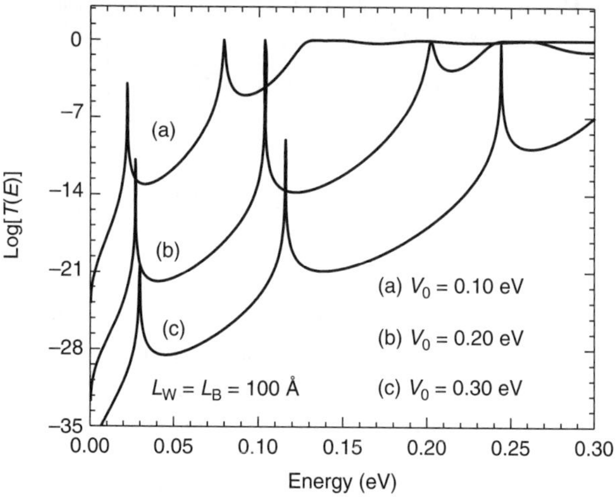

FIGURE 2.24 Transmission coefficient is plotted as function of electron energy for a fixed well width and three different barrier heights.

2.14 ENVELOPE FUNCTION APPROXIMATION

The theoretical calculations of the energy states in the semiconductor nanomaterials are very complicated and require computer analysis. One of the simplest theoretical approaches is to find the boundary conditions at the heterojunction interfaces where the wave functions are almost invariant. This approach is called envelope function approximation, which has been used successfully (Bastard 1988, Bastard *et al.* 1991) in determining the energy sates in quantum wells, superlattices and heterojunctions. Following Bastard's formalism, the envelope function approximation assumes that the constituents of the quantum wells are lattice matched with abrupt interfaces such as GaAs/AlGaAs multiple quantum wells, as shown in Fig. 2.25. This figure shows two types of multiple quantum wells. Type I is illustrated in Fig. 2.25a, where the electrons and holes exist in the same layer. An example of this type is the GaAs/AlGaAs multiple quantum well. The second type of band alignment is called type II, which is illustrated in Fig. 5.25b, where the electrons and located in one layer and the holes are located in the adjacent layer. A typical example of this band alignment is found in InAs/InGaSb superlattices.

Inside each layer of the multiple quantum wells, the wave function is expanded in the periodic part of wave functions called *Bloch functions* as

$$\psi(r) = \sum_n f_n^A(r) u_{nk_0}^A \quad \text{for } -L/2 \leq z \leq L/2 \text{ (well)}$$

and

$$\psi(r) = \sum_n f_n^B(r) u_{nk_0}^B \quad \text{for } z > L/2 \text{ or } z < -L/2 \text{ (barrier)}, \tag{2.172}$$

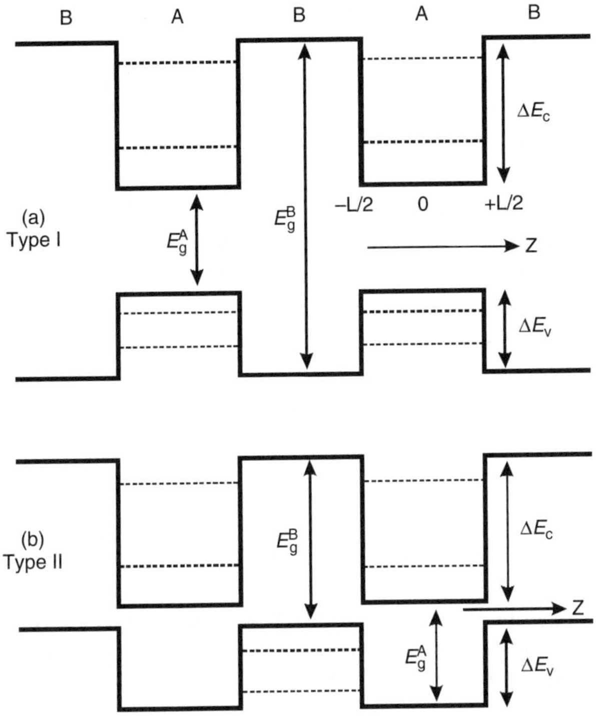

FIGURE 2.25 A sketch of the band alignment of the quantum wells for (a) type I such as GaAs/AlGaAs quantum wells and (b) type II such as InAs/InGaSb superlattices. The dashed lines represent the confined energy levels in both the conduction and valence quantum wells. The layers A and B are designated as the wells and barriers, respectively.

where $f_n^A(r)$ is the envelope function and $u_{nk_0}^A(r)$is the Block function in the well. The second equation in Equation 2.172 stands for the barrier and the summation is over all the finite energy bands. If one assumes that the Block function is the same for the well and the barrier, then Equation 2.172 becomes

$$\psi(r) = \sum_n f_n^{A,B}(r) u_{nk_0}. \tag{2.173}$$

It is thus required to determine the function $f_n^{A,B}(r)$, where $f_n^A(r)$ stands for the wave function in the well and $f_n^B(r)$ stands for the wave function in the barrier. The assumption of identical Bloch functions in the well and the barrier implies that interband matrix element is the same for the barrier and the well, which yields

$$f_n^A(r_\perp, z_0) = f_n^B(r_\perp, z_0) \tag{2.174}$$

where $r_\perp$ is the $x-y$ plane, and the subject interface along the growth axis, which is z, occurs at $z = z_0$. Since the lattice constants are assumed to be the same for both the well and the barrier, the envelope wave function can be factorized as

$$f_n^{A,B}(r_\perp, z) = \frac{1}{\sqrt{S}} e^{i(\mathbf{k}_\perp \cdot \mathbf{r}_\perp)} \chi_n^{A,B}(z), \tag{2.175}$$

where S is the sample area and $k_\perp$ is the (k_x, k_y) bidimensional wave vector, which is assumed to be the same for both A and B, and $\chi_n^{A,B}(z)$ is a slowly varying function with respect to the host's unit cells. To summarize, the heterojunction wave function $\psi(r)$ is composed of the sum of the of the product of the rapidly varying Bloch function $u_{nk_0}(r)$ and the slowly varying envelop function $f_n^{A,B}$.

Since the effective masses in A and B layers are different, the equation of motion of the envelope functions inside the well and the barrier are

$$\left(E_\mathrm{c} + V_\mathrm{c}(z) - \frac{\hbar^2}{2\mu(z)} \left(\frac{\mathrm{d}^2}{\mathrm{d}z^2} + k_x^2 + k_y^2 \right) \right) \chi_{A,B}(z) = E \chi_{A,B}(z), \tag{2.176}$$

where E_c is the conduction band edge in the well which can be set to zero, V_c(z) is zero in the well and equal to the conduction band offset in the barrier, and $\mu(z)$ is either m_A^* in the well or m_B^* in the barrier. Notice that $k_\perp = k_x^2 + k_y^2$. We assume that Equation 2.176 is written for the conduction band, but the case of the valence band is more complicated. The boundary conditions are those of BenDaniel–Duke conditions such that

$$\chi_A(\pm L/2) = \chi_B(\pm L/2) \tag{2.177}$$

and

$$\frac{1}{m_A^*} \frac{\mathrm{d}\chi_A(z)}{\mathrm{d}z} \bigg|_{z=\pm L/2} = \frac{1}{m_B^*} \frac{\mathrm{d}\chi_B(z)}{\mathrm{d}z} \bigg|_{z=\pm L/2}.$$

The effective mass mismatch leads to a discontinuity in the derivative of the envelope function at the interfaces, and $k_\perp$ adds step-like variation to the barrier $\mathrm{V_c}(z)$ since the effective potential is now $V_\mathrm{c}(z) + \frac{\hbar^2 k_\perp^2}{2\mu(z)}$. The wave functions of the bound states can be chosen for the even state as

$$\chi_\mathrm{even}(z) = A\cos(k_A z) \quad \text{for the well}$$
$$\chi_\mathrm{even}(z) = B e^{\left[-k_B\left(z - \frac{L}{2}\right)\right]} \quad \text{for the barrier} \tag{2.178}$$

and

$$\chi_\mathrm{even}(-z) = \chi_\mathrm{even}(z)$$

and for the odd state as

$$\chi_\mathrm{odd}(z) = A\sin(k_A z) \quad \text{for the well}$$
$$\chi_\mathrm{odd}(z) = B e^{\left[-k_B\left(z - \frac{L}{2}\right)\right]} \quad \text{for the barrier} \tag{2.179}$$

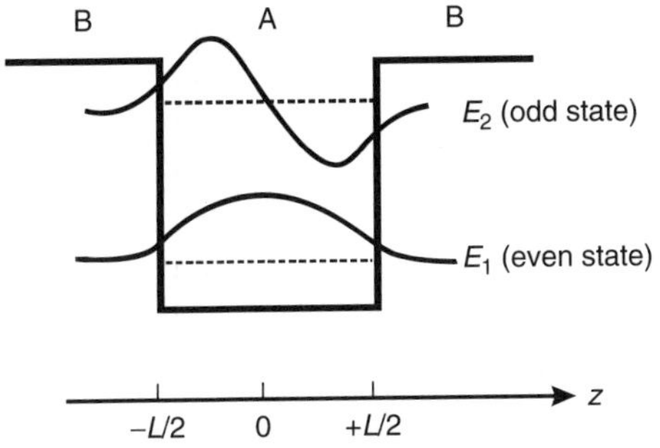

FIGURE 2.26 A sketch of the conduction quantum well plotted with the ground state (even) and the excited state (odd).

and

$$\chi_{\text{odd}}(-z) = -\chi_{\text{odd}}(z),$$

where the even and odd states are shown schematically in Fig. 2.26. The wave vectors are given as follows

$$k_A = \sqrt{\frac{2m_A^* E}{\hbar^2} - k_\perp^2}$$
$$k_B = \sqrt{\frac{2m_B^*(V_c - E)}{\hbar^2} - k_\perp^2}, \text{ for } E < V_c. \tag{2.180}$$

The BenDaniel–Duke boundary conditions give the following straightforward relations from which the energy levels are obtained:

$$\tan(k_A L/2) = \frac{m_A^* k_B}{m_B^* k_A} \quad \text{for even state} \tag{2.181}$$

and

$$-\cot(k_A L/2) = \frac{m_A^* k_B}{m_B^* k_A} \quad \text{for the odd state.} \tag{2.182}$$

The left-hand sides of Equations 2.181 and 1.182 are plotted as a function of $k_A L/2$ as shown in Fig. 2.27 with $k_\perp = 0$. The nodes in the figure indicate the intersections that correspond to the values of $k_A L/2$ from which the energy levels of the bound states are obtained. It is clear from this figure that the ratio of the effective masses plays a major role in determining the eigenvalues. The energy levels seems to reach constant values for a higher m_B^*/m_A^* ratio, which implies that $k_A L = n\pi$ for $n = 0, 1, 2, \ldots$ This result resembles the infinite potential well in which the wave function is zero at the boundary conditions. This is actually the essence of the envelope function approximation wherein the wave function vanishes at the interfaces.

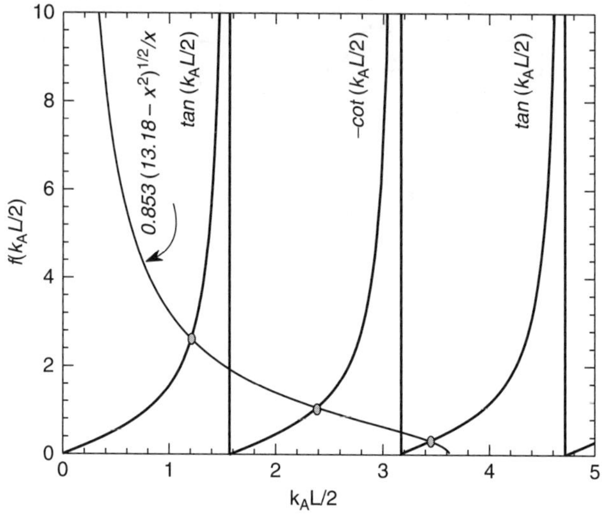

FIGURE 2.27 The left-hand sides of Equations 2.181 and 2.182 plotted as a function of $(k_A L/2)$. The nodes indicate the intersections from which the energy levels are obtained.

Let us consider a GaAs/AlGaAs quantum well with a well width of 100 Å and a barriers thickness of 400 Å. Assume that the conduction band offset is 0.3 eV, which corresponds to about 30% Al mole fraction in the AlGaAs barriers. Let the effective mass be $0.067m_0$ for the well and $0.092m_0$ for the barrier. The energy levels of the bound states in this well can be obtained graphically from Fig. 2.27 in which the wave functions in Equations 2.181 and 2.182 were plotted as a function of $k_A L/2$. For $k_\perp = 0$ Equation 2.180 is reduced to

$$k_A = \sqrt{\frac{2m_A^* E}{\hbar^2}}$$

$$k_B = \sqrt{\frac{2m_B^*(V_C - E)}{\hbar^2}}, \quad \text{for } E < V_C \tag{2.183}$$

Substituting Equation 2.183 into the right-hand side of Equations 2.181 and 2.182, we obtain

$$\frac{m_A^* k_B}{m_B^* k_A} = \frac{m_A^* \sqrt{\frac{2m_B^*}{\hbar^2}(V_c - E)}}{m_B^* \sqrt{2m_A^* E/\hbar^2}} \frac{L/2}{L/2}. \tag{2.184}$$

The value of y is calculated to be 13.18 for the effective masses, band offset, and well width mentioned above. Furthermore, Equations 2.181 and 2.182 are reduced to

$$-\cot(x) = 0.853\frac{\sqrt{13.18 - x^2}}{x} \quad \text{and} \quad \tan(x) = 0.853\frac{\sqrt{13.18 - x^2}}{x} \tag{2.185}$$

TABLE 2.1 The Energy Levels of the Bound States in a Single GaAs/AlGaAs Quantum Well Obtained From Three Different Approximation Energy Levels

Energy Levels	Graphical Presentation Using Fig. 2.27 (eV)	Propagation Matrix Method Using Fig. 2.28 (eV)	Infinite Potential Well (eV)
E_1	0.0320	0.0292	0.0474
E_2	0.1257	0.1160	0.1894
E_3	0.2681	0.2460	0.4263

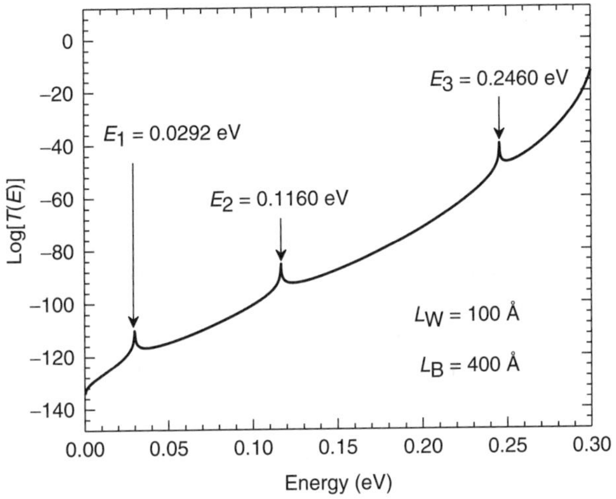

FIGURE 2.28 The transmission coefficient obtained from the matrix transfer method for a GaAs/AlGaAs quantum well.

From Fig. 2.27, the intersections of the curves occur at $k_A L/2 = 1.1863$, 2.350, and 3.432. From these values and with the help of Equation 2.183, the energy levels listed in Table 2.1 are obtained. The effective mass in the propagation matrix method and the infinite potential well calculations is taken as the average between effective masses of the electron in the well and in the barrier materials.

2.15 PERIODIC POTENTIAL

A single electron in a period potential is a difficult system to find, but the closest example is that of a free electron in a solid single crystal. Free electron here means that there is only one electron in the conduction band of the crystal. This simplistic example requires that the atoms of the single crystal be perfectly arranged in a single lattice and the electron–electron interactions are ignored. Such a *one-electron single-crystal approximation* leads to a description of allowed electronic energy levels in the crystal under the constraints of Pauli exclusion principle and Fermi–Dirac statistics. This approximation is actually the foundation of most theoretical analysis of the crystalline solids. Based on this foundation, there are other approximations such as the absence of imperfections in the single crystal, tight-binding method, and effective mass approximations. For the *one-electron single crystal approximation* to work, the periodic potential must satisfy the following relation assuming a one-dimension crystal

$$V(x) = V(x + L), \tag{2.186}$$

where L is the period of the potential. The periodic potential could be square shaped, δ-function, or any arbitrary shape such that it repeats itself periodically and has the same periodicity of the lattice. The Schrödinger equation of the one-electron single crystal can be written as

$$\frac{\partial^2 \psi(x)}{\partial x^2} + \left[\frac{2m}{\hbar^2}(E_n - V(x))\right]\psi(x) = 0. \tag{2.187}$$

If $V(x)$ is a periodic function, then $\frac{2m}{\hbar^2}(E_n - V(x))$ must be periodic. A typical periodic potential is shown in Fig. 2.29, where we plot a square periodic potential in Fig. 2.29a and a period potential due to a line of atoms in a crystal in Fig. 2.29b.

Independent electrons in a crystalline solid each obeying a one-dimensional Schrödinger equation with a periodic potential are commonly called *Bloch electrons*. A Bloch electron reduces to a free electron when the periodic potential

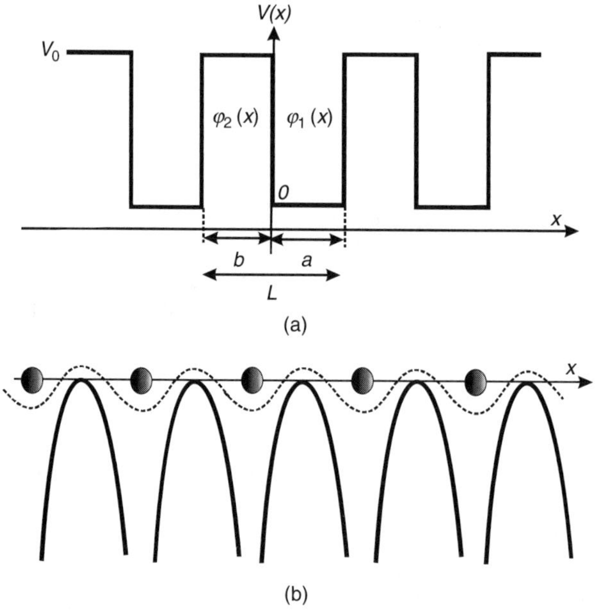

FIGURE 2.29 (a) Square periodic potential wells and (b) a typical crystalline periodic potential plotted along a line of ions. The solid lines are the potentials along the line of ions, and the dashed line is the potential along the line between planes of ions.

is zero. The discussion in this section starts by introducing Bloch's theorem. A simple model known as Kronig–Penney model is presented in which the allowed and forbidden energy bands are obtained for an electron in a periodic potential. The discussion covers other approximations, such as a Bloch electron in a weak periodic potential and an electron in a periodic δ-function potential.

2.15.1 Bloch's Theorem

The Bloch theorem was derived in 1928, which was based on the nineteenth century result of *Floquet*. This theorem states that the eigenstates of the one-electron Hamiltonian in one dimension can be written as $H = -\hbar^2\nabla^2/2m + V(x)$, where $V(x + L) = V(x)$ and L is the period of the periodic potential. The wave function can be chosen to have the form of a plane wave, e^{-ikx}, times a function of periodicity, $\varphi_k(x)$, of the primitive lattice cell such that

$$\psi_k(x) = \mathrm{e}^{ikx}\varphi_k(x) \tag{2.188}$$

where $\varphi_k(x)$ satisfies the following condition:

$$\varphi_k(x + L) = \varphi_k(x). \tag{2.189}$$

The one-dimensional propagation vector, $\boldsymbol{k}$, is introduced as a subscript. Each $\boldsymbol{k}$ may have several eigenvalues. For a three-dimensional case, x is replaced by $\boldsymbol{r}$. By combining Equations 2.188 and 2.189, we obtain

$$\psi_k(x + L) = \mathrm{e}^{ikL}\psi_k(x). \tag{2.190}$$

This equation states that the eigenstates of H can be chosen so that associated with each ψ is a wave vector $\boldsymbol{k}$ such that the following condition is satisfied:

$$\psi(x + L) = e^{ikL}\psi(x) \tag{2.191}$$

for every L in the lattice.

The proof of Bloch theorem is left as an exercise. Bloch theorem is the key to answering many of the unresolved questions posed by the free-electron theory and serves as the starting point for most of the more detailed calculations of wave functions and energy levels in crystalline solids including semiconductors and insulators.

2.15.2 The Kronig–Penney Model

Let us consider a periodic rectangular well potential as shown in Fig. 2.29a. The Schrödinger equation for this periodic potential was first solved by R. de L. Kronig and G. Penney in 1931, which lead to the well-known Kronig–Penny model. This model allows one to reach an exact solution to Schrödinger equation. While the model is a crude approximation of real crystal potentials, it illustrates explicitly most of the important characteristics of the quantum behavior of electrons in real crystalline solids such as semiconductors. Using the one-electron approximation, the wave function of Schrödinger equation can be obtained by assuming that the net force acting on the electron is regarded as derivable from the periodic potential, which has the familiar form given by Equation 2.187. The periodic potential in this equation satisfies the following conditions: $V(x) = 0$ for $0 < x < a$ and $V(x) = V_0$ for $-b < x < 0$. Thus, the lattice constant can be considered as $L = a + b$, which is the potential period. The wave functions of Schrödinger equation are given by Equation 3.3. Substituting the wave functions into Schrödinger equation gives

$$\frac{\mathrm{d}^2 \mathrm{e}^{ikx}\varphi_k(x)}{\mathrm{d}x^2} + \left[\frac{2m}{\hbar^2}(E_n - V(x))\right]\mathrm{e}^{ikx}\varphi_k(x) = 0. \tag{2.192}$$

By performing the second derivative on $\mathrm{e}^{ikx}\varphi(x)$, this equation can be rewritten as

$$\frac{\mathrm{d}^2\varphi_k(x)}{\mathrm{d}x^2} + 2ik\frac{\mathrm{d}\varphi_k(x)}{\mathrm{d}x} - \left[k^2 - \frac{2m}{\hbar^2}(E_n - V(x))\right]\varphi_k(x) = 0. \tag{2.193}$$

The square periodic potential shown in the Fig. 2.29 a requires to write two equations for $\varphi_1(x)$ and $\varphi_2(x)$ such that

$$\begin{aligned} &\frac{\mathrm{d}^2\varphi_1(x)}{\mathrm{d}x^2} + 2ik\frac{\mathrm{d}\varphi_1(x)}{\mathrm{d}x} - [k^2 - \alpha^2]\varphi_1(x) = 0 \quad \text{for } 0 < x < a \\ &\frac{\mathrm{d}^2\varphi_2(x)}{\mathrm{d}x^2} + 2ik\frac{\mathrm{d}\varphi_2(x)}{\mathrm{d}x} - [k^2 - \beta^2]\varphi_2(x) = 0 \quad \text{for } -b < x < 0, \end{aligned} \tag{2.194}$$

where $\alpha^2 = \frac{2mE_n}{\hbar^2}$ and $\beta^2 = \frac{2m(E_n - V_0)}{\hbar^2}$. The solutions to these two linear differential equations are taken as

$$\begin{aligned} &\varphi_1(x) = A\mathrm{e}^{i(\alpha-k)x} + B\mathrm{e}^{-i(\alpha+k)x} \quad \text{for } 0 < x < a \\ &\varphi_2(x) = C\mathrm{e}^{i(\beta-k)x} + D\mathrm{e}^{-i(\beta+k)x} \quad \text{for } -b < x < 0, \end{aligned} \tag{2.195}$$

where A, B, C, and D are arbitrary constants. Using the continuous boundary conditions, that is, the wave functions and their first derivatives are continuous at the boundaries, at $x = 0$ and $x = -b$, one can obtain the following four equations:

$$\begin{aligned} A + B &= C + D \\ i(\alpha - k)A - i(\alpha + k)B &= i(\beta - k)C - i(\beta + k)D \\ \mathrm{e}^{i(\alpha-k)a}A + \mathrm{e}^{-i(\alpha+k)a}B &= \mathrm{e}^{-i(\beta-k)b}C + \mathrm{e}^{i(\beta+k)b}D \\ i(\alpha - k)\mathrm{e}^{i(\alpha-k)a}A - i(\alpha + k)\mathrm{e}^{-i(\alpha+k)a}B &= i(\beta - k)\mathrm{e}^{-i(\beta-k)b} \\ &\quad C - i(\beta + k)\mathrm{e}^{i(\beta+k)b}D. \end{aligned} \tag{2.196}$$

Notice that the period function at $x = a$ is the same as $x = -b$. A trivial solution of Equation 3.11 would be to set $A = B = C = D$ 0. However, a nontrivial solution is to set the determinant of the coefficients to zero such as

$$\begin{vmatrix} 1 & 1 & -1 & -1 \\ (\alpha - k) & -(\alpha + k) & -(\beta - k) & (\beta + k) \\ e^{i(\alpha - k)a} & e^{-i(\alpha + k)a} & -e^{-i(\beta - k)b} & -e^{i(\beta + k)b} \\ (\alpha - k) & -(\alpha + k) & -(\beta - k) & (\beta + k) \\ e^{i(\alpha - k)a} & e^{-i(\alpha + k)a} & e^{-i(\beta - k)b} & e^{i(\beta + k)b} \end{vmatrix} = 0. \tag{2.197}$$

Using determinant minor technique and very tedious algebra, one can reach the following well-known result:

$$\frac{-(\alpha^2 + \beta^2)}{2\alpha\beta} \sin(\alpha a) \sin(\beta b) + \cos(\alpha a) \cos(\beta b) = \cos(ka + kb) = \cos(kL). \tag{2.198}$$

This equation is derived for the case in which the electron energy E is larger than the potential barrier height ($E_n > V_0$). This means that β^2 is a positive real quantity. For the case in which $0 < E_n < V_0$, β is a pure imaginary number. By letting $\beta = i\gamma$, we have $\beta^2 = -\gamma^2$. From the trigonometric relations, we have $\cos(ix) = \cos h(x)$ and $\sin(ix) = i \sin h(x)$. Substituting the above relations into Equation 2.198, we obtain

$$\frac{-(\alpha^2 - \gamma^2)}{2\alpha\gamma} \sin(\alpha a) \sin\text{h}(\gamma b) + \cos(\alpha a) \cos\text{h}(\gamma b) = \cos(ka + kb) = \cos(kL). \tag{2.199}$$

Additional approximations (see, e.g., Kittle) can be made to Equation 2.199. One of these approximations is based on the assumption that the periodic square wells can be replaced by δ-functions such that the product of the width and height of the δ-function remains finite. Incorporating this simplified approximation reduces the above equation to

$$P\frac{\sin(\alpha a)}{\alpha a} + \cos(\alpha a) = \cos(ka + kb) = \cos(kL) \quad \text{for } E < V_0, , \tag{2.200}$$

where $P = \frac{mbaV_0}{\hbar^2}$. The left-hand side of Equation 2.200 is plotted as a function of αa in Fig. 2.30. Notice that the function cos(kL) on the right hand side of Equation 2.200 is always within the interval $-1 \leq \cos(kL) \leq +1$ for all real values of kL. For the nonzero imaginary part of kL, we have wave functions that diverge at $\pm\infty$, which is not an acceptable solution for the one-electron approximation in periodic potential. Thus, there are ranges of energy in which no quantum states can exist. These bands are shown as the shaded regions in Fig. 2.30. The unshaded bands between ± 1 are the allowed energy bands in which energy states exist. Equation 2.200 is the dispersion relation, which gives the relation between the propagation vector $\boldsymbol{k}$ and the energy E_n for which the Schrödinger equation has a solution.

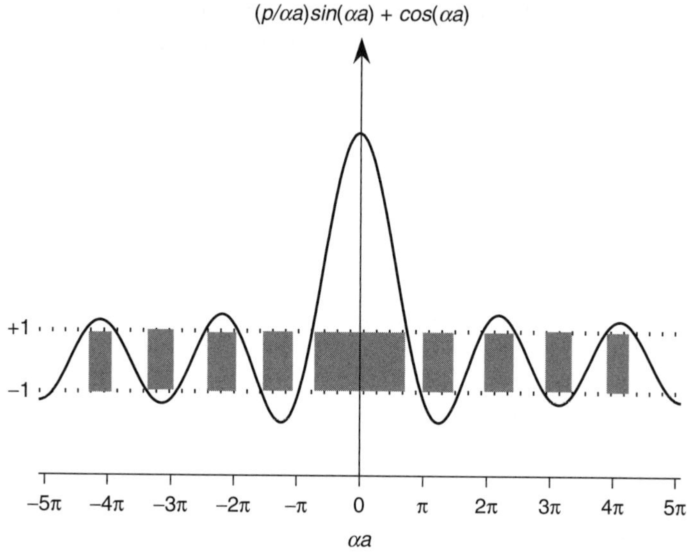

FIGURE 2.30 A plot of the left-hand side of Equation 2.200 with $P = 5$. The allowed energy bands are shown as the unshaded bands for which the function lies between ± 1. The forbidden bands are shown as gray bands.

To understand Fig. 2.30, one may consider the case in which the potential height is zero, which is the case of a free electron. Equation 3.15 is then reduced to

$$\cos(\alpha a) = \cos(KL) \quad \text{or } \alpha = K, \tag{2.201}$$

where $E = \frac{\hbar^2 K^2}{2m}$ is the free-electron energy, which is shown as the dashed parabola in Fig. 2.31. Notice that the propagation vector for the free electron is written as K to distinguish it from the propagation vector, k, of Bloch electron. The solid segment lines in Fig. 2.31 represent the allowed energy bands, where the energy is a continuous function of k. This figure illustrates the concept of allowed and forbidden bands in solids such as semiconductor materials. Notice that k is continuous in the allowed band. The discontinuities at $n\pi/a$ in Fig. 2.31, where n is a positive or negative integer, show where Bragg reflections takes place. At these places, the slope of the $E(k)$ should be zero. The Bloch electron energy bands can be presented by folding the curve segment as shown in Fig. 2.32. This is called reduced-zone representation. To understand this representation, let us consider the right-hand side of Equation 2.200, which is a periodic function and satisfies the following condition

$$\cos(kL) = \cos(kL + 2n\pi) = \cos(kL - 2n\pi) \tag{2.202}$$

where n is a positive integer. This equation is still satisfied by adding or subtracting $2n\pi$ from the cosine argument, hence one can displace the curve segments as shown in the figure.

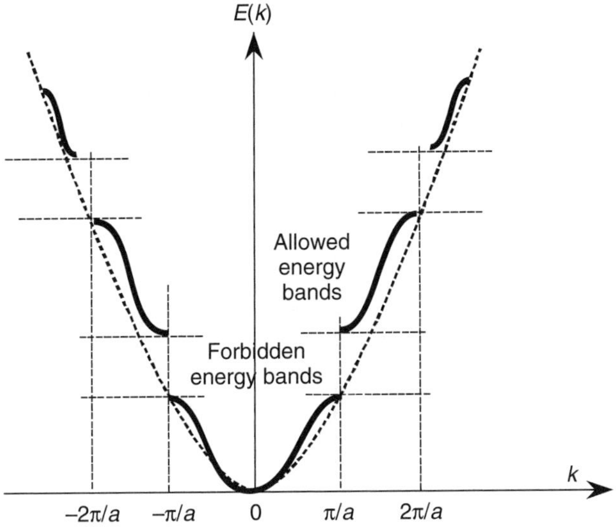

FIGURE 2.31 The electron energy, $E(k)$, versus k for both Bloch electron (segments) showing the allowed and forbidden bands according to Kronig–Penney model and the free-electron energy (dashed parabola).

2.15.3 One-Electron Approximation in a Periodic Dirac δ-Function

It is mentioned in Section 3.2 that the Kronig–Penney model can be simplified by assuming that the periodic potential can be approximated as δ-functions with a finite product of the width and height. In fact, this assumption is quite feasible since the atoms in single crystals can be considered as periodic δ-function potentials in many theoretical models. This problem is treated by Mihály and Martin. Let us assume that the atoms are arranged in a one-dimensional crystal with a lattice constant of a. Each atom is thus represented by the potential $V(x) = aV_0\delta(x)$ where V_0 is the height of the δ-function. Assume that the atoms are placed at $x = na$ where n is an integer. The Schrödinger equation between the atoms for the range $0 < x < a$ is

$$-\frac{\hbar^2}{2m}\frac{d^2\psi(x)}{dx^2} = E\psi(x), \tag{2.203}$$

and the wave function is

$$\psi(x) = Ae^{iKx} + Be^{-iKx}, \tag{2.204}$$

where $K = \sqrt{\frac{2mE}{\hbar^2}}$. For the wave function in the full range $-\infty < x < +\infty$, Bloch wave function, Equation 2.188, will be adopted and combined with

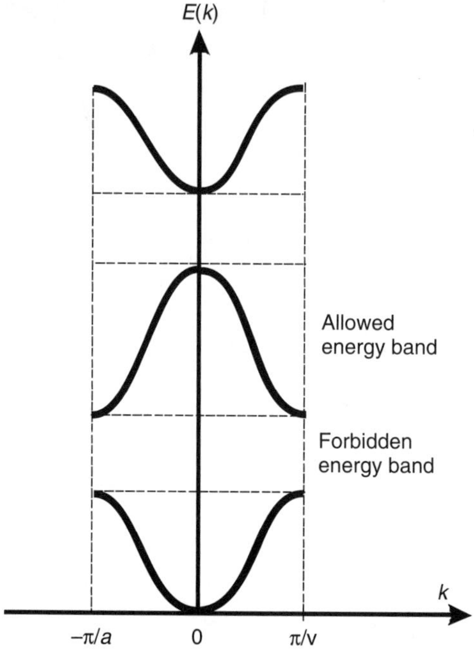

FIGURE 2.32 Reduced-zone representation of the allowed and forbidden bands. The curve segments of Bloch electron were displaced by $\pm 2n\pi$ and mirror reflected at each Bragg plane.

Equation 3.30 to give

$$\psi(x) = Ae^{iKx} + Be^{-iKx} = e^{ikx}\varphi_k(x) \tag{2.205a}$$

$$\varphi_k(x) = Ae^{i(K-k)x} + Be^{-i(K+k)x} \quad \text{for } 0 < x < a, \tag{2.205b}$$

where $\varphi_k(x)$ is a periodic function and can be generated for the whole crystal by setting $x = na$. This function is continuous, but its derivative is not as shown in a previous section. The jump (discontinuity) in the derivative of the δ-function can be found by integrating Schrödinger equation over a small range around a such that $a - \epsilon < x < a + \epsilon$

$$\psi(x)|_{x=a+\epsilon} = \psi(x)|_{x=a-\epsilon} \tag{2.206a}$$

$$\frac{d}{dx}\psi(x)\bigg|_{x=a+\epsilon} - \frac{d}{dx}\psi(x)\bigg|_{x=a-\epsilon} = \frac{2maV_0}{\hbar^2}\psi(a) \tag{2.206b}$$

Since $\varphi_k(x)$ is periodic, $\varphi_k(0) = \varphi_k(a)$, which in the limit of $\epsilon \to 0$ leads to

$$A + B = Ae^{i(K-k)a} + Be^{-i(K+k)a}. \tag{2.207}$$

The derivative of Bloch wave function gives

$$\frac{\mathrm{d}}{\mathrm{d}x}\psi(x) = ik\mathrm{e}^{ikx}\varphi_k(x) + \mathrm{e}^{ikx}\frac{\mathrm{d}\varphi_k(x)}{\mathrm{d}x} \tag{2.208}$$

Combining Equations 2.206 and 2.208, we have

$$\mathrm{e}^{ik(a+\epsilon)}\left.\frac{\mathrm{d}\varphi_k(x)}{\mathrm{d}x}\right|_{x=a+\epsilon} - \mathrm{e}^{ik(a-\epsilon)}\left.\frac{\mathrm{d}\varphi_k(x)}{\mathrm{d}x}\right|_{x=a-\epsilon} = \frac{2\mathrm{maV}_0}{\hbar^2}\mathrm{e}^{ika}\varphi_k(a)$$

or

$$\left.\frac{\mathrm{d}\varphi_k(x)}{\mathrm{d}x}\right|_0 - \left.\frac{\mathrm{d}\varphi_k(x)}{\mathrm{d}x}\right|_a = \frac{2\mathrm{maV}_0}{\hbar^2}\varphi_k(0). \tag{2.209}$$

By using the explicit form of $\varphi_k(x)$ in Equation 2.205, the first derivative yields

$$\frac{\mathrm{d}\varphi_k(x)}{\mathrm{d}x} = i(K-k)A\mathrm{e}^{i(K-k)x} - i(K+k)\mathrm{B}\ \mathrm{e}^{-i(K+k)x}. \tag{2.210}$$

Combining Equations 2.209 and 2.210 to obtain

$$\begin{aligned} &i(K-k)A - i(K+k)B - i(K-k)A\mathrm{e}^{i(K-k)a} \\ &+ i(K+k)B\mathrm{e}^{-i(K+k)a} = \frac{2maV_0}{\hbar^2}(A+B). \end{aligned} \tag{2.211}$$

One can now solve Equations 2.207 and 2.211 such that

$$\begin{pmatrix} 1-\mathrm{e}^{i(K-k)a} & 1-\mathrm{e}^{-i(K+k)a} \\ i(K-k)(1-\mathrm{e}^{i(K-k)a})-\gamma & -i(K+k)(1-\mathrm{e}^{-i(K+k)a})-\gamma \end{pmatrix}\begin{pmatrix} A \\ B \end{pmatrix} = 0, \tag{2.212}$$

where $\gamma = \frac{2maV_0}{\hbar^2}$. For a nontrivial solution, the determinant of the coefficient should be zero

$$\begin{vmatrix} 1-\mathrm{e}^{i(K-k)a} & 1-\mathrm{e}^{-i(K+k)a} \\ i(K-k)(1-\mathrm{e}^{i(K-k)a})-\gamma & -i(K+k)(1-\mathrm{e}^{-i(K+k)a})-\gamma \end{vmatrix} = 0. \tag{2.213}$$

With simple algebra, the determinant gives the following solution

$$\cos(ka) = \frac{ma^2V_0}{\hbar^2}\frac{\sin(Ka)}{Ka} + \cos(Ka), \tag{2.214}$$

which is the same form obtained for Kronig–Penney model as illustrated in Equation 2.200. The plot of this equation is similar to the plot shown in Fig. 2.30. The above equation is derived for $V_0 > 0$, where $E = \frac{\hbar^2K^2}{2m}$, but when $V_0 < 0$, we have negative energy. For this reason, we can define $K = i\rho$, which gives $E = -\frac{\hbar^2\rho^2}{2m}$, and Equation 2.214 is changed to

$$\cos(ka) = \frac{ma^2 V_0}{\hbar^2} \frac{\sin h(\rho a)}{\rho a} + \cos h(\rho a). \tag{2.215}$$

The graphical solution of Equation 2.215 shows that at least one bound sate exists for $V_0 < 0$. The proof of this case is left as an exercise.

2.15.4 Superlattices

A semiconductor superlattice is a periodic structure that can be used to illustrate the behavior of a periodic potential. This class of systems is composed of a number of semiconductor quantum wells of thickness L_w separated by barriers of thickness L_b as shown in Fig. 2.33. Generally speaking, the number of periods is ranging between 10 and 50. The superlattice here means that the quantum wells are close to each other such that an electron can tunnel through the barriers and exist in any of the wells with a nonzero probability. Thus, the quantized energy levels form what are called *minibands*. The potential energy $V(z)$ is a periodic function of z, where z is the growth direction, with a period of $L = L_w + L_b$. This can be written as

$$V(z) = \sum_{n=-\infty}^{+\infty} V(z - nL), \tag{2.216}$$

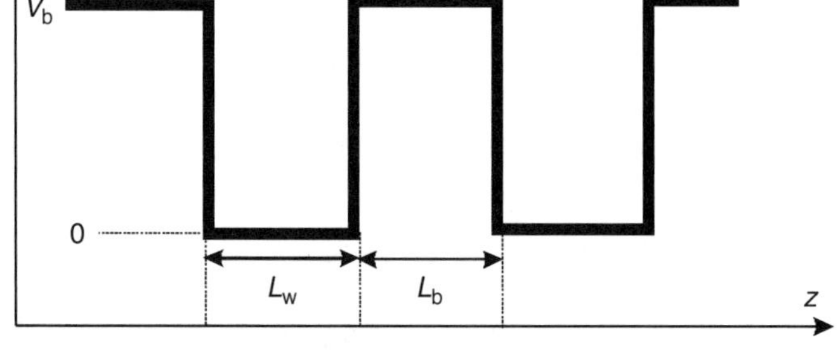

FIGURE 2.33 A segment of the potential energy profile of a GaN/AlN superlattice plotted along the z-direction.

where n is the number of periods and

$$V(z-nd)=\begin{cases}0 & \text{if } \left|z-nL\right|\leq \dfrac{L_w}{2}\\ V_b & \text{if } \left|z-nL\right|>\dfrac{L_w}{2}.\end{cases} \tag{2.217}$$

Following Bastard formalisms, the form of the wave functions solutions can be chosen as

$$\psi(z)=\begin{cases}Ae^{ik(z-nL)}+Be^{-ik(z-nL)}\\ \qquad \text{for the well, that is, } \left|z-nL\right|\leq \dfrac{L_w}{2}\\ Ce^{i\rho(z-nL-\frac{L}{2})}+De^{-i\rho(z-nL-\frac{L}{2})}\\ \qquad \text{for the barrier, that is, } \left|z-nd-\dfrac{L}{2}\right|\leq \dfrac{L_b}{2},\end{cases} \tag{2.218}$$

where k is the propagation vector in the well and ρ is the propagation vector in the barrier, which are related through the energy

$$E=\frac{\hbar^2\rho^2}{2m^*}=V_b+\frac{\hbar^2k^2}{2m^*}. \tag{2.219}$$

The parameter m^* is the effective mass of the electron in the superlattice. Since the potential function of the superlattice is periodic, the wave function expressed in Equation 2.218 must satisfied the Bloch theorem such that $\psi_q(z+nL)=\psi_q(z)$, where we introduced the subscript q to indicate that the function is a Bloch function. Again, the q-space is called the reciprocal or momentum space. Thus, the solution of Schrödinger equation can be limited to the first Brillouin zone (see Ashcroft and Mermin or Kittel for further discussions on the Brillouin zones). The continuity conditions at the interfaces labeled I and II in Fig. 2.33 give the following results for the case of $E>V_b$:

$$\begin{aligned}
e^{ikL_w/2}A+e^{-ikL_w/2}B&=e^{-i\rho L_b/2}C+e^{i\rho L_b/2}D\\
ke^{ikL_w/2}A-ke^{-ikL_w/2}B&=\rho e^{-i\rho L_b/2}C-\rho e^{i\rho L_b/2}D\\
e^{i(k-q)L_w/2}A+e^{-i(k+q)L_w/2}B&=e^{-i(\rho-q)L_b/2}C+e^{i(\rho+q)L_b/2}D\\
(k-q)e^{i(k-q)L_w/2}A-(k+q)e^{-i(k+q)L_w/2}&\\
B&=(\beta-k)e^{-i(\rho-q)L_b/2}C-(\rho+q)e^{i(\rho+q)L_b/2}D.
\end{aligned} \tag{2.220}$$

We used Bloch theorem at the interface II in a manner similar to that of the Kronig–Penney model discussed in an earlier section. Notice that q here represents the crystal momentum. To solve these four equations with four unknowns, we rely on the determinant method, which gives

$$\begin{vmatrix} e^{ikL_w/2} & e^{-ikL_w/2} & -e^{-i\rho L_b/2} & -e^{i\rho L_b/2} \\ ke^{ikL_w/2} & -ke^{-ikL_w/2} & -\rho e^{-i\rho L_b/2} & \rho e^{i\rho L_b/2} \\ e^{i(k-q)L_w/2} & e^{-i(k+q)L_w/2} & -e^{-i(\rho-q)L_b/2} & -e^{i(\rho+q)L_b/2} \\ (k-q)\, e^{i(k-q)L_w/2} & -(k+q)\, e^{-i(k+q)L_w/2} & -(\beta-k)\, e^{-i(\rho-q)L_b/2} & (\rho+q)\, e^{i(\rho+q)L_b/2} \end{vmatrix} = 0. \tag{2.221}$$

The solution of the determinant is similar to that of Equation 2.198, which yields

$$\frac{-(k^2+\rho^2)}{2k\rho}\sin(kL_w)\sin(\rho L_b) + \cos(kL_w)\cos(\rho L_b) = \cos(qL_w + qL_b) = \cos(qL). \tag{2.222}$$

Similarly, for $0 \leq E \leq V_b$ we have

$$\frac{-(k^2-\gamma^2)}{2k\gamma}\sin(kL_w)\sin h(\gamma L_b) + \cos(kL_w)\cos h(\gamma L_b) = \cos(qL_w + qL_b) = \cos(qL), \tag{2.223}$$

where $\rho = i\gamma$. For the case in which L_b is very large, Equation 2.223 diverges unless the multiplication coefficients were set to zero such that

$$\cos(kL_w) - \frac{(k^2-\gamma^2)}{2k\gamma}\sin(kL_w) = 0. \tag{2.224}$$

The form of this equation is familiar to us, which is the solution of isolated quantum wells (see section 2.9 on the quantum well with a finite barrier height).

The above analysis is for a single-electron model, and the results are still approximate. The propagation matrix method is another simple approximation that is adequate for calculating the quantized energy levels in quantum wells and superlattices. This method is discussed in more details by Levi. An example of the transmission coefficient in a few superlattices is shown in Fig. 2.34 for GaN/AlN superlattices. The conduction band offset was taken as $\Delta E_c = 0.68$ eV and the effective mass is 0.22 m_0. A detailed plot of the superlattice and the transmission coefficient are shown in Fig. 2.35 for a 10 periods of GaN/AlN superlattice. Notice the formation of the minibands, which consist of energy levels formed in the superlattices. Each well in the superlattice contributes one energy level. The ground state is highly degenerate, while the excited states are not. For example, the first excited miniband in Fig. 2.35 is composed of 9 energy levels because of the presence of 9 wells in the GaN/AlN superlattice.

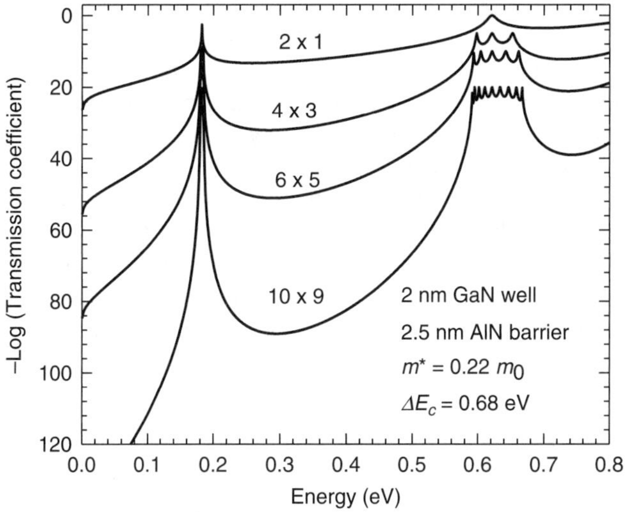

FIGURE 2.34 The transmission coefficient plotted as a function of energy for different GaN/AlN superlattice structures. The curves are displaced vertically for clarity. The ground states are highly degenerate, while the excited states form minibands.

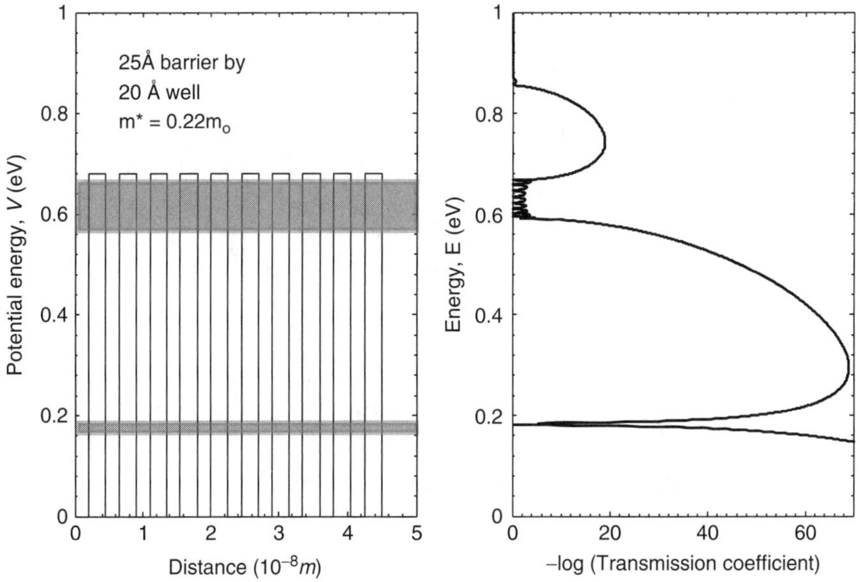

FIGURE 2.35 A plot of the potential energy as a function of distance and the corresponding transmission coefficient of a 10 periods (10 barriers and 9 wells) of 20 Å GaN/25 Å AlN superlattice. The shaded areas represent the location of the minibands in the superlattice.

2.16 EFFECTIVE MASS

When the periodic potential is zero, we have a free electron with mass "m_0". But when the periodic potential is nonzero, the electron will move in the periodic crystal with a different mass known as the *effective mass*. It is usually denoted as "$m*$". To obtain an expression for the effective mass, one can start from the free-electron case and work the problem by exerting a force on the electron. We know from previous discussions that the energy of free electron can be written as $E = \hbar\omega = \frac{\hbar^2 k^2}{2m}$. By using the duality concept, the wave packet of the electron is assumed to be moving with a group velocity $\upsilon_g = \frac{\partial\omega}{\partial k}$. When a force, F, is applied to an electron, the electron is accelerated and the motion of electron is given by the classical relation

$$\frac{\partial E}{\partial t} = F.\upsilon_g \tag{2.225}$$

On the other hand, if the energy band of the electron $E(\boldsymbol{k})$ is peaking at $\boldsymbol{k}_0$, one can expand $E(\boldsymbol{k})$ about $\boldsymbol{k}_0$ assuming that $\boldsymbol{k}$ is very close to $\boldsymbol{k}_0$. The linear term in $(\boldsymbol{k} - \boldsymbol{k}_0)$ vanishes at $\boldsymbol{k} = \boldsymbol{k}_0$, and the quadratic term will be proportional to $(\boldsymbol{k} - \boldsymbol{k}_0)^2$ according to the following relation, where $\boldsymbol{k}_0$ is assumed to be a point of high symmetry:

$$E(\boldsymbol{k}) \approx E(\boldsymbol{k}_0) + A(\boldsymbol{k} - \boldsymbol{k}_0)^2,, \tag{2.226}$$

where A is a positive quantity since E is maximum at $\boldsymbol{k}_0$. It is obvious that one can easily guess that $A = \frac{\hbar^2}{2m*}$. Furthermore, for energy levels with wave vectors near $\boldsymbol{k}_0$, we have

$$\upsilon_g = \frac{1}{\hbar}\frac{\partial E}{\partial \boldsymbol{k}} \approx \frac{\hbar(\boldsymbol{k} - \boldsymbol{k}_0)}{m^*}. \tag{2.227}$$

The acceleration, a, of the electron in the applied force is thus given by

$$a = \frac{\partial \upsilon_g}{\partial t} = \frac{\hbar}{m^*}\frac{d\boldsymbol{k}}{dt} = \frac{1}{m^*}\frac{d\hbar\boldsymbol{k}}{dt} = \frac{1}{m^*}\frac{dp}{dt} = \frac{1}{m^*}F. \tag{2.228}$$

Furthermore, the first derivative of the group velocity with respect to time can be expressed as

$$\frac{\partial \upsilon_g}{\partial t} = \frac{1}{\hbar}\frac{\partial}{\partial t}\frac{\partial E}{\partial \boldsymbol{k}} = \frac{1}{\hbar}\frac{\partial}{\partial \boldsymbol{k}}\frac{\partial E}{\partial t} = \frac{1}{\hbar}\frac{\partial}{\partial k}F \cdot \upsilon_g = \frac{1}{\hbar}\frac{\partial}{\partial \boldsymbol{k}}\frac{1}{\hbar}\frac{\partial E}{\partial \boldsymbol{k}}F = \frac{1}{\hbar^2}\frac{\partial^2 E}{\partial \boldsymbol{k}^2}F. \tag{2.229}$$

By equating Equations 2.228 and 2.229, one can find that

$$m* = \left[\frac{1}{\hbar^2}\frac{\partial^2 E}{\partial \boldsymbol{k}^2}\right]^{-1}, \tag{2.230}$$

which can be written in the tensor form as

$$M_{ij}^{-1} = \frac{1}{\hbar^2}\frac{\partial^2 E}{\partial \boldsymbol{k}_i \partial \boldsymbol{k}_j} = \frac{1}{\hbar^2}\nabla_{k_i}\nabla_{k_j}E \tag{2.231}$$

M^{-1} is called the inverse of the effective mass tensor. The above derivation is made for the electron in the conduction band. The same procedure could be followed for the hole in the valence band and the result is similar to Equation 2.231 except that the mass tensor has a minus sign from of the second derivative of the energy. From Equation 2.231, one can conclude that the curvature of the energy band is proportional to the inverse effective mass tensor. This means that the smaller the effect mass, the large the band curvature. The effective masses of charged particles in solids can be calculated from more elaborate quantum theories (see for example Chuang).

2.17 SUMMARY

In this chapter, we reviewed the basic concepts of quantum mechanics needed for the analysis of nanomaterial structures and energy quantization. Several examples, such as blackbody radiation and photoelectric effect, were briefly discussed to illustrate the limitation of classical mechanics and the need for the quantum mechanics. The concept of duality and de Broglie relation were also briefly discussed. Schrödinger equation and the concept of wave functions were presented from which a spectrum of energy levels can be obtained. The concept of energy quantization, probabilities, wave packet, Heisenberg uncertainty principle, Dirac notations, and the most important postulates of quantum mechanics were discussed. Quantum mechanics models and theories require the knowledge of mathematical tools. Thus, we briefly discussed the separation of variables, scalar product, linear operators, adjoint operator, eigenfunction operators, Dirac δ-function, and Fourier transform. There are several approximations in quantum mechanics that one needs to understand at early stages. The variational method is one among them. This approximation is encountered in many quantum mechanical treatments of solids in general and semiconductors in particular.

The periodic potentials and various approximations used to calculate the single-electron energy levels in this type of potentials are introduced. The periodic potential is considered here because it can represent a real single-crystal solid. Bloch theorem is briefly discussed, which provides a means to construct the wave function of a single electron in a periodic potential. Once the forms of the wave functions and the periodic potentials are known, the Schrödinger equation can be solved. The solution of Schrödinger equation, however, can be complicated, and in fact it is impossible to solve it exactly for a real crystal. Therefore, several approximations were introduced to construct the dispersion relations. Kronig–Penney theory is introduced here because it provides the concept of energy bands in solids.

A few examples of periodic potentials were introduced in this chapter, which include a weak periodic potential, periodic δ-function potential, and semiconductor superlattices. The concept of the effective mass of a charge carrier, such as an electron or hole, is discussed.

The envelope function approximations were discussed because of its applicability of constructing the band structure of bulk materials as well as that of quantum structures, such as quantum wells, superlattices, and quantum dots. The quantization and calculations of the energy levels in quantum wells were discussed, and BenDaniel–Duke boundary conditions were introduced. These boundary conditions are very helpful in calculating the discrete energy levels in a quantum well. Finally, the propagation matrix method is introduced to estimate the energy levels of bound states in quantum wells and short-period superlattices.

PROBLEMS

2.1. The work function of a material is the minimum energy required to remove an electron from the surface of the material. Calculate the maximum wavelength of light for the photoelectric emission from gold ($\varphi_0 = 4.90$ V) and cesium ($\varphi_0 = 1.90$ V).

2.2. Use the uncertainty relation discussed in Appendix A to evaluate the ground state of the hydrogen atom.

2.3. Calculate the de Broglie wavelength for (a) an electron with kinetic energy of 10^4 eV, (b) a proton of kinetic energy of 10^2 eV, and (c) a (150 kg) man running at a speed of 0.25 m/s.

2.4. Derive the dispersion relationship described by Equation 2.15.

2.5. If $|\psi\rangle$ can be normalized to unity and assuming that an operator $\mathbf{A} = |\psi\rangle\langle\psi|$, show that $\mathbf{A}^2 = \mathbf{A}$.

2.6. Assuming that $[X, P] = i\hbar$ show that $[X, P^2] = 2i\hbar P$, then show that $[X, P^n] = i\hbar n P^{n-1}$.

2.7. Care must be taken when working with operator. The order of the operator is very important. Assume that **A** and **B** are operators that do not commute. Show that $e^{\mathbf{A}}e^{\mathbf{B}}$, $e^{\mathbf{B}}e^{\mathbf{A}}$, and $e^{\mathbf{A}+\mathbf{B}}$ are not equal.

2.8. A series of lines in hydrogen correspond to transitions to a final state characterized by some quantum number n. If the wavelength of the radiation giving rise to the first line is 657 nm, what are the wavelengths corresponding to the next two lines. Assume that $\Delta n = 1$.

2.9. Show that the integration of a δ-function is a step function.

2.10. Derive the expression of Fourier transform function shown in Equation 2.53.

2.11. If $A = \left(\begin{smallmatrix} 1 & 0 \\ 0 & -1 \end{smallmatrix}\right)$, write an expression for e^A in a matrix form.

2.12. Show that $\sum_j |u_j\rangle\langle u_j| = 1$. This is called the closure relation.

2.13. Find the Fourier transform of the following functions:

$$\begin{aligned} \overline{\psi}(x) &= \frac{1}{a} \quad \text{for} \quad -\frac{a}{2} < x < \frac{a}{2} \\ &= 0 \quad \text{for } |x| > \frac{a}{2} \end{aligned} \tag{a}$$

$$\begin{aligned} \psi(x) &= e^{-ax} \qquad \text{for } x > 0 \\ &= 0 \qquad \text{for } x < 0 \end{aligned} \tag{b}$$

$$\psi(x) = e^{-x^2/a^2} \tag{c}$$

2.14. Use the following trial function $\psi(x) = (x^2 + a)^{-1}$, where a is a positive number, to calculate $\langle H \rangle$ for a simple harmonic oscillator as described in the variational method approach.

2.15. Derive an expression for the transmission coefficient for the potential barrier shown in Fig. P2.15. Simplify your answer for the case of $k_2 a = n\pi$, where n is an even integer.

2.16. Consider the potential well shown in Fig. P2.16. Derive an expression for the energy levels in the potential well.

2.17. Consider an infinite three-dimensional cubic potential well with a side $\boldsymbol{a}$ where $V(0) = 0$ for $0 < x < a$, $0 < y < a$, and $0 < z < a$ and infinity everywhere else. Derive an expression for the eigenvalues.

2.18. Consider the step potential barrier shown in Fig. 2.1 where an electron traveling from left to right with energy of 2.0 eV and the potential height is 2.2 eV. Determine the relative probability of finding the electron at 10 Å and 30 Å beyond the barrier.

2.19. Derive Equations 2.112 and 2.113.

2.20. The eigenvalues of Schrödinger equation for a finite well can be obtained graphically as shown in Fig. 2.13. Start from the eigen functions shown in Equation 2.114, and use the boundary conditions at $x \simeq L/2$ and $x = +L/2$ to derive Equation 2.119.

2.21. Derive the transmission coefficient for the step potential well shown in Fig. 2.3 assuming that the particle is traveling from $x = +\infty$ to $x = -\infty$ with energy $E > V_0$.

2.22. Start from Equation 2.89 to obtain Equation 2.90.

2.23. The δ-function is very useful in solving many mathematical problems. Show that the following properties of δ-function are true.

$$f(x)\delta(x) = f(0)\delta(x)$$

$$x\delta(x) = 0$$

$$\delta(ax) = \frac{1}{|a|}\delta(x)$$

$$\delta(-x) = \delta(x).$$

2.24. Solve the determinant of the one-electron approximation with the periodic δ-function potential. Start from Equation 2.213 and obtain the relationship shown in Equation 2.214.

2.25. Show graphically that Equation 2.215 has at least one bound state. When do you expect to see more than one bound state? Show your results.

2.26. Use the wave packet analysis to show that the group velocity is the gradient of the electron energy such that $\upsilon_g = \frac{1}{\hbar}\frac{\partial E}{\partial k}$. Sketch E and υ_g in the first Brillouin zone.

2.27. Consider the following periodic δ-function potential $V(x) = \frac{\hbar^2\lambda}{2ma}\sum_n \delta(x+na)$, where λ is a positive dimensionless constant, m is the mass of the electron, and a is the lattice constant. Use Equation 2.200 to derive an expression for the lowest energy level at $k=0$ and then find the band gap at $k=\pi/a$.

2.28. Use BenDaniel–Duke boundary conditions to obtain Equations 2.181 and 2.182.

2.29. Use the graphical presentation and BenDaniel–Duke boundary conditions to calculate all the energy levels of bound states in a GaN/AlN quantum well of a thickness of 100 Å and a conduction band offset of 0.68 eV. The effective masses are $m_A^* = 0.20m_0$ and $m_B^* = 0.27m_0$ for the well and barrier, respectively. Calculate the energy levels using the propagation matrix method and compare your results obtained from both methods.

BIBLIOGRAPHY

Balkanski M, Wallis RF. semiconductor physics and applications. Oxford: Oxford; 2000.

Bastard G. Wave mechanics applied to semiconductor heterostructures. New York: Halsted Press; 1988.

Bastard, G, Brum JA, Ferreira R. Electronics states in semiconductor heterostructures. Volume 44. Solid state physics. San Diego: Academic; 1991. p. 229,

BenDaniel DJ, Duke CB. Phys Rev B 1966;152:684.

Bohm D. Quantum theory. Englewood Cliffs, NJ: Prentice Hall; 1953.

Chuang SL. Physics of photonic devices. 2nd ed. New York: Wiley; 2009.

Cohen-Tannoudji C, Diu B, Laloë F. Quantum Mechanics. New York: Wiley; 1977.

Dicke RH, Witske JP. Introduction to quantum mechanics. Reading: Addison–Wesley; 1960.

Ferry DK. Quantum mechanics: an introduction for device physicists and electrical engineers. Bristol: IOP; 2001.

Gasiorowicz S. Quantum mechanics. 3rd ed. New York: Wiley; 2003.

Harrison P. Quantum wells, wires and dots. New York: Wiley; 2000.

Krane K. Modern physics. 2nd ed. New York. Wiley; 1996.

Levi AFJ. Applied quantum mechanics. Cambridge: Cambridge University Press; 2003.

Merzbacher E. Quantum mechanics. New York: Wiley; 1970.

Mihály L, Martin MC. Solid state physics: problems and solutions. New York: Wiley; 1996.

Mitin VV, Kochelap VA, Stroscio MA. Quantum heterostructures microelectronics and optoelectronics. Cambridge: Cambridge University Press; 1999.

Singh J. Electronic and optoelectronic properties of semiconductor structures. Cambridge: Cambridege University press; 2003.

Singh J. Modern physics for engineers. New York: Wiley; 1999.

Stern F. Phys Rev B 1972;5:2891.

3

DENSITY OF STATES IN SEMICONDUCTOR MATERIALS

3.1 INTRODUCTION

Semiconductor heterojunctions and nanomaterials consist of large numbers of identical particles such as electrons, atoms, holes, and harmonic oscillators. In such cases, it is impossible to try to trace the motion of each individual particle. An alternative way of looking at these large numbers of particles is to settle for knowing averages of relevant dynamical quantities over the entire range of possible system configurations. This leads to the construction of the macroscopic properties of the system and to understanding of how energy, velocity, and momentum are distributed among the particles that form the system. The branch of physics that addresses the distribution function of a system is called *statistical mechanics*, which links the microscopic properties of the system with its macroscopic domains. For physical systems such as semiconductor nanomaterials, there are constraints associated with any distribution function. For example, the number of particles is finite, or the total energy of the system is constant. These constraints usually alter the probabilities associated with possible system configurations.

The techniques of statistical mechanics have been applied to a variety of physical problems such as gases, liquids, polymers, metals, semiconductors, transport theory, DNA, adsorption, spectroscopy, and optical and electrical properties of solids among many other fields of study. Statistical thermodynamics is usually applied to a system in equilibrium. This branch of statistical mechanics links thermodynamics with molecular physics. Thermodynamics, on the other hand, provides connections between the properties of the system without supplying any

Introduction to Nanomaterials and Devices, First Edition. Omar Manasreh.

information about the magnitude of any one of them, while statistical thermodynamics assumes the existence of atoms and molecules to calculate thermodynamic quantities from a molecular point of view. Statistical thermodynamics is further divided into two areas: first, the study of systems of molecules in which molecular interaction is neglected, such as dilute gases; and second, the study of systems in which molecular interactions are of prime importance, such as liquids.

To illustrate the terminology of statistical mechanics, let us consider the energy states of a particle in a three-dimensional infinite cubic potential well. These energy states are given by

$$E = E(k_x, k_y, k_z) = \frac{\hbar^2}{2ma^2}\left(n_x^2 + n_y^2 + n_z^2\right), \tag{3.1}$$

where a is the side of the cube and n_x, n_y, and n_z are positive integers. The degeneracy, $g(E)$, is given by the number of ways that an integer $M = 2ma^2E/\hbar^2$ can be written as the sum of the squares of three positive integers. The results could be erratic and discontinuous functions for small values of n_x, n_y, and n_z, but it becomes smooth for large values of n_x, n_y, and n_z. Consider a three-dimensional space spanned by $k_x = n_x/a$, $k_y = n_y/a$, and $k_z = n_z/a$ as shown in Fig. 3.1. Equation 3.1 becomes an equation of a sphere of radius k, where $k^2 = k_x^2 + k_y^2 + k_z^2$. Now, it is possible to calculate the number of states in the range $\mathrm{d}k$. This is simply obtained by finding the volume of the shell between k and $k + \mathrm{d}k$, which is given as

$$\mathrm{d}V_k = 4\pi k^2\,\mathrm{d}k = 2\pi\left(\frac{2m}{\hbar^2}\right)^{3/2}\sqrt{E}\,\mathrm{d}E. \tag{3.2}$$

In order to calculate the density of states from Equation 3.2, one needs to consider the positive values of n_x, n_y, and n_z. Thus $\mathrm{d}V_k$ should be divided by the number 8, since the space is composed of eight quadrants, which include the negative values of n_x, n_y, and n_z. Moreover, $\mathrm{d}V_k$ in Equation 3.2 should be divided by the volume

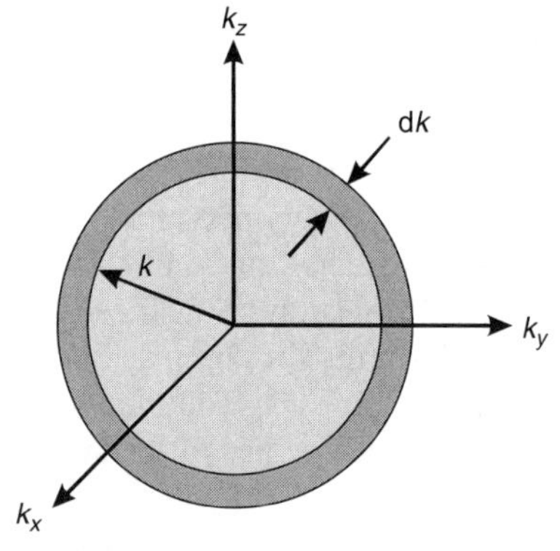

FIGURE 3.1 A spherical surface of a constant energy is plotted in the $\boldsymbol{k}$-space. The shell of thickness dk is used to calculate the density of states of a particle in a box.

of the unit cell in $\boldsymbol{k}$-space, which is $(\pi/\mathrm{a})^3$, to give the density of state in the shell with thickness dE. Including the spin degeneracy, this density of state can now be written as

$$\begin{aligned} g(E)\mathrm{d}E &= 2\frac{\mathrm{d}V_k}{8(\pi/a)^3} \\ &= \frac{1}{8}\frac{1}{(\pi/a)^3}4\pi\left(\frac{2m}{\hbar^2}\right)^{3/2}\sqrt{E}\,\mathrm{d}E \\ &= \frac{a^3}{2\pi^2}\left(\frac{2m}{\hbar^2}\right)^{3/2}\sqrt{E}\,\mathrm{d}E. \end{aligned} \tag{3.3}$$

The density of states can be thought of as the number of states per unit energy or the degeneracy of the energy levels. If one assumes that the energy $E = 3k_\mathrm{B}T/2$, where k_B is the Boltzmann's constant, $T = 300$ K, $m = 9.11 \times 10^{-31}$ kg, $a =$ 100 Å, the density of state can be easily obtained as $g(E)\mathrm{d}E = 8.40 \times 10^{21}\mathrm{d}E$. Thus, even for a system as simple as a particle in a box, the degeneracy can be very large at room temperature.

For a system consisting of noninteracting N particles, the degeneracy is extremely high. The energy of the system is

$$E = \frac{\hbar^2}{2ma^2}\sum_j^N\left(n_{xj}^2 + n_{yj}^2 + n_{zj}^2\right) = \frac{\hbar^2}{2ma^2}\sum_j^N R_j^2, \tag{3.4}$$

where $n_{xj}^2, n_{yj}^2, n_{zj}^2$, and R_j^2 are positive integers. The degeneracy of the system can be calculated by generalizing the above procedure for a one-particle system. The density of state can be written as

$$\begin{aligned} g(E)\mathrm{d}E &= \prod_j^N g_j(E)\mathrm{d}E \\ &= \frac{\sqrt{\pi}}{\Gamma(N+1)\Gamma(3N/2)}\left(\frac{a^3}{2\pi^2}\right)^N\left(\frac{2m}{\hbar^2}\right)^{3N/2} E^{(3N/2-1)}\mathrm{d}E, \end{aligned} \tag{3.5}$$

where $\Gamma(x)$ is the gamma function given by

$$\Gamma(x) = \int_0^\infty \mathrm{e}^{-t}t^{x-1}\,\mathrm{d}t, \tag{3.6}$$

and it has the following properties

$$\Gamma(x+1) = x\Gamma(x), \tag{3.7a}$$

$$\Gamma(n+1) = n! \quad \text{for } n = \text{integer}, \tag{3.7b}$$

$$\Gamma(1/2) = \sqrt{\pi}, \tag{3.7c}$$

$$\Gamma(n+1/2) = \frac{(2n)!}{2^{2n}n!}\sqrt{\pi}. \tag{3.7d}$$

The density of state $g(E)\mathrm{d}E$ is calculated to be on the order of 10^N, where N is the Avogadro's number. For $N = 10$, we obtain a density of state in the order of 10^{1089}. Thus, the concept of density of state is very important for macroscopic systems.

3.2 DISTRIBUTION FUNCTIONS

Consider a liter of salt solution. From a macroscopic point of view, we can completely specify the system by a few parameters such as volume, concentration, and temperature. Regardless of the complexity of the system, it requires only a small number of parameters to describe it. From a microscopic point of view, there are enormous numbers of quantum states associated with the fixed macroscopic properties. Gibbs introduced the concept of "ensemble," which is a virtual collection of a very large number of systems, denoted A; each is constructed to be a replica on the macroscopic level of particular system of interest. Suppose that an isolated system has a volume V, contains N molecules, and is known to have an energy E. Thus, the ensemble would have a volume AV, contains AN molecules, and has a total energy of AE. Each of the systems in this ensemble is a quantum mechanical system of N interacting molecules in a container of volume V. The values of N, V, and the interaction between molecules are sufficient to determine the energy eigenvalues, E_j of Schrödinger equation along with their associate degeneracies, $g(E_j)$. These energies are the only eigenvalues available to the system. The fixed energy of the system, E, is one of these E_j's and there is a degeneracy $g(E)$. While the systems in the ensemble are identical at the macroscopic level, they may differ at the molecular level. Nothing has been said thus far about the distribution of the member of the ensemble with respect to the degeneracy of the possible quantum states.

The ensembles are required to obey the principle of *equal priori probability*, which states that every $g(E)$ is represented an equal number of times in the ensembles. Thus, each $g(E)$ is treated equally and the number of systems in the ensembles is an integer multiple of $g(E)$. An alternative interpretation of the principle of equal *a priori* probabilities is that an isolated system is equally likely to be in any of its $g(E)$ possible quantum states.

The most commonly used ensemble is called *canonical ensemble* in which the volume, number of particles, and the temperature are constant. The occupation number means the number of systems of the ensemble occupying a specific quantum state. The occupation numbers must satisfy the conditions

$$\sum_j a_j = A, \tag{3.8}$$

where a_j is the occupation number of the jth sate, and A is the total number of the systems in the ensemble, and

$$\sum_j a_j E_j = E, \tag{3.9}$$

where E_j is the energy of the jth state and E is the total energy of the ensemble. The above two conditions mean that all the members of the ensemble are included in the calculations and the total energy of the system is fixed. The number of ways, $\Omega(\mathbf{a}) = \Omega(a_1, a_2, a_3, \ldots)$, that any particular distribution of the a_j's can be realized is the number of ways that A *distinguishable* objects can be arranged in groups, such as a_1 is in the first group, a_2 is in the second group, and so on, is

$$\Omega(a) = \frac{A!}{a_1!a_2!a_3!\ldots} = \frac{A!}{\prod_j a_j!}. \tag{3.10}$$

The distinguishable objects mean that we can label each object uniquely. The overall probability, P_j, that a system is in the jth quantum state is obtained by averaging the fraction (a_j/A) of the systems or members of the canonical ensemble in the jth state with an energy E_j, which can be written as

$$P_j = \frac{\overline{a}_j}{A} = \frac{1}{A}\frac{\sum_a \Omega(a)a_j(a)}{\sum_a \Omega(a)}, \tag{3.11}$$

where the notation $a_j(\boldsymbol{a})$ indicates that the value a_j depends on the distribution and summations over all distributions that satisfy Equations 3.8 and 3.9. Given the probability that a system is in the jth state, one can calculate the canonical ensemble average of any property from the following relation

$$\overline{M} = \sum_j M_j P_j, \tag{3.12}$$

where M_j is the value of M in the jth quantum state. For more discussion on this subject, see Huang, McQuarie, and Reif.

3.3 MAXWELL–BOLTZMANN STATISTIC

For distinguishable number of particles (N) in a container with many compartments (g_i), the number of ways that the particles can be distributed among the compartments can be written as

$$\Omega(N_1, N_2, \ldots, N_j) = \frac{N!}{\prod_{i=1}^{j} N_i!}\prod_{i=1}^{j} g_i^{N_i}, \tag{3.13}$$

where $g_i^{N_i}$ is the degeneracy of finding the N_i particle in the g_i compartment. The problem of finding what set of values for the numbers $N_1, N_2, N_3, \ldots, N_j$ that will make Ω as large as possible, subject to the constraints of a constant number of particles, $\sum_{i=1}^{j} N_i = N = \text{constant}$, and constant energy, $\sum_{i=1}^{j} \epsilon_i N_i = E = \text{constant}$, is more or less a mathematical exercise. To maximize a function of many variables with a given constraint, one can apply the Lagrange's method of undetermined multipliers as follows:

$$\frac{\partial f}{\partial x_i} + \alpha \frac{\partial g_1}{\partial x_i} + \beta \frac{\partial g_2}{\partial x_i} = 0, \tag{3.14}$$

where f is the function to be maximized, α and β are the undetermined multipliers and g_1 and g_2 are the constraints. To maximize Equation 3.13 with two constraints, one can settle for maximizing the logarithm of Ω since the logarithm of products is converted to sum, which is much easier to handle mathematically. Hence, one can write

$$\frac{\partial}{\partial N_j}\left(\ln(\Omega) + \alpha \sum_{i=1}^{j} N_i + \beta \sum_{i=1}^{j} \epsilon_i N_i\right) = 0. \tag{3.15}$$

The first derivative with respect to the N_j particle of the logarithm of Equation 3.13 is

$$\begin{aligned}\frac{\partial \ln(\Omega)}{\partial N_j} &= \frac{\partial \ln N!}{\partial N_j} + \frac{\partial \sum_{i=1}^{j} N_i \ln g_i}{\partial N_j} - \frac{\partial \sum_{i=1}^{j} \ln N_i!}{\partial N_j} \\ &= 0 + \ln g_j - \ln N_j \\ &= \ln gj - \ln N_j.\end{aligned} \tag{3.16}$$

Notice $\frac{\partial \ln N_i!}{\partial N_j} \approx \ln N_j$. It follows that the derivative of the two constraints are

$$\alpha \frac{\partial \sum_{i=1}^{j} N_i}{\partial N_j} = \alpha \quad \text{and} \quad \beta \frac{\partial \sum_{i=1}^{j} \epsilon_i N_i}{\partial N_j} = \beta \epsilon_j. \tag{3.17}$$

Substituting Equations 3.16 and 3.17 into Equation 3.15, we have

$$\ln g_j - \ln N_j + \alpha + \beta \epsilon_j = 0. \tag{3.18}$$

The quantity N_j in the above equation is the most probable number of particles to be found in the jth energy level and g_j is the number of the quantum states associated with the jth energy level. Thus, N_j/g_j is the average number of particles per quantum state at that energy level, which is by definition the distribution function $f(\epsilon_j)$. Equation 3.18 becomes

$$\frac{N_j}{g_j} = f(\epsilon_j) = \mathbf{e}^{\alpha} \mathbf{e}^{\beta \epsilon_j}. \tag{3.19}$$

Without going through the thermodynamic derivation, it is found that

$$\beta = -\frac{1}{k_{\mathrm{B}} T}, \tag{3.20}$$

where k_B is the Boltzmann constant and T is the temperature in Kelvin. The task now is to determine the multiplier α. From Equation 3.19, the number that occupies the jth quantum state can be written as

$$N_j = g_j \mathrm{e}^{\alpha} \mathrm{e}^{\beta \epsilon_j}. \tag{3.21}$$

It follows that the total number of particles in the system is

$$N = \sum_j N_j = \sum_j g_j \mathrm{e}^{\alpha} \mathrm{e}^{-\epsilon_j/k_B T} = \mathrm{e}^{\alpha} \sum_j g_j \mathrm{e}^{-\epsilon_j/k_B T}, \tag{3.22}$$

which can be solved for the quantity e^{α} such as

$$\mathrm{e}^{\alpha} = \frac{N}{\sum_j g_j \mathrm{e}^{-\epsilon_j/k_B T}}. \tag{3.23}$$

Substituting Equation 3.23 into Equation 3.21, we have

$$N_j = \frac{N g_j \mathrm{e}^{-\epsilon_j/k_B T}}{\sum_j g_j \mathrm{e}^{-\epsilon_j/k_B T}}. \tag{3.24}$$

When the system is composed of quasi-continuum eigenvalues, the degeneracy g_j can be replaced by the density of state $g(E)\mathrm{d}E$, the population N_j is replaced by the function $N(E)\mathrm{d}E$, and the summation is replaced by integration over the region of allowed energies. These above substitutions allow one to rewrite the above equations as

$$N(E)\mathrm{d}E = f(E)g(E)\mathrm{d}E = \mathrm{e}^{\alpha}\mathrm{e}^{-E/k_B T} g(E)\mathrm{d}E, \tag{3.25}$$

$$N = \int N(E)\mathrm{d}E = \mathrm{e}^{\alpha} \int \mathrm{e}^{-E/k_B T} g(E)\mathrm{d}E, \tag{3.26}$$

$$\mathrm{e}^{\alpha} = \frac{N}{\int \mathrm{e}^{-E/k_B T} g(E)\mathrm{d}E}, \tag{3.27}$$

and

$$N(E)\mathrm{d}E = \frac{N \mathrm{e}^{-E/k_B T} g(E)\mathrm{d}E}{\int \mathrm{e}^{-E/k_B T} g(E)\mathrm{d}E}. \tag{3.28}$$

For an ideal monatomic gas in a cubic container with side a, the density of states can be presented by the expression shown in Equation 3.3. By substituting this expression in Equation 3.27, we have

$$\mathrm{e}^{\alpha} = \frac{N}{\dfrac{a^3}{2\pi^2}\left(\dfrac{2m}{\hbar^2}\right)^{3/2} \displaystyle\int_0^{\infty} \mathrm{e}^{-E/k_B T}\sqrt{E}\,\mathrm{d}E}. \tag{3.29}$$

Let $x = E/k_B T$, which implies that $dE = k_B T dx$ and $\sqrt{E} = \sqrt{k_B T}\sqrt{x}$. Substituting these quantities back into Equation 3.29 yields

$$\mathrm{e}^{\alpha} = \frac{N}{\dfrac{a^3}{2\pi^2}\left(\dfrac{2mk_B T}{\hbar^2}\right)^{3/2}\displaystyle\int_0^{\infty} \mathrm{e}^{-x}\sqrt{x}\,\mathrm{d}x}. \tag{3.30}$$

The integral in the above equation is a Γ-function (Eq. 3.6) with an argument of 3/2. Γ(3/2) can be evaluated from Equation 3.7d to be $\sqrt{\pi}/2$. Substitute this quantity back in Equation 3.30 to obtain

$$\mathrm{e}^{\alpha} = \frac{\sqrt{2}N}{a^3}\left(\frac{\pi\hbar^2}{mk_B T}\right)^{3/2}. \tag{3.31}$$

Finally, substitute Equation 3.31 into Equation 3.19 to obtain the following expression for the distribution function

$$f(E) = \frac{\sqrt{2}N}{a^3}\left(\frac{\pi\hbar^2}{mk_B T}\right)^{3/2} \mathrm{e}^{-E/k_B T}. \tag{3.32}$$

This expression is known as *Maxwell–Boltzmann distribution function*, which is applicable to noninteracting particles in a system, whose density of states is defined by Equation 3.3. A plot of this function is shown in Fig. 3.2 for four different temperatures. It is clear that this function is becoming steeper as the temperature is decreased.

3.4 FERMI–DIRAC STATISTICS

The Maxwell–Boltzmann distribution function is applicable to classical systems, where the particle can be identified and labeled. For quantum systems, there is no way one can distinguish between electrons and protons, for example. Quantum systems are composed of *inherently indistinguishable* particles and therefore Maxwell–Boltzmann statistics cannot be applied. In addition to this point, Pauli exclusion principle requires taking particles' spin into consideration. These two points require a different distribution function known as *Fermi–Dirac distribution*. For indistinguishable particles, the number of ways of realizing a distribution of N_j indistinguishable particles is as follows:

$$\Omega_{\mathrm{FD}}(N_1, N_2, N_3, \ldots, N_n) = \prod_j^n \frac{g_j!}{N_j!(g_j - N_j)!}, \tag{3.33}$$

where g_j is the quantum state. The logarithm of this equation is

$$\ln \Omega_{\mathrm{FD}} = \sum_j \ln g_j! - \sum_j \ln N_j! - \sum_j \ln(g_j - N_j)!. \tag{3.34}$$

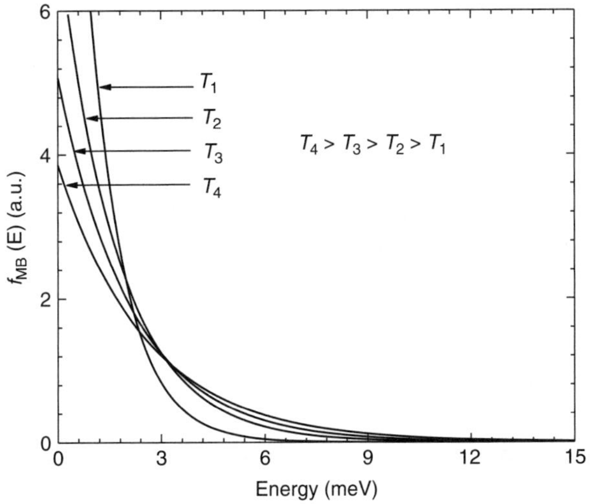

FIGURE 3.2 Maxwell–Boltzmann distribution function plotted as a function of energy for four different temperatures.

Taking the first derivative of Equation 5.34 with respect to N_i, we can write

$$\begin{aligned}\frac{\partial \ln \Omega_{\rm FD}}{\partial N_i} &= \frac{\partial \sum_j \ln g_j!}{\partial N_i} - \frac{\partial \sum_j \ln N_j!}{\partial N_i} - \frac{\partial \sum_j \ln(g_j - N_j)!}{\partial N_i} \\ &= 0 - \ln N_i + \frac{\partial \sum_j \ln(g_j - N_j)!}{\partial (g_i - N_i)} \\ &= -\ln N_i + \ln(g_i - N_i) \\ &= \ln\left(\frac{g_i}{N_i} - 1\right). \qquad (3.35)\end{aligned}$$

Lagrange's method of undetermined multipliers can now be applied to maximize Equation 3.35 by using Equations 3.14 and 3.15, which yields

$$\ln\left(\frac{g_i}{N_i} - 1\right) = -\alpha - \beta E_i. \qquad (3.36)$$

The quantity g_i/N_i is called the *Fermi–Dirac distribution function*, $f_{\rm FD}(E_i)$, and can be written as

$$\frac{g_i}{N_i} = f_{\rm FD}(E_i) = \frac{1}{1 + \mathbf{e}^{-(\alpha + \beta E_j)}}, \qquad (3.37)$$

where β is taken as the same as in Equation 3.20 and α in the Fermi–Dirac distribution function is taken as

$$\alpha = \frac{E_F}{k_{\rm B} T}, \qquad (3.38)$$

where E_F is known as *Fermi energy level*. Substituting Equations 3.20 and 3.38 into Equation 3.37 yields

$$f_{FD}(E_i) = \frac{1}{1 + \mathbf{e}^{(E_j - E_F)/k_B T}}. \tag{3.39}$$

This function, known as *Fermi–Dirac distribution function*, is plotted as a function of energy for different temperatures as shown in Fig. 3.3. Notice that at $T = 0$ K, f_{FD} becomes a step function.

For quasi-continuous energy levels in which the degeneracy is represented by a density of state function, we can write

$$N(E)\mathrm{d}E = f_{FD}(E)g(E)\mathrm{d}E = \frac{g(E)\mathrm{d}E}{1 + \mathbf{e}^{(E - E_F)/k_B T}}. \tag{3.40}$$

Substitute Equation 3.39 into Equation 3.40, and then integrate to give

$$\begin{aligned} N &= \int f_{FD}(E)g(E)\mathrm{d}E = \int \frac{g(E)\mathrm{d}E}{1 + \mathbf{e}^{(E - E_F)/k_B T}} \\ &= \frac{a^3}{2\pi^2}\left(\frac{2m}{\hbar^2}\right)^{3/2} \int_0^\infty \frac{\sqrt{E}\mathrm{d}E}{\left(1 + \mathbf{e}^{(E - E_F)/k_B T}\right)}. \end{aligned} \tag{3.41}$$

The integral in the above equation is difficult to evaluate analytically. For $(E - E_F) \gg k_B T$, one can find a solution, for this integral, of the following form

$$\mathrm{N} = \frac{a^3}{2\pi^2}\left(\frac{2m}{\hbar^2}\right)^{3/2} \frac{\mathbf{e}^{E_F/kT}\sqrt{\pi}(k_B T)^{3/2}}{2}$$

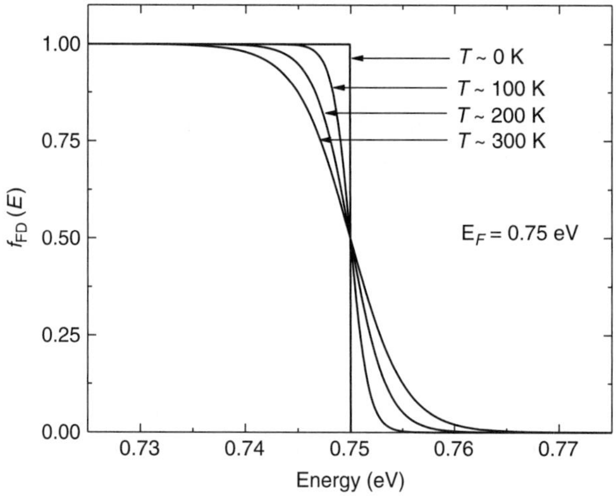

FIGURE 3.3 Fermi–Dirac function plotted as a function of energy for four different temperatures.

$$= \frac{a^3}{4}\left(\frac{2mk_BT}{\pi\hbar^2}\right)^{3/2} e^{E_F/k_BT}, \tag{3.42}$$

or

$$E_F = k_BT \ln\left(\frac{4N}{a^3}\left(\frac{\pi\hbar^2}{2mk_BT}\right)^{3/2}\right). \tag{3.43}$$

The Fermi energy of the form (Eq. 3.43) is plotted for bulk GaAs materials as a function carrier concentration with respect to the conduction band minimum as shown in Fig. 3.4. The curves obtained in this figure were plotted for different temperatures ranging from 300 to 4.0 K. The conduction band minimum was included as a function of temperature as well. The sample size was chosen as a cubic specimen with a side of 1.0 cm. It is customary to divide the density of state by the volume in real space so that the volume of the sample would not show in the final expressions of either the Fermi energy or the carrier concentrations (Eq. 3.42). As the temperature decreases, the Fermi energy is reduced because of carrier freeze out.

3.5 BOSE–EINSTEIN STATISTICS

Bose–Einstein statistics is concerned with particles that possess zero or integer spins ($S = 0, 1, 2, 3, \ldots$), which do not obey Pauli exclusion principle. These particles are still indistinguishable. The most common particles that follow

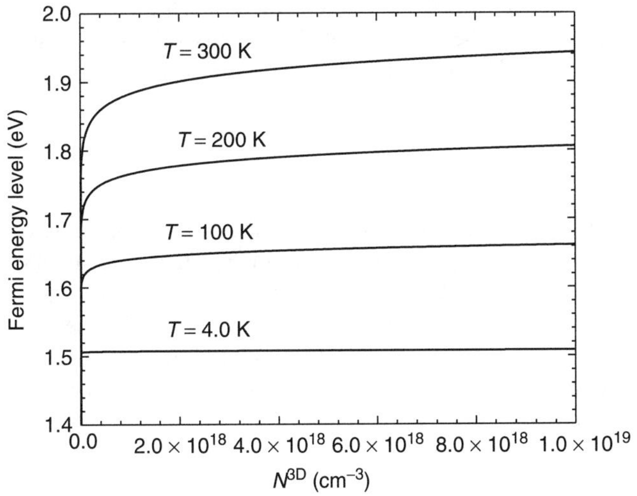

FIGURE 3.4 The Fermi energy plotted as a function of carrier concentration using Equation 5.43 for different temperatures. The Fermi energy was taken with respect to the top of the valence band minimum of GaAs assuming the band gap is 1.5 eV.

Bose–Einstein statistics are photons ($S = 1$). In this case, an arbitrary number of particles can occupy a single quantum state and hence the number of ways of arranging N particles in the system can be written as

$$\Omega_{\mathrm{BE}}(N_1, N_2, N_3, \ldots, N_n) = \prod_{j}^{n} \frac{(N_j + g_j - 1)!}{N_j!(g_j - 1)!}. \tag{3.44}$$

One now can proceed to maximize this function using Lagrange's method of undetermined multipliers as described in the previous two sections. The final results can be written as

$$f_{\mathrm{BE}}(E) = \frac{1}{\mathrm{e}^{\alpha}\mathrm{e}^{E/k_{\mathrm{B}}T} - 1}, \tag{3.45}$$

where $f_{\mathrm{BE}}(E)$ is the Bose–Einstein distribution function and α can be considered zero, since the photons can be easily created and annihilated, and therefore, the constraint having a constant number of particles can be easily discarded. The function $f_{\mathrm{BE}}(E)$ can be reduced to

$$f_{\mathrm{BE}}(E) = \frac{1}{\mathrm{e}^{E/k_{\mathrm{B}}T} - 1}. \tag{3.46}$$

The three distribution functions (Maxwell–Boltzmann, Fermi–Dirac, and Bose–Einstein) are plotted as a function of energy in Fig. 3.5 at $T = 300$ K and $T = 77$ K with the Fermi energy chosen as $E_{\mathrm{F}} = 0.2$ eV. It is obvious from this plot that these three distribution functions are in good agreement with each other when $(E - E_{\mathrm{F}}) \gg k_{\mathrm{B}}T$. The agreement is even improved as the temperature is reduced from 300 to 77 K. In many semiconductor cases, the Fermi–Dirac distribution function can be approximated as a Maxwell–Boltzmann distribution function, in particular, when $\exp\{(E - E_{\mathrm{F}})/k_{\mathrm{B}}T\} \gg 1$.

3.6 DENSITY OF STATES

To understand the energy and momentum distribution among particles in a system, one needs to answer the question of how many states are available for these particles to occupy in a particular system. For a large number of particles in a three-dimensional system, such as electrons in a crystalline semiconductor, the answer to this question can be understood by applying Bloch's theorem to the crystalline semiconductor, where the wave function exhibits periodicity within the period structure (crystal) such as

$$\begin{aligned}\psi(x, y, z) &= \psi(x + L_x, y + L_y, z + L_z) \\ &= \exp\left[i\left\{k_x(x + L_x) + k_y(y + L_y) + k_z(z + L_z)\right\}\right] \\ &= \exp\left[i\left\{k_x x + k_y y + k_z z\right\}\right]\exp\left[i\left\{k_x L_x + k_y L_y + k_z L_z,\right\}\right]\end{aligned} \tag{3.47}$$

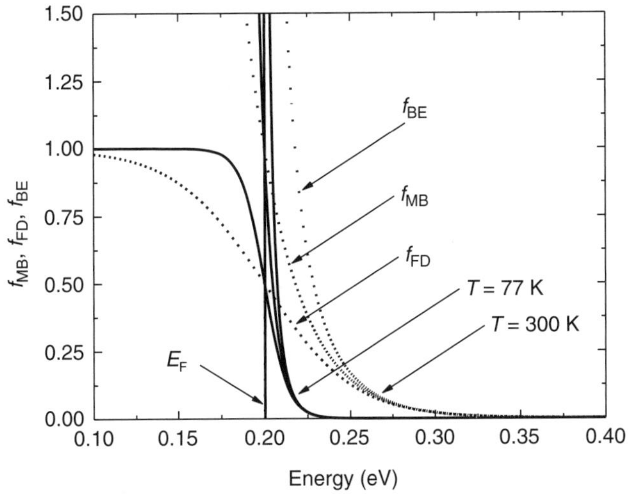

FIGURE 3.5 Fermi–Dirac (f_{FD}), Maxwell–Boltzmann (f_{MB}), and Bose–Einstein (f_{BE}) distribution functions plotted as a function of energy at $T = 300$ K (dashed lines) and at $T = 77$ K (solid lines).

where $L_x = L_y = L_z = L$ is the period of a cubic crystal. For Bloch's theorem to be valid, the second exponential of Equation 3.47 must be unity, which implies that

$$k_x L = 2\pi n_x, k_y L = 2\pi n_y, k_z L = 2\pi n_z, \tag{3.48}$$

where n_x, n_y, and n_z are integers. The volume of a unit cell (V_k^o) in the $\boldsymbol{k}$-space occupied by one state is

$$V_k^o = k_x k_y k_z = \frac{(2\pi)^3}{L^3}. \tag{3.49}$$

Other states are obtained by assuming other values for n_x, n_y, and n_z such as (000), (100), (110), (200), and (210), which gradually fill a sphere of radius $\boldsymbol{k}$. The Fermi energy is thus defined at zero temperature, where the states within the sphere of radius $\boldsymbol{k}_F$ are all occupied and states for $k > k_F$ are all empty, where $\boldsymbol{k}_F$ is the Fermi wave vector.

We can define the density of states, $g(E)$, as the number of states per unit energy per unit volume of real space (see, e.g., Harrison) such that

$$g(E) = \frac{\partial N}{\partial E}, \tag{3.50}$$

which follows that the total number of states (N) is equal the degeneracy times the volume of the sphere in $\boldsymbol{k}$-space divided by the volume occupied by one state (primitive unit cell) and divided again by the volume of real space such that

$$N = 2\frac{4\pi k^3}{3} \cdot \frac{1}{\left(\frac{2\pi}{L}\right)^3} \cdot \frac{1}{V} = 2\frac{4\pi k^3}{3(2\pi)^3}, \tag{3.51}$$

where we assume $V = L^3$. For electrons of spin 1/2, the degeneracy is 2 for spin up and spin down. The density of states can be written as

$$g(E) = \frac{\partial N}{\partial k}\frac{\partial k}{\partial E}, \tag{3.52}$$

where

$$\frac{\partial N}{\partial k} = 2\frac{4\pi k^2}{(2\pi)^3}. \tag{3.53}$$

From the effective mass approximation, the energy of the electrons is assumed to be parabolic in $\boldsymbol{k}$-space as follows:

$$E = \frac{\hbar^2 k^2}{2m^*}, \tag{3.54}$$

which yields

$$\frac{\partial k}{\partial E} = \left(\frac{2m^*}{\hbar^2}\right)^{1/2} \cdot \frac{1}{2\sqrt{E}}. \tag{3.55}$$

Substituting Equations 3.53–3.55 into Equation 3.52, we obtain

$$g(E) = \frac{1}{2\pi^2}\left(\frac{2m^*}{\hbar^2}\right)^{3/2}\sqrt{E}. \tag{3.56}$$

A plot of $g(E)$ as a function of energy is shown in Fig. 3.6, where the effective mass is assumed to be $m* = 0.067\ m_0$. The inset is the three-dimensional sphere of radius $\boldsymbol{k}_{\mathrm{F}}$ in the $\boldsymbol{k}$-space. This is a typical example of electrons in bulk semiconductor material such as GaAs or silicon.

To understand the concept of Fermi energy and electrons and holes distribution in semiconductors, let us first assume that the semiconductor is intrinsic, which means that the number of electrons in the conduction band is equal to the number of holes in the valence band. The density of states for both the conduction and valence bands can be written as follows:

$$g_e(E) = \frac{1}{2\pi^2}\left(\frac{2m_e^*}{\hbar^2}\right)^{3/2}\sqrt{E - E_{\mathrm{c}}} \quad \text{and} \quad g_h(E) = \frac{1}{2\pi^2}\left(\frac{2m_h^*}{\hbar^2}\right)^{3/2}\sqrt{E_{\mathrm{v}} - E}, \tag{3.57}$$

where the subscripts "e" and "h" stand for electrons and holes, respectively, and E_{c} and E_{v} are the bottom and the top of the conduction and valence bands, respectively. The results of plotting Equation 3.57 are shown in Fig. 3.7. The

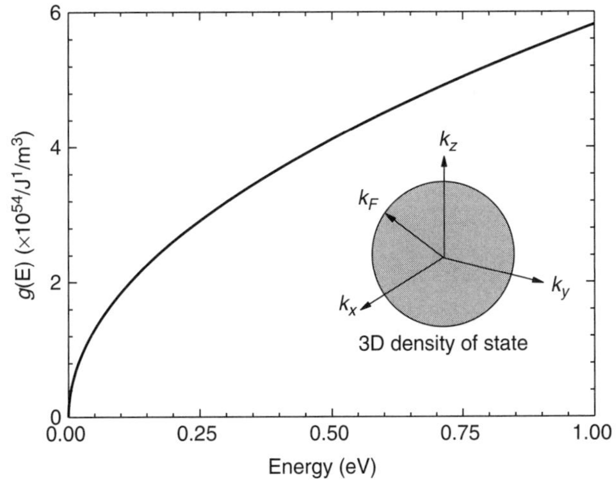

FIGURE 3.6 The density of states of a three-dimensional (3D) system, which is a typical bulk material such as semiconductor single crystals. The calculations were made for the electron effective mass in GaAs. The inset is a sphere in the $\boldsymbol{k}$-space of radius $\boldsymbol{k}_F$.

electron distribution function is given by the Fermi–Dirac function (Eq. 3.39), while the distribution function of holes can be expressed as the distribution function of unfilled states $(1 - f_{FD})$

$$\begin{aligned} f_{FD}^h(E) &= 1 - f_{FD}(E) = 1 - \frac{1}{1 + \mathbf{e}^{(E-E_F)/k_BT}} \\ &= \frac{1}{\mathbf{e}^{-(E-E_F)/k_BT} + 1}. \end{aligned} \tag{3.58}$$

The superscript "h" is introduced to refer to the hole distribution function. The electron and hole concentrations are shown as the shaded areas in Fig. 3.7. The carrier concentrations are plotted for intrinsic GaAs materials as shown in Fig. 3.7a. In this figure, the number of electrons $[n(E)]$ is equal to the number of holes $[p(E)]$. For n-type GaAs, where the material is doped with donors, the Fermi energy is shifted toward the conduction band as shown in Fig. 3.7b and $n(E) > p(E)$ as indicated by the shaded area. The Fermi–Dirac distribution function is also shifted. For a degenerate n-type semiconductor (heavily doped semiconductor), the Fermi energy is pinned above the conduction band minimum. Similarly, for a p-type semiconductor, the Fermi energy will be shifted toward the valence band and for degenerate p-type semiconductor, the Fermi energy is resonant in the valence band.

It was shown, as illustrated in Fig. 3.5, that the Fermi–Dirac distribution function can be approximated by the Maxwell–Boltzmann function for $(E_c - E_F) \gg k_BT$, where E_c is the bottom of the conduction band. This is a valid assumption for intrinsic or lightly doped semiconductors, where E_F is pinned

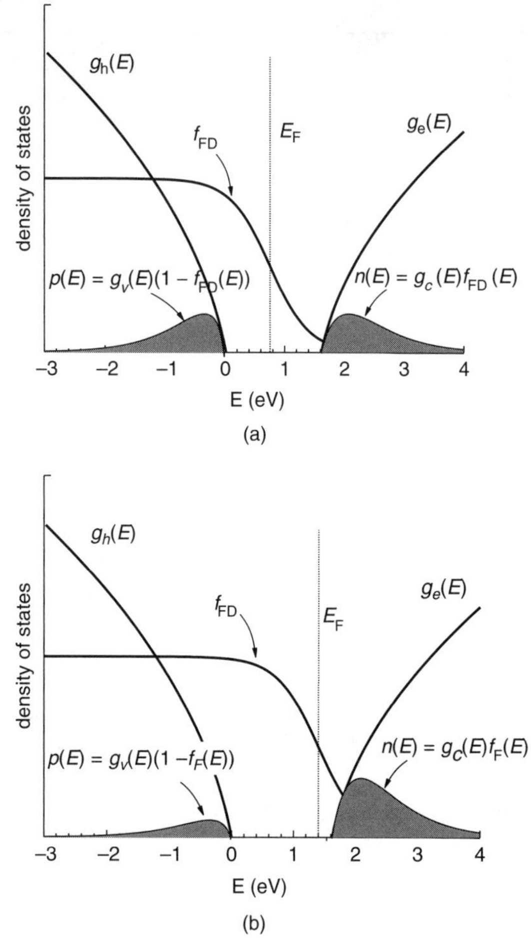

FIGURE 3.7 The distribution function, density of states, Fermi energy level, and the carrier population for (a) intrinsic and (b) n-type GaAs are sketched as a function of energy. The y-axis unit is taken as an arbitrary unit since we have several overlaid parameters.

near the mid gap. This implies that $\exp((E - E_{\mathrm{F}})/k_{\mathrm{B}}T) \gg 1$, which yields

$$f_{\mathrm{FD}}(E) = \frac{1}{1 + \mathbf{e}^{(E-E_{\mathrm{F}})/k_{\mathrm{B}}T}} \simeq \frac{1}{\mathbf{e}^{(E-E_{\mathrm{F}})/k_{\mathrm{B}}T}} = \mathbf{e}^{-(E-E_{\mathrm{F}})/k_{\mathrm{B}}T} = \mathbf{e}^{E_{\mathrm{F}}/k_{\mathrm{B}}T}\mathbf{e}^{-E/k_{\mathrm{B}}T} \tag{3.59a}$$

and

$$f_{\mathrm{FD}}^{h}(E) = 1 - \frac{1}{1 + \mathbf{e}^{(E-E_{\mathrm{F}})/k_{\mathrm{B}}T}} = \frac{1}{\mathbf{e}^{-(E-E_{\mathrm{F}})/k_{\mathrm{B}}T} + 1}$$
$$\simeq \mathbf{e}^{+(E-E_{\mathrm{F}})/k_{\mathrm{B}}T} = \mathbf{e}^{-E_F/k_{\mathrm{B}}T}\mathbf{e}^{E/k_{\mathrm{B}}T}. \tag{3.59b}$$

With this Maxwell–Boltzmann approximation, the electron and hole densities can be easily evaluated for a semiconductor at equilibrium as follows:

$$
\begin{aligned}
n_0 &= \int f_{\mathrm{FD}}(E) g(E) \mathrm{d}E = \int \frac{g(E)\mathrm{d}E}{1+\mathbf{e}^{(E-E_{\mathrm{F}})/k_{\mathrm{B}}T}} \\
&= \frac{1}{2\pi^2}\left(\frac{2m_e^*}{\hbar^2}\right)^{3/2} \int_{E_{\mathrm{c}}}^{\infty} \frac{\sqrt{E-E_c}\,\mathrm{d}E}{(1+\mathbf{e}^{(E-E_{\mathrm{F}})/k_{\mathrm{B}}T})} \\
&= \frac{1}{2\pi^2}\left(\frac{2m_e^* k_{\mathrm{B}}T}{\hbar^2}\right)^{3/2} \mathbf{e}^{(E_{\mathrm{F}}-E_{\mathrm{c}})/k_{\mathrm{B}}T} \int_0^{\infty} \mathbf{e}^{-x}\sqrt{x}\mathrm{d}x \\
&= \frac{1}{2\pi^2}\left(\frac{2m_e^* k_{\mathrm{B}}T}{\hbar^2}\right)^{3/2} \mathbf{e}^{(E_{\mathrm{F}}-E_{\mathrm{c}})/kT}.\frac{\sqrt{\pi}}{2} = \frac{1}{4}\left(\frac{2m_e^* k_{\mathrm{B}}T}{\pi\hbar^2}\right)^{3/2} \mathbf{e}^{-(E_{\mathrm{c}}-E_{\mathrm{F}})/k_{\mathrm{B}}T} \\
&= N_{\mathrm{c}}\mathbf{e}^{-(E_{\mathrm{c}}-E_{\mathrm{F}})/k_{\mathrm{B}}T},
\end{aligned}
\tag{3.60}
$$

where n_0 is the electron density and N_{c} is given by

$$
N_{\mathrm{c}} = \frac{1}{4}\left(\frac{2m_e^* k_{\mathrm{B}}T}{\pi\hbar^2}\right)^{3/2}. \tag{3.61}
$$

The density of states used in the above derivation is given by Equation 3.57. Similarly, the hole concentration (p_0) can be obtained as

$$
p_0 = N_{\mathrm{v}}\mathbf{e}^{-(E_{\mathrm{F}}-E_{\mathrm{v}})/k_{\mathrm{B}}T}, \quad \text{where } N_{\mathrm{v}} = \frac{1}{4}\left(\frac{2m_h^* k_{\mathrm{B}}T}{\pi\hbar^2}\right)^{3/2}. \tag{3.62}
$$

The mass action law, $n_0 p_0 = n_i^2$, where n_i is the intrinsic carrier concentration can now be written as

$$
\begin{aligned}
n_i = \sqrt{n_0 p_0} &= \left(\frac{1}{4}\left(\frac{2m_h^* k_{\mathrm{B}}T}{\pi\hbar^2}\right)^{3/2} \frac{1}{4}\left(\frac{2m_e^* k_{\mathrm{B}}T}{\pi\hbar^2}\right)^{3/2}\right)^{1/2} \\
&\quad \times \mathbf{e}^{-(E_{\mathrm{F}}-E_{\mathrm{v}})/2k_{\mathrm{B}}T}\mathbf{e}^{-(E_{\mathrm{c}}-E_{\mathrm{F}})/2k_{\mathrm{B}}T} \\
&= \frac{1}{4}\left(\frac{2\sqrt{m_h^* m_e^*}k_{\mathrm{B}}T}{\pi\hbar^2}\right)^{3/2} \mathbf{e}^{-(E_{\mathrm{c}}-E_{\mathrm{v}})/2k_{\mathrm{B}}T} = \frac{1}{4}\left(\frac{2\sqrt{m_h^* m_e^*}k_{\mathrm{B}}T}{\pi\hbar^2}\right)^{3/2} \mathbf{e}^{-E_g/2k_{\mathrm{B}}T},
\end{aligned}
\tag{3.63}
$$

where $E_g = E_{\mathrm{c}} - E_{\mathrm{v}}$ is the band gap energy. The intrinsic carrier concentration is thus independent of the Fermi energy level. The Fermi energy for an intrinsic semiconductor can be evaluated by equation n_0 and p_0, which yields

$$
\frac{1}{4}\left(\frac{2m_e^* k_{\mathrm{B}}T}{\pi\hbar^2}\right)^{3/2} \mathbf{e}^{-(E_{\mathrm{c}}-E_{\mathrm{F}})/k_{\mathrm{B}}T} = \frac{1}{4}\left(\frac{2m_h^* k_{\mathrm{B}}T}{\pi\hbar^2}\right)^{3/2} \mathbf{e}^{-(E_{\mathrm{F}}-E_{\mathrm{v}})/k_{\mathrm{B}}T}
$$

or

$$\mathbf{e}^{[2E_F-(E_c+E_v)]/k_BT} = \left(\frac{m_h^*}{m_e^*}\right)^{3/2}. \tag{3.64}$$

Taking the natural log of both sides and rearrange, we have

$$E_F = \frac{1}{2}(E_c + E_v) + \frac{3}{4}k_BT \ln\left(\frac{m_h^*}{m_e^*}\right). \tag{3.65}$$

Usually, the top of the valance band is taken as a reference point, which can be set as zero. The intrinsic Fermi energy at room temperature is ~0.78 eV for GaAs with $m_h{}^* = 0.45m_0, m_e{}^* = 0.067m_0$, and $E_c = 1.48$ eV.

3.7 DENSITY OF STATES OF QUANTUM WELLS, WIRES, AND DOTS

The density of states in low dimensional systems is derived in this section. To avoid the confusion about how the low dimensional systems are defined, we consider that the charge carriers have degree of freedom directions and confinement directions. For bulk materials, there are three degrees of freedom directions and zero confined directions. Thus, bulk materials are called *three-dimensional systems*. Quantum wells are considered two-dimensional systems, which mean that the charge carriers have two degrees of freedom directions and one confined direction. In this case, the growth direction is the confined direction. Quantum wires, on the other hand, have one degree of freedom direction and two confined directions. Thus, quantum wires are considered one-dimensional systems. When the charge carriers are confined in three directions, the structure is called *zero dimensional system*. We refer to this system as quantum dots.

3.7.1 Quantum Wells

The density of states in a quantum well system is restricted to the $k_x - k_y$ plane as shown in the inset of Fig. 3.8, where the electrons or holes are now confined in this plane and their motion is restricted along the growth axis (z-direction in real space, or k_z-direction in momentum space). The total number of states per unit cross-sectional area is given by the area in $\boldsymbol{k}$-space divided by the area of the unit cell in $\boldsymbol{k}$-space and divided by the area in the real space

$$\begin{aligned} N^{2D} &= 2\pi k^2 . \frac{1}{(2\pi/L)^2} . \frac{1}{L^2} \\ &= 2\frac{\pi k^2}{(2\pi)^2}, \end{aligned} \tag{3.66}$$

where the factor 2 is the spin degeneracy of the electrons and L^2 is the real-space square area, and $2\pi/L^2$ is the two-dimensional primitive unit cell in the $\boldsymbol{k}$-space.

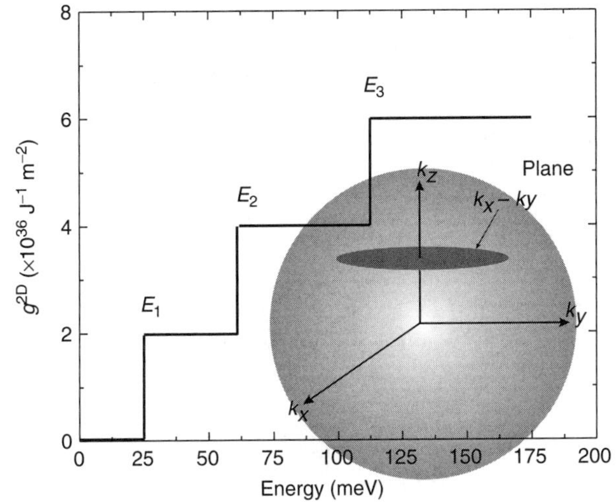

FIGURE 3.8 The density of states as a function of energy for a two-dimensional (2D) system such as GaAs/AlGaAs multiple quantum wells. The inset illustrates the 2D confinement where the charge carriers are confined in the k_x–k_y plane.

The density of state can be expressed as

$$g^{2D}(E) = \frac{\partial N^{2D}}{\partial E} = \frac{\partial N^{2D}}{\partial k}\frac{\partial k}{\partial E} = \left(\frac{k}{\pi}\right)\left(\frac{2m^*}{\hbar^2}\right)^{1/2}\frac{1}{2\sqrt{E}} = \frac{m^*}{\pi\hbar^2}, \tag{3.67}$$

where the energy, E, is defined in Equation 3.54. Notice that the density of states is independent of energy. If there are more than one confined states in the quantum well system, the density of states at a given energy is the sum over all subbands below that particular energy. Hence, Equation 3.67 can be rewritten as

$$g^{2D}(E) = \sum_{j=1}^{n} \frac{m^*}{\pi\hbar^2} Y(E - E_i), \tag{3.68}$$

where n is the total number of confined subbands below a particular energy and Y is a step function defined as

$$Y(E - E_i) = \begin{cases} 1 \text{ for} & E > E_j, \\ 0 \text{ for} & E < E_j. \end{cases} \tag{3.69}$$

The two-dimensional density of states is plotted in Fig. 3.8 for three confined subband energy levels. A typical system of two-dimensional structure is

GaAs/AlGaAs multiple quantum wells, where three energy levels can be confined in a well of thickness 200 Å and a barrier height of ~0.3 eV. The barrier height is determined by the Al mole fraction of the AlGaAs layer.

For n-type GaAs/AlGaAs multiple quantum wells, the total number of electrons (n^{2D}) within a subband is given as

$$n^{2D} = \int_{\text{subband}} g^{2D}(E) f_{FD}(E) \mathrm{d}E, \tag{3.70}$$

where $f_{FD}(E)$ is the Fermi–Dirac distribution function defined in Equation 3.39. Substituting Equations 3.39 and 3.57 into Equation 3.70 yields

$$n_j^{2D} = \int_{E_j}^{\infty} \frac{m^*}{\pi \hbar^2} \frac{\mathrm{d}E}{\left(\mathrm{e}^{(E-E_F)/k_B T} + 1\right)}. \tag{3.71}$$

To integrate this equation, we let $y = \exp[(E - E_F)/k_B T]$, which gives $\mathrm{d}y = [y/(k_B T)]\mathrm{d}E$ and $y_j = \exp[(E_j - E_F)/k_B T]$. Inserting this transformation into Equation 3.71, we have

$$n_j^{2D} = \frac{m^*}{\pi \hbar^2} \int_{y_j}^{\infty} \frac{\mathrm{d}y}{(y+1)y}. \tag{3.72}$$

The integral in the above equation can be solved using integration by part or by using Mathematica. The result of the integration gives

$$\int \frac{\mathrm{d}y}{(y+1)y} = -\ln\left(1 + \frac{1}{y}\right). \tag{3.73}$$

Substitute Equation 3.72 back into Equation 3.71 and rearrange, the density of electrons in the jth subband is finally obtained as

$$\begin{aligned} n_j^{2D} &= \frac{m^* k_B T}{\pi \hbar^2}\left(-\ln\left(1+\frac{1}{y}\right)\right)\Bigg|_{y_j}^{\infty} = \frac{m^*}{\pi \hbar^2} \ln\left(1 + \frac{1}{y_j}\right) \\ &= \frac{m^* k_B T}{\pi \hbar^2} \ln\left(1 + \mathrm{e}^{(E_F - E_j)/k_B T}\right). \end{aligned} \tag{3.74}$$

For n bound subbands in the quantum wells, we have

$$n^{2D} = \sum_{j=1}^{n} n_j^{2D} = \frac{m^* k_B T}{\pi \hbar^2} \sum_{j=1}^{n} \ln(1 + \mathrm{e}^{(E_F - E_j)/k_B T}). \tag{3.75}$$

It appears from the above equation that the Fermi energy, E_F, is explicitly independent of temperature when $(E_F - E_j) \gg k_B T$. However, for

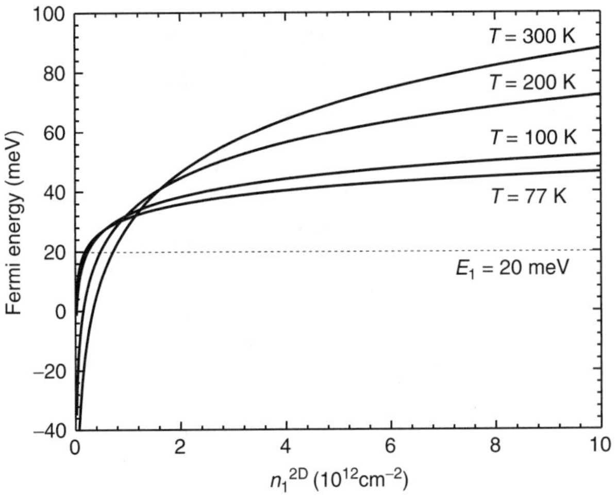

FIGURE 3.9 The Fermi energy plotted as a function of the 2D carrier concentration for the first subband energy level (n_1^{2D}) at different temperatures. The first subband energy level (E_1) was taken as 20 meV.

the limit $n^{2D}\pi\hbar^2 \ll m^* k_B T$, the Fermi energy can be approximated as $E_F \approx E_j + k_B T \ln[n^{2D}\pi\hbar^2/(m^* k_B T)]$. The Fermi energy in the latter limit is plotted as a function of the two-dimensional electron density as shown in Fig. 3.9. In this figure, the Fermi energy was plotted for different temperatures with respect to the first bound state (E_1), which was taken as 20 meV. The Fermi energy is below the bound state for population on the order of 5×10^{11} cm^{-2} and at $T > 100$ K. This can be understood by the fact the Fermi energy in the above formalisms is merely a quasi-Fermi energy, which describes the occupancy of the subband energy levels.

3.7.2 Quantum Wires

The charge carrier confinement in semiconductors can be further reduced by reducing the number of degrees of freedom in the carrier momentum. This can be accomplished through photolithography or even self-assembled epitaxial growth of what is called quantum wires. A typical example of an n-type GaAs/AlGaAs quantum wire is sketched in Fig. 3.10, where the electrons in the GaAs layer are confined in both the growth direction (z-direction) and the y-direction, but they can move along the x-direction. The z- and y-directions are called the *directions of confinements* and the x-direction is called the *degree of freedom direction*. The quantum wire is usually referred to as *one-dimensional* (1D) *system*. Thus, for quantum wells, we have one direction of confinement and two directions of degree of freedom directions. In contrast, the bulk materials have three degrees of freedom directions and zero direction of confinement.

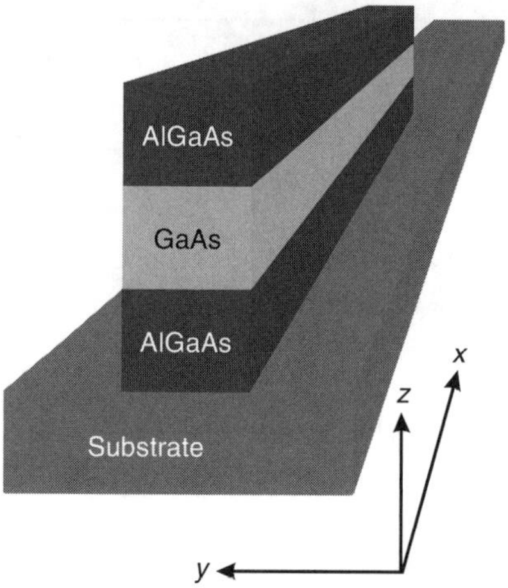

FIGURE 3.10 A schematic presentation of GaAs/AlGaAs quantum wire showing two directions of confinement (y- and z-directions) and one degree of freedom (x-direction).

The density of states of the 1D system can be obtained by assuming that the electron momenta fill states along a line as shown in Fig. 3.11. The total number of states can be obtained by dividing the total length of the quantum wire ($2k$) by the primitive unit cell and then divided by the length in the real space such as

$$n^{1D} = 2\frac{2k}{(2\pi/L)}\frac{1}{L} = \frac{2k}{\pi}, \tag{3.76}$$

where n^{1D} is the total number of states and the 2 is included to account for the electron spin degeneracy. The density of states can then be written as

$$g^{1D}(E) = \frac{\partial n^{1D}}{\partial E} = \frac{\partial n^{1D}}{\partial k}\frac{\partial k}{\partial E}, \tag{3.77}$$

where

$$\frac{\partial n^{1D}}{\partial k} = \frac{2}{\pi} \quad \text{and} \quad \frac{\partial k}{\partial E} = \left(\frac{2m^*}{\hbar^2}\right)^{1/2}\frac{1}{2\sqrt{E}}. \tag{3.78}$$

The second term of the above equation is obtained from Equation 3.54. Substituting Equation 3.78 into Equation 3.77, we have

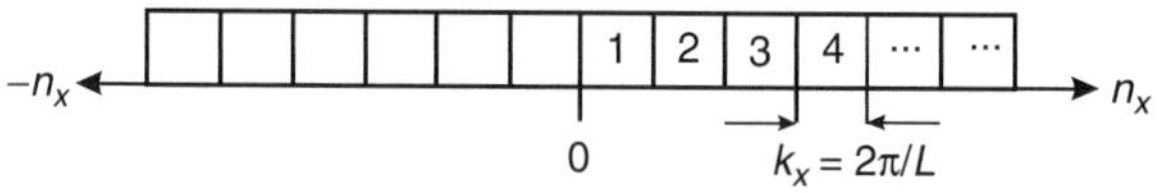

FIGURE 3.11 A quantum wire of length $2k$ divided into one-dimensional primitive unit cells of length $k_x = 2\pi/L$.

$$g^{1D}(E) = \left(\frac{2m^*}{\hbar^2}\right)^{1/2} \frac{1}{\pi\sqrt{E}}. \tag{3.79}$$

Following the same discussion for the 2D system, the total density of states of the 1D system with an n number of confined energy levels is given as

$$g^{1D}(E) = \left(\frac{2m^*}{\pi^2\hbar^2}\right)^{1/2} \sum_{j=1}^{n} \frac{1}{\sqrt{E - E_j}} Y(E - E_j), \tag{3.80}$$

where $Y(E - E_j)$ is a step function defined in Equation 3.69. A plot of the density of states of a quantum wire is shown in Fig. 3.12 for four bound states. The inset is an illustration in the $\boldsymbol{k}$-space for the two confinement directions (k_y- and k_z-directions) depicted as the two ellipses and the one degree of freedom direction depicted as the k_x line indicated by the arrow as the solid thick line. Notice that the units of the density of state is Joule^{-1} meter^{-1} = 1.602×10^{-21} /eV/cm.

The linear electron density in the quantum wire can be obtained in a manner similar to that of the two-dimensional system, where the population density for the jth subband can be expressed as

$$n_j^{1D} = \int_0^\infty g^{1D}(E) f_{FD}(E) \mathrm{d}E = \left(\frac{2m^*}{\pi^2\hbar^2}\right)^{1/2} \int_{E_j}^{\infty} \frac{\mathrm{d}E}{\sqrt{E - E_j}(\mathrm{e}^{(E-E_F)/k_BT} + 1)}. \tag{3.81}$$

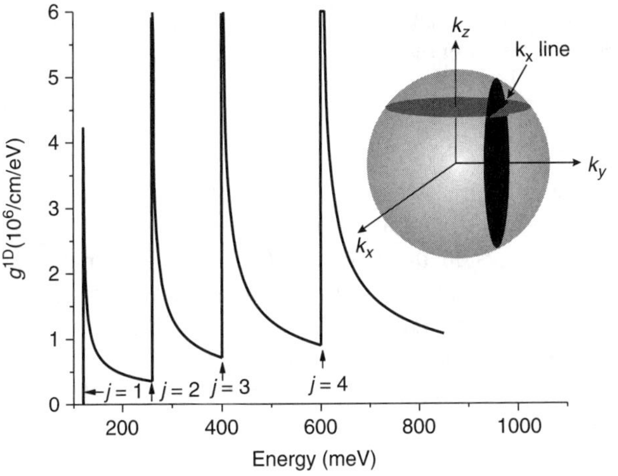

FIGURE 3.12 The density of states for a GaAs/AlGaAs quantum wire is plotted as a function of energy for four bound states labeled $j = 1, 2, 3,$ and 4. The inset is the k-space illustration of the two directions of confinement (k_y and k_z) and one degree of freedom direction (k_x).

The total linear density of states for a quantum wire with n bound states is thus

$$n^{1\text{D}} = \sum_{j=1}^{n} n_j^{1\text{D}}. \tag{3.82}$$

The integral in Equation 3.81 is difficult to solve, but can be evaluated analytically in the limit of $(E - E_\text{F})/k_\text{B}T \gg 1$ or in the limit of $(E - E_\text{F})/k_\text{B}T \ll 1$ and numerically between these two limits. For example, when $(E - E_\text{F})/k_\text{B}T \gg 1$, Equation 3.81 becomes

$$n_j^{1\text{D}} \approx \left(\frac{2m^*}{\pi^2\hbar^2}\right)^{1/2} \int_{E_j}^{\infty} \frac{\text{d}E}{\sqrt{E - E_j}(\text{e}^{(E-E_\text{F})/k_\text{B}T})} \approx \left(\frac{2m^* k_\text{B}T}{\pi^2\hbar^2}\right)^{1/2}$$
$$\text{e}^{(E_\text{F}-E_j)/k_\text{B}T} \int_0^{\infty} \frac{\text{e}^{-x}\,\text{d}x}{\sqrt{x}}$$
$$\approx \left(\frac{2m^* k_\text{B}T}{\pi^2\hbar^2}\right)^{1/2} . \text{e}^{(E_\text{F}-E_j)/k_\text{B}T} . \sqrt{\pi} = \left(\frac{2m^* k_\text{B}T}{\pi\hbar^2}\right)^{1/2} . \text{e}^{(E_\text{F}-E_j)/k_\text{B}T}, \tag{3.83}$$

where we used the transformation $x = (E - E_j)/k_\text{B}T$. The general form of this equation is similar to that shown in Equation 3.60. The Fermi energy is plotted with respect to the first bound state (E_1) as a function of the quantum wire carrier concentration for different temperatures as shown in Fig. 3.13. There is a similarity between the behavior of the Fermi energy in quantum wires and quantum wells as a function of the electron density as shown in Figs. 3.9 and 3.13.

3.7.3 Quantum Dots

The quantum dot is characterized by having three confinement directions and zero degree of freedom direction as shown in the inset of Fig. 3.14, where we sketch the confinements in the $\boldsymbol{k}$-space. The wave vector of the quantum dot is represented by the white dot where the three circles in the figures intercept as indicated by the arrow. Owing to the lack of dispersion curves, the wave vector selection rules are absent. The density of state is thus represented by the number of confined states divided by the energy interval. If the energy interval is approaching zero, then the density of states is simply a series of δ-functions centered on the confined energy levels ($E_1, E_2, E_3, \ldots$) as shown in Fig. 3.14. The energy levels are entirely discrete and are given by

$$E_{n_x,n_y,n_z} = \frac{\pi\hbar^2}{2m^*}\left(\frac{n_x^2}{L_x^2} + \frac{n_y^2}{L_y^2} + \frac{n_z^2}{L_z^2}\right), \tag{3.84}$$

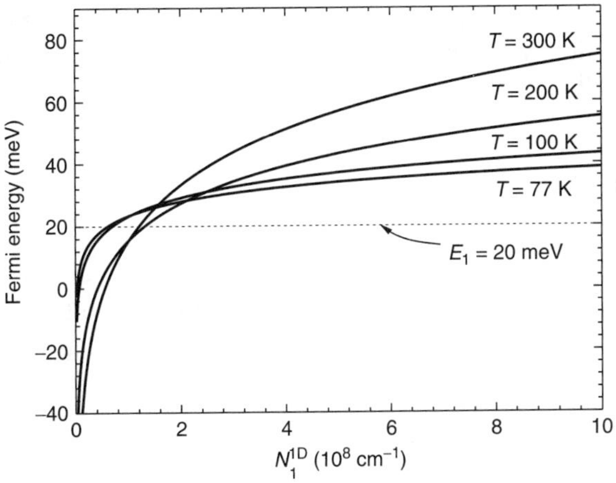

FIGURE 3.13 The Fermi energy plotted as a function of the 1D carrier concentration for the first subband energy level at different temperatures. The first subband energy level (E_1) was taken as 20 meV.

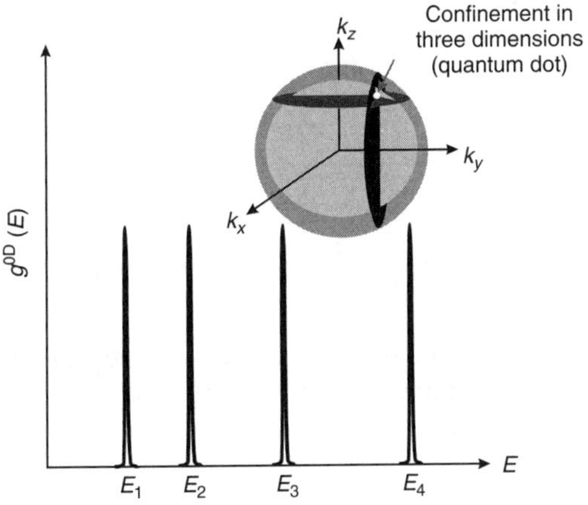

FIGURE 3.14 The density of states of a quantum dot is presented by a series of δ-functions centered on the confined energy levels. The inset is the k-space presentation of the quantum dot showing confinement in three directions. The energy dispersion is absent and is represented by the white dot where the three circles intercept.

where L_x, L_y, and L_z are the dimensions of the quantum dot and n_x, n_y, and n_z are positive integers.

3.8 DENSITY OF STATES OF OTHER SYSTEMS

This section is focused on deriving the density of states for systems that are occasionally encountered in semiconductor physics. In particular, superlattices are bulk materials under the influence of magnetic or electric fields. The density of states in quantum wells and wires under the influence of external electric fields will be briefly discussed.

3.8.1 Superlattices

Semiconductor superlattices are discussed in this chapter, where the confined energy levels are approximated by minibands. This happens when the barriers are too thin. A typical example of superlattices is InAs/InGaSb type II superlattices. One approach to estimate the density of states in superlattices is to take the general form

$$g(E) = \sum_{j=1}^{n} \delta(E - E_j), \tag{3.85}$$

where the sum is over all the eigenvalues. This general form of the density of state is actually very convenient, since the eigenvalues are quantized regardless of how small the separation is between them. The eigenvalues of the electrons in the superlattices can be expressed as (see Bastard)

$$E(n, q, k_\perp) = \frac{\hbar^2 k_\perp^2}{2m^*} + E_n(q). \tag{3.86}$$

The density of state is thus

$$g(E) = 2 \sum_{n,q,k_\perp} \delta\left(E - E_n(q) - \frac{\hbar^2 k_\perp^2}{2m^*}\right) \tag{3.87}$$

and the factor 2 is included for spin degeneracy. Convert the sum into an integral and use the δ-function definition, the density of state in the $\boldsymbol{k}$-space becomes

$$g^s(E) = 2\frac{1}{(2\pi/L)^2} \cdot \frac{Nd\, m^*}{L^2\, \hbar^2} .2 \int_0^{\pi/d} Y(E - E_n(q))\mathrm{d}q$$

$$= \frac{Nd\, m^*}{\pi^2\, \hbar^2} \int_0^{\pi/d} Y(E - E_n(q))\mathrm{d}q = \sum_n g_n^s(E), \tag{3.88}$$

where $\mathrm{Y}(E - E_n(q))$ is a step function, Nd is the length of the superlattice, and the superscript "s" was introduced to indicate that the density of states is for the superlattice, $g_n^s(E)$ is the density of states associated with the nth miniband, and the integral limits are the first Brillouin zone boundary. Since the miniband has a width such that $E_{\min} < E_n(q) < E_{\max}$, $g_n^s(E) = 0$ for $E_n(q) < E_{\min}$ and $g_n^s(E) = N\frac{m^*}{\pi\hbar^2}$ for $E_n(q) > E_{\max}$. The jump from one miniband to the next is not abrupt as in the case for quantum wells or wires. This is due to the fact that

$E_n(q) = E_n + S_n + 2T_n \cos(qd)$ is a function of the wave vector (q). The final results of the density of states, according to Bastard, can be written as

$$g^s(E) = \begin{cases} 0 & \text{for } E < (E_n + S_n - 2|T_n|) \\ \dfrac{Nm^*}{\pi\hbar^2} \text{ Arc } \cos\left(\dfrac{-E + E_n + S_n}{2|t_n|}\right) & \text{for } E - E_n - S_n < 2|T_n| \\ \dfrac{Nm^*}{\pi\hbar^2} & \text{for } E > (E_n + S_n + 2|T_n|) \end{cases} . \quad (3.89)$$

A plot of Equation 3.89 is shown for the three minibands in Fig. 3.15. Notice that the widths of the minibands are increased as the subband quantum number, n, is increased. The density of state is also multiplied by the number of the superlattice period, N.

3.8.2 Density of States of Bulk Electrons in the Presence of a Magnetic Field

Bulk electrons here are assumed to be electrons in the conduction band of bulk semiconductor materials, so that they have three degree of freedom directions. The allowed eigenvalues are quasi-continuum. In the presence of a magnetic field, each energy level splits into what is called Landau energy levels. The separation between the Landau energy levels is directly proportional to the strength of the magnetic field. The eigenvalues of an electron in a magnetic field parallel to the growth axis is given by

$$E_{n,k_z,\sigma_z} = \left(n + \frac{1}{2}\right)\hbar\omega_c + \frac{\hbar^2 k_z^2}{2m^*} + \sigma_z g^* \mu_B B, \quad (3.90)$$

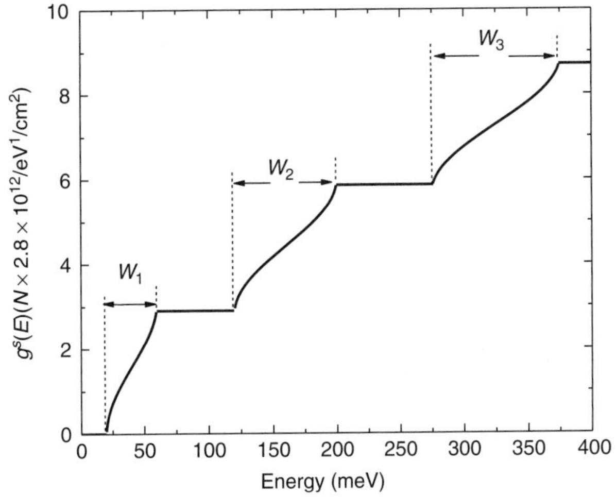

FIGURE 3.15 The density of states of a superlattice structure plotted as a function of energy. The width of the bands were indicated by W_i, where $i = 1,2,3$.

where n is the Landau quantum number, ω_c is the cyclotron frequency, which is given by $(eB)/(m^*c)$, μ_B is the Bohr magneton, g^* is the effective Landé g-factor, and σ_z is the spin eigenvalues, which is $\pm 1/2$. Using the general definition of the density of state (Eq. 3.85), we have

$$g^B(E) = \sum_{n,k_z,\sigma_s} \delta(E - E_{n,k_z,\sigma_s}), \tag{3.91}$$

where the superscript "B" is introduced to indicate the presence of magnetic field. Substitute Equation 3.90 into Equation 3.91 to obtain

$$\begin{aligned} g^B(E) &= \sum_{n,k_z,\sigma_s} \delta\left(E - (n + \tfrac{1}{2})\hbar\omega_c - \frac{\hbar^2 k_z^2}{2m^*} - \sigma_z g^* \mu_B B\right) \\ &= \sum_{n,k_z,\sigma_s} \delta\left(x - \frac{\hbar^2 k_z^2}{2m^*}\right) = \sum_{n,k_z,\sigma_s} g_n^B(E), \end{aligned} \tag{3.92}$$

where

$$x = E - (n + \tfrac{1}{2})\hbar\omega_c - \sigma_z g^* \mu_B B. \tag{3.93}$$

The degeneracy of any Landau level in one-dimensional k_z-space becomes

$$g_n^B(E) = \frac{L_x L_y}{(2\pi^2) l^2} \int_{-\infty}^{\infty} \delta\left(x - \frac{\hbar^2 k_z^2}{2m^*}\right) \mathrm{d}k_z, \tag{3.94}$$

where l is the magnetic length and given by $(\hbar c/eB)^{1/2}$, and the quantity $\frac{L_x L_y}{(2\pi^2)l^2}$ is included because of the degeneracy of any level in the k_x–k_y plane. Let $y = \frac{\hbar^2 k_z^2}{2m^*}$, which leads to $\mathrm{d}y = \frac{\hbar^2 k_z}{m^*}\mathrm{d}k_z$. Using this transformation, the density of states is further simplified, such that

$$\begin{aligned} g_n^B(E) &= 2\frac{L_x L_y}{(2\pi)^2 l^2 \hbar} \int_0^{\infty} \sqrt{\frac{m}{2y}} \delta(x - y) \mathrm{d}y \\ &= 2\frac{L_x L_y}{(2\pi)^2 l^2 \hbar}\sqrt{\frac{m}{2x}} = \frac{L_x L_y}{(2\pi)^2 l^2}\sqrt{\frac{2m}{\hbar^2}}(E - (n + \tfrac{1}{2})\hbar\omega_c - \sigma_z g^* \mu_B B)^{-1/2} \\ &= \frac{L_x L_y}{8\pi^2}\left(\frac{2m}{\hbar^2}\right)^{3/2} \hbar\omega_c (E - (n + \tfrac{1}{2})\hbar\omega_c - \sigma_z g^* \mu_B B)^{-1/2}. \end{aligned} \tag{3.95}$$

By including the spin degeneracy, the total density of state can be rewritten as

$$g^B(E) = \frac{1}{4\pi^2}\left(\frac{2m}{\hbar^2}\right)^{3/2} \hbar\omega_c \sum_n (E - (n + \tfrac{1}{2})\hbar\omega_c - \sigma_z g^* \mu_B B)^{-1/2}. \tag{3.96}$$

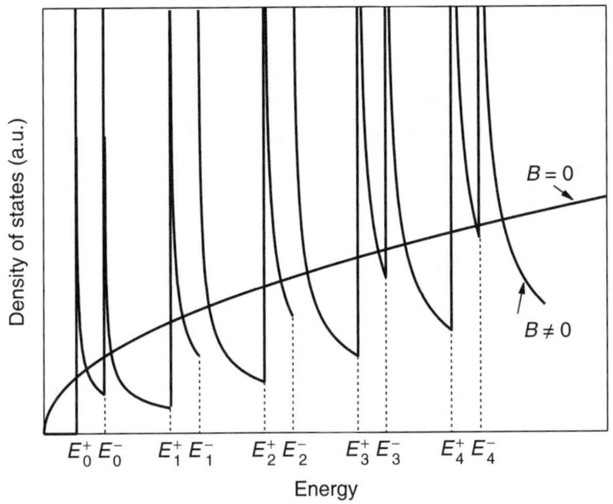

FIGURE 3.16 The density of states of electrons in the conduction band of a bulk semiconductor material plotted as a function of energy for both zero magnetic field and in the presence of the magnetic field.

Notice that we dropped $L_x L_y$ from the last expression to obtain the density of states per unit area. A plot of the density of state described in Equation 3.96 is shown in Fig. 3.16 for both $B = 0$ and $B \neq 0$. The density of states for $B \neq 0$ is zero only for $E < E_0^+$. The energies labeled E_n^+ correspond to $\sigma_z = +1/2$ and E_n^- correspond to $\sigma_z = -1/2$. The above analysis indicates that only k_z is a good quantum number. The magnetic field produces energy quantization in the x–y plane. One may imagine this situation by assuming that the electrons are trapped in the Landau circular orbits in the x–y plane generated by the magnetic field, but the electrons can move along the z-axis in a helical form. This form of motion is analogous to the confinement of electrons in a quantum wire, as it is clear from the density of states in both cases that they have the same energy dependence as shown in Equations 3.80 and 3.96.

For two-dimensional systems, such as multiple quantum wells, the confinement occurs along the growth axis (z-direction), which is not a good quantum number. By applying a magnetic field parallel to the growth direction, the x- and y-directions are no longer good quantum numbers. This implies that the electrons are confined in the three directions without any degree of freedom directions. In this case, the density of states is similar to that of the quantum dots. In other words, the density of states is a series of δ-functions as shown in Fig. 3.14.

3.8.3 Density of States in the Presence of an Electric Field

The density of states under the influence of an electric field is very complicated, and yet it is important to understand the behavior of density of states in devices that operate under applied bias voltage. In this section, we follow the analysis of

Davies (1998) and Davies *et al*. (1988). When a semiconductor is experiencing an applied bias voltage, the conduction and valence bands bend or vary in a way that the properties of the device have to be solved using self-consistent calculations. An example of how the semiconductor bands are changed under bias voltage is shown in Fig. 3.17 for a type I quantum well. It is evident that the interfaces are modified into triangular shapes similar to simple heterostructures.

If one assumes that a constant electric field is applied to a heterojunction, the electrostatic potential energy ($e\phi$) is

$$e\phi = eEz, \tag{3.97}$$

where e is the charge of the electron, E is the electric field, and z is the distance from the interface. The stationary Schrödinger equation can be written as

$$-\frac{\hbar^2}{2m^*}\frac{\mathrm{d}^2\psi(z)}{\mathrm{d}z^2} + eEz\psi(z) = E_n\psi(z). \tag{3.98}$$

The solution of this equation is expressed in terms of the Airy function

$$\psi(z) = Ai\left(\frac{e\epsilon z - E_n}{\epsilon_o}\right), \tag{3.99}$$

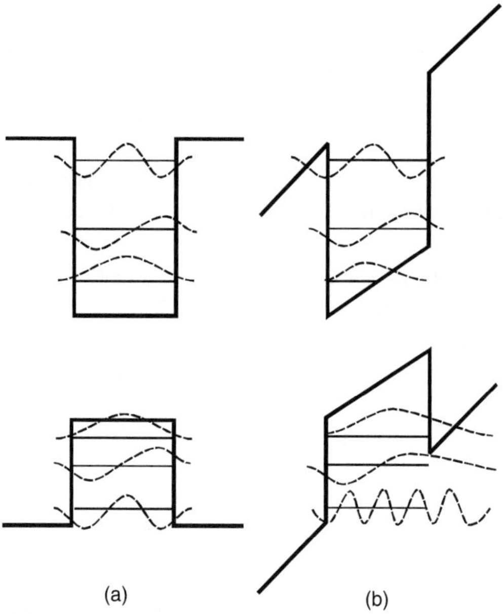

(a) (b)

FIGURE 3.17 The effect of an applied electric field on quantum wells includes the modification of the energy levels and their wave functions. (a) A quantum well is sketched in the absence of the electric field and (b) the modified band structure in the presence of an electric field.

where

$$\epsilon_0 = \left(\frac{(e\epsilon\hbar)^2}{2m^*}\right)^{1/3} = e\epsilon z_0, \text{ and } z_0 = \left(\frac{\hbar^2}{2m^* e\epsilon}\right)^{1/3}. \tag{3.100}$$

The Airy function is plotted for three energy levels in a triangular quantum well as shown in Fig. 3.18. For any particular energy level, the wave function in Fig. 3.18 exhibits propagation behavior for $E < e\epsilon z$ and tunneling behavior for $E > e\epsilon z$. This feature has a very interesting effect on the local density of states at fixed values of z. To demonstrate this effect, consider the general definition of the density of states

$$g(E, z) = \sum_k |\psi_k(z)|^2 \delta(E - E_k), \tag{3.101}$$

where the sum is over all eigenstates, labeled by k. The formalism of obtaining the density of states for bulk, quantum wells, and quantum wires under the influence of an electric field was reported by Davies and the final results are

$$g^{\epsilon}_{1\text{D}}(E, z) = \frac{2}{\hbar}\sqrt{\frac{2m^*}{\epsilon_o}} Ai^2\left(-\frac{E - e\epsilon z}{\epsilon_o}\right), \tag{3.102}$$

$$g^{\epsilon}_{2\text{D}}(E, z) = \frac{m^*}{2\pi} AiI(2^{2/3} S), \tag{3.103}$$

where

$$S = \frac{E - e\epsilon z}{\epsilon_o} \tag{3.104}$$

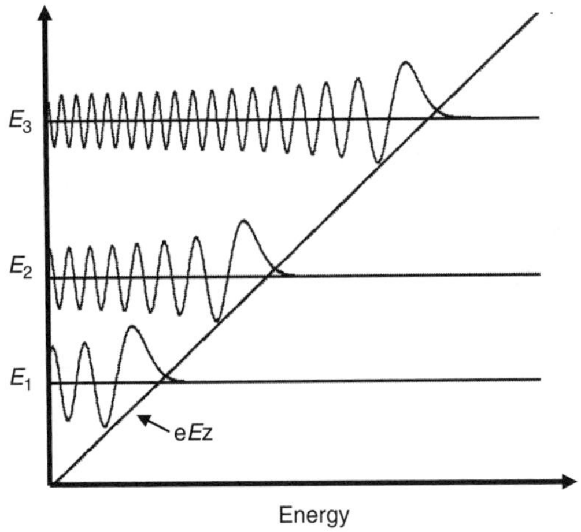

FIGURE 3.18 A triangular quantum well formed by applying a uniform electric field. Three energy levels are shown along with their wave functions. The wave functions have the form of Airy functions that satisfy the boundary conditions.

and

$$AiI(x) = \int_{x}^{\infty} Ai(y)\mathrm{d}y. \tag{3.105}$$

Finally, the density of states for a 3D system is

$$g_{3\mathrm{D}}^{\epsilon}(E, z) = \frac{m}{\pi \hbar^3}\sqrt{2m^*\epsilon_o}\{[Ai(S)]^2 - S[Ai(S)]^2\}, \tag{3.106}$$

where S is defined in Equation 3.104. The superscript "ϵ" is added to indicate the density of states under the influence of the applied electric field. Equations 3.102, 3.103, and 3.106 are plotted as a function of energy, as shown in Fig. 3.19. For simplicity, it is assumed in this figure that the ground state energy values for the

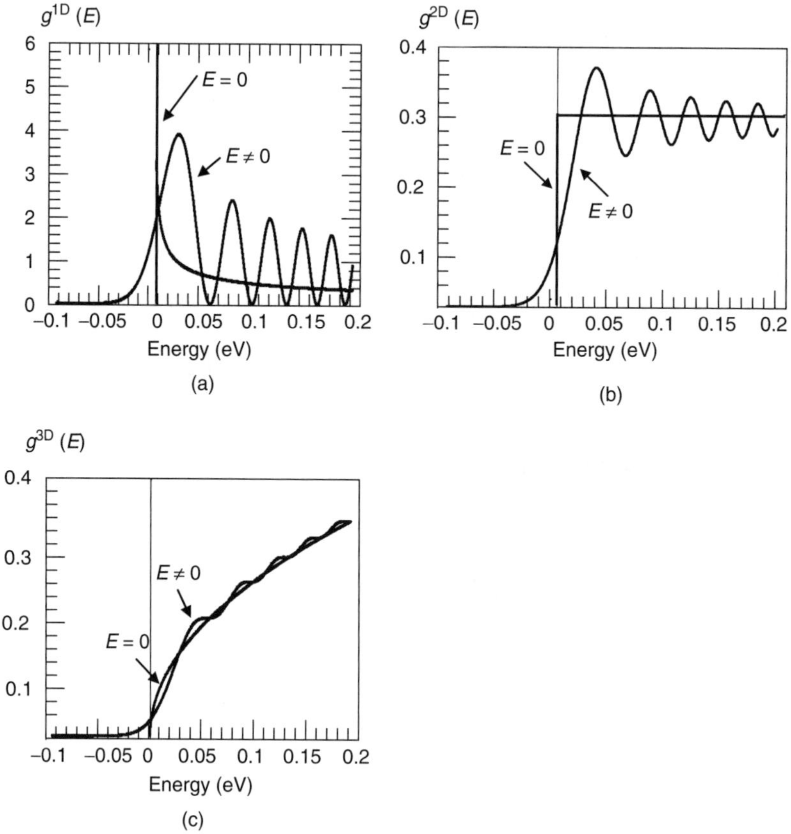

FIGURE 3.19 The density of states under the influence of a uniform electric field is plotted as a function of energy for (a) quantum wire, (b) quantum well, and (c) bulk materials (after Davies).

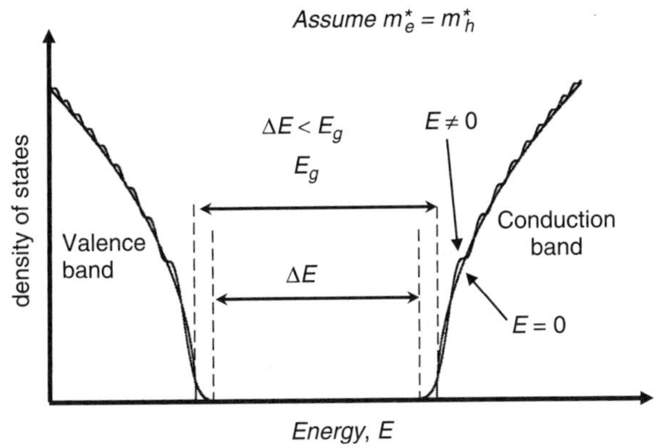

FIGURE 3.20 The densities of states is plotted as a function of energy for both the conduction and valence bands with (rippled curves) and without (smooth curves) an applied electric field. Notice that the densities of states leaked into the fundamental band gap causing an apparent decrease in E_g. For simplicity, we assumed the electrons and holes have the same effective mass.

quantum wire and quantum well are zero and the conduction band minimum is also set to zero. It can be seen from Fig. 3.19 that the density of states tunneled below the energy levels in the three cases. An interesting feature in the bulk material is that the density of states has a tail extended below the bottom of the conduction band minimum as shown in Fig. 3.19c. A similar result is obtained for the hole density of states as shown in Fig. 3.20, where we assumed that the effective mass of the electrons and holes are the same. The tunneling of states in the fundamental band gap (E_g), when a uniform electric field is applied, leads to the Franz–Keldysh effect. Photon with energy of $\Delta E < E_g$ can be absorbed. The oscillations observed in the density of state when an electric field is applied are difficult to observe by using the optical absorption technique because of the fact that most photons with energy above the fundamental band gap energy are reflected or absorbed at the edge of the conduction band. However, these oscillations can be observed using the photoreflectance technique. The absorption tail due to the tunneling of states into the band gap can be expressed as (see, for example, Mitin *et al*. and Fox)

$$\alpha(E) \propto \mathbf{e}^{\left[-\left(\frac{E_g-\hbar\omega}{\hbar\omega_F}\right)^{3/2}\right]}, \quad \text{for } \hbar\omega < E_g, \tag{3.107}$$

where

$$\omega_\text{F} = \frac{e^2}{2\hbar}\left(\frac{m_e^* + m_h^*}{m_e^* m_h^*}\right) E^{2/3}, \tag{3.108}$$

m_e^* and m_h^* are the electron and hole effective masses, respectively, and E is the constant applied electric field. The Franz–Keldysh effect has no significant applications in a bulk semiconductor, but it can wash out the desired excitonic effect.

3.9 SUMMARY

The distribution functions and density of states play a major role in the transport, electrical, and optical properties of semiconductor materials and devices. Thus, knowledge of these important parameters is necessary before proceeding. In this chapter, we presented derivations for Maxwell–Boltzmann, Fermi–Dirac, and Bose–Einstein distribution functions. The Fermi–Dirac distribution function is widely used in both bulk and low dimensional semiconductor materials, since it describes the distribution of particles with spin 1/2, such as electrons and holes. It should be pointed out that these distribution functions were derived for systems at equilibrium. For nonequilibrium cases, different analyses are applied. These analyses are presented in Chapter 7.

A fair amount of discussion in this chapter was devoted to the density of states in various systems. The density of states was derived for bulk semiconductors, and then compared to the density of states in low dimensional systems, such as quantum wells, quantum wires, and quantum dots. The density of states was also derived for semiconductor superlattices and bulk materials under the influence of a magnetic field. The electron motion in the presence of a magnetic field is confined to a two-dimensional plane. This condition is similar to confinement of electrons in quantum wires.

The density of states in bulk semiconductors, quantum wells, and quantum wires exhibits oscillatory behavior under the influence of an electric field. In addition, the density of states leaks into the fundamental band gap in case of bulk materials, causing an apparent decrease in the band gap. It also exhibits a tail below the bound energy levels in case of quantum wells and quantum wires.

The distribution functions and density of states are used to obtain the Fermi energy level in bulk semiconductor, quantum well, and quantum wire systems. The expressions for the Fermi energy levels are always easy to handle. These expressions can yield an approximate behavior of the Fermi energy levels in certain regimes, such as high or low temperature regimes. A plot of the Fermi energy level as a function of temperature or as a function of carrier concentrations is shown for these systems.

PROBLEMS

3.1. Γ–function is very useful in solving many statistical problems. Show that $\Gamma(n) = (n-1)\Gamma(n-1)$ and $\Gamma(n) = (n-1)!$

3.2. Derive Equation 3.2.

3.3. Gaussian distribution, $P(x) = \frac{1}{\sigma\sqrt{2\pi}} e^{\left(-\frac{(x-\overline{x})^2}{2\sigma^2}\right)}$, is used occasionally to describe certain properties in semiconductors. For example, the diffusion of carriers can be described by a Gaussian function. Show that $\int_{-\infty}^{\infty} P(x)\mathrm{d}x = 1$. Plot $P(x)$ as a function of x for at least three different values of σ. What type of distribution do you obtain when $\sigma \to 0$?

3.4. Plot the Fermi energy as a function of carrier concentration for different temperatures (Fig. 3.4) using Equation 3.43 for a cubic GaAs sample of an edge of 10^{-6} cm.

3.5. Show that the hole concentration in an intrinsic semiconductor is given by Equation 3.62.

3.6. Calculate the density of states for the following: (i) bulk GaAs, (ii) the lowest state of GaAs/AlGaAs quantum well, and (iii) the lowest band GaAs/AlGaAs quantum wire. Express your answer in terms of energy, centimeter, and electron volts.

3.7. Consider Fig. P3.7, where the energy level is sketched in GaAs/AlGaAs quantum well. The Fermi energy is shown to be above the bound state, E_1.

1. Calculate the Fermi Energy position for a 2D electron density of 4×10^{12} cm^{-2} at $T = 300$, 77, and 4.2 K.
2. Calculate the Fermi energy levels at $T = 300$ and 77 K for the following 2D electron density: 3×10^{11} cm^{-2}, 1×10^{12} cm^{-2}, and 5×10^{13} cm^{-2}.

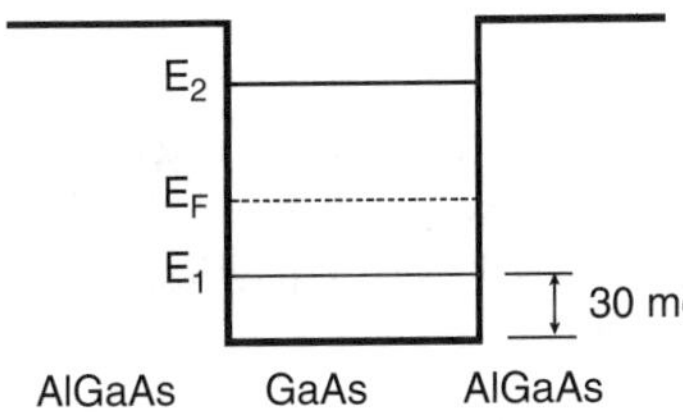

Fig. P3.7

3.8. Use the general definition of the density of state as described by the summation of δ-functions (Eq. 3.85) to derive the density of states for bulk semiconductors (3D system), quantum wells (2D system), and quantum wires (1D system).

3.9. For the Bose–Einstein distribution function (Eq. 3.45), assume that the total number of the particles, N, with spin zero and mass m in a two-dimensional system is constant. Derive an expression for the parameter α.

3.10. The electron density in bulk GaAs can be written as $n_0 = \frac{1}{2\pi^2}\left(\frac{2m_e^*kT}{\hbar^2}\right)^{3/2}\int_{E_c}^{\infty}\frac{\sqrt{E-E_c}\mathrm{d}E}{(1+\mathrm{e}^{(E-E_\mathrm{F})/kT})} = N_\mathrm{c}F_{1/2}(\eta)$, where $F_{1/2}(\eta)$ is Fermi integral and for $\eta > 1.25$ is it approximated as $F_{1/2}(\eta) = (4/3)\pi^{3/2}\eta^{3/2} + (\pi^{3/2}/6)\eta^{1/2}$, where $\eta = (E_\mathrm{F} - E_\mathrm{c})/kT$. Plot the Fermi energy as a function of electron density for $T = 300$, 200, 100, 77, and 4.2 K.

3.11. The electron thermal energy in the conduction band of GaAs can be expressed as $k_B T$. Plot the magnetic field required to split the energy levels into Landau levels as a function of temperature. From the graph, find the magnetic field required to generate Landau levels at 4.2, 77, and 300 K.

3.12. Consider a quantum dot to be a cubic quantum box with a finite potential, V_0, outside the well. For bound states in the quantum well, the energy $E \leq 0$. Assume that the density of states is 3D-like inside the well. Calculate the number of state inside the quantum dot for $V_0 = 0.6$ eV and for $L_x = L_y = L_z = 150$ Å.

3.13. Plot Equation 3.107 for several values of the electric field. When do you start to see an effect on the band-edge absorption?

BIBLIOGRAPHY

Davies JH. The physics of low-dimensional semiconductors: an introduction. Cambridge, UK: Cambridge University Press; 1998.

Davies JH, Wilkins JW. Narrow electronic bands in high electric filed: static properties. Phys Rev B 1988;38:1667.

Fox M. Optical properties of solids. Oxford: Oxford University Press; 2001.

Harrison P. Quantum wells, wires and dots. New York: Wiley; 2000.

Huang K. Statistical mechanics. New York: Wiley; 1963.

McQuarrie DA. Statistical mechanics. New York: Harper & Raw; 1973.

Mitin VV, Kochelap VA, Stroscio MA. Quantum heterostructures microelectronics and optoelectronics. Cambridge: Cambridge University Press; 1999.

Rief F. Fundamentals of statistical and thermal physics. New York: McGraw-Hill; 1965.

4

OPTICAL PROPERTIES

The optical properties of any material are the result of photon interaction with the constituents of the material. The aim of this chapter is to describe photon interaction with semiconductor materials, including nanoscale material systems, which leads to effects that are the basis for many technologies, such as detectors, emitters, optical communications, display panels, and optical oscillators. The interaction of photons with electrons in semiconductor materials is the most important interaction that gives rise to many phenomena. Electrons in semiconductor materials can absorb photons and be excited from the valence band to the conduction band. This transition is called *interband* transition. The inverse of this process occurs when electrons decay from a higher energy level, such as a conduction band, to a lower energy level, such as a valence band, and as a result photons are emitted. This is the basis for light-emitting diodes (LEDs) and laser diodes. Electrons can absorb photons and be excited from one state to another within a particular band, such as a conduction band. This transition is called an *intraband* transition. In nanoscale material systems, such as quantum wells, wires, and dots, electrons can be excited by photons and jump from one confined energy level to another. When the electrons are excited from one bound state to another in the conduction band of a quantum well, as an example, the transition is called an *intersubband* transition. These terminologies are also applied to heavy or light holes in semiconductors. These transitions are illustrated in Fig. 4.1 for both a bulk material (Fig. 4.1a) and a quantum structure (Fig. 4.1b).

The band-to-band transition in a bulk material is usually referred to as the *optical band gap*. In the case of a quantum structure, the conventional optical

Introduction to Nanomaterials and Devices, First Edition. Omar Manasreh.

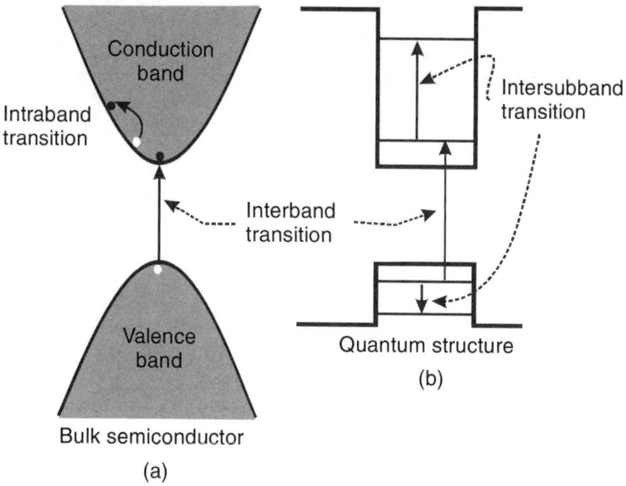

FIGURE 4.1 Illustrations of various electronic transitions are shown for (a) a bulk semiconductor material and (b) a quantum structure.

band gap is no longer allowed and the effective band gap is referred to as the *transition* from the ground state in the valence band to the ground state in the conduction band, as illustrated in Fig. 4.1b. Thus, the effective band gap in quantum structures is larger than the conventional optical band gap in bulk materials. If an electron is excited from the valence band to the conduction band of a semiconductor, it leaves behind a positively charged hole. This process is called *electron–hole pair generation*. When the electron and hole interact with each other because of Coulomb interaction, the result is called an *exciton*. The excitonic energy levels are usually formed in the fundamental band gap as shown in Fig. 4.2a. The exciton may move about the crystal. In this case, the electron–hole pair is called a *free exciton* or *Wannier–Mott exciton*. If the exciton is trapped by an impurity or an atom in the crystal, it is called a *bound exciton* or *Frenkel exciton* as shown in Fig. 4.2b. The binding energy of a free exciton is usually smaller than that of a bound exciton.

Many experimental techniques are used to probe electronic transitions in semiconductors. Essentially, the electron–photon interaction is the most dominant process in optoelectronic devices that is based on semiconductors and their nanostructures. In this chapter, we discuss various aspects of the optical properties of bulk semiconductors and nanoscale material systems.

4.1 FUNDAMENTALS

The interaction of photons with any material can be understood from Maxwell's classical electromagnetic theory. In MKS units, the four Maxwell equations that

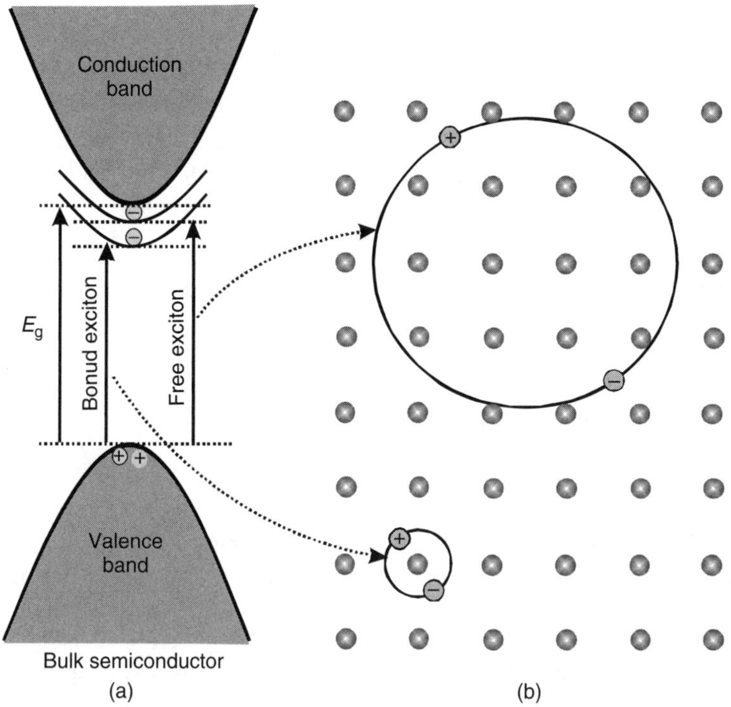

FIGURE 4.2 A schematic presentation of exciton energy levels with respect to the conduction band is shown in (a). The bound electron–hole pairs are shown in (b) for both free and bound excitons.

govern electromagnetic phenomena are

$$\nabla \cdot \boldsymbol{\mathcal{E}} = \frac{\rho}{\epsilon_0} \tag{4.1a}$$

$$\nabla \times \boldsymbol{\mathcal{E}} = -\mu_0 \frac{\partial \boldsymbol{B}}{\partial t} \tag{4.1b}$$

$$\nabla \cdot \boldsymbol{B} = 0 \tag{4.1c}$$

$$\nabla \times \boldsymbol{B} - \epsilon_0 \frac{\partial \boldsymbol{\mathcal{E}}}{\partial t} = \boldsymbol{J}, \tag{4.1d}$$

where $\boldsymbol{\mathcal{E}}$ is the electric field, $\boldsymbol{B}$ is the magnetic field, ρ is the electric charge density, $\boldsymbol{J}$ is the electric current density, ϵ_0 is the permittivity of free space constant (8.854×10^{-12} F/m), and μ_0 is the permeability of free space constant ($4\pi \times 10^{-7}$ Wb/m A). For the interaction of electromagnetic waves with electrically polarized material, we have

$$\boldsymbol{D} = \epsilon_0 \boldsymbol{\mathcal{E}} + \boldsymbol{P}, \tag{4.2}$$

where $\boldsymbol{D}$ is the electric displacement vector, and $\boldsymbol{P}$ is the polarization vector. In the linear limit, the polarization vector can be written as

$$\boldsymbol{P} = \epsilon_0 \overleftrightarrow{\chi} \cdot \boldsymbol{\mathcal{E}}, \tag{4.3}$$

where $\overleftrightarrow{\chi}$ is the dielectric susceptibility tensor. Combining Equations 4.2 and 4.3, we have

$$\boldsymbol{D} = \epsilon_0 (1 + \overleftrightarrow{\chi}) \cdot \boldsymbol{\mathcal{E}} = \epsilon_0 \overleftrightarrow{\epsilon} \cdot \boldsymbol{\mathcal{E}}, \tag{4.4}$$

where $\overleftrightarrow{\epsilon} = 1 + \overleftrightarrow{\chi}$ is the dielectric tensor. These tensors can be written in terms of the scalar quantities ϵ and χ, such that $\overleftrightarrow{I}\epsilon = 1 + \overleftrightarrow{I}\chi$, where $\overleftrightarrow{I}$ is a unit tensor.

For a conductive medium, the current density is related to the electric field according to the following relation

$$\boldsymbol{J}_{\mathrm{T}} = \sigma \boldsymbol{\mathcal{E}}, \tag{4.5}$$

where σ is the electrical conductivity, which may be a complex quantity, and $\boldsymbol{J}_T$ is the total current density composed of both the steady state and the time-dependent current densities. For the optical properties of semiconductors, we are concerned with the time-dependent contribution to the current density. Hence, the steady-state contribution can be ignored. The time-dependant current density can now be written as

$$\boldsymbol{J}_{\mathrm{T}} = \boldsymbol{J} = \frac{\partial \boldsymbol{P}}{\partial t}. \tag{4.6}$$

Substituting Equation 4.6 into Equation 4.1d and using Equation 4.2, we have

$$\nabla \times \boldsymbol{B} = \frac{\partial \boldsymbol{D}}{\partial t}. \tag{4.7}$$

By substituting Equation 4.2 into Equation 4.1a, we have

$$\nabla \cdot \boldsymbol{\mathcal{E}} = \frac{\nabla \cdot (\boldsymbol{D} - \boldsymbol{P})}{\epsilon_0} = \frac{\rho}{\epsilon_0}$$

$$\therefore \nabla \cdot \boldsymbol{D} = \nabla \cdot \boldsymbol{P} + \rho. \tag{4.8}$$

Substituting the continuity equation $\frac{\partial \rho}{\partial t} + \nabla \cdot \boldsymbol{J} = 0$ into Equation 4.8, we obtain

$$\nabla \cdot \boldsymbol{D} = 0 \tag{4.9}$$

The wave equation for nonmagnetic materials can be derived by taking the curl of Equation 4.1b

$$\nabla \times \nabla \times \boldsymbol{\mathcal{E}} = -\mu_0 \frac{\partial}{\partial t} \nabla \times \boldsymbol{B}. \tag{4.10}$$

Substituting Equations 4.4 and 4.7 into Equation 4.10, we obtain

$$\nabla \times \nabla \times \boldsymbol{\mathcal{E}} = -\mu_0 \epsilon_0 \epsilon \frac{\partial^2 \boldsymbol{\mathcal{E}}}{\partial t^2}$$

$$= -\frac{\epsilon}{c^2} \frac{\partial^2 \boldsymbol{\mathcal{E}}}{\partial t^2}, \tag{4.11}$$

where c is the speed of light, given by $c = 1/\sqrt{\mu_0 \epsilon_0}$. Recall that $\nabla \times \nabla \times \boldsymbol{\mathcal{E}} = \nabla(\nabla \cdot \boldsymbol{\mathcal{E}}) - \nabla^2 \boldsymbol{\mathcal{E}}$ and $\nabla \cdot \boldsymbol{F} = 0$, hence Equation 4.11 becomes

$$\nabla^2 \boldsymbol{\mathcal{E}} = \frac{\epsilon}{c^2} \frac{\partial^2 \boldsymbol{\mathcal{E}}}{\partial t^2}. \tag{4.12}$$

A solution of this wave equation is a plane wave with the following form:

$$\boldsymbol{\mathcal{E}}(\boldsymbol{r}, t) = \boldsymbol{\mathcal{E}}_0 \exp[i(\boldsymbol{k} \cdot \boldsymbol{r} - \omega t)], \tag{4.13}$$

where $\boldsymbol{\mathcal{E}}_0$ is the amplitude of the electric field, $\boldsymbol{k}$ is the propagation vector, and ω is the angular frequency. Substituting the above solution into the wave equation (Eq. 4.12), one can obtain the following dispersion relation:

$$c^2 k^2 = \omega^2 \epsilon. \tag{4.14}$$

The dielectric constant, ϵ, is frequency dependent and its explicit form is required to evaluate the dispersion relation. Substituting Equation 4.14 back into Equation 4.13 yields

$$\boldsymbol{\mathcal{E}}(\boldsymbol{r}, t) = \boldsymbol{\mathcal{E}}_0 \exp\left[i\omega\left(\tfrac{\sqrt{\epsilon}}{c}\widehat{\boldsymbol{k}} \cdot \boldsymbol{r} - t\right)\right]. \tag{4.15}$$

The dielectric constant is related to the refractive index, $n_r(\omega)$, according to the following relation:

$$n_{\mathrm{r}}(\omega) = \sqrt{\epsilon(\omega)}. \tag{4.16}$$

Both the refractive index and the dielectric constant are complex numbers and can be written as

$$n_r(\omega) = n_1(\omega) + i n_2(\omega) \tag{4.17a}$$

$$\epsilon(\omega) = \epsilon_1(\omega) + i\epsilon_2(\omega), \tag{4.17b}$$

where n_1 and ϵ_1 are the real parts and n_2 and ϵ_2 *are the imaginary parts*. Substituting Equations 4.16 and 4.17a into Equation 4.15, we obtain

$$\boldsymbol{\mathcal{E}}(\boldsymbol{r}, t) = \boldsymbol{\mathcal{E}}_0 \exp\left[-\frac{\omega n_2(\omega)}{c}\widehat{\boldsymbol{k}} \cdot \boldsymbol{r}\right] \exp\left[i\omega\left(\tfrac{n_1}{c}\widehat{\boldsymbol{k}} \cdot \boldsymbol{r} - t\right)\right]. \tag{4.18}$$

The intensity, I, of the electromagnetic wave is related to the electric field, according to the following relation:

$$\begin{aligned} \boldsymbol{I} &\propto |\mathcal{E}(\boldsymbol{r},t)|^2 \\ &\propto |\mathcal{E}_0|^2 \exp\left[-\frac{2\omega n_2(\omega)}{c}\widehat{\boldsymbol{k}}\cdot\boldsymbol{r}\right] \\ &\propto |\mathcal{E}_0|^2 \exp[-\alpha(\omega)\widehat{\boldsymbol{k}}\cdot\boldsymbol{r}], \end{aligned} \tag{4.19}$$

where $\alpha(\omega)$ is the optical absorption coefficient and is defined according to the above equation as

$$\begin{aligned} \alpha(\omega) &= \frac{2\omega n_2(\omega)}{c} \\ &= \frac{\omega\epsilon_2(\omega)}{cn_1(\omega)}. \end{aligned} \tag{4.20}$$

The optical absorption coefficient can also be obtained by using Beer's law:

$$I(z) = I_0 \exp[-\alpha(\omega)z], \tag{4.21}$$

where $I(z)$ is the electromagnetic radiation intensity at a distant z inside the media and I_0 is the intensity at $z = 0$.

4.2 LORENTZ AND DRUDE MODELS

The classical Lorentz model is applicable to solids with band gaps. This model is analogous to the quantum mechanically treated interband transitions. The Lorentz model assumes that the electron is bound to the nucleus, similar to a mass attached to a spring. The motion of the jth electron in a solid can be described according to the following equation of motion:

$$m^*\frac{\mathrm{d}^2x_j}{\mathrm{d}t^2} + m^*\Gamma\frac{\mathrm{d}x_j}{\mathrm{d}t} + m^*\omega_0^2x_j = -e\mathcal{E}, \tag{4.22}$$

where m^* is the effective mass of the electron, Γ is a damping constant, and $\mathcal{E}$ is the electric field. The second term on the right-hand side of this equation represents various dampings, such as collisions, and the third term is Hooke's law restoring force. The time-dependent x and $\mathcal{E}$ can be taken as

$$\begin{aligned} x_j &= x_{oj}\exp[-i\omega t] \\ \mathcal{E} &= \mathcal{E}_0\exp[-i\omega t]. \end{aligned} \tag{4.23}$$

The solution of Equation 4.22 is thus given as follows:

$$x_j = \frac{e\mathcal{E}}{m^*[(\omega^2 - \omega_0^2) + i\Gamma\omega]}. \tag{4.24}$$

The induced dipole moment, $\boldsymbol{P}$, per unit volume, V, is given by

$$\boldsymbol{P} = -e\sum_j \frac{x_j}{V} = -\frac{Ne^2\mathcal{E}}{m^*[(\omega^2 - \omega_0^2) + i\Gamma\omega]}, \tag{4.25}$$

where N is the electron concentration. Using Equations 4.2 and 4.4, one can obtain the dielectric constant as

$$\epsilon(\omega) = 1 - \frac{Ne^2}{\epsilon_0 m^*[(\omega^2 - \omega_0^2) + i\Gamma\omega]}. \tag{4.26}$$

For electrically neutral solids with free electrons, there exists a plasma with equal concentrations of positive and negative charges. If the damping and Hooke's force are ignored in Equation 4.22, the plasma frequency can be obtained as

$$\omega_p^2 = \frac{Ne^2}{\epsilon_0 m^*} \tag{4.27}$$

Interband electronic transitions in semiconductors contribute to the dielectric constants and this contribution, labeled ϵ_∞, should be included in Equation 4.26. The final expression for the complex dielectric constant is

$$\epsilon(\omega) = \epsilon_\infty\left[1 - \frac{\omega_p^2}{(\omega^2 - \omega_0^2) + i\Gamma\omega}\right], \tag{4.28}$$

where ω_p is redefined as $\omega_\mathrm{p}^2 = \frac{Ne^2}{\epsilon_0\epsilon_\infty m^*}$. The real and imaginary parts of $\epsilon(\omega)$ can now be evaluated and given by the following expressions:

$$\epsilon_1(\omega) = \epsilon_\infty\left[1 - \frac{\omega_\mathrm{p}^2(\omega^2 - \omega_0^2)}{(\omega^2 - \omega_0^2)^2 + \Gamma^2\omega^2}\right] \tag{4.29a}$$

$$\epsilon_2(\omega) = \epsilon_\infty\frac{\omega_\mathrm{p}^2\Gamma\omega}{(\omega^2 - \omega_0^2)^2 + \Gamma^2\omega^2} \tag{4.29b}$$

Notice that $\sqrt{\epsilon_\infty} \approx n_r$ for $\omega \gg \omega_\mathrm{p}$. The absorption coefficient defined in Equation 4.20 can be rewritten as

$$\alpha(\omega) = \frac{\epsilon_\infty\omega^2\,\omega_\mathrm{p}^2\Gamma}{cn_1(\omega)(\omega^2 - \omega_0^2)^2 + \Gamma^2\,\omega^2}. \tag{4.30}$$

The absorption coefficient has a Lorentzian lineshape. In the Lorentz model, the optical absorption is derived for band-to-band transitions. This is a simplistic form of the optical absorption of interband transitions. In a real semiconductor material, there are many effects that have to be included when deriving this coefficient. For example, momentum matrix elements or the oscillator strengths need to be considered.

For $\Gamma \ll \omega$ and $\omega_0 = 0$, the absorption coefficient is reduced to

$$\alpha(\omega) \approx \frac{\epsilon_\infty \omega_p^2 \Gamma}{c n_1(\omega)\omega^2}. \tag{4.31}$$

This absorption coefficient expression is actually the result of the Drude model. Notice that $\alpha(\omega) \propto \omega^{-2}$, which is the characteristic behavior of free electron absorption. The electrical conductivity can be obtained according to the Drude model by setting the restoring force (Hooke's law) in Equation 4.22 to zero. The solution of the equation of motion becomes

$$x_j = \frac{e\mathcal{E}}{m^*\omega(\omega + i\Gamma)}. \tag{4.32}$$

By taking the first derivative of x_j with respect to time, we have $\upsilon_j = \frac{\partial x_j}{\partial t} = -i\omega x_j$, where υ_j is the electron velocity. Substituting this derivative into Equation 4.32, one obtains

$$\upsilon_j = -\frac{ie\mathcal{E}}{m^*(\omega + i\Gamma)}. \tag{4.33}$$

On the other hand, the conduction current density is $J = -Ne\upsilon_j$. Multiply Equation 4.33 by the electron change and by the electron concentration to obtain

$$J = \frac{iNe^2\mathcal{E}}{m^*(\omega + i\Gamma)}. \tag{4.34}$$

One can see that the conductivity, $\sigma(\omega)$, is

$$\sigma(\omega) = \frac{iNe^2}{m^*(\omega + i\Gamma)}. \tag{4.35}$$

The real and imaginary parts of the dielectric constant in the Drude model can be obtained by setting $\omega_0 = 0$ in Equation 4.29. For $\omega = 0$, the conductivity is reduced to its DC value of

$$\sigma(\omega) = \frac{Ne^2}{m^*\Gamma} = \frac{Ne^2\tau}{m^*}, \tag{4.36}$$

where $\tau = \Gamma^{-1}$ and is designated as the scattering time.

4.3 THE OPTICAL ABSORPTION COEFFICIENT OF THE INTERBAND TRANSITION IN DIRECT BAND GAP SEMICONDUCTORS

A direct band gap semiconductor is characterized by having its valence band maximum and conduction band minimum at the same $\boldsymbol{k}$-value in reciprocal space, or momentum space, as shown in Fig. 4.3. It is customary to assume that this $\boldsymbol{k}$-value is zero, which is designated as the center of the first Brillouin zone and in group theory is labeled as Γ-point symmetry. Many authors have taken different approaches for the calculation of the optical absorption of interband transitions in semiconductors. In our case, we follow the steps taken by Balkanski *et al*. The absorption coefficient is defined according to Beer's law, as shown in Equation 4.21. By taking the first derivative of the light intensity with respect to z, we have

$$\frac{\mathrm{d}I(z)}{\mathrm{d}z} = -I(z)\alpha(\omega). \tag{4.37}$$

For a sample with a cross-sectional area of A, the rate of energy absorption is

$$\frac{\mathrm{d}E}{\mathrm{d}t} = -A\mathrm{d}I = I\alpha(\omega)A\mathrm{d}z, \tag{4.38}$$

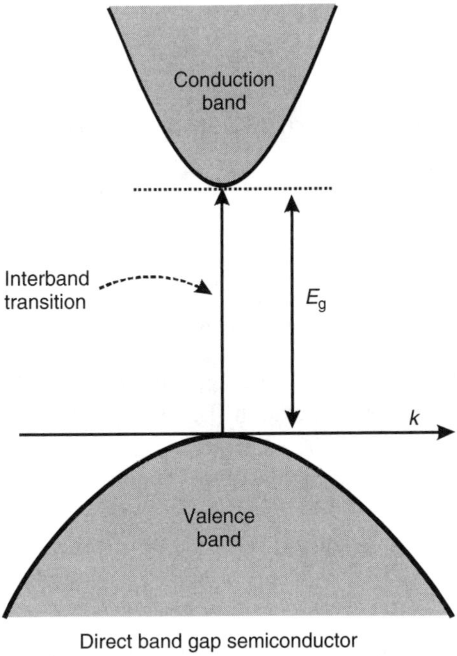

FIGURE 4.3 A sketch of the direct band gap energy showing the vertical interband transition.

where dI is the change in the light intensity after passing through the sample. The rate of energy absorption can also be written as

$$\frac{\mathrm{d}E}{\mathrm{d}t} = \hbar\omega W_{\mathrm{vc}}, \tag{4.39}$$

where $\hbar\omega$ is the photon energy and $W_{\mathrm{vc}}^{\mathrm{t}}$ is the total transition probability of an electron transition from a valence band state to a conduction band state. The absorption coefficient can be rewritten by combining Equations 4.38 and 4.39 as

$$\alpha(\omega) = \frac{\hbar\omega W_{\mathrm{vc}}^{\mathrm{t}}}{I A \mathrm{d}z}. \tag{4.40}$$

The two major issues that need to be determined in this equation are the $I(z)$ and W_{vc}. The intensity can be obtained by assuming that it is the mean value of the Poynting vector $\boldsymbol{S} = \boldsymbol{\mathcal{E}} \times \boldsymbol{B}$. The electric and magnetic fields can be written in terms of the vector potential $\boldsymbol{A}$ as $\epsilon = -\frac{\partial A}{\partial t}$ and $\mu_0 \boldsymbol{B} = \nabla \times \boldsymbol{A}$. Let $\boldsymbol{A} = \boldsymbol{A}_0 \cos(\boldsymbol{k} \cdot \boldsymbol{r} - \omega t)$. The electric and magnetic fields can now be written as

$$\boldsymbol{\mathcal{E}} = -\omega \boldsymbol{A_0} \sin(\boldsymbol{k} \cdot \boldsymbol{r} - \omega t) \tag{4.41a}$$

$$\mu_0 B = -\boldsymbol{k} \times \boldsymbol{A_0} \sin(\boldsymbol{k} \cdot \boldsymbol{r} - \omega t). \tag{4.41b}$$

The Poynting vector takes the following form:

$$\boldsymbol{S} = \frac{\omega}{\mu_0} \boldsymbol{A_0} x (kx\boldsymbol{A_0} \sin^2(\boldsymbol{k} \cdot \boldsymbol{r} - \omega t)). \tag{4.42}$$

Thus, the intensity I can be written as

$$I = \langle \boldsymbol{S} \rangle = \frac{\omega^2 n_{\mathrm{r}}(\omega)}{2\mu_0 c} |\boldsymbol{A_0}|^2, \tag{4.43}$$

where $\langle \boldsymbol{S} \rangle$ is the time average of $\boldsymbol{S}$ over one period. In reaching the form shown in Equation 4.43, the dispersion relation (Eq. 4.14) and the vector analysis identity $\boldsymbol{A} \times (\boldsymbol{B} \times \boldsymbol{C}) = (\boldsymbol{A} \cdot \boldsymbol{C})\boldsymbol{B} - (\boldsymbol{A} \cdot \boldsymbol{B})\boldsymbol{C}$ were used assuming that $\boldsymbol{k}$ and $\boldsymbol{A_0}$ are orthogonal. Substituting Equation 4.41 into Equation 4.40, we have

$$\alpha(\omega) = \frac{2\mu_0 c\hbar W_{\mathrm{vc}}^{\mathrm{t}}}{\omega n_{\mathrm{r}} |\boldsymbol{A_0}|^2 V} = \frac{2\hbar^2}{\epsilon_0 c n_{\mathrm{r}} |\boldsymbol{A_0}|^2 V} \frac{1}{\hbar\omega} W_{\mathrm{vc}}^{\mathrm{t}}. \tag{4.44}$$

The next step is to evaluate $W_{\mathrm{vc}}^{\mathrm{t}}$. The approach here is to evaluate the transition probability (W_{vc}) from one Bloch state in the valence band to another Bloch state in the conduction band. In obtaining the transition probability, the photon–electron interaction can be treated as perturbation in the following Hamiltonian

$$H = \frac{(\boldsymbol{p} + e\boldsymbol{A})^2}{2m_0} + V(\boldsymbol{r})$$
$$= \frac{\boldsymbol{p}^2}{2m_0} + \frac{e}{2m_0}(\boldsymbol{p}.\boldsymbol{A} + \boldsymbol{A}.\boldsymbol{p}) + \frac{e}{2m_0}\boldsymbol{A}^2 + V(\boldsymbol{r}), \quad (4.45)$$

where $\boldsymbol{p}$ is the momentum, $\boldsymbol{A}$ is the vector potential, and $V(\boldsymbol{r})$ is the electron potential energy. For low light intensity, $\boldsymbol{A}^2$ can be neglected. Because the photon momentum is negligible, the term rising from $\boldsymbol{p}$ acting on $\boldsymbol{A}$ can also be neglected. Notice that we use the free electron mass instead of the electron effective mass. The photon–electron interaction Hamiltonian can be written as

$$H' = \frac{e}{2m_0}\boldsymbol{A}.\boldsymbol{p}. \quad (4.46)$$

The transition probability of an electron from the valence band to the conduction band is given by the Fermi golden rule:

$$W_{\mathrm{vc}} = \frac{2\pi}{\hbar}|\langle k'c|H'|kv\rangle|^2\delta(E_{k'c} - E_{kv} - \hbar\omega), \quad (4.47)$$

where $\langle k'c|$ *and* $|kv\rangle$ are Bloch states in the conduction and valence bands, respectively. The δ-function is included to conserve the energy. If $\boldsymbol{A}$ has the following form $\boldsymbol{A} = \boldsymbol{A_0}\cos(\boldsymbol{k} \cdot \boldsymbol{r} - \omega t)$, the interband matrix element of H' can be written as

$$\langle k'c|H'|kv\rangle = \frac{e}{2m_0}\langle k'c|\boldsymbol{A_0}.\boldsymbol{p}|kv\rangle. \quad (4.48)$$

Substitute Equation 6.48 into Equation 6.47 to obtain

$$W_{\mathrm{vc}} = \frac{\pi e^2}{2\hbar m_0^2}|\boldsymbol{A_0}|^2|\langle k'c|\boldsymbol{p}_A|kv\rangle|^2\delta(E_{k'c} - E_{kv} - \hbar\omega), \quad (4.49)$$

where $\boldsymbol{p}_A$ is the momentum component along the $\boldsymbol{A}$ direction. In order for Equation 6.49 to be evaluated, the Bloch form of the valence and conduction bands are used, such as

$$\langle k'c|\boldsymbol{p}_A|kv\rangle = \int_{\mathrm{crystal}} \mathrm{e}^{-ik'\cdot r}\varphi^*_{k'c}(\boldsymbol{r})\boldsymbol{p}_A\mathrm{e}^{\boldsymbol{ik}.\boldsymbol{r}}\varphi_{kv}(\boldsymbol{r})d^3\boldsymbol{r}. \quad (4.50)$$

Taking the integral over one primitive unit cell and summing over all the unit cells, the above equation can be rewritten as

$$\begin{aligned}\langle k'c|\boldsymbol{p}_A|kv\rangle &= \sum_l \int_l \mathrm{e}^{-ik'.r}\varphi^*_{k'c}(\boldsymbol{r})\boldsymbol{p}_A \mathrm{e}^{ik.r}\varphi_{kv}(\boldsymbol{r})\mathrm{d}^3\boldsymbol{r} \\ &= \sum_l \int_l \mathrm{e}^{i(k-k').r}\varphi^*_{k'c}(r)(\hbar k_A + \boldsymbol{p}_A)\varphi_{kv}(\boldsymbol{r})\mathrm{d}^3 r,\end{aligned} \tag{4.51}$$

where the summation is over all unit cells and the integration is over one unit cell labeled l. The term $\hbar \boldsymbol{k}_A$ is the result of the $\boldsymbol{p}_A$ operation on the exponential part of the wave function and $\boldsymbol{k}_A$ is the component of the wave vector in the $\boldsymbol{A}$ direction. Utilizing the periodicity of the crystal, where $\boldsymbol{r} = \boldsymbol{R}(l) + \boldsymbol{r}'$ and $\varphi^*_{nk}(\boldsymbol{r}) = \varphi^*_{nk}(\boldsymbol{R}(l) + \boldsymbol{r}') = \varphi^*_{nk}(\boldsymbol{r}')$, Equation 4.51 can be written as

$$\langle k'c|p_A|kv\rangle = \sum_l \mathrm{e}^{i(k-k').R(l)} \int_{\text{cell } 0} \mathrm{e}^{i(k-k')\cdot \boldsymbol{r}'}\varphi^*_{k'c}(\boldsymbol{r}')(\hbar \boldsymbol{k}_A + \boldsymbol{p}_A)\varphi_{kv}(\boldsymbol{r}')\mathrm{d}^3\boldsymbol{r}', \tag{4.52}$$

where the sum is over all unit cells in the crystal and the integral is over the unit cell labeled “0.” The sum over all unit cells obviously gives the number of cells in the crystal. The wave functions $\varphi_{k'c}(\boldsymbol{r})$ and $\varphi_{kv}(\boldsymbol{r})$ are orthogonal, which means that the first term of the integral in Equation 4.52 is zero. The momentum matrix element can now be written as

$$\langle k'c|\boldsymbol{p}_A|kv\rangle = \mathrm{N}\delta_{k,k'} \int_{\text{cell } 0} \varphi^*_{kc}(\boldsymbol{r}')\boldsymbol{p}_A\varphi_{kv}(\boldsymbol{r}')\mathrm{d}^3\boldsymbol{r}', \tag{4.53}$$

where N is the total number of unit cells in the crystal. The momentum matrix element can now be written as

$$W_{\text{vc}} = \frac{\pi e^2}{2\hbar m_0^2}|\boldsymbol{A}_0|^2 P^2 \delta(E_{k'c} - E_{kv} - \hbar\omega)\delta_{k,k'}, \tag{4.54}$$

where

$$P = \mathrm{N} \int_{\text{cell } 0} \varphi^*_{kc}(\boldsymbol{r}')p_A\varphi_{kv}(\boldsymbol{r}')\mathrm{d}^3\boldsymbol{r}'. \tag{4.55}$$

The quantity P is a number known for many semiconductors. For example, $2P^2/m_0 \approx 25.7$ (eV) for GaAs, ≈ 20.9 (eV) for InP, and ≈ 22.2 (eV) for InAs (Singh, 2003). The δ-function, $\delta_{k,k'}$, in Equation 4.54 gives the selection rules for the direct transition from the valence band to the conduction band.

The total transition probability from the valence band to the conduction band over all $\boldsymbol{k}$ and $\boldsymbol{k}'$ is obtained summing over all $\boldsymbol{k}$ and $\boldsymbol{k}'$,

$$W_{\rm vc}^{\rm t} = 2\sum_{\rm k}\sum_{\rm k'} W_{\rm vc} f_{\rm FD}^{\boldsymbol{k}v}(1 - f_{\rm FD}^{\boldsymbol{k}'c}), \tag{4.56}$$

where $f_{\rm FD}^{\boldsymbol{k}v}$ and $(1 - f_{\rm FD}^{\boldsymbol{k}'c})$ are Fermi–Dirac distribution functions for a full valence band and an empty conduction band, respectively. The factor 2 is added to account for the electron spin degeneracy. For a two-band model at $T = 0$ K, we have $f_{\rm FD}^{\boldsymbol{k}v} = 1$, $f_{\rm FD}^{\boldsymbol{k}'c} = 0$ and

$$\begin{aligned} E_{\rm kc} - E_{\rm kv} &= E_{\rm g} + \frac{\hbar^2 k^2}{2m_{\rm c}^*} + \frac{\hbar^2 k^2}{2m_v^*} \\ &= E_{\rm g} + \frac{\hbar^2 k^2}{2m_{\rm r}^*}, \end{aligned} \tag{4.57}$$

where m_r^* is the reduced mass of the electron and hole system given by

$$\frac{1}{m_{\rm r}^*} = \frac{1}{m_{\rm c}^*} + \frac{1}{m_{\rm v}^*}. \tag{4.58}$$

Putting all these together, Equation 4.56 can be rewritten as (Balkanski *et al.*)

$$W_{\rm vc}^{\rm t} = \frac{e^2 V}{8\pi^2 \hbar m_0^2} |\boldsymbol{A}_0|^2 P^2 \int \delta(E_{kc} - E_{\rm k}v - \hbar\omega) {\rm d}^3 k, \tag{4.59}$$

where V is the volume of the semiconductor sample. The integral in Equation 4.59 can be evaluated using spherical coordinates as

$$\begin{aligned} \int \delta(E_{kc} - E_{kv} - \hbar\omega){\rm d}^3 k &= 4\pi \int_0^\infty k^2 \delta\left(E_{\rm g} - \hbar\omega + \frac{\hbar^2 k^2}{2m_{\rm r}^*}\right) {\rm d}k \\ &= 4\pi \sqrt{\frac{2{\rm m}_{\rm r}^*}{\hbar^2}} \int_0^\infty \sqrt{E}\delta\left(E_{\rm g} - \hbar\omega + \frac{\hbar^2 k^2}{2m_{\rm r}^*}\right) k{\rm d}k \\ &= 4\pi \sqrt{\frac{2{\rm m}_{\rm r}^*}{\hbar^2}\frac{m_{\rm r}^*}{\hbar^2}} \int_0^\infty \sqrt{E}\delta(E_{\rm g} - \hbar\omega + E)\,{\rm d}E \qquad (4.60) \\ &= 4\pi \sqrt{\frac{2{\rm m}_{\rm r}^*}{\hbar^2}\frac{m_{\rm r}^*}{\hbar^2}} \sqrt{\hbar\omega - E_{\rm g}} \\ &= 2\pi \left(\frac{2m_{\rm r}^*}{\hbar^2}\right)^{3/2} \sqrt{\hbar\omega - E_{\rm g}}. \end{aligned}$$

This expression is valid for $\hbar\omega \geq E_g$ and it is zero for $\hbar\omega \leq E_g$. Substituting Equations 4.60 and 4.59 into Equation 4.44, we obtain the following expression for the optical absorption coefficient of direct interband transition in a bulk semiconductor:

$$\alpha(\omega) = \frac{2\mu_0 c\hbar W_{vc}^{t}}{\omega n_r |\boldsymbol{A_0}|^2 V} = \frac{e^2\hbar}{2\pi\epsilon_0 c n_r m_0^2} P^2 \left(\frac{2m_r^*}{\hbar^2}\right)^{3/2} \frac{1}{\hbar\omega}\sqrt{\hbar\omega - E_g}. \qquad (4.61)$$

A plot of the absorption coefficient given by Equation 4.61, using GaAs parameters, is shown in Fig. 4.4. The energy is plotted for $\hbar\omega \geq E_g$. The optical absorption coefficient can be expressed in terms of the oscillator strength f_{vc}, which is defined as $f_{vc} = \frac{2P^2}{m_0\hbar\omega}$. The maximum value of f_{vc} can be obtained from the sum rule (see, for example, Wooten) as

$$f_{vc} \approx \begin{cases} \left|1 - \dfrac{m_0}{m_e^*}\right| & \text{for electron} \\ 1 + \dfrac{m_0}{m_h^*} & \text{for hole.} \end{cases} \qquad (4.62)$$

When the direct interband transition is forbidden at $\boldsymbol{k} = 0$, but allowed at $\boldsymbol{k} \neq 0$, the optical absorption coefficient depends on the photon energy as $\alpha(\omega) \propto \frac{(\hbar\omega - E_g)^{3/2}}{\hbar\omega}$ (Pankove). It should be noted that Fig. 4.4 does not include the absorption from either excitons or absorption from other valleys.

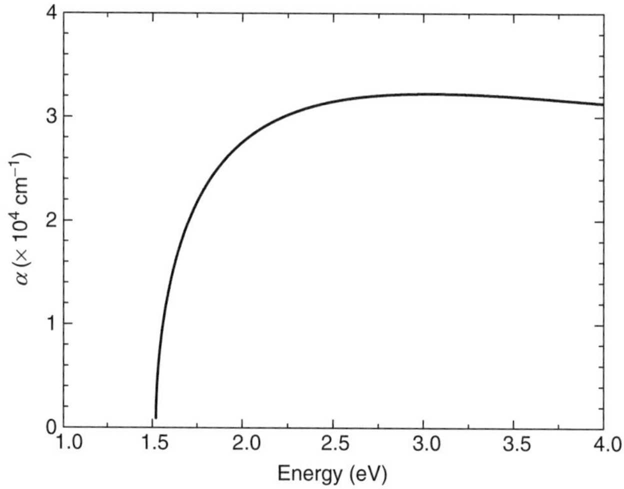

FIGURE 4.4 The optical absorption coefficients as a function of the photon energy for GaAs bulk material. The band gap was chosen at 1.52 eV.

4.4 THE OPTICAL ABSORPTION COEFFICIENT OF THE INTERBAND TRANSITION IN INDIRECT BAND GAP SEMICONDUCTORS

The interband transition in an indirect band gap semiconductor such as Si occurs between the valence band maximum and conduction band minimum that are located at different $\boldsymbol{k}$-values, as shown in Fig. 4.5. In a direct band gap semiconductor, the interband transition is excited by only electron–photon interaction. The interband transition of an indirect band gap semiconductor requires electron–photon and electron–phonon interactions. A phonon is the quanta of lattice vibrations. Thus, momentum and energy conservation require that

$$\begin{aligned} \boldsymbol{k}_{\mathrm{c}} &= \boldsymbol{k}_{\mathrm{v}} \pm \boldsymbol{q} \\ \hbar\omega &= E_{\mathrm{c}} - E_{\mathrm{v}} \pm \hbar\omega_{\mathrm{p}}, \end{aligned} \tag{4.63}$$

where $\boldsymbol{q}$ the phonon wave vector and $\hbar\omega_{\mathrm{p}}$ is the phonon energy. The "plus" and "minus" signs are for emission or absorption of phonons, respectively. The optical absorption coefficient can be derived in a manner similar to that of the direct band gap semiconductor. However, owing to the electron–phonon interaction, several steps have to be modified. For example, the electron–phonon matrix element must be included in the analysis. Furthermore, the argument of the δ-function must include the phonon energy. The number of phonons (from Bose–Einstein

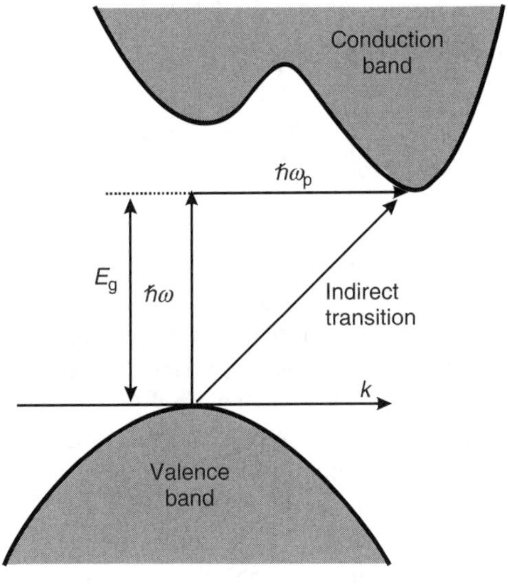

FIGURE 4.5 A sketch of an indirect band gap semiconductor showing the indirect interband transition.

statistics) must also be included. The final result is

$$\alpha(\omega) = \frac{A(\hbar\omega + \hbar\omega_{\mathrm{p}} - E_{\mathrm{g}})^n}{\exp(\hbar\omega_{\mathrm{p}}/kT) - 1} + \frac{B\exp(\hbar\omega_{\mathrm{p}}/kT)(\hbar\omega - \hbar\omega_{\mathrm{p}} - E_{\mathrm{g}})^n}{\exp(\hbar\omega_{\mathrm{p}}/kT) - 1}, \quad (4.64)$$

where A and B are constants, $n = 2$ when a vertical transition is allowed, and $n = 3$ when vertical transitions are not allowed (Wooten).

4.5 THE OPTICAL ABSORPTION COEFFICIENT OF THE INTERBAND TRANSITION IN QUANTUM WELLS

A typical example of an interband transition in a type I quantum well is shown in Fig. 4.1b, where the electrons are excited from the bound ground state in the valence band to the bound ground state in the conduction band. The steps used to calculate the interband transition in quantum wells are similar to those discussed above for optical absorption of the interband transition in direct band gap bulk semiconductor materials. There are, however, a few modifications that must be included as discussed below.

(i) The density of states appears explicitly in Equation 4.61. This form is given by Equation 4.59 except that the effective mass is replaced by the reduced mass. Hence, the notation *reduced density of states* is introduced. For a two-dimensional system, such as a multiple quantum well, the reduced density of states $\left[\frac{1}{2\pi^2}\left(\frac{2m_{\mathrm{r}}^*}{\hbar^2}\right)^{3/2}\sqrt{\hbar\omega - E_{\mathrm{g}}}\right]$ needs to be replaced by the reduced two-dimensional density of states

$$\frac{g_{\mathrm{cv}}^{\mathrm{2D}}}{L} = \frac{m_{\mathrm{r}}^*}{\pi\hbar^2 L}\sum_{m.n}\langle g_{\mathrm{v}}^m \mid g_{\mathrm{c}}^n\rangle\Theta(E_{nm} - \hbar\omega), \quad (4.65)$$

where

$$E_{nm} = E_{\mathrm{g}} + E_{\mathrm{c}}^n + E_{\mathrm{v}}^m, \quad (4.66)$$

$\langle g_{\mathrm{v}}^m \mid g_{\mathrm{c}}^n\rangle$ is the overlap integral between the z-dependent envelop functions of the conduction band and the valence band, L is the width of the quantum well, and Θ is the Heaviside step function. Notice that E_{nm} is the photon energy required to excite the interband transition in the quantum well. The energies E_{c}^n and E_{v}^m correspond to the energies of the ground bound states in the conduction and valence bands, respectively. The width of the well is introduced in Equation 4.65 to account for the transformation of the momentum matrix element (Eq. 4.55) as it goes from the three-dimensional system to the two-dimensional system.

(ii) The number of wells, N_{w}, should be included in the final expression of the optical absorption coefficient.

(iii) The absorption coefficient is calculated for the well only.

(iv) The overlap integral defined in (1) provides the selection rules for the transition. Let us assume that the envelope function has the following form

$$F_{nk_\perp}(r) = e^{ik_\perp \cdot r_\perp} \chi_n(z). \tag{4.67}$$

The overlap integral can now be written as

$$\langle g_v^m \mid g_c^n \rangle = \langle m\mathbf{k}_\perp \mid n\mathbf{k}'_\perp \rangle = \delta_{\mathbf{k}_\perp,\mathbf{k}'_\perp} \int_{-L/2}^{L/2} \chi_m^h(z)\chi_n^e(z)\, dz. \tag{4.68}$$

Thus, the overlap integral is nonzero if and only if $\chi_m^h(z)$ and $\chi_n^e(z)$ are both odd parity or both even parity.

By considering all these modifications, the optical absorption coefficient of the interband transition in type I multiple quantum wells can be written as

$$\alpha(\omega) = \frac{e^2 N_w m_r^*}{2\epsilon_0 c n_r \hbar L m_0^2} \frac{P^2}{\hbar\omega} \sum_{n,m} \Theta(E_{nm} - \hbar\omega). \tag{4.69}$$

We assumed that the square of the overlap integral (Eq. 4.68) is unity and, therefore, it was not included in the absorption coefficient expression. Equation 4.69 is valid for both heavy holes and light holes. The only difference is that the reduced mass, m_r^*, is different. Notice that $\hbar\omega$ in the denominator is the minimum photon energy needed to cause an electronic transition from the valence band to the conduction band within the quantum well. If the definition of the oscillator strength, $f_{vc} = \frac{2P^2}{m_0\hbar\omega}$, where its maximum value is given by Equation 4.62, then the optical absorption coefficient can be rewritten as

$$\alpha(\omega) = \frac{e^2 N_w m_r^*}{4\epsilon_0 c n_r \hbar L m_0} f_{vc} \sum_{n,m} \Theta(E_{nm} - \hbar\omega). \tag{4.70}$$

For example, the oscillator strength for an electronic transition in a GaAs/AlGaAs quantum well is $f_{vc} \approx |1 - 1/0.067| = 13.925$. This quantity is comparable to the value obtained from $f_{vc} = \frac{2P^2}{m_0\hbar\omega} = 25.7/1.75 = 14.68$ for $\hbar\omega = 1.75$ eV. A plot of the absorption coefficient depicted in Equation 4.70 is shown in Fig. 4.6. The ladderlike behavior of the optical absorption is due to the step function, which is the characteristic signature of the reduced density of states in the quantum well. An example of the absorbance spectrum of interband transitions in multiple quantum wells is shown in Fig. 4.7, where the spectrum was recorded at 4.2 K for InSb/InAlSb multiple quantum wells. The sharpest peak at 0.234 eV is due to the Hh1–CB1 transition.

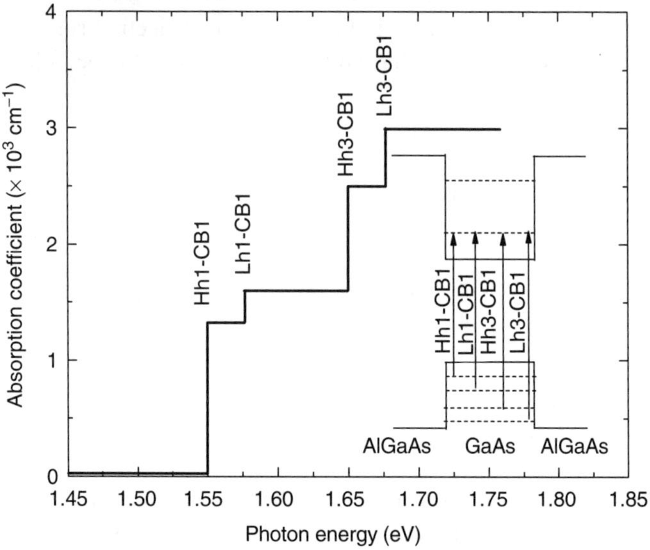

FIGURE 4.6 The optical absorption coefficient in 30 Å GaAs/AlGaAs quantum well plotted as a function of photon energy.

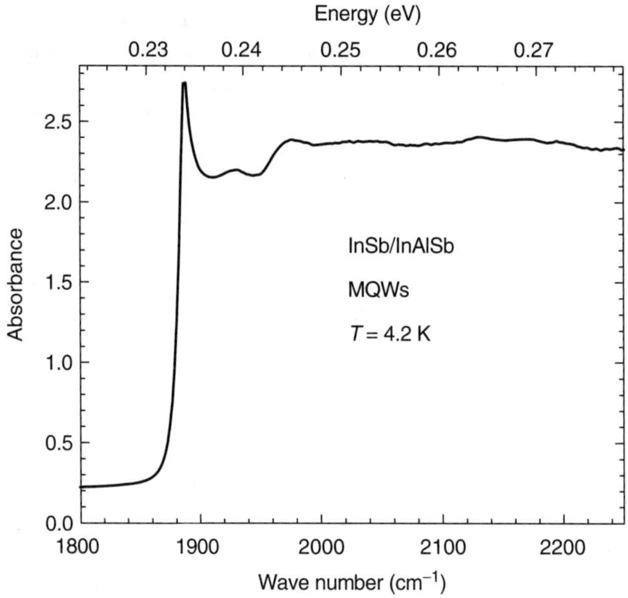

FIGURE 4.7 Absorbance spectra of InSb/InAlSb multiple quantum well measured at 4.2 K. The sharp peak at 0.234 eV is a transition from the heavy hole round state (Hh1) to the electron ground state (CB1).

4.6 THE OPTICAL ABSORPTION COEFFICIENT OF THE INTERBAND TRANSITION IN TYPE II SUPERLATTICES

A typical example of a type II superlattice is an InAs/InGaSb structure as shown in Fig. 4.8. In this figure, we plotted the conduction band as a thick line and the valence band as a thin line. The bound states are shown as the dotted lines and the wave functions for the ground bound states in the conduction band are plotted as the dashed lines. The barrier material is grown thin enough to allow the wave functions to overlap, forming what are called minibands for both the holes and the electrons. The wave functions for the holes are not shown. The intriguing property of this system is that the interband transitions are indirect in real space as indicated by the arrows in the figure, but the system exhibits a direct band gap in $\boldsymbol{k}$-space. The energy dependence of the optical absorption of the band-to-band transitions can thus be described by the Pankove expression, $\alpha(\omega) \propto (\hbar\omega - E_g)^{3/2}/\hbar\omega$. This energy dependence is valid at least near the band edge.

The analysis of the optical absorption coefficient for the type II superlattice is more complicated than interband transitions in type I superlattices. This is due to the fact that the overlap integral is no longer unity. In this section, we simply report (for full derivation, see Bastard, 1988) the absorption coefficient of the interband transition between the ground state of the heavy hole in an InGaSb layer and the ground state of the electron in the conduction band of an InAs layer as follows:

$$\alpha(\omega) = \frac{e^2 m_r^* P^2 P_b(E_1)}{\pi \epsilon_0 n_r \, c m_0^2 \omega \, \hbar^2 L}\left[\frac{-x}{1+x^2} + \arctan(x)\right], \tag{4.71}$$

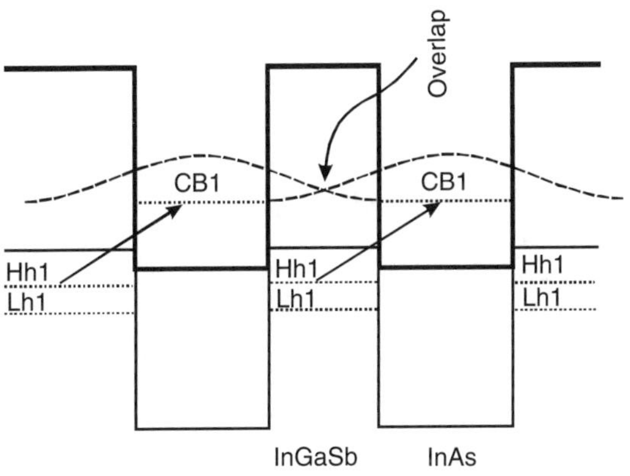

FIGURE 4.8 A sketch of the band alignment of the InAs/InGaSbsuperlattice is shown. Minibands are formed because of the overlap of the wave functions as indicated by the dashed line.

where $P_b(E_1)$ is the probability of finding the electron in an InAs layer while in the CB1 state (Fig. 4.8) and is given by

$$P_b(E_1) = \frac{B_c^2}{k_c}. \tag{4.72}$$

The parameters B_c and k_c are the amplitude of the envelope wave function of the electron in the barrier and the corresponding propagation vector, respectively. In other words, the electron envelope wave function in the barrier (InGaSb) is given by $\chi_1^e(z) = B_c \exp[-k_c(z - L/2)]$. The subscript "1" indicates the wave function for the ground state (CB1). The parameter x in Equation 4.71 is given by the following expression:

$$x = \sqrt{\frac{2m_h^*}{\hbar^2}} \frac{\sqrt{\hbar\omega - E_g^{InAs} + \Delta E_v - E_{CB1}}}{k_c}, \tag{4.73}$$

where ΔE_v is the valence band offset, E_g^{InAs} is the InAs band gap, and E_{CB1} is the electron ground state in the conduction band. The rest of the parameters in Equation 4.71 were defined previously. The lineshape of the optical absorption coefficient is defined by the behavior of the quantity in the square brackets, which is plotted in Fig. 4.9. The onset of the optical absorption profile occurs at $\hbar\omega_0 = E_g^{\mathrm{InAs}} - \Delta E_v + E_{\mathrm{CB1}}$. Above this onset, the lineshape appears to depend on the photon energy according to the following relation $\alpha(\omega) \sim (\hbar\omega - E_g)^{3/2}$.

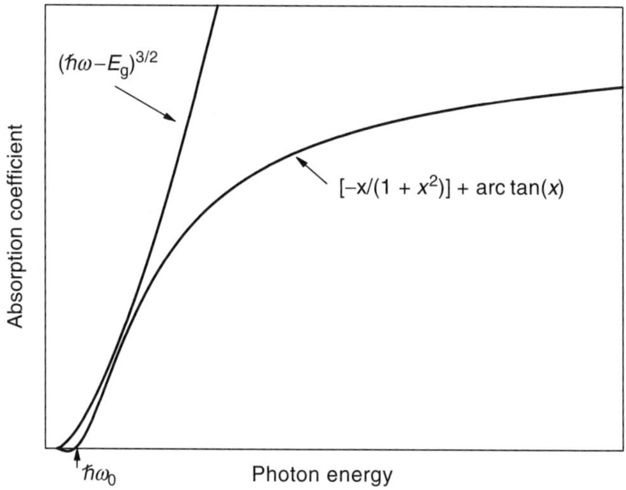

FIGURE 4.9 The lineshape of the optical absorption coefficient defined by Equation 4.71 is plotted as a function of photon energy ($\hbar\omega$). The lineshape defined by $(\hbar\omega—\epsilon_g)^{3/2}$ is also shown.

This energy dependence implies that the direct electronic transition is forbidden. The latter relation is plotted in Fig. 4.9 for comparison purposes.

4.7 THE OPTICAL ABSORPTION COEFFICIENT OF THE INTERSUBBAND TRANSITION IN MULTIPLE QUANTUM WELLS

Intersubband transitions in low dimensional quantum structures have been investigated for their infrared application as detectors and lasers. The intersubband transitions are generated in n- or p-type quantum well structures with at least one bound state as shown in Fig. 4.10. In this figure, we have shown a bound-to-bound transition, where both the ground state and the first excited stat are bound (Fig. 4.10a), a bound to continuum, where the ground state is bound while the first excited state is resonant in the conduction band (Fig. 4.10b), the transitions between the states depicted in $\boldsymbol{k}$-space (Fig. 4.10c), the optical absorption lineshape for the bound-to-bound transition (Fig. 4.10d), and the optical absorption profile for bound-to-continuum transition (Fig. 4.10e).

As indicated in Fig. 4.10d and e, the optical absorption profile of the electrons that undergo the intersubband transition from bound to bound is different from that of bound-to-continuum transition. Let us first obtain the optical absorption coefficient for the bound-to-bound transition, which has a Lorentzian lineshape. The envelope wave function for the two bound states in Fig. 4.10a can be written as

$$\psi_{nk_\perp}(\boldsymbol{r}) = \mathrm{e}^{i\boldsymbol{k}_\perp\cdot\boldsymbol{r}_\perp}\chi_n(z), \tag{4.74}$$

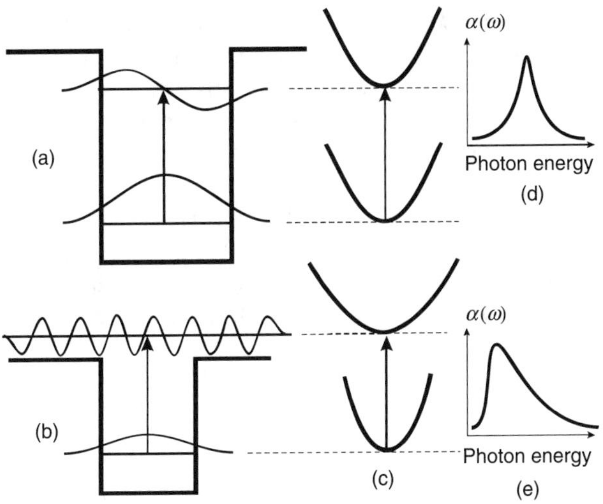

FIGURE 4.10 Illustration of the intersubband transition in n-type quantum well structures is shown for bound-to-bound and bound-to-continuum configurations. The transitions are shown in $\boldsymbol{k}$-space along with their optical absorption coefficient profiles.

where

$$\chi_n(z) = \sqrt{\frac{2}{L}}\cos(n\pi z/L), \quad \text{for } n = \text{odd integer} \tag{4.75a}$$

$$\chi_n(z) = \sqrt{\frac{2}{L}}\sin(n\pi z/L), \quad \text{for } n = \text{even integer.} \tag{4.75b}$$

L is the width of the thickness of the quantum well, $\boldsymbol{k}_\perp = \sqrt{k_x^2 + k_y^2}$, and $\boldsymbol{r}_\perp = x\widehat{\boldsymbol{x}} + y\widehat{\boldsymbol{y}}$. Following the procedure discussed in Section 4.3, the optical absorption coefficient can be written as

$$\alpha(\omega) = \frac{2\pi e^2 N_{\mathrm{w}}\hbar}{\epsilon_0 c n_{\mathrm{r}} m *^2 V}\frac{1}{\hbar\omega}\sum_{i,j}|\langle i|p_z|j\rangle|^2\delta(E_j - E_i - \hbar\omega)(f_{\mathrm{FD}}^i - f_{\mathrm{FD}}^j), \tag{4.76}$$

where f_{FD} is the Fermi–Dirac occupation functions and i and j are the initial and final states. The factor 2 is added for the electron spin degeneracy. There are a few approximations in this equation. First, the free electron mass is replaced by the electron effective mass. Second, the number of the quantum wells is included to account for the absorption from all quantum wells. Third, the momentum component is taken along the z-direction, which is the growth direction. Fourth, i and j are used, but they stand for $(n, \boldsymbol{k}_\perp)$ and $(n', \boldsymbol{k}'_\perp)$, respectively. The momentum matrix element can be expressed as

$$\langle n\boldsymbol{k}_\perp|p_z|n'\boldsymbol{k}'_\perp\rangle = \delta_{k_\perp,k'_\perp}\langle n|p_z|n'\rangle. \tag{4.77}$$

The wave vectors $\boldsymbol{k}_\perp$ and $\boldsymbol{k}'_\perp$ were removed from the bracket since they are not good quantum numbers for p_z to operate on them. The d-function was introduced to conserve the momentum. By using the wave functions described in Equation 4.80, one can find that the nonvanishing matrix elements of p_z are those associated with the following selection rule

$$n - n' = \text{odd integer}, \tag{4.78}$$

which means that only transitions between subbands with opposite parity are allowed. For n is odd and n' is even, the matrix element can be written as

$$\begin{aligned}\langle n\boldsymbol{k}_\perp|p_z|n'\boldsymbol{k}'_\perp\rangle &= \int_{-L/2}^{L/2}\sqrt{2/L}\cos(n\pi z/L)\left(-i\hbar\frac{\mathrm{d}}{\mathrm{d}z}\right)\sqrt{2/L}\sin(n'\pi z/L)\,\mathrm{d}z\\ &= -\frac{2i\hbar n'}{\mathrm{L}}\left\{\frac{\sin[(n'+n)\pi/2]}{n'+n} + \frac{\sin[(n'-n)\pi/2]}{n'-n}\right\}.\end{aligned} \tag{4.79}$$

For $n = 1$ and $n = 2$, the momentum matrix element is $-i8h/3L$ and the absorption coefficient is

$$\alpha(\omega) = \frac{2\pi e^2 N_\mathrm{w}\hbar}{\epsilon_0 c n_\mathrm{r} m *^2 V} \frac{1}{\hbar\omega} \left(\frac{8\hbar}{3L}\right)^2 \sum_{k_\perp,n,n'} \delta(E_{n'} - E_n - \hbar\omega)(f_\mathrm{FD}^{n,k_\perp} - f_\mathrm{FD}^{n',k'_\perp}). \tag{4.80}$$

If the quantum wells are doped and the Fermi energy is above the ground state $(n = 1, \boldsymbol{k}_\perp)$, and if the excited state $(n' = 2, \boldsymbol{k'}_\perp)$ is completely empty, then one can write

$$2\sum_{k_\perp} f_\mathrm{FD}^{n,k_\perp} = N_1 \text{ and } \sum_{k_\perp^i} f_\mathrm{FD}^{n',k_\perp} = 0. \tag{4.81}$$

The factor 2 is for electron spin degeneracy. Substituting Equation 4.81 into Equation 4.80, we have

$$\alpha(\omega) = \frac{n_1\pi e^2 N_\mathrm{w}\hbar}{\epsilon_0 c n_\mathrm{r} m^{*2} l} \frac{1}{\hbar\omega} \left(\frac{8\hbar}{3L}\right)^2 \delta(\Delta E - \hbar\omega), \tag{4.82}$$

where l is the total thickness of the quantum wells and $\Delta E = E_2 - E_1$. In this expression, we substituted $n_1 = N_1/\text{area}$, where n_1 is the electron sheet density, or the two-dimensional electron gas density. The oscillator strength can now be defined as

$$f_{01} = \frac{2P^2}{m^*\hbar\omega} = \frac{2}{m^*\hbar\omega}\left(\frac{8\hbar}{3L}\right)^2, \tag{4.83}$$

where the subscript "01" stands for the electronic transition from the ground state to the first excited state and P is the value of the momentum matrix element along the z-direction. An equivalent definition of the oscillator strength is $f_{01} = \frac{2m^*\omega}{\hbar}|\langle n\boldsymbol{k}_\perp|z|n'\boldsymbol{k'}_\perp\rangle|^2$. Notice that $\hbar\omega$ is the photon energy required to excite a transition from the ground to the excited state. For a GaAs/AlGaAs quantum well of thickness 100 Å, and $\hbar\omega = 150\,\mathrm{meV}$, $f_{01} \approx 1.08$. A typical example of the intersubband transition in GaAs/AlGaAs multiple quantum wells (MQWs) is shown in Fig. 4.11, where the solid lines represent experimental measurements at 300 and 77 K and the dashed lines represent Lorentzian lineshape fits for both spectra. The peak position shift of the intersubband transition is explained in terms of many-body effect (Manasreh, 1991). Since the experimental measurements show broadening because of several effects, the δ-function in Equation 4.82 can now be replaced by a Lorentzian lineshape. By inserting Equation 4.83 into Equation 4.82, we obtain

$$\alpha(\omega) = \frac{n_1\pi e^2 N_\mathrm{w}\hbar}{\epsilon_0 c n_\mathrm{r} m * l} f_{01} \frac{\Gamma}{\pi((\hbar\omega - \Delta E)^2 + \Gamma^2)}, \tag{4.84}$$

where Γ is the half width at half maximum.

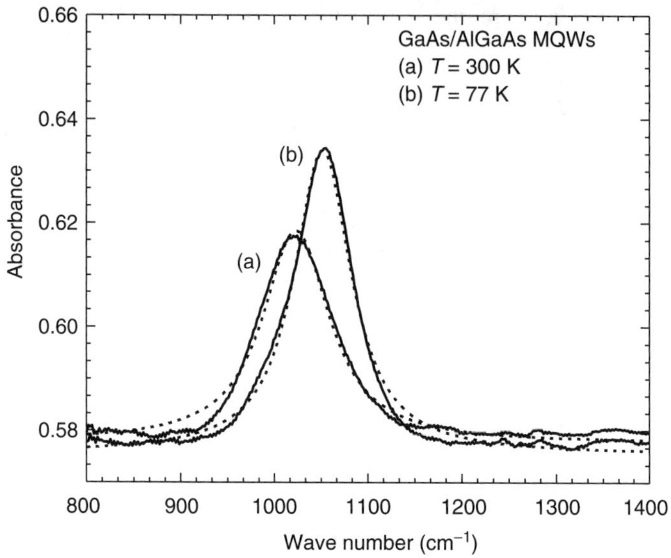

FIGURE 4.11 Absorbance of the intersubband transition in 75 Å GaAs/AlGaAs MQWs measured at 300 and 77 K (solid lines). The dashed lines are Lorentzian fits of the data.

Intersubband transitions in n-type semiconductor quantum wells were found experimentally to be excited by photon with an electric component parallel to the growth axis (z-axis). Thus, the momentum matrix element should contain the following factor:

$$\boldsymbol{\mathcal{E}} \cdot \widehat{z} \approx \sin\theta, \tag{4.85}$$

where $\boldsymbol{\mathcal{E}}$ is the unit vector of the light polarization, θ the internal incident angle, and $\widehat{z}$ is the unit vector along the z-direction as shown in the figure. If the light is polarized in the $x-y$ plane, the electron–photon coupling is zero. But if the light has a component parallel to the growth axis (z-axis), then the electron–photon coupling is nonzero and the intersubband transition can be observed. If the light reaches the sample at an angle, then the light intensity has to be scaled by a factor of $\cos\theta$ The absorption coefficient for the intersubband transition in n-type quantum wells can now be rewritten as

$$\alpha(\omega) = \frac{n_1 \pi e^2 N_w \hbar}{\epsilon_0 c n_r m * l} f_{01} \frac{\sin^2\theta}{\cos\theta} \frac{\Gamma}{\pi((\hbar\omega - \Delta E)^2 + \Gamma^2)}. \tag{4.86}$$

If θ is 45°, then the factor $(\sin^2\theta/\cos\theta)$is ~0.71. One can now state the polarization selection rule for n-type quantum wells: *The electron–photon coupling for a spherically symmetric band in a quantum well is nonzero for photons polarized along the growth direction of the quantum well.*

For bound-to-continuum intersubband transition, the calculation of the momentum matrix element is more complicated, since the excited state is a propagating plane wave as shown in Fig. 4.10b. This problem has been discussed by Choi, who derived the optical absorption lineshape as

$$\alpha(\omega) \propto \frac{\sqrt{\hbar\omega + E_1 - \Delta E_c}}{1 + C^2(\hbar\omega + E_1 - \Delta E_c)(\hbar\omega - (E_2 - E_1))^2}, \tag{4.87}$$

where ΔE_c is the conduction band offset, E_1 and E_2 are the ground and first excited states, respectively, and C is a constant. For this expression to be valid, the excited state should be above ΔE_c. A similar expression was derived by Liu [1996].

Owing to the small thickness of the quantum wells, the measured optical absorption of the intersubband transition is usually very small. One way to increase the absorption intensity is to fabricate a waveguide, where the light will make multiple passes. Figure 4.12 shows the measurements of the absorbance of the intersubband transition for both Brewster's angle and waveguide configuration. Notice that the absorption coefficient can be obtained by dividing the absorbance by the total thickness of the active region in multiple quantum wells. For 75 Å well and 50 periods, the total thickness of the active region is 0.375 μm for a single pass. The effective optical active region in the waveguide configuration is 0.374 times the number of passes the photons will make before exiting the sample. It is clear from Fig. 4.12 that the signal obtained from using

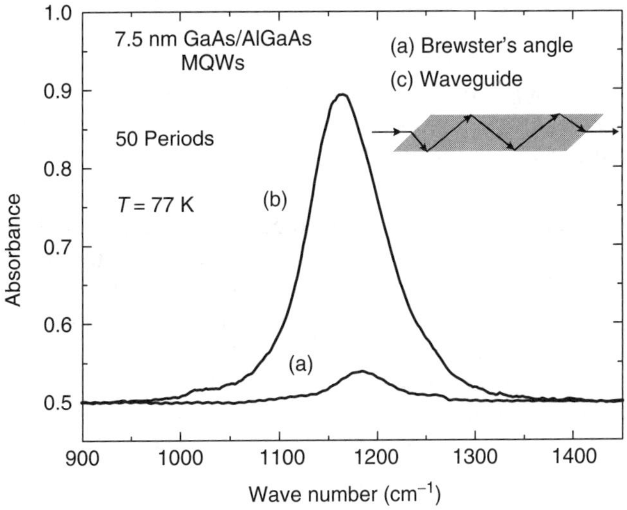

FIGURE 4.12 Intersubband transition in 75 Å GaAs/AlGaAs multiple quantum wells measured at $T = 77$ K using (a) Brewster's angle and (b) waveguide configuration. The inset is a sketch of a waveguide with multiple passes.

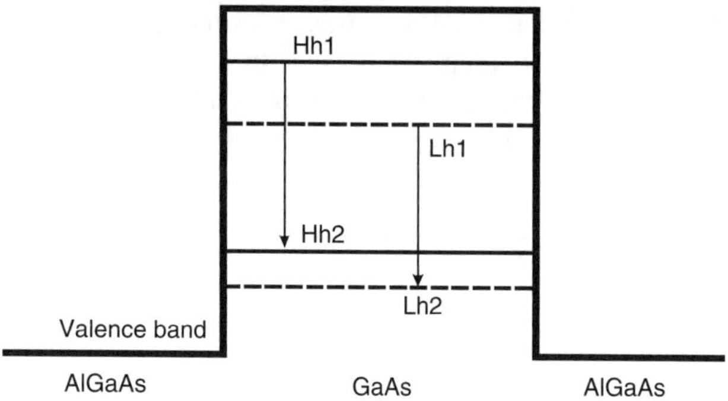

FIGURE 4.13 Intersubband transitions in p-type GaAs/AlGaAs multiple quantum wells shown for transitions for heavy hole bound states (solid lines) and for light hole bound state (dashed lines).

the waveguide configuration is much larger than that of the Brewster's angle configuration due to multiple passes that the photons make inside the waveguide.

The above discussion is directed toward n-type multiple quantum wells, such as GaAs/AlGaAs, where the well is doped with a donor such as Si. It is quite possible, however, that intersubband transition can be observed in p-type multiple quantum wells where the dopant is an accepter such as Be. In p-type multiple quantum wells, quantum confinement is in the valence band, as shown in Fig. 4.13. The momentum matrix elements show that normal incident photon–electron coupling is possible because of the bands mixing as detailed by Brown and Szmulowicz. Transitions between heavy holes and light holes are also possible.

4.8 THE OPTICAL ABSORPTION COEFFICIENT OF THE INTERSUBBAND TRANSITION IN GaN/AlGaN MULTIPLE QUANTUM WELLS

III-Nitride semiconductors have been studied for their applications in the visible and ultraviolet spectral regions. They can also be used for infrared applications by investigating the intersubband transitions in quantum structures such as GaN/AlGaN multiple quantum wells and quantum dots. The most common crystallographic structure of III-nitride materials is the wurtzite (hexagonal) structure. A large spontaneous polarization oriented along the c-axis occurs because of the lack of inversion symmetry and the large ionicity associated with the covalent nitrogen bond (Bernardini *et al.*). The electrostatic charge densities associated with the piezoelectric polarization field influence the carrier distributions, electric field, and consequently, a wide range of optical and electronic properties of nitride materials and devices. The total polarization at the GaN/AlGaN interface is the sum of the effective piezoelectric polarization and the difference spontaneous polarization (Morkoç and Yu). A typical value for the total polarization is

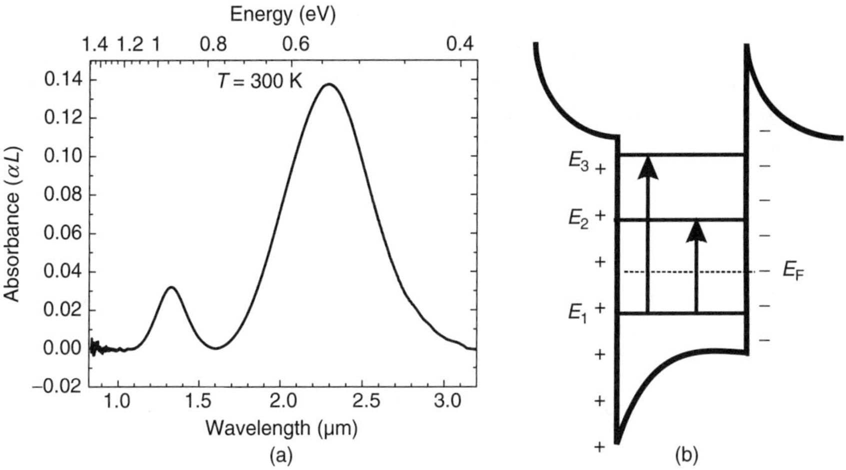

FIGURE 4.14 (a) Absorbance spectrum of intersubband transitions in GaN/AlN multiple quantum well and (b) the structure was designed to generate two transitions as shown by the vertical arrows.

of the order of $-0.096\,x$ C/m^2, where x is the Al mole fraction in the AlGaN (Morkoç). For $x = 0.3$, the total polarization is -0.0288 C/m^2. Notice that 1C/m^2 $= 6.24 \times 10^{14}$ electrons/cm^2. Thus, one AlGaN/GaN interface can produce a total polarization charge of the order of $\sim 1.80 \times 10^{13}$ electrons/cm^2. A test of the polarization-induced charges is to measure the intersubband transition in certain GaN/AlN multiple quantum well structures, as shown in Fig. 4.14a, where the absorbance spectrum was measured at room temperature for a superlattice structure with two intersubband transitions as shown in Fig. 4.14b.

III-Nitride semiconductor materials have a wide range of applications and their use extends from the ultraviolet and visible spectrum to near the infrared spectral region as in Fig. 4.14. In addition to their optoelectronic applications, the III-nitride materials have been used for high power modulation doped field effect transistors.

4.9 ELECTRONIC TRANSITIONS IN MULTIPLE QUANTUM DOTS

The intersubband transitions in multiple quantum wells discussed in the previous section show that there is a selection rule that permits electron–phonon coupling at a certain angle of the incident light. The maximum electron–phonon coupling occurs at the Brewster's angle, while the waveguide configuration allows one to obtain stronger absorption intensity because of the multiple passes that the light makes before exiting the sample. In the multiple quantum well case, the wave functions of the ground and excited states are highly symmetrical (s-type) and the transition is determined by the overlap integral or by the selection rules determined from the momentum matrix elements. In the case of quantum dots, there is a strong p-type mixture in the conduction band states (Singh, 2003). For an eight-band **k**·**p** model, the wave functions of the electronic energy levels in quantum dots can be written according to Singh as

$$\psi_n(x) = \sum_{j=1}^{8} \phi_{nj}(\boldsymbol{r}) u_j(x), \tag{4.88}$$

where ϕ_{nj} is the envelope part and u_j is the central cell part. The momentum matrix element can now be written as

$$p_{fi} = \sum_{jj'} \{ \langle \phi_{fj'} |\boldsymbol{p}| \phi_{ij} \rangle \langle u_{j'} | u_j \rangle + \langle u_{j'} |\boldsymbol{p}| u_j \rangle \langle \phi_{fj'} | \phi_{ij} \rangle \}. \tag{4.89}$$

The absorption coefficient can now be written for bound-to-bound transition with a Lorentzian lineshape as

$$\alpha(\omega) = \frac{n_1 \pi e^2 N_q \hbar}{\epsilon_0 c n_r m *^2 l_{av}} \frac{1}{\hbar\omega} (\widehat{\epsilon} \cdot \boldsymbol{p}_{fi})^2 \frac{\Gamma}{\pi \left((\hbar\omega - \Delta E)^2 + \Gamma^2 \right)}, \tag{4.90}$$

where n_1 is the number of electrons per unit area in each quantum dot layer, N_q is the number of quantum dot layers, l_{av} is the average quantum dot layer thickness, and $\widehat{\epsilon}$ is the direction of the polarization unit vector. A typical example of a multiple quantum dot structure is depicted in Fig. 4.15a. In this figure, we sketched $In_{0.3}$ $Ga_{0.7}As$ triangular-shaped quantum dots grown by the molecular beam epitaxy technique using the *Stranski–Krastanow* mode with GaAs being the barrier. The wetting layer and the average quantum dot height are shown. Two samples were designed such that the intersubband transition is bound to bound (Fig. 4.15b) in one sample and bound to continuum in the other (Fig. 4.15c). Waveguides were made from these two samples to allow the light to make multiple passes to increase the optical length. The average thicknesses (l_{av}) of the bound-to-bound and bound-to-continuum samples were 25 and 15 monolayers (MLs), respectively. An ML of $In_{0.3}$ $Ga_{0.7}As$ is approximately 2.588 Å.

Typical optical absorbance spectra for bound-to-bound and bound-to-continuum intersubband transitions obtained for multiple quantum dots are shown in Fig. 4.16. The solid lines are the experimental spectra and the dashed lines are theoretical spectra. For the bound-to-bound spectrum (a), a Lorentzian lineshape was used as given by Equation 4.84. The bound-to-continuum spectrum (b) was fitted with a lineshape described by Equation 4.87. Bound-to-continuum transition exhibits an asymmetrical lineshape because of the fact that the transition occurs between the bound ground state and all available states in the continuum, including the resonant state, which has the propagation property shown in Fig. 4.15c. Notice that the optical absorption coefficient of the intersubband transition contains the electron effective mass. This is due to the

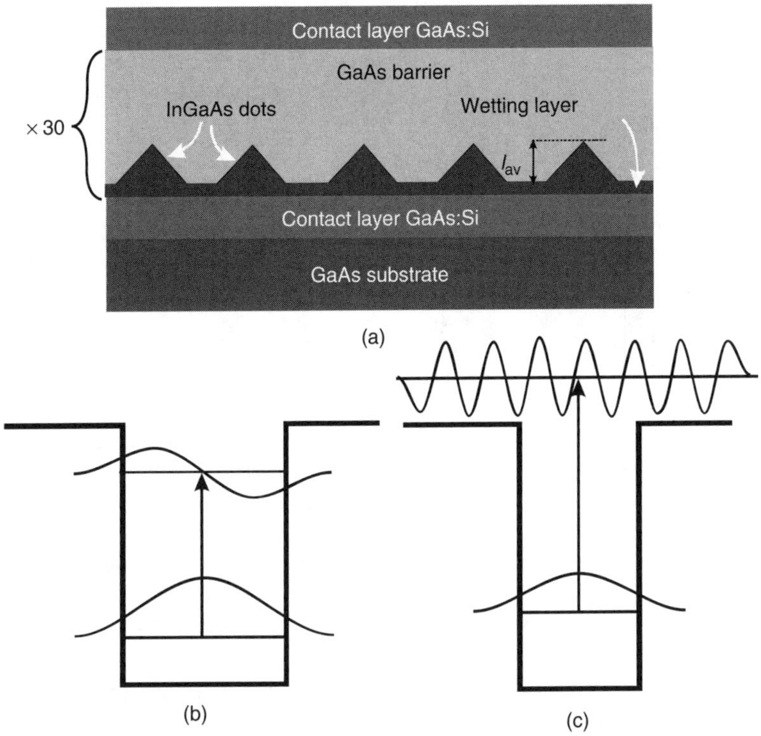

FIGURE 4.15 (a) A sketch of an InGaAs/GaAs multiple quantum dot structure used for optical absorption coefficient measurements showing the wetting layers, triangular shape quantum dots, and contact layers, (b) bound-to-bound transition, and (c) bound-to-continuum transition.

higher order terms arising from the canonical transformations of the effective mass theory (Wallis).

The sketch shown in Fig. 4.15a indicates that the dot size is the same for all dots in the structure. This figure, however, is very simplistic since the self-assembled quantum dot shows that there is a variation in size, shape, and strain causing a variation in the energy levels, which leads to an inhomogeneous broadening in the quantum dot ensemble properties. An example of this effect is the interband transitions in quantum dot ensembles, where the distribution of the quantum dot size is assumed to be a Gaussian of the form

$$G(a) = \frac{1}{\sqrt{2\pi}\sigma_a} \exp\left[-\frac{(a-a_0)^2}{2\sigma_a^2}\right]. \tag{4.91}$$

where σ_a is the standard deviation and is given by $\sigma_a = \sqrt{\langle a - a_0 \rangle^2}$. The quantum dots are assumed to be cubic in shape with an average side of a_0 (Wu *et al.*). The same analysis can be applied to spherical quantum dots with an average radius

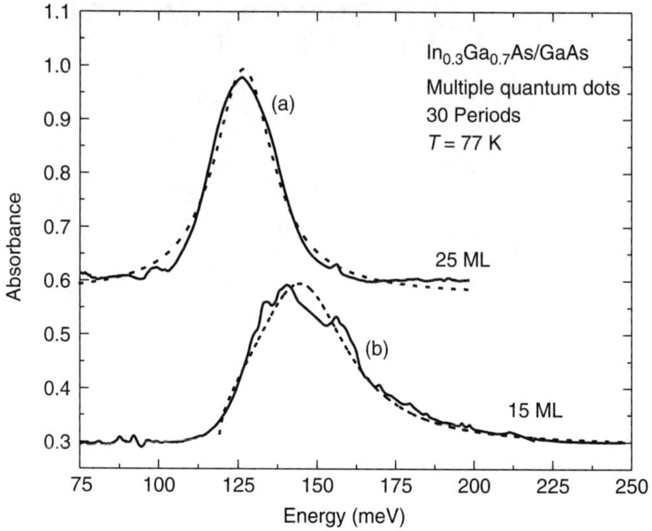

FIGURE 4.16 Optical absorbance spectra obtained for two multiple quantum dot structures with average dots height of (a) 25 monolayer (ML) and (b) 15 ML. Solid lines represent the experimental spectra and the dashed lines represents the theoretical spectra.

r_0. The optical absorption coefficient for the interband transition in a quantum dot can be written as

$$\alpha(\omega) = \frac{2\pi m_r^* e^2 \hbar}{\epsilon_0 c n_r m_0^2 a^3} \frac{|P_n|^2}{\hbar\omega} \sum_{n,l} (2l+1)\delta\left(\hbar\omega - E_g - \frac{\pi^2\hbar^2 n^2}{2m_r^* a^2}\right), \tag{4.92}$$

where $\sum_l (2l+1)$ is the degeneracy of the energy level, P_n is the momentum matrix element, and E_g is the band gap. The selection rules dictate that $\Delta n = 0$, which means the allowed transitions are those between HH_1 and E_1, HH_2 and E_2, and so on. The reduced effective mass, m_r^*, has been defined previously (Eq. 4.58). The convolution of Equations 4.91 and 4.92 gives

$$\alpha(\omega) = \frac{2\pi m_r^* e^2 \hbar}{\epsilon_0 c n_r m_0^2 a^3} \frac{|P_n|^2}{\hbar\omega} \frac{1}{\sqrt{2\pi}\sigma_a} \sum_l (2l+1)$$

$$\int_0^\infty \delta\left(\hbar\omega - E_g - \frac{\pi^2\hbar^2 n^2}{2m_r^* a^2}\right) \frac{1}{a^3} \exp\left(-\frac{(a-a_0)^2}{2\sigma_a^2}\right) \mathrm{da}. \tag{4.93}$$

Let $x^2 = \frac{2m_r^* a_0^2}{\pi^2\hbar^2}(\hbar\omega - E_g)$, $\xi = \sigma_a/a_0$, and $A = \frac{2\pi e^2 \hbar}{\epsilon_0 c n_r m_0^2 a^3} \frac{|P_n|^2}{\hbar\omega} \frac{m_r^*}{\sqrt{2\pi}\pi^2\hbar^2}$, we have

$$\alpha(\omega) = \frac{A}{a_0} \sum_{n,l} \frac{2l+1}{\xi n^2} \exp\left(-\frac{(n/x-1)^2}{2\xi^2}\right). \tag{4.94}$$

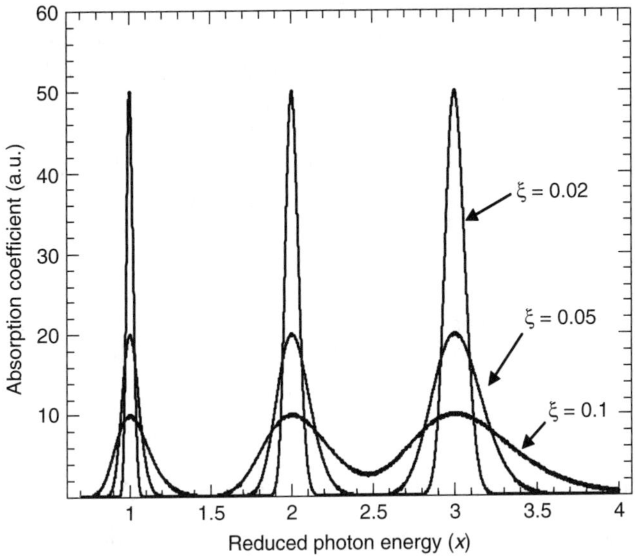

FIGURE 4.17 Absorption coefficient of interband transitions in quantum dot ensembles having a Gaussian distribution plotted as a function of reduced photon energy. The spectra are shown for three different standard deviations.

A plot of Equation 4.94 is shown in Fig. 4.17 for different values of ξ assuming that the parameter A is the same for all transitions. The absorption peak position energies can be expressed as

$$\hbar\omega = E_{\mathrm{g}} + x^2 \frac{\pi^2\hbar^2}{2m_{\mathrm{r}}^* a_0^2}, \tag{4.95}$$

where E_{g} is the band gap of the bulk material and x can be read directly from Fig. 4.17. If one ignores the small red shift of the peak position energy due to the broadening effect, x is taken as an integer as shown in the figure. It is clear from Equation 4.95 that the peak positions are determined solely by the size of the quantum dots for known electron and hole effective masses.

4.10 SELECTION RULES

4.10.1 Electron–Photon Coupling of Intersubband Transitions in Multiple Quantum Wells

It was found experimentally that the intersubband transitions in n-type multiple quantum wells can be observed when the incident light has a polarization component in the z-direction or the growth direction. The light has electrical ($\boldsymbol{\mathcal{E}}_x$) and magnetic ($\boldsymbol{B}_y$) components, which are orthogonal to each other and to the propagation direction, as illustrated in Fig. 4.18. For normal incident light, as illustrated in Fig. 4.18a, the electrical component is perpendicular to the z-axis and, therefore, the electric component along the z-axis is zero. However, when the incident light reaches the surface of the sample at an angle φ from the normal, as illustrated in Fig. 4.18b, the electric field has a component in the z-axis such that $\mathcal{E}_x.\widehat{z} = \mathcal{E}_x \cos\phi = \mathcal{E}_x \sin\theta$. The maximum electron–photon coupling occurs when φ is the Brewster's angle. For GaAs, with a refractive index of 3.27, φ is 73°. Thus, $\theta = \sin^{-1}[(\sin 73)/3.27] = 17^\circ$. The electron–photon coupling selection rule for a spherically symmetric band in a quantum well is that this coupling is nonzero for photons polarized along the growth direction of the quantum well. For intersubband transitions in p-type multiple quantum wells, this selection rule is no longer valid and the photon can be absorbed at normal incident because of heavy hole and light hole wave functions mixing.

4.10.2 Intersubband Transition in Multiple Quantum Wells

The envelope functions of the bound states in the conduction quantum well are given by Equation 4.75 for the even and odd states. The momentum matrix element is given by Equation 4.77. Since the momentum operator, $p_z = -i\hbar d/\mathrm{dz}$, changes the parity of the wave function, the wave functions in the integral $\langle n|p_z|n'\rangle$ must have opposite parities for a nonzero momentum matrix element. In other words, $|n\rangle$ and $|n'\rangle$ must have different parity, which leads to the selection rule $(n'-n) = \textit{odd integer}$. The same conclusion can be reached if the dipole matrix element is used instead of the momentum matrix element such that $\langle n|z|n'\rangle$. In this case, z is an odd function and, therefore, $|n\rangle$ and $|n'\rangle$ must have different parities for the integral to have a nonzero value.

4.10.3 Interband Transition

The selection rules of interband transitions in multiple quantum wells can be understood by examining the wave functions of the valence and conduction bands, which can be written as

$$|i\rangle = \frac{1}{\sqrt{V}} u_{\mathrm{v}}(\boldsymbol{r})\varphi_{nh}(z)\exp(i\boldsymbol{k}_\perp.\boldsymbol{r}_\perp) \tag{4.96a}$$

$$|f\rangle = \frac{1}{\sqrt{V}} u_{\mathrm{c}}(\boldsymbol{r})\varphi_{n'e}(z)\exp(i\boldsymbol{k}'_\perp.\boldsymbol{r}'_\perp), \tag{4.96b}$$

where $|k_\perp| = \sqrt{k_x^2 + k_y^2}$, $|r_\perp| = \sqrt{x^2 + y^2}$, $u_{\mathrm{v}}(\boldsymbol{r})$, and $u_{\mathrm{c}}(\boldsymbol{r})$ are the envelope functions for the valence and conduction bands, $\varphi_{nh}(z), \varphi_{n'e}(z)$ are the wave functions for the bound states in the valence and conduction bands, and the exponentials are the plane waves for free motion in the $x-y$ plane. Either the momentum or the dipole matrix element can be used to determine the allowed

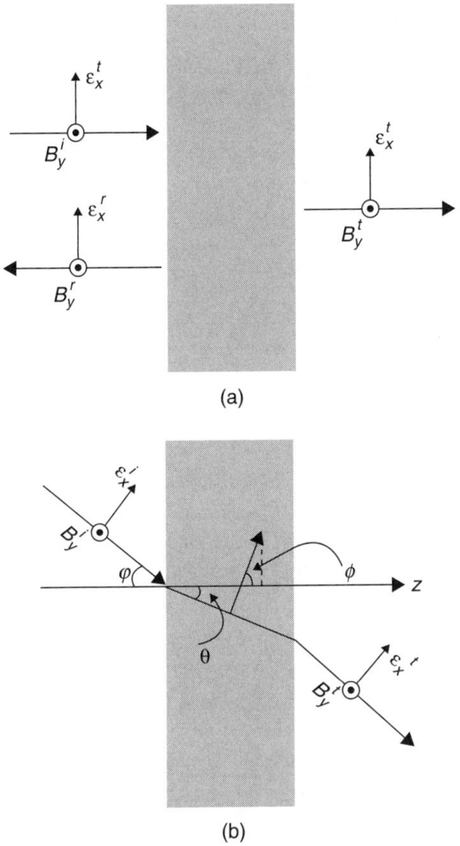

FIGURE 4.18 Reflection and transmission of an electromagnetic wave showing the electric (E_x) and magnetic (B_y) fields with respect to the direction of propagation. (a) Normal incidence gives a zero component of the electric field along the propagation direction. (b) Incidence at an angle ϕ from the normal yields a nonzero component of the electric field along the z-direction.

interband transitions. The conservation of momentum allows one to set $\boldsymbol{k}_\perp = \boldsymbol{k}'_\perp$ since the photon momentum is very small compared to the electron momentum. The matrix element, M, can now be written as

$$M = \langle f|\boldsymbol{r}|i\rangle. \tag{4.97}$$

For quantum wells, we have

$$\langle f|x|i\rangle = \langle f|y|i\rangle \neq \langle f|z|i\rangle. \tag{4.98}$$

Since the interband transition is in a plane perpendicular to the growth axis or z-direction, we are concerned in evaluating the matrix element along the x- or

y-direction, which yields

$$
\begin{aligned}
M = \langle f|x|i\rangle &= \frac{1}{V}\int\int u_c^*(\boldsymbol{r})\varphi_{n'e}^*(z)xu_v(\boldsymbol{r})\varphi_{nh}(z)d^3\boldsymbol{r}\mathrm{d}z \\
&= \frac{1}{V}\int u_c^*(\boldsymbol{r})xu_v(\boldsymbol{r})d^3\boldsymbol{r}\int \varphi_{n'e}^*(z)\varphi_{nh}(z)\mathrm{d}z \\
&= \frac{1}{V}\langle u_c|x|u_v\rangle\langle n'e|nh\rangle = M_{cv}M_{nn'},
\end{aligned} \tag{4.99}
$$

where

$$
\frac{1}{V}\langle u_c|x|u_v\rangle = M_{cv} \tag{4.100}
$$

and

$$
\langle n'e|nh\rangle = M_{nn'}, \tag{4.101}
$$

where $M_{nn'}$ is known as the *electron–hole overlap*. If one assumes that the wave functions of the bound states in the valance and the conduction bands have the following forms

$$
\begin{aligned}
\varphi_{nh}(z) &= \sqrt{\frac{2}{L}}\cos\left(\frac{n\pi z}{L} + \frac{n\pi}{2}\right) \\
\varphi_{n'e}(z) &= \sqrt{\frac{2}{L}}\cos\left(\frac{n'\pi z}{L} + \frac{n'\pi}{2}\right)
\end{aligned} \tag{4.102}
$$

then the electron–hole overlap integral is

$$
\begin{aligned}
M_{nn'} &= \frac{2}{L}\int_{-L/2}^{L/2}\cos\left(\frac{n\pi z}{L} + \frac{n\pi}{2}\right)\cos\left(\frac{n'\pi z}{L} + \frac{n'\pi}{2}\right)\mathrm{d}z. \\
&= \delta_{\mathrm{nn'}}.
\end{aligned} \tag{4.103}
$$

From this equation, we have the following selection rule $n = n'$ or $\Delta n = 0$.

4.11 EXCITONS

A brief discussion of the excitons in bulk semiconductors and low dimensional systems is presented in the following subsections. Excitons in GaN thin films are discussed as an example for bulk materials. Then, excitons in quantum wells and quantum dots are discussed.

4.11.1 Excitons in Bulk Semiconductors

Excitons are quasi-particles used to describe electron–hole pairs coupled by Coulomb interaction in a manner similar to the hydrogen atom. As mentioned in the introduction of this chapter, there are two types of excitons: free and bound excitons as illustrated in Fig. 4.2. Excitons in semiconductors are stable so long as their binding energy is smaller than the thermal energy (k_BT). The optical absorption and photoluminescence (PL) emission of excitons affect the optical properties of the band edge of semiconductors and their heterojunctions. Exciton absorption is profound at low temperatures in most direct band gap semiconductor materials and it can even be observed at room temperature in semiconductors, such as GaN, where the binding energy of the exciton is slightly larger than the room temperature thermal energy. Figure 4.19 illustrates how the free exciton affects the band edge absorption of a pure semiconductor materials at low temperature. The dashed line in this figure depicts the band edge absorption of a

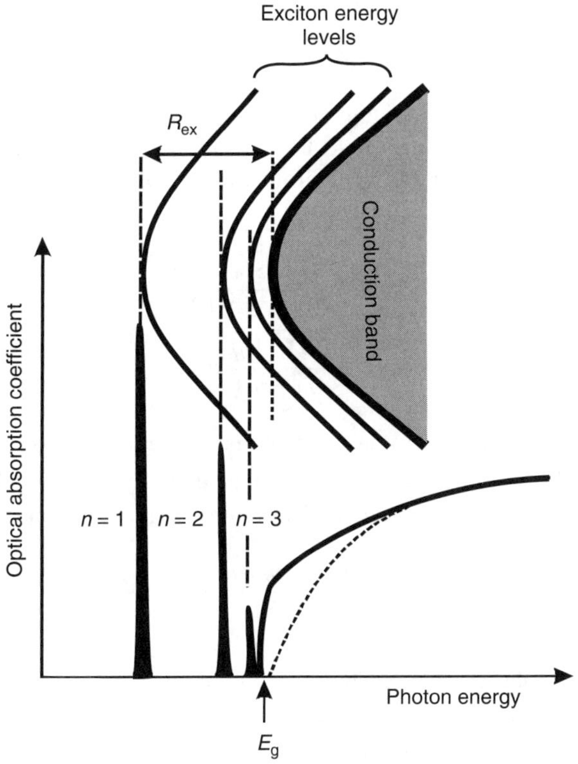

FIGURE 4.19 Illustration of the band edge absorption of a direct semiconductor in the absence (dashed curve) and in the presence of exciton (solid lines). The exciton energy levels (n = 1, 2, and 3) are shown. R_{ex} is the exciton binding energy in bulk semiconductors.

direct band gap semiconductor without the exciton effect. The solid curve is the band edge absorption with the exciton effect included. The lines labeled $n = 1$, 2, and 3 are the excitonic energy levels.

To obtain the exciton energy levels, one needs to solve the Schrödinger equation for a two-body problem. By considering the relative motion of the electron–hole system and ignoring the motion of the center of mass (the kinetic energy of the center of mass which is translation invariant), the Schrödinger equation can be written as

$$\left[-\frac{\hbar^2 k^2}{2\mu^*} - \frac{e^2}{4\pi\epsilon\epsilon_0 r}\right]\psi_{\text{ex}} = E_n\psi_{\text{ex}}, \tag{4.104}$$

where the first term is the relative motion of the electron–hole system (kinetic energy), the second term is the Coulomb interaction energy between the electron and hole, ϵ is the dielectric constant of the material, r is the distance between the electron and hole, and m^* is the exciton reduced effective mass $[1/\mu^* = (1/m_{\text{e}}^*) + (1/m_{\text{h}}^*)]$. The exciton wave function can be written as

$$\psi_{\text{ex}} \propto \chi(r)\phi_{\text{c}}(r_e)\phi_{\text{v}}(r_h), \tag{4.105}$$

where $\chi(r)$ is the envelope function and $\phi_{\text{c}}(r_{\text{e}})$ and $\phi_{\text{v}}(r_{\text{h}})$ are Wannier functions that represent the electron and hole band edge states. Equation 4.104 can be solved in a manner similar to the hydrogen atom, where the energy levels can be written as

$$\begin{aligned} E_n &= -\frac{\mu^* e^4}{2(4\pi\epsilon\epsilon_0)^2\hbar^2 n^2} \\ &= -\frac{\mu^*}{m_0\epsilon^2}\frac{R_H}{n^2}, \end{aligned} \tag{4.106}$$

where m_0 is the free electron mass and R_{H} is the hydrogen atom Rydberg constant given by $R_{\text{H}} = \frac{m_0 e^4}{2(4\pi\epsilon_0)^2\hbar^2} = 13.60\text{eV}$. The quantity $R_{\text{ex}} = \frac{\mu^* R_{\text{H}}}{m_0\epsilon^2}$ can now be called the *exciton Rydberg constant*. The radius of the electron–hole orbit can be written as

$$r_n = \frac{4\pi\epsilon\epsilon_0\hbar^2 n^2}{\mu^* e^2} = \frac{m_0\epsilon n^2}{\mu^*}a_{\text{H}} = n^2 a_{\text{ex}}, \tag{4.107}$$

where a_H is the Bohr radius of the hydrogen atom given by $a_{\text{H}} = \frac{4\pi\epsilon_0\hbar^2}{m_0 e^2} = 0.5293$ Å, and a_{ex} is the exciton Bohr radius. The exciton binding energy can be taken as R_x or the energy of the ground state. For example, the binding energy of a free exciton in GaAs is found to be 4.35 meV, assuming that the electron and heavy hole effective masses are $0.067m_0$ and $0.54m_0$, respectively, and the dielectric constant is taken as 13.6. The Bohr radius of the free exciton in GaAs is calculated to be 12.07 nm. The equivalent temperature to the free exciton binding energy is ~50 K. Thus, the free exciton in GaAs is stable at temperatures below 50 K. In a highly pure GaAs sample with high mobility, the free exciton is observed at temperatures as high as 180 K. As a comparison, we calculated the binding energy of the free exciton in GaN material to be ~23.4 meV for electron and heavy hole effective masses of $0.20\,m_0$ and $0.60\,m_0$, respectively, and for a dielectric constant of 9.2. The Bohr radius is found to be ~3.24 nm. The equivalent temperature of the exciton in GaAs is ~271 K, which means that free excitons can be observed at room temperature in relatively pure GaN samples. The exciton binding energy and radius were calculated for a few semiconductor materials and are given in Table 4.1. The exciton binding energy increases as the band gap increases. On the other hand, the exciton radius decreases with increasing band gap.

Excitons are mostly observed at the high symmetry points in the Brillouin zone such as the Γ-point (the center of the Brillouin zone). At these points, the slopes of the energy bands are zero and the group velocities of electrons and holes are the same, which is a necessary condition to observe excitons. The excitonic energy levels in a direct band gap semiconductor can be written as

$$E_n = E_g - \frac{\mu^* e^4}{2(4\pi\epsilon\epsilon_0)^2\hbar^2 n^2} = E_g - \frac{R_{ex}}{n^2}. \tag{4.108}$$

If the motion of the center of mass of the exciton is included in the Schrödinger equation, the excitonic energy levels become

$$E_n = E_g + \frac{\hbar^2 k_{ex}^2}{2M} - \frac{R_{ex}}{n^2}, \tag{4.109}$$

where $\boldsymbol{k}_{ex}$ is the exciton wave vector and $M = m_e^* + m_h^*$. The exciton center of mass in the above equation behaves like a particle with mass M and a wave

TABLE 4.1 Several Well-Known Direct Band Gap Semiconductor Materials Shown with their Band Gaps (E_g), Electron (m_e^*), and Heavy Hole (m_h^*) Effective Mass in Units of Free Electron Mass (9.11 x 10^{-31} kg), Dielectric Constant (ϵ), Exciton Binding Energy (E_{ex}^{bulk}) and Exciton Radius (a_{ex}^{bulk})

Material	E_g (eV)	m_e^*	m_h^*	ϵ	E_{ex}^{bulk} (meV)	a_{ex}^{bulk} (Å)
InSb	0.23	0.013	0.40	16.8	0.61	706.2
InAs	0.35	0.027	0.40	15.15	1.50	317.0
GaSb	0.75	0.042	0.40	15.69	2.10	218.5
GaAs	1.52	0.067	0.54	13.18	4.67	117.0
InP	1.35	0.073	0.64	12.56	5.65	101.4
CdTe	1.48	0.086	0.60	10.6	10.02	67.8
ZnTe	2.39	0.12	1.30	8.7	19.90	41.9
GaN	3.44	0.20	0.60	9.50	22.60	33.5
ZnO	3.28	0.24	0.78	8.1	33.00	23.4

vector $\boldsymbol{k}_{ex}$. The translational energy of the exciton center of mass, which is usually very small, can be dropped from Equation 4.109. The driving force of exciton generation is the Coulomb interaction between the electron–hole pair. If this interaction is zero, the exciton energy levels will vanish. The exciton functions are hydrogen atomlike functions. For example, the ground state wave function can be written as

$$\phi^{100}(\boldsymbol{r}) = \frac{1}{\sqrt{\pi a_{ex}^3}} e^{-r/a_{ex}}. \tag{4.110}$$

The free exciton radius in many semiconductors that have a band gap in the range of 1–2 eV is of the order of 100 Å, which means that the exciton is spread over many unit cells as shown in Fig. 4.2. In wide band gap materials, such as GaN and ZnO, the exciton radius is smaller, but the binding energy is larger and free excitons are observable even at room temperature. The stability of the exciton at room temperature is very important for exciton-based device applications.

The optical absorption spectra of excitons have been reported for many direct band gap semiconductor materials. For example, optical absorption measurements on wurtzite GaN thin films, grown on sapphire, exhibit three free excitons, as shown in Fig. 4.20. Room temperature spectrum show excitonic behavior near the band edge absorption. Upon cooling down the sample to 10 K, the spectrum

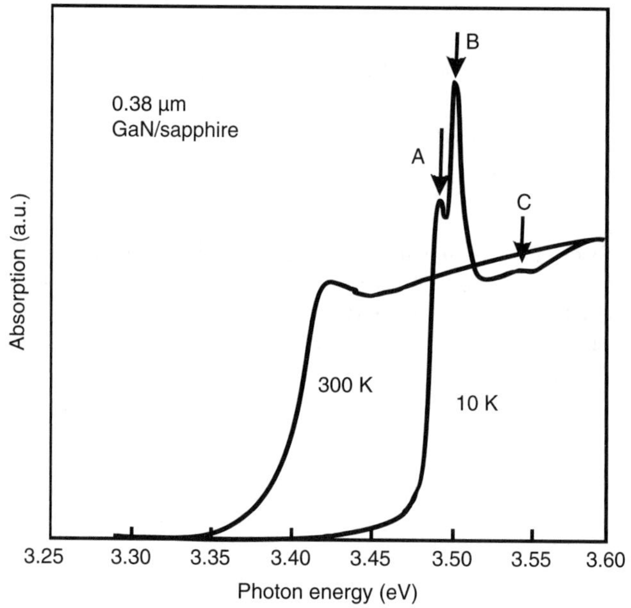

FIGURE 4.20 The optical absorption coefficient spectra of GaN thin film measured at 300 and 10 K. The three exciton lines, A, B, and C are clearly visible in the spectrum measured at 10 K.

shows the three excitons labeled A, B, and C. These excitons are usually observed in epitaxially grown thin films with thicknesses ranging from 0.1 to 1.0 μm. Within this thickness range, absorption above the band gap is possible since thick layers tend to absorb and/or reflect light just above the band edge. An alternative technique used to measure the absorption coefficient above the band gap is ellipsometry. The absorption coefficient can then be obtained from the imaginary part of the dielectric constant.

The origin of the A, B, and C excitons can be understood by examining Fig. 4.21. In this figure, the band structure is sketched at the center of the Brillouin zone, where the wurtzite structure has a nondegenerate energy level for the conduction band (CB) and a degenerate energy level for the valence band (VB). The valence band energy level splits into two energy levels (Γ_1, Γ_5) under the action of the axial crystal field (Δ_{cr}). The spin–orbit interaction (Δ_{so}) causes a similar effect on the valence band. The combined actions of Δ_{cr} and Δ_{so} result in splitting of the valence band into three energy levels labeled Γ_9, Γ_7 and Γ_7. The transitions from the three valence energy levels to a single conduction band energy level dominate the optical absorption near the band edge of GaN. The three exciton transitions are labeled E_g^A, E_g^B, and E_g^C, while E_g^0 is the band gap transition in the absence of the exciton effect.

The excitonic energy levels in GaN have been investigated by many techniques. GaN thin films are epitaxially grown on lattice mismatched substrates, such as sapphire or SiC. The interfaces are usually plagued by dislocations and extended defects. The structural property of the thin film usually improves as the thickness of the layer increases. Conversely, optical absorption above the band edge becomes difficult as the layer thickness increases. However, the photoreflectance technique is useful in this case, where the excitonic bound states are probed. Figure 4.22 is a typical photoreflectance spectrum measured at 10 K for

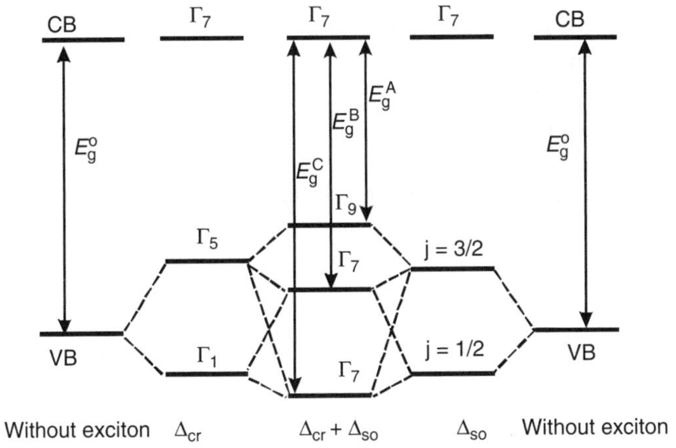

FIGURE 4.21 A sketch of the band structure near the fundamental band edge in wurtzite GaN showing the effect of the crystal field and spin–orbit interactions on the valence band.

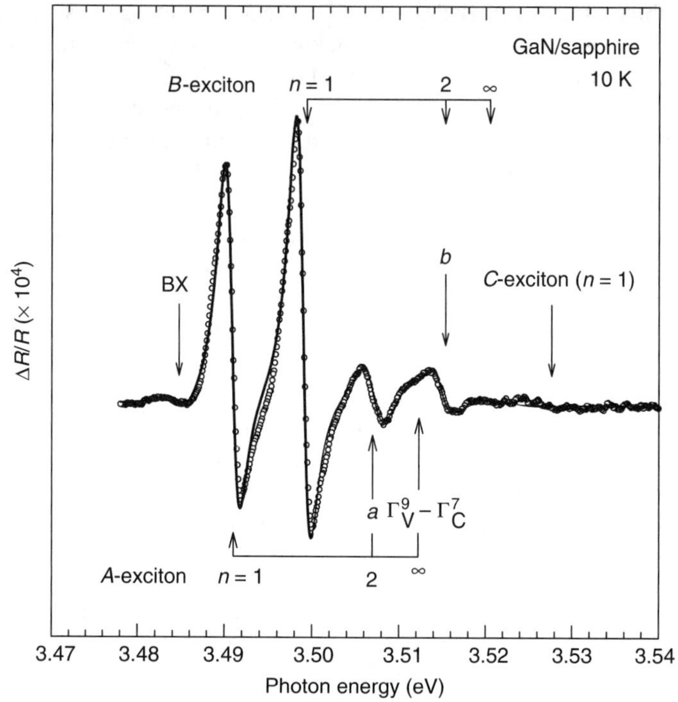

FIGURE 4.22 Photoreflectance spectrum of 7.2-μm thick GaN grown by MOCVD on (0001) sapphire substrate is shown as a function of photon energy. The excitonic energy levels ($n = 1$ and 2) are shown for excitons A and B (Schmidt *et al.*).

a 07.2 μm thick GaN thin film grown on sapphire. The exciton transition for $n = 1, 2$, and ∞ are shown for excitons A and B. The reflectance from exciton C is very weak due to the fact that its energy is way above the bottom of the conduction band. In addition to the splitting of the valence band under the crystal field and spin–orbit interactions, a fine structure splitting of the exciton lines (of the order of 1 meV) due to electron–hole exchange interaction may occur. This fine structure, however, has not been observed yet.

The optical absorption of the exciton in bulk semiconductors was derived by Elliot, and is given by

$$\alpha_{\mathrm{ex}} = \begin{cases} \alpha_0 \dfrac{2\pi\sqrt{a_{\mathrm{ex}}}}{\sqrt{\hbar\omega - E_{\mathrm{g}}}} & \text{for } \hbar\omega \approx E_{\mathrm{g}} \\ \alpha_0 \dfrac{\pi\zeta\exp(\pi\zeta)}{\sinh(\pi\zeta)} & \text{for } \hbar\omega > E_{\mathrm{g}}, \end{cases} \tag{4.111}$$

where α_0 is the optical absorption in the absence of Coulomb interaction (Eq. 4.70) and $\zeta = \frac{a_{\mathrm{ex}}}{\sqrt{\hbar\omega - E_{\mathrm{g}}}}$.

4.11.2 Excitons in Quantum Wells

The Hamiltonian of the exciton in a quantum well does not have a simple analytical solution, but the problem can be solved using the variational method described in Chapter 1. If the variational wave function is assumed to have the following form:

$$\phi_n^{11}(\boldsymbol{r}) = \sqrt{\frac{2}{\pi a_{\rm ex}^2}}\,{\rm e}^{-r/a_{\rm ex}}, \tag{4.112}$$

where $\boldsymbol{r}$ is the relative coordinate of the electron and hole in the x-y plane, then the Hamiltonian in polar coordinate can be written as

$$H = -\frac{\hbar^2}{2\mu^*}\left[\frac{1}{r}\frac{\partial}{\partial r}\left(r\frac{\partial}{\partial r}\right) + \frac{1}{r^2}\frac{\partial^2}{\partial \theta^2}\right] - \frac{e^2}{4\pi\epsilon\epsilon_0 r}. \tag{4.113}$$

Using the variational method, one can write the energy expectation value as

$$\langle E\rangle = \frac{\int \phi_n^{11*}(\boldsymbol{r},\theta)H\phi_n^{11}(\boldsymbol{r},\theta)\boldsymbol{r}\,{\rm d}\boldsymbol{r}\,{\rm d}\theta}{\int \phi_n^{11*}(\boldsymbol{r},\theta)\phi_n^{11}(\boldsymbol{r},\theta)\boldsymbol{r}\,{\rm d}\boldsymbol{r}\,{\rm d}\theta}. \tag{4.114}$$

To simplify the solution of Equation 4.114, let us set $\xi = 1/a_{\rm ex}$, where ξ is the variational parameter. Equation 4.114 can be rewritten as

$$\begin{aligned}\langle E\rangle &= \int \exp(-r\xi)\left\{-\frac{\hbar^2}{2\mu^*}\left[\frac{1}{r}\frac{\partial}{\partial r}\left(r\frac{\partial}{\partial r}\right)+\frac{1}{r^2}\frac{\partial^2}{\partial \theta^2}\right]-\frac{e^2}{4\pi\epsilon\epsilon_0 r}\right\}\exp(-r\xi)r\,dr\,d\theta \\ &= \left[\frac{\hbar^2\xi^2}{2\mu^*} - \frac{2e^2\xi}{4\pi\epsilon\epsilon_0}\right].\end{aligned} \tag{4.115}$$

The energy can now be maximized as follows. Take the first derivative of $\langle E\rangle$ with respect to ξ and equate the results to zero to obtain

$$\zeta = \frac{2\mu^* e^2}{\hbar^2 4\pi\epsilon\epsilon_0}. \tag{4.116}$$

Substituting Equation 4.116 back into Equation 4.115, the exciton ground state energy, $E_0^{\rm 2D}$, is

$$\begin{aligned}E_0^{\rm 2D} &= -\frac{2\mu^* e^4}{\hbar^2(4\pi\epsilon\epsilon_0)^2} \\ &= -\frac{4\mu^* e^4}{2\hbar^2(4\pi\epsilon\epsilon_0)^2} = -4E_0^{\rm 3D},\end{aligned} \tag{4.117}$$

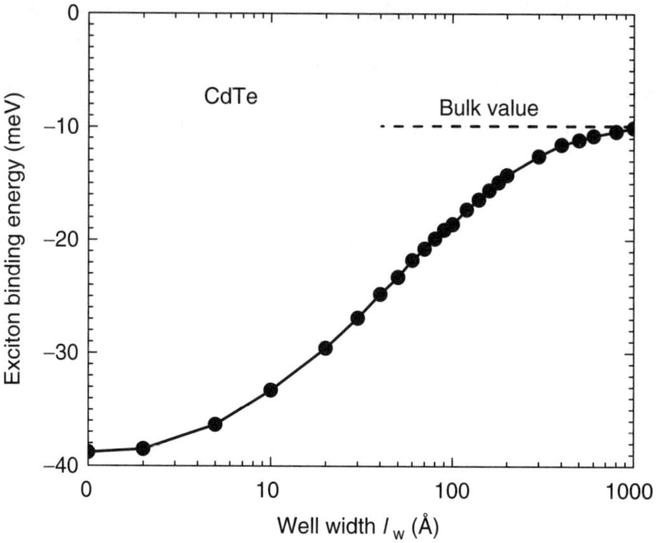

FIGURE 4.23 Exciton binding energy in an infinitely CdTe deep quantum well (Harrison).

where E_0^{3D} is the exciton ground state energy of the bulk material. One can generalize this result according to the following expression:

$$E_n^{2D} = -\frac{R_{ex}}{(n - \frac{1}{2})^2}, \quad n = 1, 2, 3 \ldots \tag{4.118}$$

The exciton binding energy in quantum wells is equal to $4R_{ex}$. The Bohr radius of the exciton in a quantum well (a_{ex}^{2D}) is the inverse of the quantity given by Equation 4.116, which is one-half the exciton radius (a_{ex}^{3D}) in bulk material ($a_{ex}^{2D} = (a_{ex}^{3D}/2)$. The theoretical value of the exciton binding energy in a quantum well presented in Equation 4.117 is the upper limit, which is a difficult limit to reach experimentally. Experimentally, the exciton binding energy in quantum wells is ~2.5 E_0^{3D}. This is still a substantial enhancement of the exciton binding energy, which is very important for many optoelectronic device applications. An example of exciton binding energy in quantum wells is shown in Fig. 4.23, where the binding energy is calculated as a function of the well width of an infinite CdTe quantum well (Harrison). The binding energy limits in this figure satisfy the bulk limit (−10.1 meV) when the well thickness is too large (1000Å) and the quantum well theoretical limit [~4 × (−10.1) meV] when the well thickness approaches zero. Notice that the energy is negative, which implies that the exciton ground state is a bound state.

4.11.3 Excitons in Quantum Dots

The calculation of exciton binding energy in quantum dots is very complicated. The most common approximation used is the variational method, which requires knowledge of a trial function and a Hamiltonian as discussed in the previous subsection. The analysis for quantum dots is more complicated because of the variation of the size and shape of the quantum dots. Generally speaking, the exciton binding energy is much higher in the case of quantum wells and dots as compared to bulk materials due to electron and hole confinement in quantum structures. Accompanying the increase in exciton binding energy is a reduction in the exciton Bohr radius. While reports in the open literature indicate a variety of results for various quantum dot shapes and sizes, a general consensus is that the binding energy in quantum dots increases as the size of the dot decreases. An example is reported by Grundmann *et al*. and shown in Fig. 4.24, where the exciton binding energy is plotted as a function of the base length of an InAs/GaAs pyramidal quantum dot size. The exciton binding energy (R_{ex}) in bulk GaAs and InAs are indicated in the figure. The behavior of the exciton binding energy in quantum dots shown in Fig. 4.24 is the trend that most theoretical calculations exhibit. For example, the change in the binding energy in an ionic semiconductor spherical quantum dot was shown recently to have the following form (Stenger *et al*.)

$$\Delta E_{ex} \propto \frac{\hbar^2 \pi^2}{2\mu^* R^2} + E_{ex}^{\text{Bulk}}, \tag{4.119}$$

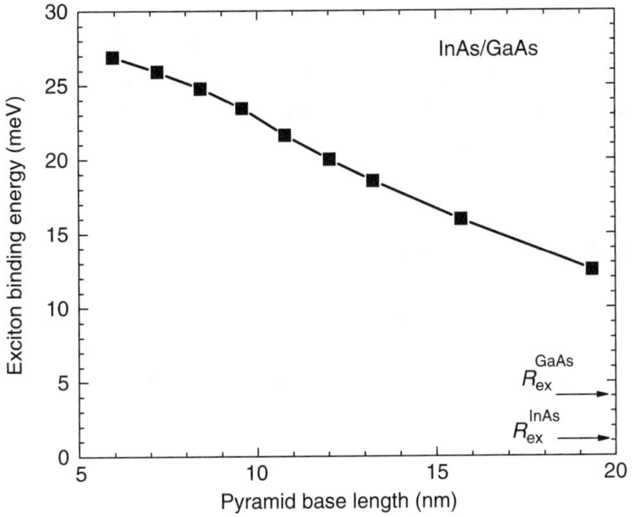

FIGURE 4.24 Exciton binding energy as a function of InAs/GaAs pyramid base length (Grundmann *et al*.).

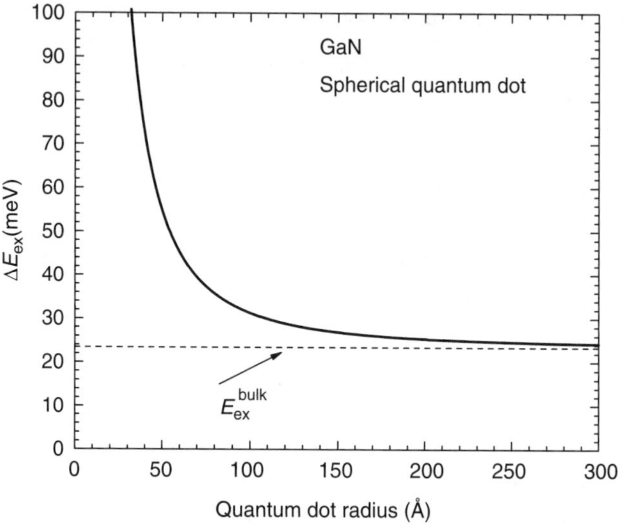

FIGURE 4.25 Variation of change in exciton binding energy as a function of dot size for GaN (after Strenger *et al.*).

where R is the radius of the sphere and E_{ex}^{Bulk} is the exciton binding energy in bulk materials. A plot of this equation is shown in Fig. 4.25 for a GaN spherical quantum dot. The dashed line in Fig. 4.25 represents the exciton binding energy in a bulk GaN material and the solid line represents the change in the exciton binding energy as a function of the quantum dot radius. Again, the behavior of the binding energy in Fig. 4.25 seems to be universal for most quantum dot materials. While optical absorption from excitons in quantum dots has not been reported, perhaps because of the significantly small optical length, theoretical results indicate that excitons in quantum dots could have large oscillator strengths (see, e.g., Bimberg *et al*. and references therein).

4.12 CYCLOTRON RESONANCE

The cyclotron resonance technique has been used to determine the effective masses of charge carriers in high purity bulk semiconductors as well as heterojunctions and quantum wells. This technique requires both electric and magnet fields. Early experiments used microwave radiation in conjunction with the magnetic field. Then, the technique was developed to incorporate an infrared laser light instead of microwave radiation. With this configuration, the magnetic field is swept in a specific range to obtain the cyclotron resonance spectrum. Most recently, infrared light was used instead of laser light. The latter configuration provided a quick determination of the effective masses since the magnetic field

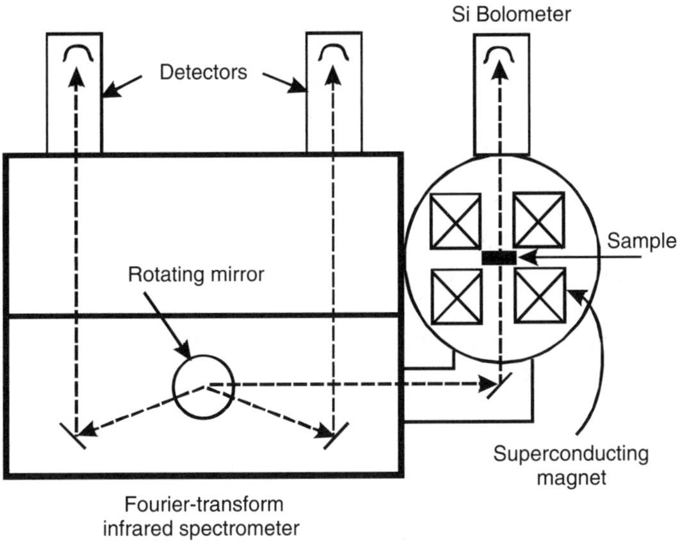

FIGURE 4.26 Cyclotron resonance experimental setup for the measurement of charge charier effective masses.

is fixed and the infrared radiation is scanned using Fourier-transform infrared spectroscopy, as shown in Fig. 4.26.

An electron with a charge e and velocity υ, under the influence of a magnetic field $\boldsymbol{B}$, will experience a force, $\mathcal{F}$ (Lorentz Force), given by

$$\boldsymbol{F} = e\boldsymbol{\upsilon} \times \boldsymbol{B}. \tag{4.120}$$

When υ is perpendicular to $\boldsymbol{B}$, the magnitude of this force is $|\mathcal{F}| = |e|\upsilon B$. On the other hand, the centripetal force, $\mathcal{F}_{\mathrm{c}}$, due to a uniform circular motion of a particle with mass, m, and acceleration υ^2/r, is given by

$$\mathcal{F}_{\mathrm{c}} = m\frac{\upsilon^2}{r}. \tag{4.121}$$

By equating the two forces in Equations 4.120 and 4.121, one can obtain the radius, r, and the period, T, of the circular orbit as

$$r = \frac{m\upsilon}{eB} \quad \text{and} \quad T = \frac{2\pi r}{\upsilon}. \tag{4.122}$$

The angular frequency of the particle is called the *cyclotron frequency*, which is given by

$$\omega_{\mathrm{c}} = \frac{2\pi}{T} = \frac{\upsilon}{r} = \frac{|e|B}{m}. \tag{4.123}$$

This equation can be modified for charge carriers in semiconductors as

$$\omega_c = \frac{|e|B}{m^*}, \tag{4.124}$$

where m^* is the charged particle effective mass. For cyclotron resonance to work, the electric field, $\mathcal{E}$, of the radiation should have a nonzero component in the plane of the cyclotron motion. The cyclotron resonance condition occurs when the radiation energy is equal to the energy needed for the charge carrier to make a transition between adjacent Landau energy levels. To determine the cyclotron resonance condition, the equation of motion of a free charged particle under the influence of electromagnetic and magnetic fields can be written as

$$m^* \left(\frac{d\boldsymbol{v}}{dt} + \frac{\boldsymbol{v}}{\tau} \right) = e(\boldsymbol{\mathcal{E}} + \boldsymbol{v} \times \boldsymbol{B}), \tag{4.125}$$

where the first term in the left-hand side is due to the particle acceleration and the second term is due to collisions, which are characterized by the relaxation time (τ) of the carriers. The velocity υ is the drift velocity under the influence of the electric field, $\mathcal{E}$. The magnetic field associated with the electromagnetic radiation is too small compared to the applied magnetic field, and hence, it is ignored.

Let us take the polarization of the electric field along the x-direction and the magnetic field along the z-direction. Since the particle motion is in the x–y plane, the drift velocity has two components in the x- and y-directions. Thus, we can write the velocity, electric, and magnetic fields as

$$\begin{aligned} \boldsymbol{\mathcal{E}} &= \mathcal{E}_x \exp(-i\omega t) \\ \boldsymbol{B} &= B_z \\ \boldsymbol{v} &= (\upsilon_x \widehat{x} + \upsilon_y \widehat{y}) \exp(-i\omega t). \end{aligned} \tag{4.126}$$

Substituting Equation 4.126 into Equation 4.125, we have

$$m^* \left(-i\omega + \frac{1}{\tau} \right) \upsilon_x = e\mathcal{E}_x + e\upsilon_y B_z \tag{4.127a}$$

$$m^* \left(-i\omega + \frac{1}{\tau} \right) \upsilon_y = -e\upsilon_x B_z \tag{4.127b}$$

Solving these two equations for υ_x, we have

$$\upsilon_x = \frac{e\mathcal{E}_x}{m^*} \frac{(-i\omega + \tau^{-1})}{(-i\omega + \tau^{-1})^2 + \omega_c^2}, \tag{4.128}$$

where ω_c is defined by Equation 4.124. The current density in the x-direction can be written as

$$j_x = eNv_x = \sigma(\omega)\mathcal{E}_x, \tag{4.129}$$

where N is the number of carriers. Combine Equations 4.129 and 4.128 to yield

$$\begin{aligned} j_x &= \frac{e^2 N\mathcal{E}_x}{m^*}\frac{(-i\omega+\tau^{-1})}{(-i\omega+\tau^{-1})^2+\omega_c^2} \\ &= \frac{e^2 N\tau\mathcal{E}_x}{m^*}\frac{(1-i\omega\tau)}{(\omega_c^2-\omega^2)\tau^2+1-2i\omega\tau}. \end{aligned} \tag{4.130}$$

This leads to the following expression for the conductivity

$$\begin{aligned} \sigma(\omega) &= \frac{e^2 N\tau}{m^*}\frac{(1-i\omega\tau)}{(\omega_c^2-\omega^2)\tau^2+1-2i\omega\tau} \\ &= \sigma_0\frac{(1-i\omega\tau)}{(\omega_c^2-\omega^2)\tau^2+1-2i\omega\tau}, \end{aligned} \tag{4.131}$$

where σ_0 is the DC conductivity given by $\sigma_0 = \frac{e^2 N\tau}{m^*}$. The frequency-dependent conductivity is a complex quantity given by

$$\sigma(\omega) = \sigma_1(\omega) + i\sigma_2(\omega), \tag{4.132}$$

where $\sigma_1(\omega)$ and $\sigma_2(\omega)$ are the real and imaginary parts of the conductivity and are given by

$$\sigma_1(\omega) = \sigma_0\frac{1+(\omega_c^2+\omega^2)\tau^2}{[1+(\omega_c^2-\omega^2)\tau^2]^2+4\omega^2\tau^2} \tag{4.133a}$$

$$\sigma_2(\omega) = \sigma_0\frac{2\omega\tau-\omega\tau[1+(\omega_c^2-\omega^2)\tau^2]}{[1+(\omega_c^2-\omega^2)\tau^2]^2+4\omega^2\tau^2} \tag{4.133b}$$

In the Faraday configuration, where the electric field is perpendicular to the magnetic field, the power absorbed by the carriers is given by

$$P(\omega) = \mathrm{Re}(j_x\mathcal{E}_x) = \sigma_0|\mathcal{E}_x|^2\frac{1+(\omega_c^2+\omega^2)\tau^2}{[1+(\omega_c^2-\omega^2)\tau^2]^2+4\omega^2\tau^2}. \tag{4.134}$$

For $\omega = \omega_c$ and $\omega_c\tau \gg 1$, the power is reduced to

$$P(\omega_c) = \frac{1}{2}\sigma_0|\mathcal{E}_x|^2. \tag{4.135}$$

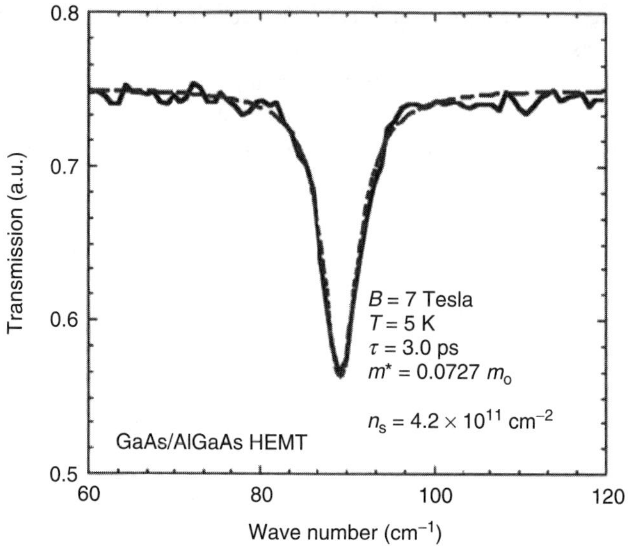

FIGURE 4.27 Cyclotron resonance spectrum (solid line) obtained for GaAs/AlGaAs high electron mobility transistor (HEMT) structure. The dashed line is the theoretical fit using Equation 4.134.

A plot of the relative power absorbed $(P(\omega)/P_0)$, where $P_0 = \sigma_0|\mathcal{E}_x|^2$, is shown in Fig. 4.27 for a GaAs/AlGaAs high electron mobility transistor structure. This figure shows the transmission spectrum obtained from the Fourier-transform infrared spectroscopy set up shown in Fig. 4.26. The dashed line is a fit of the experimental spectrum using Equation 4.134. The effective mass was determined from the cyclotron frequency, which was used as a fitting parameter. Notice that the electron effective mass of $0.0727m_0$ is slightly larger than the electron effective mass in bulk materials. The larger effective mass in heterostructures has been observed in many semiconductor quantum structures. Notice that Fig. 4.27 displays the transmission spectrum. Taking the negative of $(P(\omega)/P_0)$ is the proper form for fitting the transmission experimental results.

The cyclotron resonance is due to energy level quantization in the presence of electric and magnetic fields. These energy levels are known as *Landau levels* and can be obtained by solving the Schrödinger equation. Neglecting the crystal potential, the Hamiltonian can be written as

$$H = \frac{1}{2m^*}(P_x^2 + (Py - e\boldsymbol{A})^2 + P_z^2) + g^*\mu_{\text{B}}\sigma_z B, \tag{4.136}$$

where $\boldsymbol{A}$ is the vector potential given in Landau gauge as $\boldsymbol{A} = (0, \text{Bx}, 0)$, μ_B is the Bohr magneton given by $eh/(2m^*)$, g^* is the effective g-factor, and σ_z is the electron spin quantum number given by $\pm 1/2$. Notice that the magnetic induction (H) is given by $\boldsymbol{H} = \mu_0 \boldsymbol{B} = \nabla \times \boldsymbol{A}$. The wave function is an envelope function, which can be written as

$$\psi(x, y, z) = \exp(ik_y y + ik_z z)u(x). \tag{4.137}$$

Schrödinger equation can be written as

$$\left[\frac{1}{2m^*}(P_x^2 + (P_y - e\mathbf{A})^2 + P_z^2) + g^*\mu_B\sigma_z B\right]\psi(x, y, z) = E\psi(x, y, z). \tag{4.138}$$

Use momentum operators to give

$$\frac{\partial^2 u(x)}{\partial x^2} - \left(k_y - \frac{eBx}{\hbar}\right)^2 u(x) + \frac{2m^*}{\hbar^2}E'u(x) = 0, \tag{4.139}$$

where

$$E' = E - \frac{\hbar^2 k_z^2}{2m^*} - g^*\mu_B\sigma_z B. \tag{4.140}$$

Equation 4.138 can be rewritten as

$$-\frac{\hbar^2}{2m^*}\frac{\partial^2 u(x)}{\partial x^2} + \frac{m^*}{2}\left(\frac{eBx}{m^*\hbar} - \frac{\hbar k_y}{m^*}\right)^2 u(x) = E'u(x). \tag{4.141}$$

This equation is a one-dimensional harmonic oscillator equation with frequency ω_c and energy given by

$$E' = \left(n + \frac{1}{2}\right)\hbar\omega_c. \tag{4.142}$$

Substituting Equation 4.140 into Equation 4.142, we have

$$E_n = \left(n + \frac{1}{2}\right)\hbar\omega_c + \frac{\hbar^2 k_z^2}{2m^*} + g^*\mu_B\sigma_z B. \tag{4.143}$$

As this equation indicates, the electronic energy level, E, splits under the influence of magnetic field into Landau energy levels, with n being the Landau quantum number, separated by $\hbar\omega_c$. Furthermore, each Landau level splits into two levels because of the inclusion of electron spin. Thus, the electronic energy levels are quantized in the x–y plane (the plane perpendicular to the magnetic field) and have translational energy $[\hbar^2 k_z^2/(2m^*)]$ along the z-direction (along the magnetic field direction).

4.13 PHOTOLUMINESCENCE

With the increasing importance of nanostructures in optoelectronics, PL becomes a powerful technique that is used to characterize semiconductor micro- and nanostructures. This is because it provides information on many fundamental properties of semiconductors and nanostructures such as crystalline order, strain, composition, doping, surface carrier depletion depth, crystal damage, quality of interfaces, layer thickness, extended defects, microscopic defects, surface quality, and many other parameters. Thus, this is one of the most important and versatile techniques for investigating compound semiconductors and their nanostructures. The interband optical absorption process in a semiconductor involves the excitation of an electron from the valence band to the conduction band after absorbing the photon. The reverse radiative process, where the photoexcited electron decays from the conduction band to the valence band, is called *photoluminescence*. In this process, the electron emits energy (photon) as it drops from the conduction band to the valence band. Luminescence can also be observed by injecting electrons into the semiconductor material, in which the injected electrons decay to the valence band by emitting photons. This process is called *electroluminescence*. Photon emission is more complicated than photon absorption in a semiconductor, but the emission results are easier to analyze. A comparison between absorption and emission is shown in Fig. 4.28, where we present the optical absorption and PL spectra for $In_{0.52}Ga_{0.48}As/In_{0.52}Al_{0.48}As$ multiple quantum wells. The optical absorption spectrum threshold occurs at ~1.14 μm (~1.088 eV), while

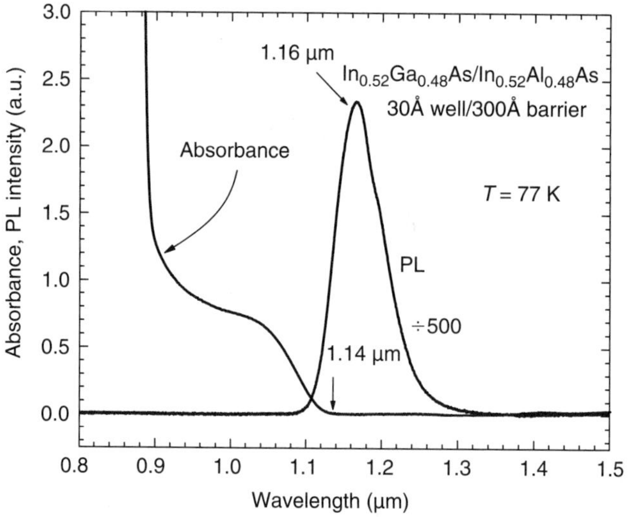

FIGURE 4.28 Absorption and photoluminescence (PL) spectra of $In_{0.52}Ga_{0.48}As/In_{0.52}Al_{0.48}As$ multiple quantum wells plotted as a function of the wavelength. The spectra were measured at 77 K.

the PL peak occurs at ~1.16 μm (1.069 eV). The optical absorption threshold and the PL peak are expected to be identical since the band gap is the same at a constant temperature. The reason for the difference between the absorption and emission is due to electron–phonon coupling. Electron–phonon coupling in a semiconductor involves extensive theoretical analysis. The simplest model used to explain electron–phonon coupling is the configuration coordinate model.

The configuration coordinate model is illustrated in Fig. 4.29. Let us consider the interband transition in a direct band gap semiconductor material. In reality, the atoms vibrate in a solid and the total energy of the electron is the sum of electronic and vibronic energies. The total energy of an electron in the valence band can be expanded in a Taylor series about a coordinate minimum Q_0 such that

$$E(Q) = E(Q_0) + \frac{dE}{dQ}(Q - Q_0) + \frac{1}{2}\frac{d^2E}{dQ^2}(Q - Q_0)^2 + \cdots \qquad (4.144)$$

Since the expansion is made about an extrema, the first derivative in Equation 4.144 is zero. Thus, the valence band can be presented by a parabola

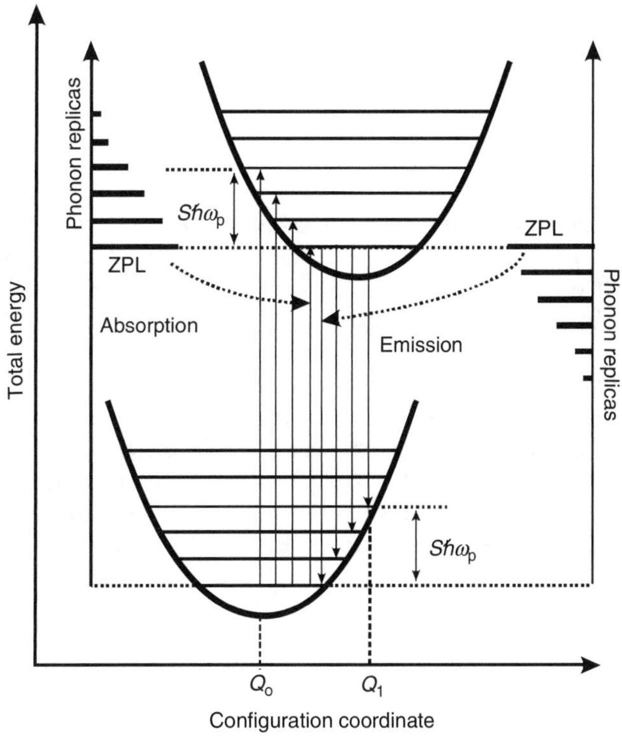

FIGURE 4.29 Schematic of the configuration coordinate interband transition in a direct band gap semiconductor in the presence of electron–phonon coupling.

around Q_0 as shown in Fig. 4.29. The conduction band can be presented in a similar manner, with a minimum at $Q_1 > Q_0$. The electron–phonon coupling is zero when $Q_1 = Q_0$. Each parabola represents a simple harmonic oscillator with a quantized energy

$$E_n^{Q_0,Q_1}(\omega_p) = \left(n + \frac{1}{2}\right)\hbar\omega_p, n = 0, 1, 2, \ldots, \quad (4.145)$$

where ω_p is the phonon angular frequency and $E_n^{Q_0,Q_1}(\omega_p)$ is the phonon energy associated with the valence and conduction bands. The optical absorption of the interband transition is presented by a series of arrows pointing upward in the figure, where the electron are excited from the phonon ground state $E_0^{Q_0}(\omega_p)$ in the valence band parabola to the phonon ground and excites states in the conduction band parabola $E_n^{Q_1}(\omega_p)$. On the other hand, the PL or emission transitions are presented by the arrows pointing downward, where the electrons decay from the phonon ground state, $E_0^{Q_1}(\omega_p)$, in the conduction band to the phonon levels, $E_n^{Q_0}(\omega_p)$, in the valence band. The transition (either absorption or emission) between $E_0^{Q_0}(\omega_p)$ and $E_0^{Q_1}(\omega_p)$ is called the *zero phonon line* (ZPL), which means it is a pure electronic transition with zero electron–phonon coupling. The electronic transitions between $E_n^{Q_1}(\omega_p)$ and $E_n^{Q_0}(\omega_p)$ for $n \neq 0$ are called *phonon replicas*. The difference between the absorption (E_a) and emission (E_e) energies is called the *Stokes shift* and can be written as

$$E_a - E_e = 2S\hbar\omega_p, \quad (4.146)$$

where S is a dimensionless parameter called the *Huang–Rhys factor*. It is a measure of strong (large value of S) or weak (small value of S) electron–phonon coupling. Half of the Stokes shift is called the *Franck–Condon shift*, which is commonly referred to as $d_{F-C} = S\hbar\omega_p$.

Notice that phonon replicas are observed at the higher energy side of the ZPL in the case of the absorption spectrum, which indicates that phonons are absorbed by the electrons. In the case of emission (PL), the phonon replicas occur at the lower energy side of the ZPL, which means that the electrons emits phonons as they decay from the conduction band to the valence band. Two examples of ZPLs and their phonon replicas are shown in Fig. 4.30. The first example is the optical absorption of the ZPL associated with the EL2 defect in GaAs (Fig. 4.30a). The ZPLs occur at 8378 cm^{-1} (1.0387 eV) and the replicas are those of the TA phonon mode (~10 meV) in GaAs (recall that 1.0 eV = 8065.46 cm^{-1}). The second example is the PL zero phonon mode observed in an InAs/GaAs single layer quantum dots (Fig. 4.30b). The ZPL peak is indicated as ZPL. The stronger peak around 0.934 eV is also a ZPL because of the fact that the quantum dots have two dominant sizes. The ripples below 0.9 eV are due to phonon replicas separated by an average energy of ~36.4 meV. This phonon energy is most likely to be the optical phonon mode generated at the InAs/GaAs interface.

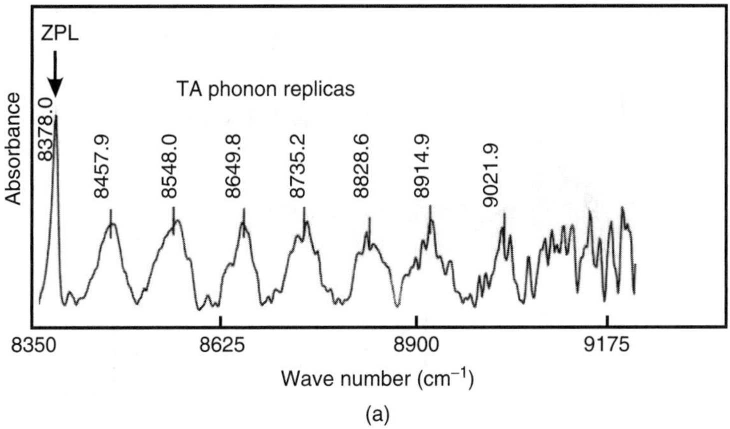

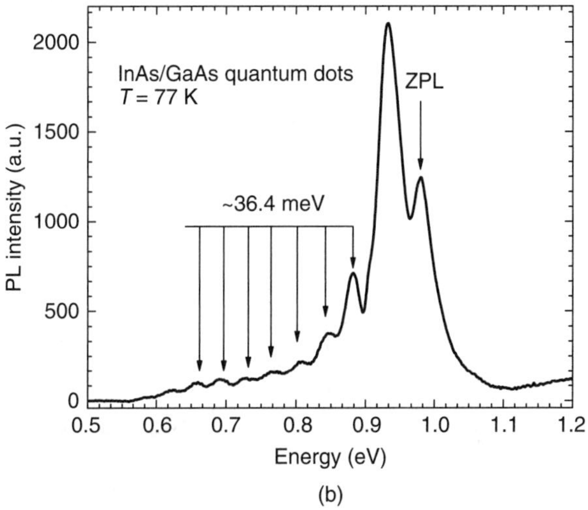

FIGURE 4.30 Zero phonon lines and their replicas in (a) EL2 defect in GaAs (Manasreh *et al.*, 1987) and (b) InAs/GaAs single layer quantum dots grown by the MBE (molecular beam epitaxy) technique.

PL can be observed in semiconductors and their nanostructures if electrons and holes are generated by optical excitation followed by radiation emission. If the electrons recombine with the holes without emitting radiation, the transition is called *nonradiative*. The PL technique is currently a standard technique in both industry and academia. It is a used to calibrate the epitaxial growth rate and growth quality as illustrated in Fig. 4.31. A test of the epitaxial growth would be to grow a few quantum wells with different thicknesses, as shown in Fig. 4.31a, where GaAs/AlGaAs quantum wells grown on semi-insulating GaAs substrate were chosen as an example. The barrier thickness is usually chosen to be thick enough to prevent tunneling between wells. As the quantum well

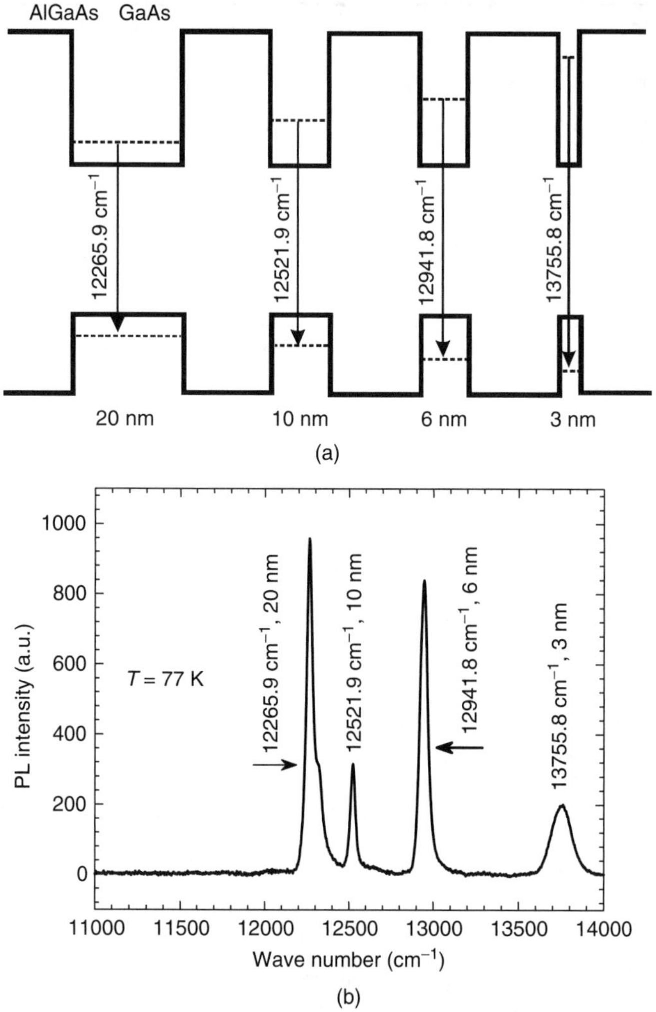

FIGURE 4.31 (a) A sketch of four GaAs/AlGaAs quantum wells grown by the molecular beam epitaxy technique on GaAs semi-insulating substrate. (b) Photoluminescence spectrum measured at $T = 77$ K for the structure described in (a).

thickness is reduced from 20 to 3 nm, the bound states are squeezed outward and the interband transition energy is increased, as shown in the figure. The corresponding PL spectrum is shown Fig. 4.31b, where the PL intensity is plotted as a function of photon energy. The PL energy (E_{PL}) can be written as

$$E_{\mathrm{PL}}(L_z) = E_{\mathrm{g}} + E_{\mathrm{CB1}}(L_z) + E_{\mathrm{HH1}}(L_z) - E_{\mathrm{ex}}(L_z), \qquad (4.147)$$

where Eg is the fundamental band gap of bulk GaAs material which is taken as 1.50 eV (12098 cm^{-1}) at $T = 77$ K, E_{CB1} is the ground bound state in the conduction band, E_{HH1} is the bound ground state of the heavy hole in the valence band, E_{ex} is the exciton binding energy, and L_z is the well width. The PL peak observed for the 20-nm quantum well exhibits a structure with a shoulder at the higher energy side. This is due to the presence of bound and free excitons in the quantum well, which is an indication of high structural interfaces.

The excess electron concentration (N) created by the laser excitation in the PL experiment is equal to the excess hole concentration, which is given by the rate equation as

$$\frac{\mathrm{d}N}{\mathrm{d}t} = -\frac{N}{\tau_r}, \tag{4.148}$$

where τ_r is the lifetime for the carriers that undergo radiative recombination, which is the inverse of the Einstein coefficient for the spontaneous emission rate. Integration of this equation yields

$$N(t) = N_0 \exp(-t/\tau_r), \tag{4.149}$$

where N_0 is the excess electron concentration at $t = 0$. The radiative recombination rate (R_r) is defined as

$$R_r = \frac{\mathrm{d}N}{\mathrm{d}t} = -\frac{N}{\tau_r}. \tag{4.150}$$

When the nonradiative recombination rate (R_n) is considered, the total spontaneous recombination rate (R_s) can be written as

$$R_s = R_r + R_n. \tag{4.151}$$

For exponential decay, the internal quantum efficiency (η_i) is given by the carrier lifetime as

$$\eta_i = \frac{\tau_r^{-1}}{\tau_r^{-1} + \tau_n^{-1}} = \frac{1}{1 + \tau_r/\tau_n}, \tag{4.152}$$

where τ_n is the lifetime of the carriers that undergo nonradiative recombination. The internal quantum efficiency of the interband transition is equal to unity when τ_n is zero.

It is possible to estimate the radiative recombination lifetime from carrier concentration in direct band gap semiconductors. The relationship between the lifetime and the carrier concentration is left as an exercise.

4.14 BASIC CONCEPTS OF PHOTOCONDUCTIVITY

The electrical conductivity due to the excess carrier generated by illuminating the sample with photon is called *photoconductivity*. The photoconductivity is measured by attaching electrodes to the sample as shown in Fig. 4.32a. The device that absorbed photons and generate electric current is called *photodetector*. For an undoped semiconductor photodetector, the DC short circuit photocurrent is given as

$$I_{\text{ph}} = e\eta wl\Phi_e G, \tag{4.153}$$

where Φ_e is the optical irradiant flux, G is the photoconductive gain, and wl is the active detector area shown in Fig. 4.32a. Since the incident photons generate electron–hole pairs in semiconductor materials, the photoconductivity is a two-carrier process, where the photoconductive current can be written as

$$I_{\text{ph}} = ewd(\mu_n \Delta n + \mu_p \Delta p)\frac{V_b}{l}, \tag{4.154}$$

where wd is the detector cross section, V_b is the bias voltage, l is distance between the two electrodes, μ_n and μ_p are the mobility of the electrons and holes, respectively. The excess carrier concentration Δn and Δp are given by

$$n = n_0 + \Delta n \quad \text{and} \quad p = p_0 + \Delta p, \tag{4.155}$$

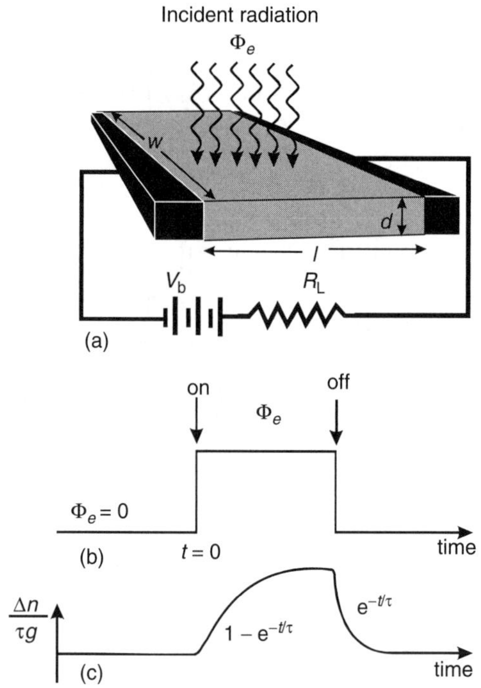

FIGURE 4.32 (a) Geometry of a biased photodetector showing the incident photons and the load resistance. (b) A radiation input pulse as a function of time. (c) The detector output signal as a function of time.

where n_0 and p_0 are the thermal equilibrium carrier concentrations.

The transport properties of photodetectors are usually dominated by electrons. The current balance equation can be constructed from the photogeneration, recombination, drift, and diffusion processes as follows:

$$\frac{\partial \Delta n}{\partial t} = \mathcal{G} - \frac{\Delta n}{\tau} - \frac{1}{e}\nabla - \boldsymbol{J}, \tag{4.156}$$

where $\mathcal{G}$ is the generation rate (in unit number/cm^{-3}.s^{-1}), τ is the recombination time, and $\boldsymbol{J}$ is the current density due to drift and diffusion. The diffusion and drift current can be neglected as compared to the generation recombination terms. Thus, Equation 4.156 can be reduced to

$$\frac{\partial \Delta n}{\partial t} = \mathcal{G}\frac{\Delta n}{\tau}. \tag{4.157}$$

The generation rate can be written as

$$\mathcal{G} = \frac{\Phi_e \eta}{d}. \tag{4.158}$$

For the steady state case, we have

$$\frac{\partial \Delta n}{\partial t} = 0 = \mathcal{G} - \frac{\Delta n}{\tau} \Rightarrow \tau = \frac{\Delta n}{\mathcal{G}} = \frac{\Delta n d}{\eta \Phi_e}. \tag{4.159}$$

In addition, the photoconductive gain can be obtained by equating Equations 4.153 and 4.154

$$G = \frac{d\mu_{\mathrm{n}} \Delta n V_{\mathrm{b}}}{\eta \Phi_e l^2}. \tag{4.160}$$

Substituting Equation 4.159 into Equation 4.160 yields

$$G = \frac{\tau \mu_{\mathrm{n}} V_{\mathrm{b}}}{l^2} = \frac{\tau}{l^2/(\mu_{\mathrm{n}} V_{\mathrm{b}})} = \frac{\tau}{\tau_{\mathrm{tr}}}, \tag{4.161}$$

where τ_{tr} is the transit time of electrons between the two electrodes, which is given as $\tau_{tr} = l/\upsilon_d = l/(\mu_n \mathcal{E}) = l^2/\mu_n V_b$. Thus, the photoconductive gain is the ratio of the recombination time to the transit time (Fig. 4.32b).

When the excess carrier density is time dependent, then the steady-state solution is not valid any more. The solution of Equation 4.157 can be obtained by multiplying both sides of the equation with $\mathrm{e}^{t/\tau}$ as follows:

$$\mathrm{e}^{t/\tau}\left(\frac{\partial \Delta n}{\partial t} + \frac{\Delta n}{\tau}\right) = \mathcal{G}\mathrm{e}^{t/\tau} \quad \text{or} \quad \frac{\partial}{\partial t}(\Delta \mathrm{n} \mathrm{e}^{t/\tau}) = \mathcal{G}\mathrm{e}^{t/\tau}. \tag{4.162}$$

Integrating the above equation to obtain

$$\Delta n e^{t/\tau} = \int_0^t \mathcal{G} e^{t/\tau} \partial t = \tau \mathcal{G}(e^{t/\tau} - 1) \Rightarrow \Delta n = \tau \mathcal{G}(1 - e^{-t/\tau}). \qquad (4.163)$$

A plot of the excess carrier as a function of time is sketched in Fig. 4.32c.

When the incident photon flux is turned off, Equation 4.157 is reduced to the following:

$$\frac{\partial \Delta n}{\partial t} = -\frac{\Delta n}{\tau}. \qquad (4.164)$$

The solution of this equation is

$$\Delta n = C_1 e^{-t/\tau}, \qquad (4.165)$$

where C_1 is a constant, which is determined from the initial conditions. If one assumes that the photon flux is turned off when Δn is maximum (i.e., $\Delta n = \tau \mathcal{G}$) as shown in Fig. 4.32c, then $C_1 = \tau \mathcal{G}$. By taking the Fourier transform of Equation 4.165, one can express the device spectral current responsivity, $\mathcal{R}_i(\omega)$, in the modulation frequency domain as

$$\mathcal{R}_i(\omega) = \frac{\mathcal{G}_i(0)}{\sqrt{(1 + \omega^2 \tau^2)}}, \qquad (4.166)$$

where $\mathcal{G}_i(0)$ is the responsivity at zero frequency given by

$$\mathcal{R}_i(0) = \frac{e\eta\lambda}{hc} G, \qquad (4.167)$$

where η is the quantum efficiency and λ is the incident photon wavelength. The expression for the photoconductive gain, G, is shown in Equation 4.162, which depends on the bias voltage. Thus, the responsivity is a function of the wavelength, quantum efficiency, and the photoconductive gain. The simplistic responsivity model discussed above excludes many effects, such as surface recombination and sweep-out effects. Furthermore, the responsivity depends on the product of the quantum efficiency and the photoconductive gain. It is difficult to measure them separately.

The internal quantum efficiency, η_0, is usually close to unity because of the fact that most of the photons are absorbed and contribute to the photoconductivity. On the other hand, the external quantum efficiency, η, depends on the reflection coefficients of both the top and bottom detector surfaces. It also depends on the optical absorption coefficient, α. The derivation of the external quantum efficiency depends on the detector type and configuration. For example, the external quantum efficiency derived for infrared charge transfer devices is given by (Nelson)

$$\eta = \frac{\eta_0(1 - r_1)(1 + r_2 e^{-2\alpha d})(1 - e^{-\alpha d})}{(1 - r_1 r_2 e^{-2\alpha d})}, \tag{4.168}$$

where d is the detector thickness (Fig. 4.32a) and r_1 and r_2 are the reflection coefficients of the top and bottom surfaces. If $r_1 = r_2 = r$ and if the absorption coefficient is sufficiently large, the above expression can be rewritten as

$$\eta \approx \eta_0(1 - r). \tag{4.169}$$

If the photodetector is made such that the reflection coefficient from the top surface is zero and the reflection coefficient from the back side is unity, the quantum efficiency is reduced to

$$\eta \approx \eta_0(1 + e^{-2\alpha d})(1 - e^{-\alpha d}). \tag{4.170}$$

For highly absorptive detector material, the external quantum efficiency described by Equation 4.170 is reduced to the internal quantum efficiency. This is an ideal design, where the external quantum efficiency approaches unity. In general, the external quantum efficiency can take different forms depending on the type of the detectors and basic physical principles of operation.

4.15 SUMMARY

The optical properties of bulk semiconductors and their low dimensional quantum structures are discussed in this chapter. We started the chapter by defining the difference between the bulk and quantum well materials with the emphasis on interband and intersubband transitions. Bound and free excitons were illustrated in crystalline structures. The basic electromagnetic formalism is introduced with a fundamental discussion regarding the refractive index, dielectric constant, and linear optical absorption coefficient. The optical absorption coefficients of interband transitions in direct and indirect semiconductor materials were derived using the Fermi golden rule. The formalisms were extended to derive the optical absorption coefficients of interband transitions in type I and type II quantum wells. Detailed discussions on the optical absorption coefficient of the intersubband transitions in quantum wells and quantum dots are presented for both bound-to-bound and bound-to-continuum cases. An example of intersubband transition is presented for GaN/AlGaN multiple quantum wells, where the piezoelectric doping is significant. A complete section on the selection rules of both interband and intersubband transitions is presented.

Excitons in both bulk and quantum structures play a major role in the optoelectronic devices. Detailed analysis of the exciton binding energy and radius is presented for bulk semiconductors, quantum wells, and quantum dots. An attractive feature of semiconductor low dimensional quantum structures is that the binding energies of the excitons are much higher than those of the excitons in bulk materials. Selected techniques used to optically characterize the semiconductor quantum structures, such as cyclotron resonance, PL, and photoconductivity are discussed.

PROBLEMS

4.1. The electric and magnetic fields can be written in terms of vector ($\boldsymbol{A}$) and scalar (ϕ) potentials such as $\boldsymbol{\mathcal{E}} = -\frac{\partial \mathbf{A}}{\partial t} - \nabla\phi$ and $\boldsymbol{B} = \frac{1}{\mu_0}\nabla x\boldsymbol{A}$. Rewrite the four Maxwell equations in terms of these two potentials.

4.2. Show that Equation 4.9 is valid.

4.3. Use the complex definition of the refractive index and the dielectric constant to show that the optical absorption coefficient can be expressed as $\alpha(\omega) = \frac{\omega\epsilon_2(\omega)}{cn_1(\omega)}$.

4.4. Derive expressions for the real (n_1) and imaginary (n_2) parts of the refractive index in terms of the real (ϵ_1) and imaginary (ϵ_2) parts of the dielectric constants. Plot n_1, n_2, ϵ_1, and ϵ_2 as a function of ω. Assume $N = 10^{17}\ m^{-3}$ and $m^* = 0.067m_0$.

4.5. The oscillator strength of the interband transition can be defined as $f_{vc} \approx \frac{2P^2}{m_0(E_{kc} - E_{kv})}$. In addition, the sum rule for a solid can be written as $\sum_{m \neq n} f_{mn} = \left|1 - \frac{m_0}{m_h^*}\right|$, where we sum the oscillator strength of an electronic transition from all m states to n state at with the same $\boldsymbol{k}$-value. Use these expressions to calculate the absorption coefficient of the interband transition in GaAs for $(\hbar\omega - E_g) = 0.1$ eV.

4.6. Calculate the optical absorption coefficient of the interband transition in 20 Å GaAs AlGaAs quantum well. Assume that the transition occurs from the ground state of the heavy hole in the valence band to the ground state of the electron in the conduction band. Assume that the number of periods is 50 and the photon energy required to excite this transition is 1.75 eV.

4.7. Calculate the oscillator strength and the optical absorption coefficient of a bound-to-bound intersubband transition in GaAs/AlGaAs multiple quantum wells. Assume that the number of wells is 50, the well width is 75 Å, the half width at half maximum is 7 meV, the electron density is 5×10^{11} cm^{-2}, and the photon energy required to excite the transition is 180 meV.

4.8. Show that the Brewster's angle of GaAs is 73°. What does this angle mean?

4.9. The following figure (Fig. P4.9) is a waveguide made of GaAs with a thickness of 0.4 mm. The photons enter the sample at 45° from the formal as shown in the figure. Finish the design of this waveguide such as three passes will be made by the photons before they exit the sample. The GaAs refractive index is 3.4. What would be the length of the waveguide?

Fig. P4.9

4.10. Use the definition of the oscillator strength of intersubband transition in GaAs/AlGaAs multiple quantum wells as $f_{01} = \frac{2m^*\omega}{\hbar}|\langle n\mathbf{k}_\perp|z|n'\mathbf{k}'_\perp\rangle|^2$, where $\langle n\mathbf{k}_\perp|z|n'\mathbf{k}'_\perp\rangle$ is known as the *overlap integral*. Calculate the oscillator strength for a 100 Å thick well, where photon energy needed to excite the transition is $\hbar\omega = 0.15$ eV.

4.11. A time-dependent quantum operator can be written as $\frac{d\boldsymbol{M}(t)}{dt} = \frac{i}{\hbar}(H_0\boldsymbol{M} - \boldsymbol{M}H_0)$. Use the dipole matrix element to show that the oscillator strength can be written as $f_{01} = \frac{2m^*\omega}{\hbar}|\langle n\mathbf{k}_\perp|z|n'\mathbf{k}'_\perp\rangle|^2$.

4.12. Calculate the optical absorption band edge associated with Hh1-E1 transition of InAs/GaAs quantum dot at room temperature. Assume that the quantum dot has a cubic shape of side 6.5 nm. Compare your results to the band gap of bulk InAs material.

4.13. Calculate the dipole matrix element of $1 \to 2$, $1 \to 4$, and $2 \to 3$ transitions in an infinite GaAs quantum well. The well width is of 10 nm, the effective mass is $0.067m_0$, and the wave functions of the bound states can be expressed as $\varphi_n(z) = \sqrt{2/L}\sin(n\pi z/L)$. Calculate the corresponding wavelengths of these three transitions.

4.14. Search the literature for the electron and heavy hole effect masses, band gaps, and dielectric constants of five direct band gap semiconductor bulk materials other than those listed in Table 4.1. Calculate and plot the exciton binding energy and the exciton radius in these materials as a function of their band gaps.

4.15. Start from Schrödinger equation and a trial function of the following form $\psi = \exp\left(-\frac{r_\perp}{a_{2D}}\right)$, where $r_\perp = \sqrt{x^2 + y^2}$ and a_{2D} is the exciton radius in quantum wells. Show that the exciton binding energy in quantum well can be written as $E_{\text{ex}}^{\text{2D}} = -4E_{\text{ex}}^{\text{3D}}$ and $a_{\text{2D}} = 0.5a_{\text{3D}}$, where 2D and 3D indicate two-dimensional (quantum wells) and three-dimensional (bulk material) systems, respectively.

4.16. A GaAs/AlGaAs quantum well has an effective electron mass of $0.072m_0$. A peak in the cyclotron resonance spectrum was observed at 20 meV. Calculate the magnetic field used to generate this peak. Estimate the splitting of Landau levels due to the electron spin. Assume $g^* = 1.75$ and $\mu_B = 9.27 \times 10^{-24}$ J/T.

4.17. In the photoluminescence experiment, the carriers' spontaneous recombination rate can be written as $R_s = Bnp$, where B is a constant, n is the electron concentration, and p is the hole concentration. On the other hand, this rate can be written as the sum of the spontaneous rates at thermal equilibrium and in the presence of the excess carriers. Derive an expression for the lifetime of the excess carriers and express your answer for the two cases of high and low injection rates.

4.18. A band-to-band photoluminescence transition was observed for InGaAs thin film at 1825.68 nm. Calculate the In composition needed to produce this peak. Repeat the same process to obtain the Al fraction in AlGaAs thin film, which has a PL peak at 689.62 nm. Search the literature to obtain the band gap of $In_xGa_{1-x}As$ and $Al_xGa_{1-x}As$ as a function of the mole fraction (x).

BIBLIOGRAPHY

Balkanski M, Wallis RF. Semiconductor physics and applications. Oxford: Oxford;2000.

Bastard G. Wave mechanics applied to semiconductor heterostructures. New York: Halsted Press; 1988.

Bernardini F, Fiorentini V, Vanderbilt D. Phys Rev B 1997; 56:R10024.

Bimberg D, Grundmann M, Ledentsov NN. Quantum dot heterostructures. New York: Wiley; 1999.

Brown GJ, Szmulowicz F. In: Razeghi M, editors. Long wavelength infrared detectors. New York: Taylor and Francis; 1996.

Choi KK. In: Feng ZC, editor. Semiconductor interfaces, microstructures and devices. Philadelphia: IOP; 1993.

Elliot RJ. Phys Rev 1957; 108:1384.

Grundmann M, Stier O, Bimberg D. Phys Rev B 1995; 52:11969.

Harrison P. Quantum wells, wires and dots. New York: Wiley;2000.

Liu HC. In: Razeghi M, editor. Long wavelength infrared detectors. New York: Taylor and Francis; 1996. Chapter 1.

Manasreh MO, Szmulowicz F, Vaughan T, Evans KR, Stutz CE, Fischer DW. Phys Rev B 1991; 43:9996.

Morkoç H, Cingolani R, Bernard Gil. Mater Res Innovat 1999; 3:97.

Pankove JI. Optical processes in semiconductors. New York: Dover Publications, Inc.; 1971.

Schmidt TJ, Song JJ. In: Manasreh MO, Jiang HX, editors. III-Nitride semiconductors: optical properties II. Volume 14, New York: Taylor and Francis; 2002. Chapter 2.

Singh J. Electronic and optoelectronic properties of semiconductor structures. Cambridge: Cambridge University press; 2003.

Stenger RT, Bajaj KK. Phys Rev B 2003; 68:45313.

Wallis RF. J Phys Chem Solids 1958; 4:101.

Wooten F. Optical properties of solids. New York: Academic Press; 1972.

Wu WY, Schulman JN, Hsu TY, Efron U. Appl Phys Lett 1987; 51:710.

Yu ET. In: Yu ET, Manasreh MO, editors. III-V Nitride semiconductors: applications and devices. Volume 16. New York: Taylor and Francis; 2003.

5

ELECTRICAL AND TRANSPORT PROPERTIES

5.1 INTRODUCTION

Electrical currents in semiconductors are due to the net flow of electrons and holes under bias voltages, and transport is the process that describes the motion of the charged particles. The two major transport processes are drift and diffusion. Drift mechanism is basically the movement of charged carriers under the influence of applied electric fields, and diffusion mechanism is the flow of charged particles due to density variation. Transport properties in semiconductors can be very complicated, depending on the actual size of the samples. Thus, it is worth discussing the classical and quantum limitations and regimes.

In order to define the limits of various transport regimes, one may scale the size of the sample against the de Broglie wavelength. This de Broglie wavelength, λ, can be expressed, as an example, for an electron traveling with a thermal kinetic energy in a semiconductor as

$$\lambda = \frac{h}{p} = \frac{h}{\sqrt{2m^* E}} = \lambda_0 \sqrt{\frac{m_0}{m^*}}, \tag{5.1}$$

where λ_0 is the de Broglie wavelength of a free electron and m^* is the electron effective mass in the semiconductor. The room temperature de Broglie wavelength of a free electron is ~76 Å, and that of an electron in GaAs is 295 Å. The de Broglie wavelength for a selection of semiconductor materials is plotted as a function of the electron effective mass, as shown in Fig. 5.1. The range of de

Introduction to Nanomaterials and Devices, First Edition. Omar Manasreh.

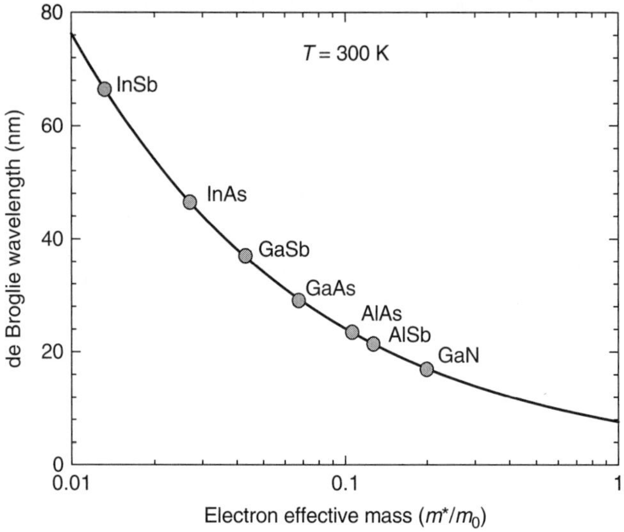

FIGURE 5.1 The de Broglie wavelength plotted as a function of electron effective mass in several semiconductor materials. The electron energy is assumed to be the thermal energy at room temperature.

Broglie wavelengths in this figure is 660–167 Å. For temperatures as low as 4.2 K, the upper limit of de Broglie wavelength increases to a fraction of a micron. This implies that the wavelength is comparable with the size of semiconductor structures and devices in the nanostructure limit. Hence, a quantum mechanical treatment of the transport properties in nanostructures must be considered.

When electrons in semiconductors lose their wavelike behavior, they can be treated classically. This could happen when the electron scattering from impurities and imperfections of the host crystal is dominant. Another reason why electrons lose their wavelike behavior is related to finite temperature and electron statistics. The electron scattering process in semiconductors and heterojunctions is dominated by scattering from impurities (including dopants), native defects, phonons, and interfaces. The scattering processes can further be divided into elastic scattering, in which the particle energy is conserved, while the momentum changes, and inelastic scattering, in which both the momentum and energy of the particle change. In elastic scattering, the motion of the electron remains coherent. The time, τ_e, between two successive elastic collisions is called *the mean free time* and can be used to define the mean free path, l_e, between scattering events, such that $l_e = \tau_e \upsilon$, where υ is the electron group velocity ($\upsilon = p/m$). The wavelike properties of the electrons are coherent when they travel a distance l_e. In addition, for elastic scattering, the electron wavelike properties remain coherent even for distances larger than l_e. For inelastic scattering, the electron wave functions have different energies and the probability of finding the electron in any state is time dependent. The distance between inelastic collisions, l_i, in which the electrons preserve their coherent properties, is called the *inelastic scattering length*. Generally speaking, l_i is larger than l_e, which means that the electrons undergo several collisions before losing their energy. The inelastic scattering length can be written as $l_i = \sqrt{D\tau_i}$, where τ_i is the time between inelastic collisions, D is the diffusion coefficient given by $D = \upsilon^2\tau_e/\alpha$, and $\alpha = 3$ for bulk, $\alpha = 2$ for quantum wells, and $\alpha = 1$ for quantum wires.

The temperature effect can cause the destruction of quantum coherence of electrons in semiconductors. As discussed in this chapter, the Fermi–Dirac distribution function is broadened as the temperature increases. If the thermal energy (k_BT) is much smaller than the Fermi energy, wave functions of the electrons maintain their amplitudes, but the phase varies slightly. If the variation in the phase is sufficiently small, temperature broadening does not break the quantum coherence properties of the electrons. For temperatures high enough such that electrons with different energies participate in the transport process, the wave function phase is spread, which leads to the destruction of the quantum coherence. The phase spreading time because of the temperature effect, τ_T, can be estimated from the uncertainty principle as $\tau_T = \hbar/(k_BT)$. The corresponding thermal diffusion length, l_T, is obtained as

$$l_T = \sqrt{D\tau_T} = \sqrt{D/(k_BT)}. \qquad (5.2)$$

This length is the distance that electrons travel before their quantum coherence is destroyed. The thermal dephasing of electrons occurs for both elastic and inelastic scattering.

The dephasing effects caused by inelastic collisions and thermal spreading can occur simultaneously. The coherence length, l_ϕ, is thus determined by the smaller value of either the inelastic scattering length or the thermal diffusion length. The superposition of the electron wave functions determines the transport properties in heterojunctions and nanostructures. The coherence length defines the limit below which the electrons have wavelike characteristics. This leads to the definition of *mesoscopic* systems, which are characterized by physical dimensions smaller than the coherence length. Mesoscopic devices cannot be characterized by macroscopic transport material parameters such as conductivity and drift velocity. Mesoscopic device and system behavior is determined by wavelike phenomena and is strongly dependent on the geometry of the sample, contacts, and position of the scatterers.

Let us assume that the dimensions of the device with contacts described in Fig. 5.2 are $L_x \ll L_y \ll L_z$. The various regimes can be defined as shown in Table 5.1. On the basis of a comparison between device dimensions and the de Broglie wavelength, λ, one can define the bulk device such that L_x, L_y, and L_z are all much larger than λ. For quantum well devices, L_z is of the order of λ, but L_x and L_y are much larger than λ. For quantum wire devices, L_x and L_z are of the order of λ, while L_y is much larger than λ. Finally, for a quantum box (dot), all the dimensions are of the order of λ.

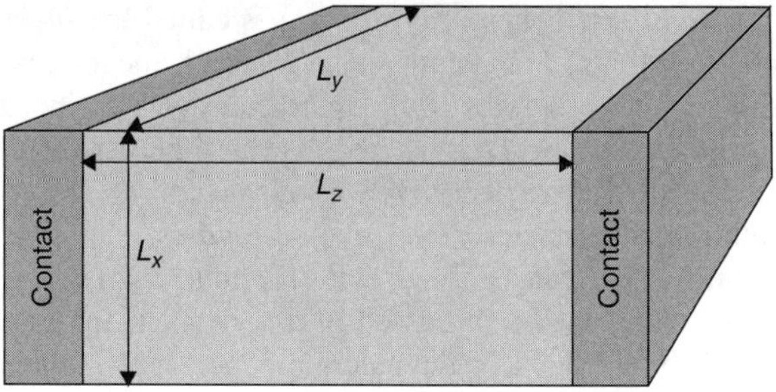

FIGURE 5.2 The device geometry with contacts used to define various transport regimes.

TABLE 5.1 The Transport Regimes in Semiconductor Devices in Terms of the Device Dimensions

Quantum regime	L_z is comparable with the electron wavelength. ($L_z \sim \lambda$)
Mesoscopic regime	$L_z \leq l_\varphi$, where l_φ is coherence length, which is also known as the dephasing length
Classical regime	$L_z > l_\varphi$
Classical ballistic regime	The mean free path of elastic collisions is larger than $L_z (l_e \geq L_z)$
Classical transverse size effect	• Effects related to the mean free path: both L_x and L_y are of the order of l_e. • Effects related to diffusion: both L_x and L_y are of the order of l_i, where l_i is the inelastic scattering length.

The wavelength, λ, is taken as the de Broglie wavelength and the interconnect distance is L_z.

The other aspects of transport properties are time and frequency. The time between successive collisions is defined as the lifetime, or free-flight time, which was previously labeled as τ_e. This time is usually much larger than the scattering duration time, τ_s. In the classical regime, the relationship between lifetime and the size of the device is very important. For example, the transit time, $t_{tr} = L_z/\upsilon$, determines the speed at which the signal propagates through the device, where υ is the electron drift velocity. The inverse of the transit time determines the ultimate frequency at which the device can operate. For further discussion, see Mitin *et al*.

The analysis of transport properties for quantum structures, such as quantum wells and dots, is more complicated than that for bulk materials. For example, three transport mechanisms can be distinguished in multiple quantum well structures, as shown in Fig. 5.3. The first mechanism (A), depicted by the dashed arrows, is the parallel transport, in which the electron motion is along the y-axis. The second mechanism (B), represented by the solid arrows, is the

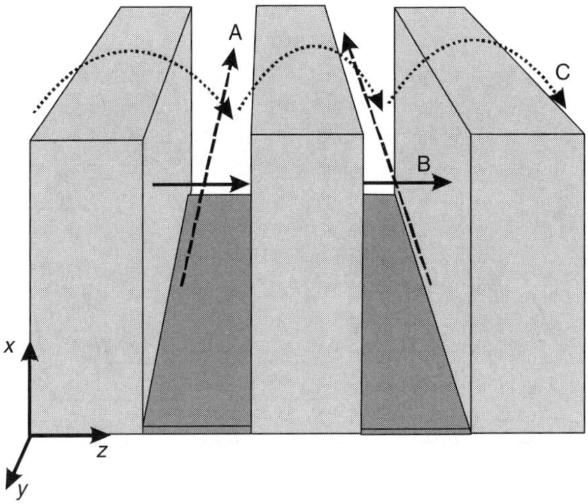

FIGURE 5.3 Transport mechanisms in quantum wells are shown by the arrows. (a) The dashed arrows represent parallel transport, (b) the solid arrows represent vertical transport by tunneling, and (c) the dotted arrows represent the vertical transport after photoexcitation or thermionic process.

tunneling through barriers, in which electron transport is along the growth axis, or z-direction. This transport is called *vertical* or *perpendicular transport*. The third mechanism (C), represented by the dotted arrows, is vertical transport resulting from the excitation of carriers to higher energy levels that are close to the top of the barriers or resonant in the continuum. The excitation of carriers can be accomplished by electron–photon coupling, as is the case for intersubband transitions, or by thermionic emission of the electrons over the barriers. Processes (A) and (C) can be analyzed classically or quantum mechanically, while process (B) is purely a quantum process.

5.2 THE HALL EFFECT

Historically, Hall effect measurements have been used extensively in determining majority carrier concentrations and their mobilities in bulk and thin film materials. Two-dimensional electron gas (2DEG) formed in quantum wells and at heterojunction interfaces have been investigated using this technique. Electric and magnetic fields are essential to observe this effect. A sketch of the sample configuration is shown in Fig. 5.4. The motions of the electrons and holes under the influence of the electric and magnetic forces are shown. The configuration in this figure is constructed such that the electric current follows along the x-axis, while the magnetic field is in the z-direction. The force on both electrons and holes is in the -y-direction. In an n-type semiconductor, in which the majority

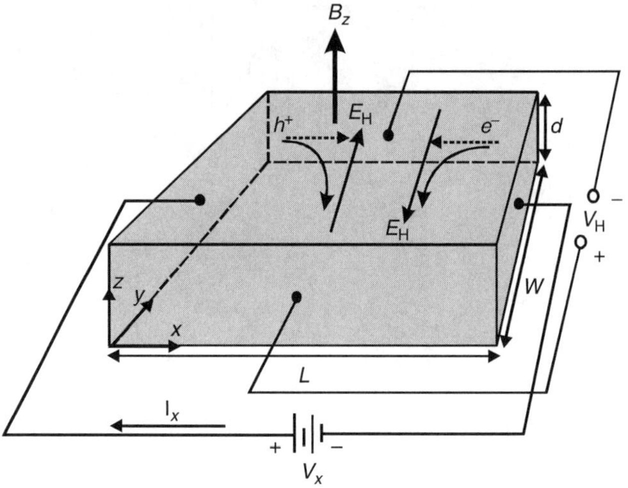

FIGURE 5.4 A sketch of a sample under the influence of electric and magnetic fields. This configuration is called the *Hall bar*.

carriers are electrons, there is a build up of negative charges at the $y = 0$ surface. For p-type material, positive charge build up is also at the $y = 0$ surface. The net change produces an electric field in the $+y$-direction.

In the steady-state case, the magnetic force is balanced by the electric force such that the net force is zero and can be expressed as

$$\begin{aligned} \boldsymbol{F} &= 0 = e[\boldsymbol{\mathcal{E}} + \upsilon \times \boldsymbol{B}] \\ &= e[\mathcal{E}_x\widehat{\boldsymbol{x}} + \mathcal{E}_y\widehat{\boldsymbol{y}} + \mathcal{E}_z\widehat{\boldsymbol{z}} - \upsilon_x \mathcal{B}_z\widehat{\boldsymbol{y}}]. \end{aligned} \tag{5.3}$$

This equation yields

$$\mathcal{E}_y = \upsilon_x \mathcal{B}_z, \tag{5.4}$$

where υ_x is the drift velocity in the x-direction. The electric field along the y-direction expressed in Equation 5.4 is called the *Hall field*, which produces the following voltage across the width (W) of the sample:

$$V_{\mathrm{H}} = \mathcal{E}_y W = \mathcal{E}_H W, \tag{5.5}$$

where $\mathcal{E}_y = \mathcal{E}_{\mathrm{H}}$ is called the *Hall field*. This voltage, V_{H}, is called the *Hall voltage*. It is negative for n-type semiconductors and positive for p-type semiconductors. Thus, the polarity of the voltage is used to determine whether the material is n-type or p-type. For n-type semiconductors, the Hall voltage can be obtained by substituting Equation 5.4 into Equation 5.5 to give

$$V_{\mathrm{H}} = \upsilon_x B_z W. \tag{5.6}$$

In addition, the drift velocity can be expressed as

$$\upsilon_x = -\frac{J_x}{en_{\mathrm{H}}} = -\frac{I_x}{en_{\mathrm{H}}A} = -\frac{I_x}{en_{\mathrm{H}}Wd}, \tag{5.7}$$

where A is the area of sample surface at $x = L$ in Fig. 5.4, which is given by the product of the sample's width (W) and thickness (d), and n_H is the Hall electron concentration, which means the concentration obtained by Hall measurements. By substituting Equation 5.7 into Equation 5.6, the Hall voltage can be rewritten as

$$V_{\mathrm{H}} = -\frac{I_x B_z}{edn_{\mathrm{H}}}. \tag{5.8}$$

The Hall voltage and current can be measured experimentally. Hence, Equation 5.8 can be used to determine the electron concentration

$$n_{\mathrm{H}} = -\frac{I_x B_z}{eV_{\mathrm{H}}d}. \tag{5.9}$$

Similarly, the hole concentration in p-type semiconductors can be obtained as

$$p = \frac{I_x B_z}{eV_{\mathrm{H}}d}. \tag{5.10}$$

Hall mobility can now be obtained from the following relation:

$$I_x = J_x Wd = en_{\mathrm{H}}\mu_{\mathrm{n}}\mathcal{E}_x Wd = \frac{en_{\mathrm{H}}\mu_{\mathrm{n}}V_x Wd}{L}, \tag{5.11}$$

where L is the length of the sample and V_x is the applied voltage (Fig. 5.4). From Equation 5.11, one can obtain the electron Hall mobility as

$$\mu_{\mathrm{n}} = \frac{I_x L}{eV_x n_{\mathrm{H}} Wd} = \frac{GL}{en_{\mathrm{H}}Wd}, \tag{5.12}$$

where G is the sample conductance. The hole Hall mobility can be obtained in a similar manner as

$$\mu_{\mathrm{p}} = \frac{I_x L}{eV_x pWd} = \frac{GL}{epWd}. \tag{5.13}$$

Another parameter that is often discussed is the Hall coefficient, R_H, which is defined as

$$R_{\mathrm{H}} = \frac{r\mathcal{E}_y}{J_x B_z} = -\frac{r}{n_{\mathrm{H}}e}, \tag{5.14}$$

where r is the Hall factor, which is close to unity. For example, r is in the range of 1.0–1.3 for GaAs.

Generally speaking, the geometry of the sample plays a significant role in the concentration and mobility results obtained from Hall effect measurements.

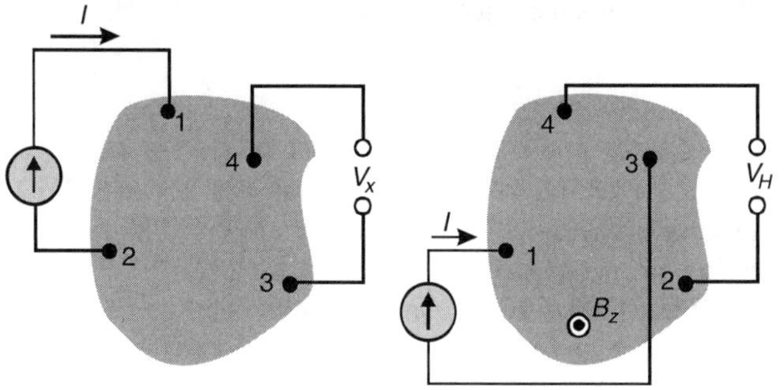

FIGURE 5.5 The van der Pauw configuration is shown for a sample with arbitrary shape. The configurations are for (a) resistivity and (b) Hall effect measurements.

The most common geometrical shape used for Hall effect measurements is the van der Pauw geometry shown in Fig. 5.5. This geometry does not require a knowledge of the sample geometry for the measurement of sheet resistance or sheet carrier concentration. The thickness of the sample, however, should be known for *volume* resistivity and carrier concentrations. The validity of the van der Pauw configuration requires that the sample has a flat, homogenous, and isotropic surface.

The relationship between current, I, and voltage, V_x, in Fig. 5.5a is determined by mapping the arbitrary shape sample geometry onto a geometry that is more regular. The Laplace equation is then solved for the simpler geometry. The final results can be obtained as follows. The resistance between points i and j can be expressed as

$$R_{ij,kl} \equiv \frac{V_{kl}}{I_{ij}}, \tag{5.15}$$

where the current enters contact i and leaves contact j, and V_{kl} is the voltage difference between contact k and contact l. For $B_z = 0$, the resistivity, ρ, is given by

$$\rho = \frac{\pi d}{\ln 2}\left[\frac{R_{21,34} + R_{32,41}}{2}\right] f, \tag{5.16}$$

where d is the sample thickness and f is determined from the following equation:

$$\frac{Q-1}{Q+1} = \frac{f}{\ln 2} \text{arccos h}\left\{\frac{1}{2}\exp\left(\frac{\ln 2}{f}\right)\right\}, \tag{5.17}$$

where $Q = R_{21,34}/R_{32,41}$ if this ratio is greater than unity, otherwise $Q = R_{32,41}/R_{21,34}$ (Look). The factor f is usually close to unity for small values of Q and is of the order of 0.3 for large values of Q. Another useful approximation is to first obtain Q, then calculate α from

$$Q = \frac{\ln(0.5 - \alpha)}{\ln(0.5 + \alpha)}, \tag{5.18}$$

and then calculate f from

$$f = \frac{\ln(0.25)}{\ln(0.5 + \alpha) + \ln(0.5 - \alpha)}, \tag{5.19}$$

The resistivity measurements can even be made more accurate when averaging ρ by including the two contact permutations and by reversing the current for all four permutations such that

$$\rho = \frac{\pi d}{8\ln(2)}\left[\{R_{21,34} - R_{12,34} + R_{32,41} - R_{23,41}\}f_A + \{R_{43,12} - R_{34,12} + R_{14,23} - R_{41,23}\}f_B\right], \tag{5.20}$$

where f_A and f_B are determined from Q_A and Q_B, respectively, by applying either Equation 5.17 or Equation 5.19. The quantities Q_A and Q_B are given by

$$Q_A = \frac{R_{21,34} - R_{12,34}}{R_{32,41} - R_{23,41}} \tag{5.21a}$$

$$Q_B = \frac{R_{43,12} - R_{34,12}}{R_{14,23} - R_{41,23}}. \tag{5.21b}$$

The Hall voltage between contacts 4 and 2 can be written as

$$V_{H42} = \frac{\rho \mu_n B_z I}{d}, \tag{5.22}$$

and the Hall coefficient is obtained by averaging V_{H42} and V_{H31}

$$R_H = \frac{d}{B_z}\left[\frac{R_{31,42} + R_{42,13}}{2}\right]. \tag{5.23}$$

It is also useful to average the Hall coefficient over current and magnetic field polarities. Doing so minimizes the magnetoresistance and many other effects, such as contact resistance.

5.3 QUANTUM HALL AND SHUBNIKOV-de HAAS EFFECTS

Quantum transport in low dimensional semiconductor systems is very interesting and offers the investigation of remarkable properties, such as the quantum Hall effect (QHE), the Shubnikov-de Haas (SdH) effect, ballistic transport, and the fractional QHE. For example, the SdH effect allows one to precisely measure the carrier concentrations formed at heterojunction interfaces. The investigation of two-dimensional systems in a perpendicular magnetic field provides quantization of Hall resistance (Klitzing *et al.*), which results from the quantization of energy in a series of Landau levels. The Landau magnetic length, l_H, (also known as the *cyclotron radius of the lowest Landau energy level*) assumes the role of wavelength in the QHE, which is given by

$$l_H = \sqrt{\frac{\hbar}{eB}}. \tag{5.24}$$

For $B = 10$ T, the magnetic length is $l_H \sim 8.12$ nm.

The original QHE device geometry used by Klitzing *et al.* is shown in Fig. 5.6a. QHE measurements are made by probing the Hall voltage across points 1 and 2, while the SdH measurements are made by probing the voltage across points 1 and 3. The device is symmetrical such that the QHE can be measured across points 3 and 4 and SdH measurements can be obtained across points 2 and 4. The initial QHE measurements were made on Si metal oxide–semiconductor field-effect transistor as schematically shown in Fig. 5.6b. A 2DEG is formed in the channel underneath the oxide layer as the gate voltage is applied. To create the channel, the gate voltage needs to be larger than the threshold voltage of ~0.7 V. The formation of the channel is very essential for the observation of both QHE and SDH effect. The band bending at the oxide–Si interface is formed by applying a gate voltage larger than 0.7 V, as shown in Fig. 5.6c. The density of the 2DEG depends on the gate voltage, as well as the drain–source voltage.

Device geometry similar to that shown in Fig. 5.6a has been applied to many semiconductor heterojunctions and quantum wells. The quantum Hall and SdH measurements from a device with such a geometry are shown in Fig. 5.7 for an InAs/AlGaSb single quantum well. A gate, in this case, is not needed since the 2DEG is formed in the quantum well because of the quantization of the energy levels in the two-dimensional nature of the quantum well. The InAs/AlGaSb single quantum well was chosen because of its high electron mobility and large band offset, which provide good carrier confinement. Furthermore, this system exhibits a large spin splitting because of the large effect g^* value (~−7.6) in InAs, as compared to other systems, such as GaAs/AlGaAs quantum wells ($g^* \sim -0.2$). The g^* is obtained from the following expression, which is derived from the fourth-order effective mass theory (Palik *et al.*):

$$g^* = 2\left[1 - \left\{\frac{1-x}{2+x}\right\}\left\{\frac{1-y}{y}\right\}\right], \tag{5.25}$$

where $x = 1/(1 + \Delta/E_g)$, $y = m*/m_0$, and Δ is the spin–orbit splitting energy in the valence band.

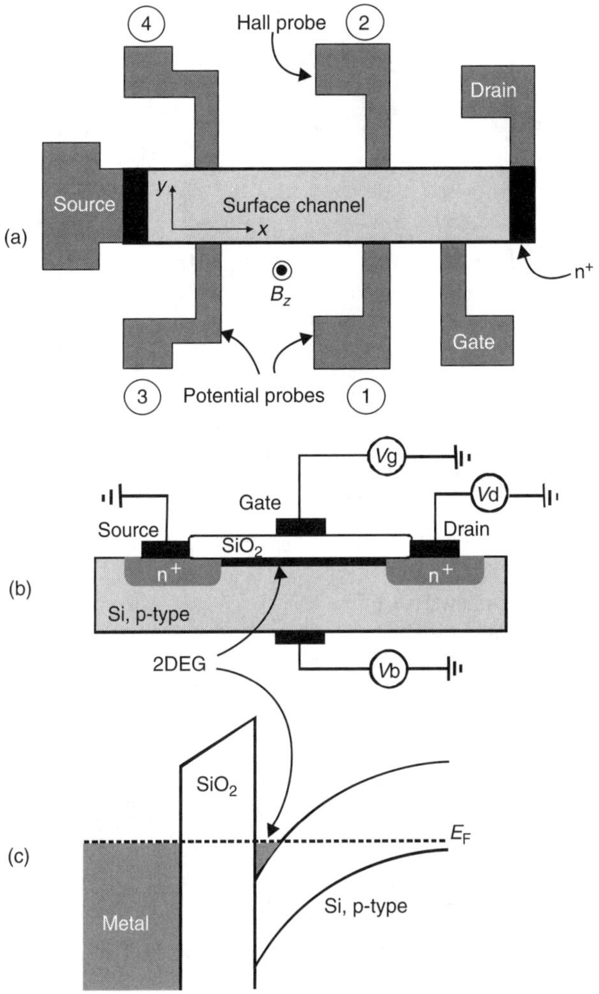

FIGURE 5.6 (a) A sketch of the device geometry used for both quantum Hall effect and Shubnikov-de Haas measurements. (b) A cross section of the n-type MOSFET device showing the channel underneath the oxide (SiO_2) layer. (c) The band bending near the oxide–Si interface showing the 2DEG.

5.3.1 Shubnikov-de Haas Effect

This effect manifests itself in the oscillations of the parallel resistivity, ρ_{xx}, obtained for the 2DEG in an InAs/AlGaSb single-quantum well system in the presence of a high magnetic field, as shown in Fig. 5.7. The oscillations observed in ρ_{xx} are periodic as a function of $(1/B_z)$ in two-dimensional systems because of the constant density of states for Landau levels. The periodicity of ρ_{xx} can be used to extract the 2DEG carrier density. Since the resistivity is expected to

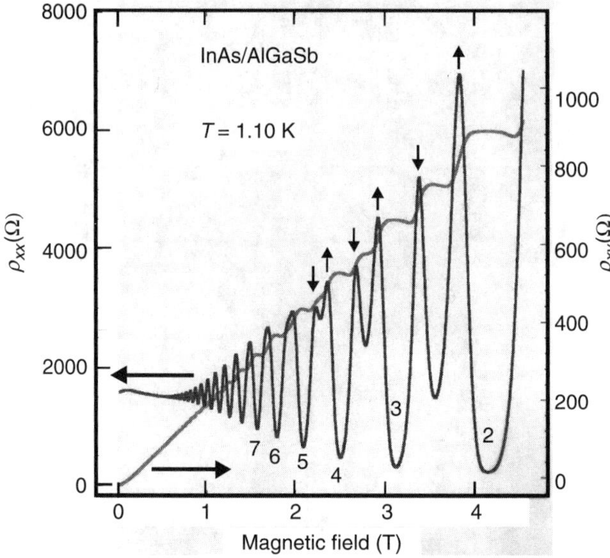

FIGURE 5.7 The quantum Hall effect resistivity, ρ_{xy}, observed as a function of the magnetic field. The parallel resistivity, ρ_{xx}, represents the Shubnikov-de Haas effect. The vertical arrows indicate electron spin up or spin down, and the integer numbers represent the filling factor. Notice that ρ_{xy} and ρ_{xx} are sheet resistivities and their unit is Ohm.

be minimum when the Fermi levels lie between two Landau levels, where the density of states is the smallest, one can define the Landau level filling factor, ν, as

$$\nu \equiv \frac{n_s h}{e B_z}, \tag{5.26}$$

where n_s is the density of the 2DEG and the filling factor, ν is an integer (1, 2, 3, ...). This equation assumes degenerate spin and valley Landau levels. Thus, for adjacent Landau levels, we have

$$n_s = \frac{e}{h} \cdot \frac{1}{\Delta(1/B_z)}. \tag{5.27}$$

An accurate measurement of n_s is obtained for larger ν (small values of B_z) where the spin splitting is minimum, as shown in Fig. 5.7. The sheet carrier density obtained by this method is more accurate than conventional Hall effect measurements. This is mainly due to the fact that the conventional Hall effect does not distinguish between two-dimensional and three-dimensional carriers, but the results from both techniques are usually very close in value. The carrier concentration can also be obtained by plotting B_z^{-1} against the consecutive minima of ρ_{xx}, n, as shown in Fig. 5.8. The slope of the plot is related to the 2DEG density through the following relation, $n_s = e/(\text{slope} \times h)$.

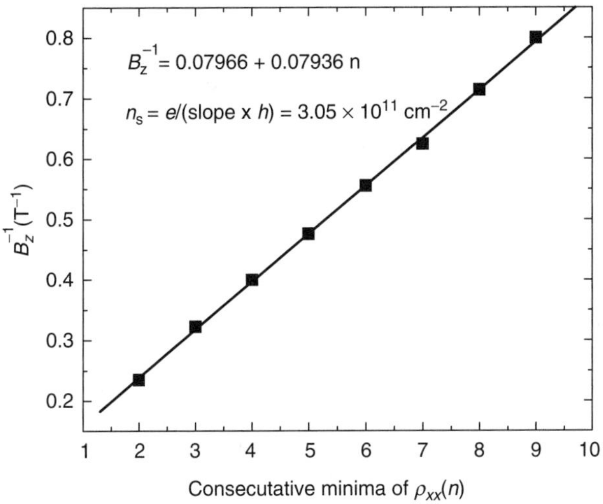

FIGURE 5.8 The inverse of the magnetic field plotted as a function of the consecutive minima obtained from ρ_{xx} in Fig. 5.7. The line is a linear fit to the data. The slope of the line is used to calculate the density of the 2DEG.

While the effective mass is not included in the SdH oscillations, it can be determined by investigating the oscillation amplitudes as a function of temperature and magnetic field of the low field oscillatory conductivity expression derived by Ando *et al*.

$$\rho_{xx}^{-1} = \sigma_{xx} = \frac{n_s e^2 \tau_f}{m^*} \frac{1}{1 + (\omega_c \tau_f)^2} \left[1 - \frac{2(\omega_c \tau_f)^2}{1 + (\omega_c \tau_f)^2} \frac{2\pi^2 k_B T}{\hbar \omega_c} \right]$$
$$\times \cos\hbar \left(\frac{2\pi^2 k_B T}{\hbar \omega_c} \right) \cos\left(\frac{2\pi E_F}{\hbar \omega_c} \right) \exp\left(-\frac{\pi}{\omega_c \tau_f} \right), \tag{5.28}$$

where E_F is the Fermi energy given by

$$E_F = \frac{\hbar^2 k_F^2}{2m^*} = \frac{2\pi \hbar^2 n_s}{m^*}. \tag{5.29}$$

τ_f is the scattering time corresponding to the dephasing of the Landau state, ω_c is the cyclotron angular frequency given by $\omega_c = \frac{|e|B_z}{m^*}$, k_B is Boltzmann's constant, and T is temperature. The scattering time, τ_f, can also be extracted from Equation 5.28. Both the effective mass and scattering time values can be quite different from the values obtained from Hall effect and cyclotron resonance measurements.

The origin of the oscillations in ρ_{xx} can be understood by examining Fig. 5.9. An n-type doped InAs/AlGaSb single quantum well is sketched in Fig. 5.9a where we assume two bound states, E_1 and E_2, exist with the Fermi energy, $E_F(0)$, at zero magnetic field is assumed to be between E_1 and E_2. By applying a magnetic field along the growth axis (z-direction), each electronic energy level splits into an n number of Landau levels with energy described in Chapter 4, Section 4.2, Equation 4.140. The separation between Landau levels is $\hbar\omega_c$. The density of states per unit area of each Landau level is obtained from the following relation:

$$(\hbar\omega_c)\left(\frac{m*}{2\pi\hbar^2}\right) = \frac{m^*\omega_c}{2\pi\hbar} = \frac{eB_z}{h}. \tag{5.30}$$

As discussed in this chapter, the density of states for bulk semiconductor material under the influence of a magnetic field resembles the quantum wire density of states because of the confinement of the electron in Landau orbits in the x–y plane. In quantum wells and heterojunctions, the electrons are confined in the z-direction. By applying a static magnetic field along the z-direction, the electrons are further confined in the x–y plane as shown in Fig. 5.9b, leading to a zero degree of freedom (confinement in the three directions). Thus, the density of states of each Landau energy level is simply a δ-function with a degeneracy of eB_z/h. This is shown in Fig. 5.9c as the solid vertical lines labeled "*no broadening*." In reality, however, the impurities, alloy fluctuations, interface roughness, and crystal imperfections will broaden the δ-function density of states. This broadening of Landau levels is depicted as Gaussian lineshape (see the curves labeled "with broadening" in Fig. 5.9c). The dashed-dotted line in Fig. 5.9c is the two-dimensional density of states of the electronic energy levels, E_1, in the absence of the magnetic field. The states at the tails of the Landau levels are called *localized states*, and they play an important role in QHE. As the magnetic filed is increased, Landau levels are swept across the Fermi levels, giving rise to the observed oscillations in ρ_{xx}.

5.3.2 Quantum Hall Effect

As a starting point, it is beneficial to understand the Drude classical model of magnetoresistance in semiconductors. The classical equation of motion of an electron in the presence of magnetic ($\boldsymbol{B}$) and electric ($\boldsymbol{\mathcal{E}}$) fields can be expressed as

$$m^*\frac{d\boldsymbol{v}}{dt} + m^*\frac{\boldsymbol{v}}{\tau} = -e[\boldsymbol{\mathcal{E}} + \boldsymbol{v} \times \boldsymbol{B}], \tag{5.31}$$

where υ is the drift velocity and τ is the scattering time. The magnetic field is applied along the z-axis, and $\boldsymbol{\mathcal{E}}$ and υ are assumed to vary with time as $\exp(-\omega t)$. This equation can be expressed in its three components as

$$\begin{aligned} m^*\frac{d\upsilon_x}{dt} + m^*\frac{\upsilon_x}{\tau} &= -e\mathcal{E}_x - e\upsilon_y B_z \\ m^*\frac{d\upsilon_y}{dt} + m^*\frac{\upsilon_y}{\tau} &= -e\mathcal{E}_y + e\upsilon_x B_z \\ m^*\frac{d\upsilon_z}{dt} + m^*\frac{\upsilon_z}{\tau} &= -e\mathcal{E}_z. \end{aligned} \tag{5.32}$$

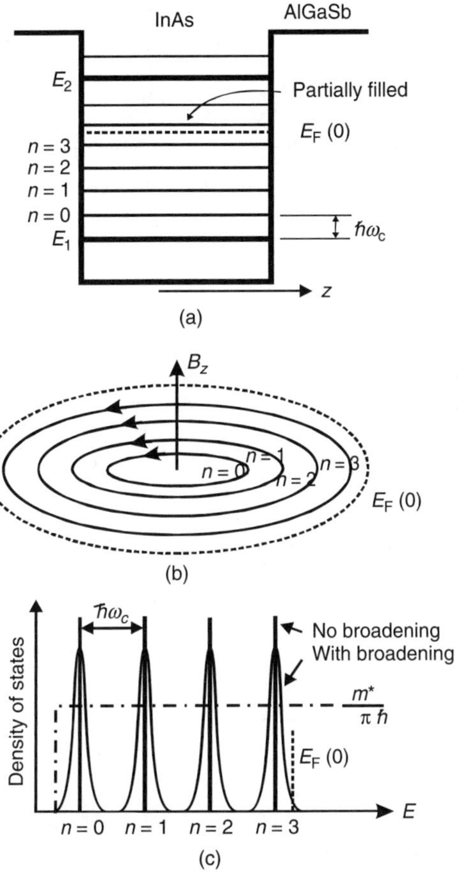

FIGURE 5.9 (a) A sketch of an InAs/AlGaSb single quantum well showing two bound sates (E_1 and E_2), Fermi energy level, $E_F(0)$, and Landau levels. (b) Landau levels are filled up to the Fermi energy level, which contains all allowed states when the magnetic field is zero. (c) Energy representation of Landau levels and Fermi level. Landau levels are broadened because of various scattering mechanisms.

By multiplying Equation 5.32 by the carrier concentration (n_s) and the electron charge ($-e$) and comparing the results with the following relation:

$$j = \overleftrightarrow{\sigma} \cdot \mathcal{E}, \tag{5.33}$$

where $\overleftrightarrow{\sigma}$ is the conductivity tensor, one can obtain the components of the conductivity tensor as

$$\sigma_{xx} = \sigma_{yy} = \frac{\sigma_0(1 - i\omega\tau)}{1 - (\omega^2 - \omega_c^2)\tau^2 - 2i\omega\tau}$$

$$\sigma_{zz} = \frac{\sigma_0}{1 - i\omega\tau}$$

$$\sigma_{xy} = -\sigma_{yx} = \frac{\sigma_0 \omega_c \tau}{1 - (\omega^2 - \omega_c^2)\tau^2 - 2i\omega\tau}$$

$$\sigma_{xz} = \sigma_{zx} = \sigma_{yz} = \sigma_{zy} = 0, \tag{5.34}$$

where $\sigma_0 = n_s e^2 \tau / m^*$ is the conductivity in the absence of the magnetic field. For the steady-state case, where $dv/dt = 0$, the conductivity tensor can be written as

$$\overleftrightarrow{\sigma} = \frac{\sigma_0}{1 + (\omega_c \tau)^2} \begin{pmatrix} 1 & -\omega_c \tau & 0 \\ \omega_c \tau & 1 & 0 \\ 0 & 0 & 1 + (\omega_c \tau)^2 \end{pmatrix}. \tag{5.35}$$

Thus, the conductivity in a two-dimensional system in the presence of a magnetic field applied along the z-direction can be expressed as

$$\overleftrightarrow{\sigma} = \frac{\sigma_0}{1 + (\omega_c \tau)^2} \begin{pmatrix} 1 & -\omega_c \tau \\ \omega_c \tau & 1 \end{pmatrix}, \tag{5.36}$$

and the resistivity tensor is related to the conductivity tensor as

$$\overleftrightarrow{\rho} = \overleftrightarrow{\sigma}^{-1}. \tag{5.37}$$

The resistivity tensor can now be written as

$$\overleftrightarrow{\rho} = \frac{1}{\sigma_{xx}^2 + \sigma_{xy}^2} \begin{pmatrix} \sigma_{xx} & -\sigma_{xy} \\ \sigma_{xy} & \sigma_{xx} \end{pmatrix}. \tag{5.38}$$

The condition $\omega_c \tau > 1$ implies that the carriers are collisionless. By applying this condition to Equation 5.34, one can obtain $\sigma_{xx} \approx 0$ and $\sigma_{xy} \approx -n_s e/B$. In the presence of collisions, where $\omega_c \tau > 1$, we have

$$\sigma_{xx} = \frac{n_s e}{B_z} \frac{\omega_c^2 \tau^2}{1 + \omega_c^2 \tau^2}$$

$$\sigma_{xy} = -\frac{n_s e}{B_z} - \frac{\sigma_{xx}}{\omega_c \tau}, \tag{5.39}$$

where these conductivity components are simply the sum of the collision and collisionless parts. When the Fermi energy level is between Landau levels labeled n and $n+1$, no elastic scattering can occur at low temperature ($T \leq 4.2$ K) and the energy separation between consecutive Landau levels is $\hbar\omega_c$. This case is thus equivalent to the condition $\omega_c \tau > 1$, which gives $\sigma_{xx} \approx 0$, and σ_{xy} is given by its classical collisionless value. From the density of states per Landau level, eB/h, one can write the carrier density, n_s, as $n_s = neB/h$, where n is the nth Landau level. The Hall conductivity, σ_{xy} can be expressed as

$$\sigma_{xy} = \frac{n_s e}{B_z} = \frac{e}{B_z} \frac{neBz}{h} = n\frac{e^2}{h} \quad \text{and} \quad \rho_{xy} = \frac{1}{n} \frac{h}{e^2}. \tag{5.40}$$

This equation shows that the Hall resistivity takes quantized values of 25812.87/n whenever the Fermi energy level lies between filled-broadened Landau levels, as illustrated by the plateaus in Fig. 5.7. This is called *QHE*.

QHE is observed for integer filling factors as described in Equation 5.26. However, at low temperature ($T \ll 5.2$ K), a fractional value of the filling factor, ν, has been observed for the lowest Landau level in many heterojunction systems with high mobility. In this case, ν can take values of p/q where p and q are integers. This is called the *fractional QHE* (Tsui *et al.*). Laughlin provided an explanation of the fractional QHE based on the condensation of electrons or holes into a collective ground state due to electron–electron or hole–hole interactions. This ground state is separated from the nearest excited state by an energy of $0.03e^2/l_H$, where l_H is the Landau magnetic length. The possibility of a repulsive interaction between carriers of the same charge, leading to a condensation, is related to the two-dimensional character of the system. The condensed phase consists of quasi-particles called *anyons*, of fractional charge $2/l$, where $l = 3, 5, 7, \ldots$, that follow statistics intermediate between Fermi–Dirac and Bose–Einstein formalisms.

5.4 CHARGE CARRIER TRANSPORT IN BULK SEMICONDUCTORS

As discussed in Section 5.1, there are several mechanisms that impact charge transport in bulk and low dimensional systems. For example, tunneling, which has been discussed in previous chapters, is a quantum effect that cannot be explained in terms of classical theory. In this section, various transport properties of bulk semiconductors are discussed.

5.4.1 Drift Current Density

The resultant movement of the electrons and holes in semiconductors under the influence of an applied electric field is called *drift*, which gives rise to drift currents. The equation of motion of an electron with mass m^* under the influence of an electric field, $\mathcal{E}$, is given by

$$m * \frac{\mathrm{d}\upsilon_\mathrm{d}(t)}{\mathrm{d}t} = -e\mathcal{E}, \tag{5.41}$$

where $\upsilon_\mathrm{d}(t)$ is the drift velocity, which, after integration, is given as

$$\upsilon_\mathrm{d}(t) = -\frac{e\mathcal{E}}{m^*}t. \tag{5.42}$$

The drift velocity increases linearly with time between collisions. The mean value of the drift velocity is

$$\langle \upsilon_\mathrm{d} \rangle = \int_0^\infty \upsilon_\mathrm{d}(t)\mathcal{P}(t)\mathrm{d}t$$

$$= -\frac{e\mathcal{E}}{m^*}\int_0^\infty t\mathcal{P}(t)\mathrm{d}t$$

$$= -\frac{e\tau}{m^*}\mathcal{E}, \tag{5.43}$$

where τ is the time that it takes for a carrier to undergo two successive collisions and $\mathcal{P}(t)$ is the probability that a carrier has *not* made a collision at time t and is given by

$$\mathcal{P}(t) = \frac{1}{\tau}\exp\left(-\frac{t}{\tau}\right). \tag{5.44}$$

From Equation 5.43, the electron mobility can be expressed as $\mu = \frac{|e|\tau}{m^*}$. The current density can now be written as

$$j_\mathrm{e} = -nev_\mathrm{e} = ne\mu_\mathrm{e}\mathcal{E}, \tag{5.45}$$

where n is the electron density and the subscript "e" stands for electrons. For holes, the current density is

$$j_\mathrm{h} = pev_\mathrm{h} = pe\mu_\mathrm{h}\mathcal{E}, \tag{5.46}$$

where p is the hole density. The above current densities assume that the drift velocity is linearly dependent on the electric field and that the mobility is independent of the electric field. This may not be the case for high electric fields ($\mathcal{E} > 10^4$ V/cm). In the case of the high electric field regime, the relaxation time, drift velocity, and mobility can all be dependent on the electric field. For additional discussion on the saturation of the drift velocity, see Look and Sze. For mixed conduction, in which both electrons and holes are present, the total current density is given by the sum of Equations 5.45 and 5.46, which gives a total conductivity of

$$\sigma = ne\mu_\mathrm{e} + pe\mu_\mathrm{h}$$

$$= \mathrm{e}^2\left[\frac{n\tau_\mathrm{e}}{m_\mathrm{e}^*} + \frac{p\tau_\mathrm{h}}{m_\mathrm{h}^*}\right]. \tag{5.47}$$

The mobility in this equation is called *conductivity mobility*. The carrier mobility can be determined by different methods, such as the Hall effect and magnetoresistance techniques, which may lead to different values for mobility. The mobility determined from various techniques depends, however, on the scattering time or the relaxation time, which was defined previously as the time between two successive collisions or scattering events. The determination of the scattering time depends on several effects that take place as the charge carrier is drifting from one end of a material to the other under an applied electric field. For example, the scattering mechanisms in GaAs include defect scattering, such as intrinsic defects, charged and neutral impurities, and alloying; carrier–carrier scattering; and lattice scattering. Lattice scattering may be due to intervalley scattering (acoustical and/or optical phonons) and intravalley scattering (phonons, deformation potential, piezoelectric, etc.). The scattering time, τ, can be written as

$$\frac{1}{\tau} = \frac{1}{\tau_1} + \frac{1}{\tau_2} + \frac{1}{\tau_3} + \ldots, \tag{5.48}$$

where the subscripts indicate different types of scattering. Consequently, the mobility of electrons can be expressed as

$$\frac{1}{\mu} = \frac{1}{\mu_1} + \frac{1}{\mu_2} + \frac{1}{\mu_3} + \ldots \tag{5.49}$$

For example, the mobility due to lattice scattering was shown to depend on $T^{-3/2}$, where T is the temperature (Smith), and the mobility due to impurity scattering varies as $T^{3/2}/N_{\mathrm{i}}$, where N_{i} is the total impurity concentration.

The mobility also depends on the effective mass of the charge carriers. When the effective mass is obtained from the conductivity measurements, it is called *mobility effective mass*. The values of the mobility effective mass may differ from those obtained from cyclotron resonance and SdH experiments.

For low values of applied electric field, the drift velocity of charge carriers in semiconductor materials and devices exhibits a linear relationship as a function of the electric field. However, many devices operate at high electric fields ($\mathcal{E} \sim$ 1–100 kV/cm) where the drift velocity is no longer linear with $\mathcal{E}$. An example of the drift velocity under the influence of a high electric field is shown in Fig. 5.10, where the drift velocity becomes almost independent of the electric field for Si and Ge. The drift velocity saturation, that is independent of the electric field, in this figure is due to the fact that electrons (holes) gain high energy (*hot electrons or holes*) from the electric field and their scattering rates are increased, leading to a reduction in the scattering time (scattering lifetime). The reduction in lifetime causes the mobility to decrease.

The curves related to GaAs and InP in Fig. 5.10 exhibit a negative differential mobility at high electric fields, which produces a negative differential resistance. This characteristic, however, is useful in the design of oscillators and low power microwave devices. The drift velocity–electric field behavior in GaAs, as well as many direct band gap materials, can be explained in terms of the conduction valley occupancy (Singh 2003). As shown in Fig. 5.11, the electrons move in the high mobility $(\mu_\Gamma)_\Gamma$-valley at low electric field, where the effective mass is $0.067m_0$. The velocity peaks at around $4–5 \times 10^5$ V/cm where most of the electrons are still in the Γ-valley. At higher electric fields, the electrons gain enough energy to transfer to the L-valley, where the electron effective mass is much greater ($\sim 0.22m_0$) and the mobility (μ_{L}) is lower. The transfer of the electrons from the Γ-valley to the L-valley is the cause of the negative differential mobility, which leads to the negative differential resistance. To observe the negative differential resistance, the energy separation between the L and Γ valleys

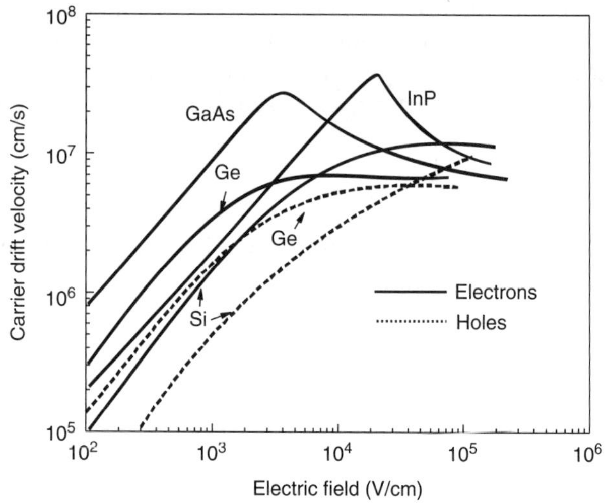

FIGURE 5.10 Carrier drift velocities as a function of the electric field for Si, Ge, GaAs, and InP.

should be much larger than $k_B T$ so that the L-valley will not be thermally populated with electrons. In the case of GaAs, this separation energy is ~0.32 eV. An additional condition to observe the negative differential resistance is that the separation between the L- and Γ-valleys should be less than the band gap of the semiconductor to avoid populating the L-valley by exciting carriers from the valence band to this valley through mechanisms such as impact ionizations.

When electrons are injected into a semiconductor by applying an electric field, they undergo several collisions in a certain period (several picoseconds) before they reach a steady-state distribution. If electrons are injected into the upper valley, where the effective mass is high, the injected electrons may have velocities lower that the steady-state velocity for a short period. This leads to what is called *velocity undershoot*. Velocity overshoot is when the electrons are injected ballistically into the sample and stay in the Γ-valley with velocities higher than the steady-state velocity. Eventually, the electrons suffer scattering, and their velocity decreases in time to the steady-state velocity.

The negative differential resistance is very useful in microwave devices and oscillators. The negative slope region in the drift velocity versus electric field curve usually occurs when a high electric field ($ll5 \times 10^5$ V/cm) is applied to semiconductor materials, such as GaAs or InP. In this region, instability can arise and current oscillation can occur. These oscillations were first observed by Gunn in 1963 and are called *Gunn oscillations*. These oscillations are observed in thin samples (of the order of 10 μm) under the influence of an electric field higher than the critical field ($\mathcal{E}_c$), shown in Fig. 5.11. The frequency of the oscillations is found to be equal to the electron drift velocity divided by the length of the sample. The origin of Gunn oscillations is that there is a fluctuation called the

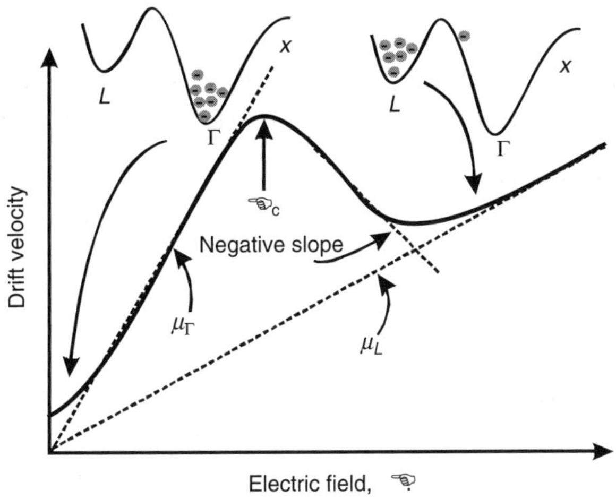

FIGURE 5.11 Illustration of the electron transfer from the Γ-valley to the L-valley in the conduction band of GaAs as the applied electric field is increased. The associated drift velocity behavior as a function of the electric field is shown with a negative slope on the right-hand side of the peak.

electric field domain formed near the cathode, as shown in Fig. 5.12a, where the carriers pile up on the left-hand side of the domain, while the carriers on the right-hand side of the domain are depleted, as shown in Fig. 5.12b. Owing to the negative differential resistance for $\mathcal{E} \gg \mathcal{E}_c$, the increase in the field inside the domain causes further slowing down of the electrons inside the domain, which leads to more charge pileup. The pileup process continues until most of the applied field is across the domain.

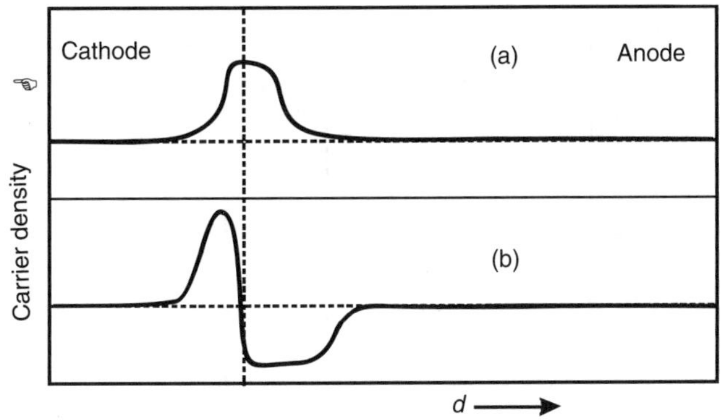

FIGURE 5.12 (a) Electric field and (b) carrier density as a function of distance for a domain moving from the cathode to the anode in a GaAs thin sample under the influence of high electric field (> 10^5 V/cm).

Only one domain can exist inside the sample at one time. The domain drifts across the sample from the cathode toward the anode at the saturation velocity under the influence of an applied bias voltage. The domain disappears once it reaches the anode, and a new domain is formed, giving rise to current oscillation. If the saturation velocity is 10^7 cm/s and the length of the sample is 10^{-4} cm, the oscillation frequency is 10 GHz. This frequency is in the microwave region. Thus, Gunn diodes are known as *microwave generators* and have applications in radar and communications.

5.4.2 Diffusion Current Density

When there is a spatial variation in carrier concentration in semiconductors, the carriers move from regions of high concentration to regions of low concentration. The movement of the carriers results in what is called *diffusion current*. The carrier diffusion is governed by Fick's law, which states that the carrier flux, $\mathcal{F}_n$, is proportional to the concentration gradient. For electrons, Fick's law has the following form:

$$\mathcal{F}_{\rm n} = -D_{\rm e}\frac{{\rm d}n}{{\rm d}x}, \tag{5.50}$$

where D_e is the electron diffusion coefficient and n is the electron concentration. Fick's law can be verified by assuming that the electron concentration in a semiconductor at a constant temperature varies along the x-axis such that $n(x)$ is described by the curve in Fig. 5.13. The average electrons flux, $\mathcal{F}_1$, crossing the concentration profile from the left can be expressed as follows:

$$\mathcal{F}_{\rm l} = \frac{n(-l).l}{2\tau} = \frac{n(-l).\upsilon_{\rm th}}{2}, \tag{5.51}$$

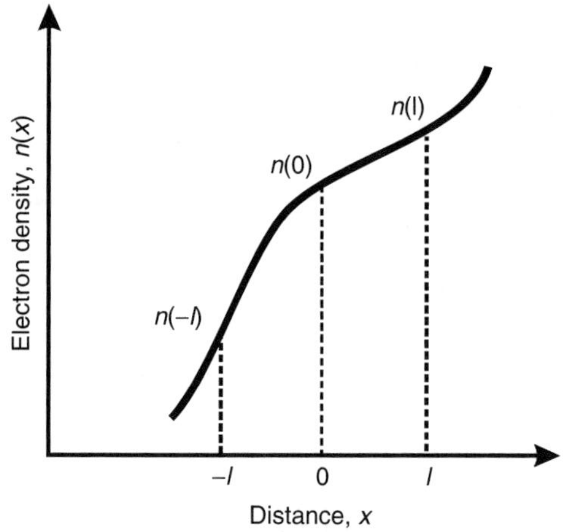

FIGURE 5.13 An example of electron concentration variation as a function of distance used to illustrate Fick's law.

where τ is the mean free time between collisions, l is the mean free path, and υ_{th} is the electron thermal velocity ($\upsilon_{\text{th}} = l/\tau$). Similarly, the average electron flux, $\mathcal{F}_2$, crossing from right to left is

$$\mathcal{F}_2 = \frac{n(l).\upsilon_{\text{th}}}{2}. \tag{5.52}$$

The net carrier flow from left to right is thus the difference between the two fluxes

$$\mathcal{F}_{\text{n}} = \mathcal{F}_1 - \mathcal{F}_2 = \frac{\upsilon_{\text{th}}}{2}[n(-l) - n(l)]. \tag{5.53}$$

One can now expand the carrier concentration at $x = \pm l$ by using the Taylor series to the first order to obtain

$$\begin{aligned}\mathcal{F}_{\text{n}} &= \frac{\upsilon_{\text{th}}}{2}\left[n(0) - l\frac{\mathrm{d}n}{\mathrm{d}x} - n(0) - l\frac{\mathrm{d}n}{\mathrm{d}x}\right] \\ &= -\upsilon_{\text{th}} l\frac{\mathrm{d}n}{\mathrm{d}x} = -D_{\text{e}}\frac{\mathrm{d}n}{\mathrm{d}x},\end{aligned} \tag{5.54}$$

where $D_{\text{e}} = \upsilon_{\text{th}} l$. The diffusion current density for conduction electrons can now be expressed as

$$J_{\text{e}} = eD_{\text{e}}\frac{\mathrm{d}n}{\mathrm{d}x}. \tag{5.55}$$

Similarly, the hole diffusion current is

$$J_{\text{h}} = -eD_{\text{h}}\frac{\mathrm{d}p}{\mathrm{d}x}, \tag{5.56}$$

where D_{h} is the hole diffusion constant and p is the hole concentration. When both the electric field and concentration gradient are present, the current densities for electrons and holes can be written as

$$J_{\text{e}} = ne\mu_{\text{e}}\mathcal{E} + eD_{\text{e}}\frac{\mathrm{d}n}{\mathrm{d}x} \tag{5.57a}$$

$$J_{\text{h}} = pe\mu_{\text{h}}\mathcal{E} - eD_{\text{h}}\frac{\mathrm{d}p}{\mathrm{d}x}. \tag{5.57b}$$

For mixed conduction in three dimensions, the total current density, which consists of the drift and diffusion components for both electrons and holes, can be generalized as

$$J = ne\mu_{\text{e}}\text{E} + pe\mu_{\text{h}}\text{E} + eD_{\text{e}}\nabla n(r) - eD_{\text{h}}\nabla p(r). \tag{5.58}$$

For a semiconductor at equilibrium, the current density of each type of carrier must be zero. For electrons, Equation 5.57a can be written as

$$ne\mu_{\text{e}}\mathcal{E} = -eD_{\text{e}}\frac{\mathrm{d}n}{\mathrm{d}x}. \tag{5.59}$$

Furthermore, the electric field is related to the electric potential, $V(r)$, according to the following relation:

$$\mathcal{E} = -\nabla V(r). \tag{5.60}$$

Substituting Equation 5.60 into Equation 5.59, we have

$$n\mu_{\mathrm{e}}\nabla V(r) = D_{\mathrm{e}}\nabla n(r). \tag{5.61}$$

The carrier concentration under nondegenerate conditions can be written as

$$n(r) = N_{\mathrm{c}} \exp\left[\frac{(E_{\mathrm{c}} - eV(r) - E_{\mathrm{F}})}{k_{\mathrm{B}}T}\right], \tag{5.62}$$

where the conduction band edge is modified in the presence of the applied voltage $[V(r)]$. By taking the gradient of Equation 5.62, one can obtain

$$\nabla n(r) = \frac{en(r)\nabla V(r)}{k_{\mathrm{B}}T}. \tag{5.63}$$

Substituting Equation 5.63 into Equation 5.61 gives

$$D_{\mathrm{e}} = \frac{k_{\mathrm{B}}T\mu_{\mathrm{e}}}{e}. \tag{5.64}$$

Similarly, the hole diffusion coefficient is

$$D_{\mathrm{h}} = \frac{k_{\mathrm{B}}T\mu_{\mathrm{h}}}{e}. \tag{5.65}$$

The above two equations are known as the *Einstein relations*. Substituting Equations 5.64 and 5.65 into Equation 5.57, we obtain

$$J_{\mathrm{e}} = \mu_{\mathrm{e}}[ne\mathcal{E} + k_{\mathrm{B}}T\nabla n] \tag{5.66a}$$

$$J_{\mathrm{h}} = \mu_{\mathrm{h}}[pe\mathcal{E} - k_{\mathrm{B}}T\nabla p]. \tag{5.66b}$$

It is clear from these equations that the current density is proportional to the mobility even in the presence of carrier diffusion.

In high frequency electronic components, an additional current density contribution, called the *displacement current density*, becomes important. Assume that the electrons in a semiconductor material are subject to an AC electric field of the form

$$\mathcal{E} = \mathcal{E}_0 \exp[-i\omega t]. \tag{5.67}$$

The displacement current density, $\boldsymbol{J}_{\mathrm{d}}$ is given by

$$\boldsymbol{J}_{\mathrm{d}} = \frac{\partial \mathbf{D}}{\partial t}, \tag{5.68}$$

where $\mathcal{D}$ is the electric displacement given by

$$\mathbf{D} = \epsilon\epsilon_0\boldsymbol{\mathcal{E}}, \tag{5.69}$$

where ϵ is the dielectric constant and ϵ_0 is the permittivity of space (8.85×10^{-12} F/m). Combining Equations 5.67–5.69 yields

$$\boldsymbol{J}_{\mathrm{d}} = -i\omega\epsilon\epsilon_0\boldsymbol{\mathcal{E}}. \tag{5.70}$$

By combining Equation 5.70 with the static contribution of the current density derived above, we have

$$\boldsymbol{J} = (\sigma - i\omega\epsilon\epsilon_0)\boldsymbol{\mathcal{E}}. \tag{5.71}$$

Thus, the electrical conductivity is composed of DC and AC components. Again, the AC component is significant and cannot be neglected in case of high frequency devices.

5.4.3 Generation and Recombination

For a semiconductor at thermal equilibrium and zero bias voltage, the product of electron (n) and hole (p) concentrations is given by $np = n_{\mathrm{i}}^2$, where n_{i} is the intrinsic carrier concentration. Many electronic devices, such as bipolar transistors and p–n junction diodes, operate on the principle of carrier injection. For excess carriers, the semiconductor is no longer at equilibrium and $np > n_{\mathrm{i}}^2$. The generation of excess carriers can be accomplished by several techniques, but the most common method is either applying a bias voltage or illuminating the sample with photons. The thermionic process is when electrons gain enough thermal energy to allow them to make transitions to higher energy levels. The introduction of excess carriers is called *carrier generation*. When the system is in nonequilibrium, a process exists to restore the system back to equilibrium. This mechanism is called *recombination*. For example, when a semiconductor sample is illuminated with light, electrons absorb the photons to make the transition from the valence band to the conduction band, leaving behind holes with positive charges. The excited electrons recombine with holes in the valence band, releasing energy in the form of photons (luminescence) or phonons (thermal energy). If photons are emitted as a by-product of the recombination, the process is called *radiative recombination. Nonradiative recombination* occurs when the energy of the electron is absorbed by the lattice. When the excited electrons recombine directly with holes in the valence band, the process is called *direct recombination*. If the recombination process takes place through centers with energy levels lying in the fundamental band gap, the process is called *indirect recombination*.

Direct recombination is common in direct band gap materials such as GaAs and GaN. Figure 5.14 illustrates the generation and recombination processes in a direct band gap semiconductor. The quantity g_{l} is the light generation rate, g_{t} is the thermal generation rate, and $\mathcal{R}$ is the recombination rate. The unit of

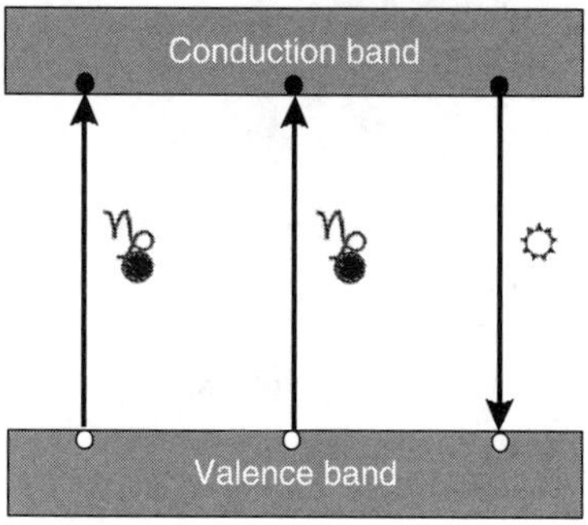

FIGURE 5.14 Direct generation and recombination of electron–hole pairs during illumination of the sample with photons.

these rates is/cm^3/ sec. For thermal equilibrium, g_l is zero and $g_t = \mathcal{R}$. For direct recombination, where the bottom of the conduction band and the top of the valence band are lined up, the recombination rate is given by

$$\mathcal{R} = \alpha_r p n, \tag{5.72}$$

where α_r is the recombination rate proportionality constant. For the nonequilibrium case when an n-type semiconductor specimen is subject to illumination by light, the recombination rate can be written as

$$\mathcal{R} = \alpha_r p_n n_n = \alpha_r (n_n^0 + \Delta n)(p_n^0 + \Delta p), \tag{5.73}$$

where n_n is the total majority carrier concentration, p_n is the total minority concentration, n_n^0 is the equilibrium majority carrier concentration, p_n^0 is the equilibrium minority carrier concentration, and Δn and Δp are the excess carrier concentrations defined as

$$\Delta n = n_n - n_n^0 \quad \text{and} \quad \Delta p = p_n - p_n^0. \tag{5.74}$$

To maintain charge neutrality, Δn and Δp must be equal. The total generation rate, Γ, is the sum of the thermal and light generation rates. Thus, the net rate of change of the hole concentration can be expressed as

$$\frac{\mathrm{d}p_n}{\mathrm{d}t} = \mathcal{G} - \mathcal{R} = g_l + g_t - \mathcal{R}. \tag{5.75}$$

For the steady-state case, the left-hand side of Equation 5.75 is zero and

$$g_l = \mathcal{R} - g_t. \tag{5.76}$$

Thus, g_l can be considered as the net recombination rate. For thermal equilibrium, we have

$$g_t = \mathcal{R} = \alpha_r p_n^0 n_n^0. \tag{5.77}$$

Substituting Equations 5.73 and 5.77 into Equation 5.76 and taking $\Delta n = \Delta p$, we obtain for the net recombination rate, the following expression

$$\begin{aligned} g_{\mathrm{l}} &= \mathcal{R} - g_{\mathrm{t}} = \mathcal{R} - \alpha_{\mathrm{r}} p_{\mathrm{n}}^0 n_{\mathrm{n}}^0 \\ &= \alpha_{\mathrm{r}} p_{\mathrm{n}} n_{\mathrm{n}} - \alpha_{\mathrm{r}} p_{\mathrm{n}}^0 n_{\mathrm{n}}^0 = \alpha_{\mathrm{r}}(n_{\mathrm{n}}^0 + \Delta n)(p_{\mathrm{n}}^0 + \Delta p) - \alpha_{\mathrm{r}} p_{\mathrm{n}}^0 n_{\mathrm{n}}^0 \\ &= \alpha_{\mathrm{r}} \Delta p [n_{\mathrm{n}}^0 + p_{\mathrm{n}}^0 + \Delta p]. \end{aligned} \tag{5.78}$$

For $p_{\mathrm{n}}^0 \ll n_{\mathrm{n}}^0$ and $\Delta p \ll n_{\mathrm{n}}^0$, the net recombination rate becomes

$$g_{\mathrm{l}} = \alpha_{\mathrm{r}} \Delta p n_{\mathrm{n}}^0 = \frac{p_{\mathrm{n}} - p_{\mathrm{n}}^0}{\tau_{\mathrm{p}}}, \tag{5.79}$$

where τ_{p} is the excess minority lifetime ($\tau_{\mathrm{p}} = 1/(\alpha_{\mathrm{r}} n_{\mathrm{n}}^0)$. This equation describes the net recombination rate when the sample is subject to light illumination with photon energy larger than the fundamental band gap energy. For indirect semiconductors such as Si, the derivation of the net rate is left as an exercise (see Problem 5.8).

In addition to direct and indirect recombinations, there are other recombination mechanisms. An example of these mechanisms is surface recombination. Since the lattice structure of any semiconductor materials at the surface contains a large number of dangling bonds, called *surface states*, the recombination rate may be enhanced at the surface. Another example of recombination is *Auger recombination*, in which the energy released from electron–hole recombination is released to a third particle (either an electron or a hole). The third particle loses its energy to the lattice, rendering Auger recombination as a nonradiative recombination process. There are several types of Auger recombinations, but they are not discussed in this chapter.

5.4.4 Continuity Equation

The continuity equation combines the drift, diffusion, generation, and recombination processes into a single equation. To construct the continuity equation in one dimension (along the x-axis), consider an element of thickness dx in a sample of cross-sectional area A, as shown in Fig. 5.15. The overall rate of change in the number of electrons in this slice can be written as the sum of the electrons entering at x minus the number of electrons leaving at dx+x and the electron generation rate minus the electron recombination rate. This can be written as

$$\frac{\partial n}{\partial t} A\mathrm{d}x = \left(\frac{J_{\mathrm{n}}(x)A}{-e} - \frac{J_{\mathrm{n}}(x + \mathrm{d}x)A}{-e} \right) + (\mathcal{G}_{\mathrm{n}} - \mathcal{R}_{\mathrm{n}})A\,\mathrm{d}x, \tag{5.80}$$

where A dx is the volume of the slice, $\mathcal{G}_{\mathrm{n}}$ is the electron generation rate, $\mathcal{R}_{\mathrm{n}}$ is the electron recombination rate, and $-e$ is the charge of the electron. Expanding $J_{\mathrm{n}}(x + \mathrm{d}x)$ in terms of a Taylor series and retaining the first-order terms

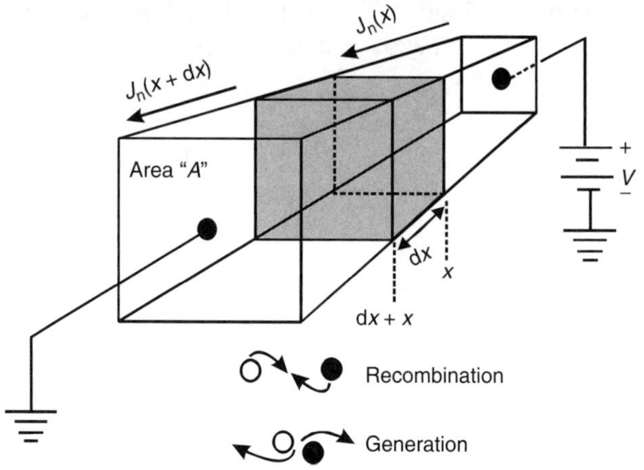

FIGURE 5.15 A sketch of a sample used to illustrate the derivation of the continuity equation. The four processes that occurred in the segment with thickness $\mathrm{d}x$ are recombination, generation, flow-in current density $J_{\mathrm{n}}(x)$, and flow-out current density $J_{\mathrm{n}}(x+\mathrm{d}x)$.

$[J_{\mathrm{n}}(x+\mathrm{d}x) = J_{\mathrm{n}}(x) + \frac{\partial J_{\mathrm{n}}(x)}{\partial x}\mathrm{d}x + \cdots]$, the continuity equation becomes

$$\frac{\partial n}{\partial t} = \frac{1}{e}\frac{\partial J_{\mathrm{n}}(x)}{\partial x} + (\mathcal{G}_{\mathrm{n}} - \mathcal{R}_{\mathrm{n}}). \tag{5.81}$$

A similar expression can be obtained for the holes as

$$\frac{\partial p}{\partial t} = -\frac{1}{e}\frac{\partial J_{\mathrm{p}}(x)}{\partial x} + (\mathcal{G}_{\mathrm{p}} - \mathcal{R}_{\mathrm{p}}), \tag{5.82}$$

where $\mathcal{G}_{\mathrm{p}}$ is the hole generation rate, $\mathcal{R}_{\mathrm{p}}$ is the hole recombination rate, and the minus sign of the first term on the right-hand side is due to the positive charge of the holes. Substituting Equation 5.57a and b into Equations 5.81 and 5.82, we obtain

$$\frac{\partial n}{\partial t} = n\mu_{\mathrm{n}}\frac{\partial \boldsymbol{\mathcal{E}}}{\partial x} + \mu_{\mathrm{n}}\boldsymbol{\mathcal{E}}\frac{\partial n}{\partial x} + D_{\mathrm{n}}\frac{\partial^2 n}{\partial x^2} + \mathcal{G}_{\mathrm{n}} - \frac{n}{\tau_{\mathrm{n}}} \tag{5.83}$$

$$\frac{\partial p}{\partial t} = -p\mu_{\mathrm{n}}\frac{\partial \boldsymbol{\mathcal{E}}}{\partial x} - \mu_{\mathrm{n}}\boldsymbol{\mathcal{E}}\frac{\partial p}{\partial x} + D_{\mathrm{p}}\frac{\partial^2 p}{\partial x^2} + \mathcal{G}_{\mathrm{p}} - \frac{p}{\tau_{\mathrm{p}}}, \tag{5.84}$$

where the regeneration rate is obtained from Equation 5.72. For minority carriers, we have the following expression for electrons (n_{p}) in p-type material and holes (p_{n}) in n-type material:

$$\frac{\partial n_{\mathrm{p}}}{\partial t} = n\mu_{\mathrm{n}}\frac{\partial \boldsymbol{\mathcal{E}}}{\partial x} + \mu_{\mathrm{n}}\boldsymbol{\mathcal{E}}\frac{\partial n_{\mathrm{p}}}{\partial x} + D_{\mathrm{n}}\frac{\partial^2 n_{\mathrm{p}}}{\partial x^2} + \mathcal{G}_{\mathrm{n}} - \frac{n_{\mathrm{p}} - n_{\mathrm{p}}^0}{\tau_{\mathrm{n}}} \tag{5.85}$$

$$\frac{\partial p_{\rm n}}{\partial t} = -p\mu_{\rm p}\frac{\partial \mathcal{E}}{\partial x} - \mu_{\rm p}\mathcal{E}\frac{\partial p_{\rm n}}{\partial x} + D_{\rm p}\frac{\partial^2 p_{\rm n}}{\partial x^2} + \mathcal{G}_{\rm p} - \frac{p_{\rm n} - p_{\rm n}^0}{\tau_{\rm p}}. \quad (5.86)$$

Notice that $\Delta n = (n_{\rm p} - n_{\rm p}^0)$ and $\Delta p = (p_{\rm n} - p_{\rm n}^0)$ are the excess minority carriers. Also, $n_{\rm p}^0$ and $p_{\rm n}^0$ are constants, and the derivatives of Δn and Δp are simply the derivatives of $n_{\rm p}$ and $p_{\rm n}$, respectively. Moreover, the first derivative of the electric field with respect to x can be written in terms of the charge density inside the semiconductor, $\rho_{\rm s}$, through Poisson's equation

$$\frac{\partial \mathcal{E}}{\partial x} = \frac{\rho_{\rm s}}{\epsilon\epsilon_0}, \quad (5.87)$$

where ϵ is the dielectric constant of the semiconductor and ϵ_0 is the permittivity of space. Assuming that the acceptors and donors in the semiconductors are totally ionized, the charge density can be expressed as

$$\rho_{\rm s} = e(p - n + N_{\rm d} - N_{\rm a}), \quad (5.88)$$

where $N_{\rm d}$ and $N_{\rm a}$ are the ionized donor and acceptor concentrations, respectively.

The continuity Equations 5.85–5.87 can be solved with imposed boundary conditions and physical approximations. Let us assume the following:

(i) If the charge neutrality condition is imposed, we have

$$\Delta p = (p_{\rm n} - p_{\rm n}^0) = \Delta n = (n_{\rm p} - n_{\rm p}^0). \quad (5.89)$$

(ii) The generation rates of the electrons and holes are equal:

$$\mathcal{G}_{\rm n} = \mathcal{G}_{\rm p} = \mathcal{G}. \quad (5.90)$$

(iii) The recombination rates of the electrons and holes are equal:

$$\mathcal{R}_{\rm n} = \frac{\Delta n}{\tau_{\rm n}} = \mathcal{R}_{\rm p} = \frac{\Delta p}{\tau_{\rm p}} = \mathcal{R}. \quad (5.91)$$

(iv) The minority carriers are equal, $n_{\rm p} \approx p_{\rm n}$.

With these assumptions, Equations 5.85 and 5.86 become

$$\frac{\partial n_{\rm p}}{\partial t} = n\mu_{\rm n}\frac{\partial \mathcal{E}}{\partial x} + \mu_{\rm n}\mathcal{E}\frac{\partial n_{\rm p}}{\partial x} + D_{\rm n}\frac{\partial^2 n_{\rm p}}{\partial x^2} + \mathcal{G} - \mathcal{R} \text{ and} \quad (5.92)$$

$$\frac{\partial n_{\rm p}}{\partial t} = -p\mu_{\rm p}\frac{\partial \mathcal{E}}{\partial x} - \mu_{\rm p}\mathcal{E}\frac{\partial n_{\rm p}}{\partial x} + D_{\rm p}\frac{\partial^2 n_{\rm p}}{\partial x^2} + \mathcal{G} - \mathcal{R}. \quad (5.93)$$

Multiply Equation 5.92 by $\mu_p p$ and Equation 5.93 by $\mu_n n$, add the two equations, and then divide by $(\mu_n n + \mu_p p)$ to obtain the following:

$$\frac{\partial n_p}{\partial t} = \mu \mathcal{E} \frac{\partial n_p}{\partial x} + D \frac{\partial^2 n_p}{\partial x^2} + \mathcal{G} - \mathcal{R}, \tag{5.94}$$

where

$$\mu = \frac{\mu_n \mu_p (p - n)}{\mu_n n + \mu_p p}, \tag{5.95}$$

and

$$D = \frac{\mu_n n D_p + \mu_p p D_n}{\mu_n n + \mu_p p} = \frac{D_p D_n (n + p)}{D_n n + D_p p}. \tag{5.96}$$

Equations 5.95 and 5.96 are called the *ambipolar mobility* and *diffusion coefficients*, respectively. For low level carrier injection, Equations 5.95 and 5.96 are reduced to the following expressions:

$$\mu = \begin{cases} \mu_n & \text{for p-type} \\ -\mu_p & \text{for n-type} \end{cases} \tag{5.97}$$

and

$$D = \begin{cases} D_n & \text{for p-type} \\ D_p & \text{for n-type.} \end{cases} \tag{5.98}$$

It is clear from the above relations that the ambipolar mobility and ambipolar diffusion coefficient are reduced to minority mobility and diffusion coefficient. Thus, the behavior of the excess majority carriers is determined by the minority carrier parameters, which is the general characteristic of semiconductor devices, such as p–n junction diodes and bipolar transistors, based on the principle of carrier injection.

The continuity equation can be solved for a steady-state system in which the time-first derivative of the carrier concentration is zero. Another way of solving the continuity equation for a system in thermal equilibrium is by assuming that both the concentration gradient and electric field are absent. A more complicated problem is encountered when a constant electric filed is applied across the sample and carrier diffusion is present. Let us assume that a mechanism exists such that a finite number of electron–hole pairs are generated in an n-type semiconductor sample at $t = 0$ and $x = 0$, where t is the time and x is the spatial coordinate. Since the electric field ($\mathcal{E}$) is constant, the gradient of $\mathcal{E}$ is zero. Assume that the generation rate $\mathcal{G}$ is zero at $t \gg 0$. The continuity equation of the minority carriers (holes) can now be expressed as

$$\frac{\partial p_n}{\partial t} = -\mu_p \mathcal{E} \frac{\partial p_n}{\partial x} + D_p \frac{\partial^2 p_n}{\partial x^2} - \frac{p_n - p_n^0}{\tau_p}. \tag{5.99}$$

A possible solution for this equation is

$$\Delta p = p_n - p_n^0 = p_n'(x,t)\exp(-t/\tau_p). \tag{5.100}$$

Substituting this solution into the continuity Equation 5.99 yields

$$\frac{\partial p_n'(x,t)}{\partial t} = -\mu_p \mathcal{E}\frac{\partial p_n'(x,t)}{\partial x} + D_p \frac{\partial^2 p_n'(x,t)}{\partial x^2} - \frac{p_n'(x,t)}{\tau_p}. \tag{5.101}$$

This equation can now be solved using the Laplace transformation technique, which yields a solution of the Gaussian form

$$p_n'(x,t) = \frac{1}{\sqrt{4\pi D_p t}}\exp\left(-\frac{(x-\mu_p \mathcal{E} t)^2}{4D_p t}\right). \tag{5.102}$$

Substituting Equation 5.102 into 5.100, we obtain the final solution as

$$\Delta p(x,t) = p_n - p_n^0 = \mathcal{N}\frac{\exp(-t/\tau_p)}{\sqrt{4\pi D_p t}}\exp\left(-\frac{(x-\mu_p \mathcal{E} t)^2}{4D_p t}\right), \tag{5.103}$$

where $\mathcal{N}$ is the number of electrons or holes generated per unit area. A plot of this solution is shown in Fig. 5.16 with and without an applied electric field. As the excess minority carriers (holes) are generated at $t = 0$ and $x = 0$ when the electric field is zero, as shown in Fig. 5.16a, the holes start to diffuse in both $+x$ and $-x$ directions. The excess majority carriers (electrons) generated during the process diffuse exactly at the same rate as the holes. As time passes, the diffused electrons and holes start to combine, leading to the disappearance of the holes as t approaches infinity. Thus, diffusion and recombination occur at

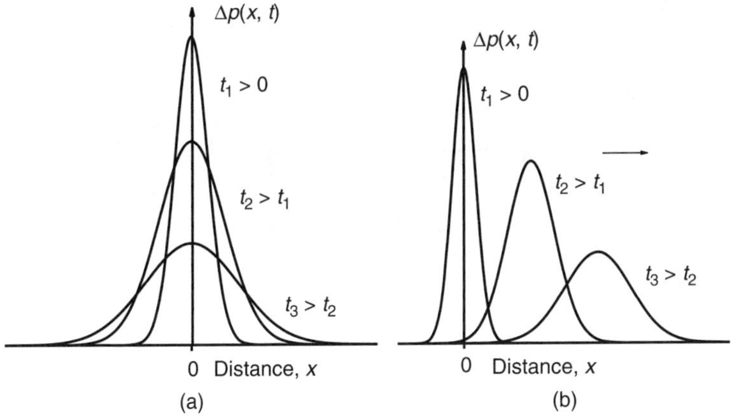

FIGURE 5.16 Equation 5.104 is plotted as a function of time and distance (a) without applying electric field and (b) by applying electric field.

the same time. When the electric field is not zero, the excess minority carriers drift in the same direction, as shown in Fig. 5.16b, since the holes have positive charge. Since the recombination is present as the holes drift in the direction of the electric field, the excess electrons seem to drift in the same direction, even though their charge is negative.

The first experiment to measure excess carrier behavior was reported by Haynes and Shockley in 1951. The basic concept of the Haynes and Shockley experiment is illustrated in Fig. 5.17. A rectangular input pulse, as shown in Fig. 5.18, is introduced at point A at time $t = 0$. The excess carriers drift along the semiconductor when an applied electric field, $\mathcal{E}_1$, is applied to the sample, producing an output voltage signal at point B after time t_1 is lapsed, as shown in Fig. 5.18. If the electric field is reduced to $\mathcal{E}_2 < \mathcal{E}_1$, the signal received at point B will arrive at time $t_2 > t_1$, as shown in the figure. This is because the drift velocity is smaller for lower electric field values. During this longer time period, there is more diffusion and recombination and the excess carrier pulse is smaller, as shown in Fig. 5.18.

The Haynes–Shockley experiment allows one to measure the mobility, diffusion coefficient, and relaxation time of the minority carriers. However, the most accurate parameter that can be extracted from this experiment is the mobility of the excess minority carriers.

5.5 BOLTZMANN TRANSPORT EQUATION

When charge carriers, such as electrons in semiconductors, are at equilibrium (absence of external perturbations), their statistical distribution obeys the Fermi–Dirac distribution function, f_k^0, given by

$$f_k^0 = \frac{1}{e^{(E_k - E_F)/k_B T} + 1}, \tag{5.104}$$

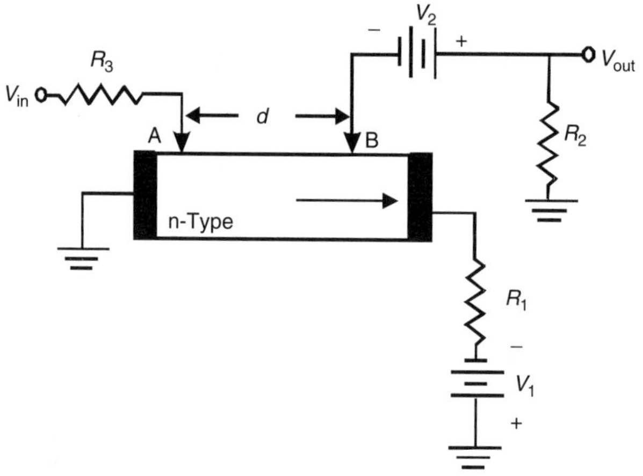

FIGURE 5.17 An illustration of the Haynes–Shockley experiment.

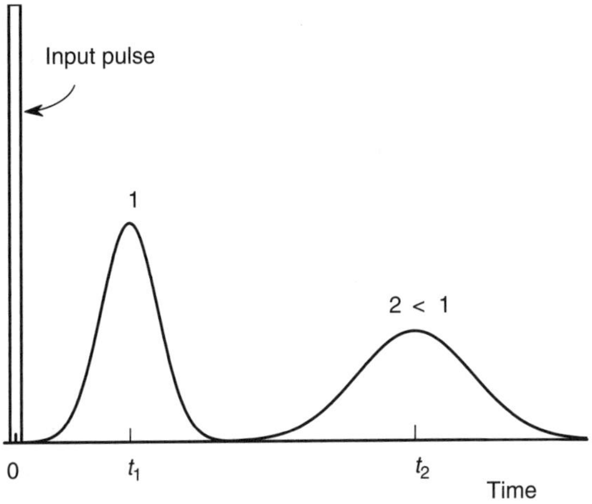

FIGURE 5.18 Carrier diffusion in the Haynes–Shockley experiment in which the input signal is a rectangular narrow pulse applied at point A in Fig. 5.17. The minority carrier pulse is received at point B in Fig. 5.17 at two different electric field values.

where E_F is the Fermi energy level and k_B is the Boltzmann constant. When electrons are subjected to an external perturbation, such as applied electric field, diffusion, or scattering, their distribution function is no longer described by the Fermi–Dirac function, but by a function, f_k, that depends on time, space, and momentum. The Boltzmann approach is used to evaluate the behavior of the nonequilibrium distribution function, f_k, with time. The evolution of f_k as a function of time because of scattering, diffusions, and external field can be written as

$$\frac{df_k}{dt} = \left.\frac{\partial f_k}{\partial t}\right|_{\text{scatterings}}. \tag{5.105}$$

This equation is known as the *Boltzmann equation*. Since f_k is a function of time, $\boldsymbol{r}$ and $\boldsymbol{k}$, the total derivative can be expanded as follows:

$$\frac{\mathrm{d}f_k}{\mathrm{d}t} = \frac{\partial f_k}{\partial t} + \frac{\partial f_k}{\partial x}\frac{\partial x}{\partial t} + \frac{\partial f_k}{\partial y}\frac{\partial y}{\partial t} + \frac{\partial f_k}{\partial z}\frac{\partial z}{\partial t} + \frac{\partial f_k}{\partial k_x}\frac{\partial k_x}{\partial t} + \frac{\partial f_k}{\partial k_y}\frac{\partial k_y}{\partial t} + \frac{\partial f_k}{\partial k_z}\frac{\partial k_z}{\partial t}. \tag{5.106}$$

Since

$$\upsilon = \frac{\partial x}{\partial t}\boldsymbol{x} + \frac{\partial y}{\partial t}\boldsymbol{y} + \frac{\partial z}{\partial t}\boldsymbol{z} \text{ and } \frac{\partial k_x}{\partial t}\boldsymbol{k}_x + \frac{\partial k_y}{\partial t}\boldsymbol{k}_y + \frac{\partial k_z}{\partial t}\boldsymbol{k}_z = \frac{\mathcal{F}}{\hbar}, \tag{5.107}$$

where υ is the electron velocity and $\mathcal{F}$ is the external force acting on the system. Substituting Equation 5.107 into Equation 5.106 and knowing that

$$\frac{\partial f_k}{\partial x}\boldsymbol{x} + \frac{\partial f_k}{\partial y}\boldsymbol{y} + \frac{\partial f_k}{\partial z}\boldsymbol{z} = \nabla_r f_k \text{ and } \frac{\partial f_k}{\partial k_x}\boldsymbol{k}_x + \frac{\partial f_k}{\partial k_y}\boldsymbol{k}_y + \frac{\partial f_k}{\partial k_z}\boldsymbol{k}_z = \nabla_k f_k, \tag{5.108}$$

we have

$$\frac{\partial f_k}{\partial t} + \upsilon \cdot \nabla_r f_k + \frac{1}{\hbar}\mathcal{F} \cdot \nabla_k f_k - \left.\frac{\partial f_k}{\partial t}\right|_{\text{scattering}} = 0. \tag{5.109}$$

This equation is known as the *Boltzmann transport equation*. The term labeled "*scattering*" represents the distribution function due to scattering between electrons and their surrounding, which can be defined as

$$\left.\frac{\partial f_k}{\partial t}\right|_{\text{scattering}} = -\int [f_k(1 - f_{k'})W_{k,k'} - f_{k'}(1 - f_k)W_{k',k}]\mathrm{d}k', \tag{5.110}$$

where the term $(1 - f_k)$ represents the probability of having a vacancy in the state k, the term $(1 - f_{k'})$ represents the probability of having a vacancy in the state k', and $W_{k,k'}$ and $W_{k',k}$ are the rates at which the electron makes a transition from state k to state k' and from state k' to state k, respectively. These rates are also called *transition matrix elements*. There is a whole subfield of transport theory devoted to the calculation of these matrix elements for various scattering mechanisms. While the Boltzmann transport equation provides a very useful description of many transport processes in semiconductors, it is still a strictly classical method of describing the transport properties. This is because the distribution function is specified in terms of position, momentum, and time. The simultaneous description of position and momentum is in contradiction to the Heisenberg uncertainty principle and, therefore, the Boltzmann transport equation is not a valid description of quantum effects.

Since the Boltzmann transport equation includes various nonequilibrium mechanisms, such as scattering, recombination, generation, drift, and diffusion, an exact solution for this equation is extremely difficult to obtain. Even approximate solutions require sophisticated numerical analyses, such as the Monte Carlo and drift–diffusion methods. One possible approximation is the relaxation time method, which assumes that the scattering term in Equation 5.110 can be replaced by a constant relaxation term. This approximation reduces Equation 5.109 to a regular differential equation. Thus, the right-hand side of 5.105 can be replaced by

$$\left.\frac{\partial f_k}{\partial t}\right|_{\text{collisions}} = -\frac{f_k - f_k^0}{\tau}. \tag{5.111}$$

This equation indicates that it will take the system a characteristic time, τ, called the *relaxation time* to relax from the nonequilibrium state to the equilibrium state. With this approximation, the Boltzmann transport equation can be rewritten as

$$\frac{\partial f_k}{\partial t} + \upsilon \cdot \nabla_r f_k + \frac{1}{\hbar}\mathcal{F} \cdot \nabla_k f_k = -\frac{f_k - f_k^0}{\tau}. \tag{5.112}$$

Let us assume that the system is in steady state, where $\frac{\partial f_k}{\partial t} = 0$, and that the distribution function is spatially uniform such that the spatial gradient of f_k is zero, that is, $\nabla_r f_k = 0$. Moreover, let us assume that the external force $\mathcal{F}$ is only due to a constant applied electric field ($\boldsymbol{\mathcal{E}}$) such as $\mathcal{F} = -e\boldsymbol{\mathcal{E}}$. Finally, the term $\frac{1}{\hbar}\nabla_k$ is simply $\frac{1}{m^*}\nabla_\upsilon$, where m^* is the electron effective mass. With these approximations, the Boltzmann transport equation is reduced to

$$-\frac{e}{m^*}\boldsymbol{\mathcal{E}} \cdot \nabla_\upsilon f_k = -\frac{f_k - f_k^0}{\tau}. \quad (5.113)$$

If we assume that the electric field is along the x-axis, the above equation becomes

$$\frac{e\tau\mathcal{E}_x}{m^*}\frac{\partial f_k}{\partial \upsilon_x} = f_k - f_k^0. \quad (5.114)$$

Another useful approximation is that if f_k is assumed to be not far from f_k^0, the derivative of these two functions with respect to the velocity is approximately the same

$$\frac{\partial f_k}{\partial \upsilon_x} \approx \frac{\partial f_k^0}{\partial \upsilon_x}. \quad (5.115)$$

The equilibrium distribution function is assumed to be a Fermi–Dirac distribution function given by Equation 5.104. For simplicity, let us assume that our reference point is the Fermi energy, which can be set to zero in Equation 5.104. Let us also assume that $\mathrm{e}^{E_k/k_\mathrm{B}T} > 1$. Thus, f_k^0 can be written as

$$f_k^0 = \mathrm{e}^{-E_k/k_\mathrm{B}T}, \quad (5.116)$$

which is the form of the Maxwell–Boltzmann distribution function. The energy, E_k, can be taken as $\frac{1}{2}m^*\upsilon_x^2 x$. The derivative of Equation 5.116 with respect to velocity is obtained as follows:

$$\frac{\partial f_k^0}{\partial \upsilon_x} = \frac{\partial}{\partial \upsilon_x}(e^{-m^*\upsilon_x^2/2k_\mathrm{B}T}) = -\frac{m^*\upsilon_x}{k_\mathrm{B}T}\mathrm{e}^{-m^*\upsilon_x^2/2k_\mathrm{B}T} \approx \frac{\partial f_k}{\partial \upsilon_x}. \quad (5.117)$$

Substituting Equation 5.117 into Equation 5.114 and rearranging yields

$$f_k = f_k^0\left(1 - \frac{e\tau\mathcal{E}_x\upsilon_x}{k_\mathrm{B}T}\right). \quad (5.118)$$

A plot of the nonequilibrium distribution function expressed in Equation 5.118 as a function of the drift velocity is shown in Fig. 5.19 for a free electron. For this plot, the temperature is assumed to be 300 K, the applied electric field is 5×10^5 V/cm, and the relaxation time (τ) is 0.4 ps. The x-component of the velocity can be taken as $\upsilon\cos\theta$. Notice that f_k^0 is centered at $\upsilon_x = 0$ and f_k is slightly shifted toward the left. If the minus sign in the parenthesis in Equation 5.118 is positive, f_k will shift to the right.

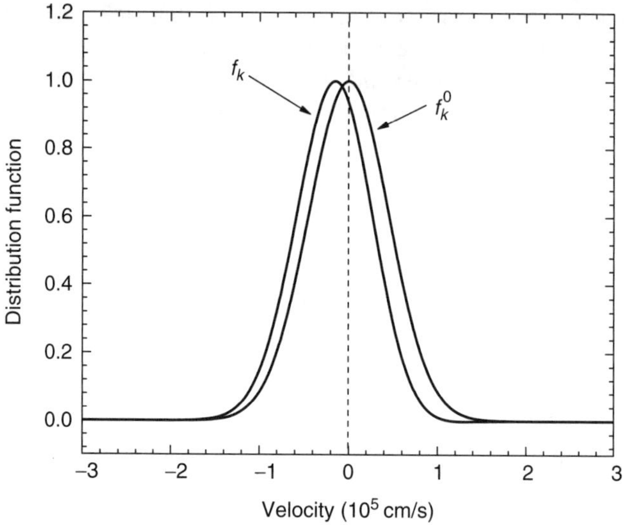

FIGURE 5.19 The distribution function plotted as a function of carrier velocity for equilibrium (f_k^0) and nonequilibrium (f_k) cases. Notice that the peak of f_k is shifted from $\upsilon_x = 0$.

The nonequilibrium distribution function can also be realized by shifting the wave vector $\boldsymbol{k}$ by $e\tau\mathcal{E}/\hbar$. This can be accomplished by considering the distribution function for electrons in a parabolic band at equilibrium, which is given by Equation 5.104. By setting the Fermi energy to zero, $E_k = \frac{\hbar^2 k^2}{2m}$, and by replacing $\boldsymbol{k}$ with $\boldsymbol{k} - e\tau\mathcal{E}/\hbar$, one can obtain the distribution function for the nonequilibrium case as shown in Fig. 5.20. If the applied electric field is along the x-direction, the distribution will shift only for k_x. In equilibrium, there is a net cancellation between positive and negative momenta, but when an electric field is applied, there is a nonzero net shift in the electron momenta given by $\delta\boldsymbol{p} = \hbar\delta\boldsymbol{k} = -e\tau\boldsymbol{\mathcal{E}}$.

5.6 DERIVATION OF TRANSPORT COEFFICIENTS USING THE BOLTZMANN TRANSPORT EQUATION

Many of the transport coefficients can be derived from the Boltzmann transport equation in the framework of a relaxation time approximation as described in Equation 5.111. The relaxation time depends on various scattering mechanisms. This relaxation time depends on the energy and mass of the scattered particles (e.g., electron) according to the following relation:

$$\tau = \tau_0 (m^*)^{\alpha} (E)^{\beta}, \tag{5.119}$$

where τ_0 is a constant and α and β are constants that characterize the scattering mechanism, which depend on the type of scattering mechanism. For example,

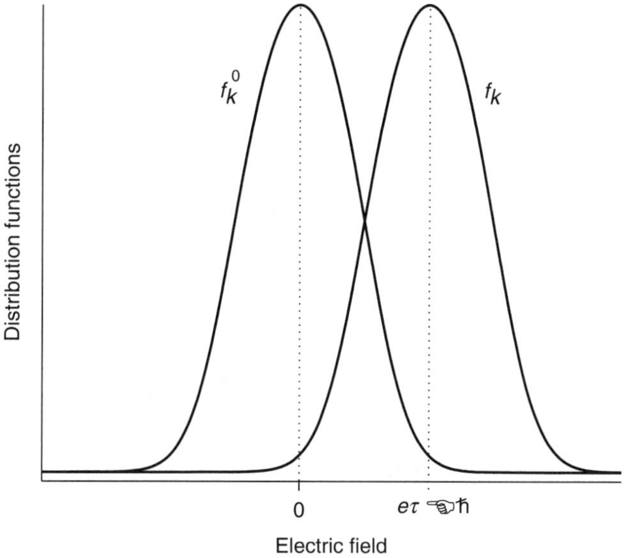

FIGURE 5.20 The displaced distribution function shows the effect of an applied electric field.

α 1/2 and β 3/2 for electron-ionized impurity scattering, while alloy scattering yields $\alpha = -1/2$ and $\beta = -3/2$.

For simplicity, let us assume that we have an n-type semiconductor in which an applied electric field, magnetic field, and temperature gradient are present. The Boltzmann transport equation for the steady-state case can now be written as

$$-\upsilon \cdot \nabla_r f_k + \frac{e}{\hbar}(\boldsymbol{\mathcal{E}} + \boldsymbol{\upsilon} \times \boldsymbol{B}) \cdot \nabla_k f_k = \frac{f_k - f_k^0}{\tau}, \tag{5.120}$$

where $\boldsymbol{\mathcal{E}}$ is the applied electric field and $\boldsymbol{B}$ is the applied magnetic field. Using $m^* \upsilon = \hbar k$ in Equation 5.120, we obtain the Boltzmann transport equation in the following form:

$$-\upsilon \cdot \nabla_r f_k + \frac{e}{m^*}(\boldsymbol{\mathcal{E}} + \boldsymbol{\upsilon} \times \boldsymbol{B}) \cdot \nabla_\upsilon f_k = \frac{f_k - f_k^0}{\tau}. \tag{5.121}$$

An analytical solution for this equation is difficult to obtain without additional approximations. One good approximation is to assume that the solution of the function f_k can be written in terms of f_k^0 and a first-order correction term such that

$$f_k = f_k^0 - \boldsymbol{\upsilon} \cdot \boldsymbol{Q}(E) \frac{\partial f_k^0}{\partial E}, \tag{5.122}$$

where $\boldsymbol{Q}(E)$ is an unknown vector function that depends only on the energy of the electron (E). For the small perturbation case, in which $(f_k - f_k^0) < 1$, the expressions in Equation 5.121 can be approximated as

$$\upsilon \cdot \nabla_r f_k \approx \upsilon \cdot \nabla_r f_k^0 = \upsilon \cdot (\nabla_r T)\left(\frac{(E_F - E)}{T}\frac{\partial f_k^0}{\partial E}\right) \tag{5.123a}$$

$$\mathcal{E} \cdot \nabla_\upsilon f_k \approx \mathcal{E} \cdot \nabla_\upsilon f_k^0 = \mathcal{E} \cdot (\nabla_\upsilon E)\frac{\partial f_k^0}{\partial E} = \mathcal{E} \cdot (m^* \upsilon)\frac{\partial f_k^0}{\partial E} \tag{5.123b}$$

$$(\upsilon \times \boldsymbol{B}) \cdot \nabla_\upsilon f_k \approx -\upsilon \cdot (\boldsymbol{B} \times \boldsymbol{Q}(E))\frac{\partial f_k^0}{\partial E}. \tag{5.123c}$$

Substituting Equations 5.122 and 5.23 into Equation 5.121, we obtain

$$-e\tau(\mathcal{E} \cdot \upsilon) + \frac{e\tau}{m^*}\upsilon \cdot (\boldsymbol{B} \times \boldsymbol{Q}(E)) + \tau\frac{(E_F - E)}{T}\upsilon \cdot (\nabla_r T) - \upsilon \cdot \mathcal{Q}(E) = 0. \tag{5.124}$$

The velocity in this equation can be factored out to obtain the Boltzmann transport equation for the steady-state case under applied electric and magnetic field and a temperature gradient. To obtain a solution for $\mathcal{Q}(E)$, let us assume that the applied electric field and temperature gradient lie in the x–y plane, while the magnetic field is applied along the z-direction. The x and y components of $\mathcal{Q}(E)$ are

$$\boldsymbol{Q}_x(E) = \frac{\tau\left(-e\mathcal{E}_x + \frac{(E_F-E)}{T}\frac{\partial T}{\partial x}\right) - \omega_c\tau^2\left(-e\mathcal{E}_y + \frac{(E_F-E)}{T}\frac{\partial T}{\partial y}\right)}{1+\omega_c^2\tau^2} \tag{5.125a}$$

$$\boldsymbol{Q}_y(E) = \frac{\tau\left(-e\mathcal{E}_y + \frac{(E_F-E)}{T}\frac{\partial T}{\partial y}\right) + \omega_c\tau^2\left(-e\mathcal{E}_x + \frac{(E_F-E)}{T}\frac{\partial T}{\partial x}\right)}{1+\omega_c^2\tau^2}, \tag{5.125b}$$

where ω_c is the cyclotron angular frequency (eB_z/m^*). By knowing $\boldsymbol{Q}(E)$, one can use Equation 5.122 to obtain the transport parameters of a semiconductor in the nonequilibrium case.

5.6.1 Electrical Conductivity and Mobility in n-type Semiconductors

Let us assume that the electric field is applied in the x- and y-directions in an n-type semiconductor. Let the magnetic field and temperature gradient be zero. With these assumptions, Equation 5.25 isreduced to the following:

$$\boldsymbol{Q}_x(E) = -\tau e\mathcal{E}_x \tag{5.126a}$$

$$\boldsymbol{Q}_y(E) = -\tau e\mathcal{E}_y. \tag{5.126b}$$

The general expression for the electron current density is

$$J_x = -en\upsilon_x = -e\int_0^\infty \upsilon_x f(E) g^{3D}(E)\mathrm{d}E, \tag{5.127}$$

where $f(E)$ is given by

$$f(E) = (f_k - f_k^0) = -\upsilon \cdot \boldsymbol{Q}(E)\frac{\partial f_k^0}{\partial E}, \tag{5.128}$$

and $g^{3D}(E)$ is the density of sates per unit volume in bulk semiconductor and is given by

$$g^{3D}(E) = \frac{1}{2\pi^2}\left(\frac{2m}{\hbar^2}\right)^{3/2}\sqrt{E}, \tag{5.129}$$

assuming that the bottom of the conduction band energy is the reference point, which can be set to zero. By combining Equations 5.126 through 5.129, the current density can be written as

$$\begin{aligned} J_x &= -en\upsilon_x = e\int_0^\infty \upsilon_x^2 Q(E)\frac{\partial f_k^0}{\partial E}g^{3D}(E)\mathrm{d}E = -e^2\boldsymbol{\mathcal{E}}_x\int_0^\infty \tau\upsilon_x^2\frac{\partial f_k^0}{\partial E}g^{3D}(E)\mathrm{d}E \\ &= \frac{2e^2\boldsymbol{\mathcal{E}}_x}{3m^*k_\mathrm{B}T}\int_0^\infty \tau E g^{3D}(E) f_k^0 \mathrm{d}E, \end{aligned} \tag{5.130}$$

where we assumed that $\upsilon_x^2 = \upsilon_y^2 = \upsilon_z^2 = \frac{2E}{3m^*}$. The Fermi–Dirac distribution function, f_k^0, is approximated as a Maxwell–Boltzmann function given by Equation 5.116, and its derivative is given by

$$\frac{\partial f_k^0}{\partial E} = -\frac{f_k^0}{k_\mathrm{B}T}. \tag{5.131}$$

The electrical conductivity can now be written as

$$\begin{aligned} \sigma &= \frac{\mathrm{J}_x}{\boldsymbol{\mathcal{E}}_x} = \frac{2e^2}{3m^*k_\mathrm{B}T}\int_0^\infty \tau E g^{3D}(E) f_k^0 \mathrm{d}E = \frac{2ne^2}{3m^*k_\mathrm{B}Tn}\int_0^\infty \tau E g^{3D}(E) f_k^0 \mathrm{d}E \\ &= \frac{ne^2}{m^*}\frac{\int_0^\infty \tau E g^{3D}(E) f_k^0 \mathrm{d}E}{\int_0^\infty \left(\frac{3k_\mathrm{B}T}{2}\right) g^{3D}(E) f_k^0 dE} = \frac{ne^2}{m^*}\frac{\int_0^\infty \tau E^{3/2} f_k^0 \mathrm{d}E}{\int_0^\infty E^{3/2} f_k^0 \mathrm{d}E} = \frac{ne^2\langle\tau\rangle}{m^*}, \end{aligned} \tag{5.132}$$

where the average total kinetic energy is given by $E = \frac{3}{2}k_\mathrm{B}T$, the average relaxation time is given by

$$\langle\tau\rangle = \frac{\int_0^\infty \tau E^{3/2} f_k^0 \mathrm{d}E}{\int_0^\infty E^{3/2} f_k^0 \mathrm{d}E}, \tag{5.133}$$

and the electron density is given by

$$n = \int_0^\infty g^{3D}(E) f_k^0 \mathrm{d}E. \tag{5.134}$$

In general, the relaxation time is a function of electron energy for most scattering processes in semiconductors as expressed in Equation 5.119. If we assume that α is zero, that is, the relaxation time is independent of the electron effective mass, Equation 5.119 is reduced to

$$\tau(E) = \tau_0 E^{\beta}. \tag{5.135}$$

For a nondegenerate semiconductor and for Maxwell–Boltzmann statistics, the relaxation time is

$$\langle\tau\rangle = \tau_0 \frac{\int_0^{\infty} E^{\beta+3/2} \mathrm{e}^{-(E-E_\mathrm{F})/k_\mathrm{B}T} \mathrm{d}E}{\int_0^{\infty} E^{3/2} \mathrm{e}^{-(E-E_\mathrm{F})/k_\mathrm{B}T} \mathrm{d}E} = \tau_0 (k_\mathrm{B}T)^{\beta} \frac{\Gamma(\frac{5}{2}+\beta)}{\Gamma(\frac{5}{2})}, \tag{5.136}$$

where Γ is the Γ-function described in this chapter. Substituting Equation 5.136 into Equation 5.132, we obtain the following expression for the electrical conductivity:

$$\sigma = \frac{ne^2\tau_0}{m^*} (k_\mathrm{B}T)^{\beta} \frac{\Gamma(\frac{5}{2}+\beta)}{\Gamma(\frac{5}{2})}, \tag{5.137}$$

from which the electron mobility is obtained as

$$\mu_n = \frac{e\tau_0}{m^*} (k_B T)^{\beta} \frac{\Gamma(\frac{5}{2}+\beta)}{\Gamma(\frac{5}{2})}. \tag{5.138}$$

For β 0, the expression for mobility is reduced to that obtained when the system is in equilibrium. Various scattering mechanisms are discussed in the next section.

The electrical conductivity and mobility discussed above were derived for a single-valley model with a spherical constant energy surface for the conduction band. These requirements are usually encountered in III–V semiconductor materials such as GaAs, GaSb, InP, etc. The electron effective mass in Equation 5.137 is usually isotropic. However, for semiconductors with multivalley conduction bands, such as Si, the effective mass is anisotropic, even though the electrical conductivity remains isotropic because of the cubic crystal structure. For multivalley semiconductors, the effective mass is given by

$$m_{\sigma}^* = \left\{\frac{1}{3}\left(\frac{1}{m_l} + \frac{1}{m_t}\right)\right\}^{-1}, \tag{5.139}$$

where m_l and m_t are the electron longitudinal and transverse effective masses along the two main axes of the ellipsoidal energy surface near the conduction band edge. The subscript "σ" in introduced in Equation 5.139 to indicate that the mass is the conductivity effective mass.

In addition, the Hall mobility described in Equation 5.138 is not required to be equal to the drift mobility where the magnetic field is zero. These mobilities are related according to the following relation:

$$\mu_{\mathrm{H}} = \frac{\langle \tau^2 \rangle}{\langle \tau \rangle^2} \mu, \tag{5.140}$$

where μ_{H} is the Hall mobility ($B_z \neq 0$) and μ is the drift mobility in the absence of magnetic field.

5.6.2 Hall Coefficient, R_H

The Hall coefficient for a nondegenerate n-type semiconductor with a single-valley spherical energy band can be derived by assuming that the magnetic field in the Hall measurement is weak, where $(\omega_c \tau)^2 \gg 1$. For the steady-state case in the absence of a temperature gradient, the vector function, $\mathcal{Q}$(E), given by Equation 5.125, isreduced to the following

$$\mathcal{Q}_x(E) = -\tau e \boldsymbol{\mathcal{E}}_x + \omega_c \tau^2 e \boldsymbol{\mathcal{E}}_y \tag{5.141a}$$

$$\mathcal{Q}_y(E) = -\tau e \boldsymbol{\mathcal{E}}_y - \omega_c \tau^2 e \boldsymbol{\mathcal{E}}_x. \tag{5.141b}$$

By following the same procedure as in the previous section and using the above expressions for $\boldsymbol{Q}(E)$, the current densities can be written as

$$\begin{aligned} J_x = -env_x &= e \int_0^\infty v_x^2 \mathcal{Q}(E) \frac{\partial f_k^0}{\partial E} g^{3D}(E) \mathrm{d}E \\ &= \frac{2e^2}{3m^* k_{\mathrm{B}} T} \int_0^\infty \tau E [\boldsymbol{\mathcal{E}}_x - \omega_c \tau \boldsymbol{\mathcal{E}}_y] g^{3D}(E) f_k^0 \mathrm{d}E \end{aligned} \tag{5.142}$$

and

$$\begin{aligned} J_y = -env_y &= e \int_0^\infty v_y^2 \mathcal{Q}(E) \frac{\partial f_k^0}{\partial E} g^{3D}(E) \mathrm{d}E \\ &= \frac{2e^2}{3m^* k_{\mathrm{B}} T} \int_0^\infty \tau E [\boldsymbol{\mathcal{E}}_y + \omega_c \tau \boldsymbol{\mathcal{E}}_x] g^{3D}(E) f_k^0 \mathrm{d}E. \end{aligned} \tag{5.143}$$

From the definition of the electron Hall coefficient in Equations 5.14, 5.142, and 5.143, we can write

$$R_{\mathrm{H}} = \left. \frac{E_y}{J_x B_z} \right|_{J_y=0} = -\left(\frac{3k_{\mathrm{B}} T}{2e} \right) \frac{\int_0^\infty \tau^2 E g^{3D} F(E) f_k^0 \mathrm{d}E}{\left[\int_0^\infty \tau E g^{3D} F(E) f_k^0 \mathrm{d}E \right]^2} = -\frac{1}{en} \frac{\langle \tau^2 \rangle}{\langle \tau \rangle^2}. \tag{5.144}$$

The Hall factor, r, shown in Equation 5.14 can now be defined as

$$r = \frac{\langle \tau^2 \rangle}{\langle \tau \rangle^2}. \tag{5.145}$$

The expression for the Hall coefficient for the holes is the same as that for the electrons except that the *minus* sign in Equation 5.144 is replaced by a *plus* sign.

Many other transport parameters can be analytically derived from the Boltzmann transport equation, depending on the conditions and the initial assumptions. For example, the Seebeck coefficient can be derived in the steady state in the absence of a magnetic field, but in the presence of a temperature gradient. The analysis can be accomplished as follows: First, set $\omega_c = 0$ in Equation 5.126a, which gives

$$\mathcal{Q}_x(E) = \tau \left(-e\boldsymbol{\mathcal{E}}_x + \frac{(E_F - E)}{T} \frac{\partial T}{\partial x} \right). \tag{5.146}$$

Substitute Equation 5.146 into the current density, J_x, shown in 5.127; then set J_x to zero to give

$$J_x = -env_x = -e \int_0^{\infty} v_x^2 \tau \left(-e\boldsymbol{\mathcal{E}}_x + \frac{(E_F - E)}{T} \frac{\partial T}{\partial x} \right) g^{3D}(E) \mathrm{d}E = 0. \tag{5.147}$$

Rearrange to obtain

$$S_n = \frac{\boldsymbol{\mathcal{E}}_x}{(\partial T / \partial x)} = -\frac{1}{eT} \left[\frac{\int_0^{\infty} \tau E^2 g^{3D}(E) \frac{\partial f_k^0}{\partial E} \mathrm{d}E}{\int_0^{\infty} \tau E g^{3D}(E) \frac{\partial f_k^0}{\partial E} \mathrm{d}E} - E_F \right] = -\frac{1}{eT} \left[\frac{\langle \tau E \rangle}{\langle \tau \rangle} - E_F \right], \tag{5.148}$$

where S_n is the Seebeck coefficient. Notice that we used $v_x^2 = \frac{2E}{3m^*}$. Other parameters that can be easily derived from the Boltzmann transport equation are the transverse magnetoresistance, Nernst coefficient, Ettingshausen coefficient, and Righi–Leduc coefficient.

5.7 SCATTERING MECHANISMS IN BULK SEMICONDUCTORS

There are several scattering mechanisms that play major roles in the determination of carrier mobilities and conductivities in semiconductors. These mechanisms are summarized in Fig. 5.21. This figure lists the scattering relaxation time as a function of the energy and the effective mass of the charge carrier. The mobility as a function of the sample temperature and the effective mass of the carrier is also shown in the figure. This figure does not include electron–electron or electron–hole scattering. The carrier–carrier scatterings become important in heavily doped semiconductors in which impact ionization and Auger processes

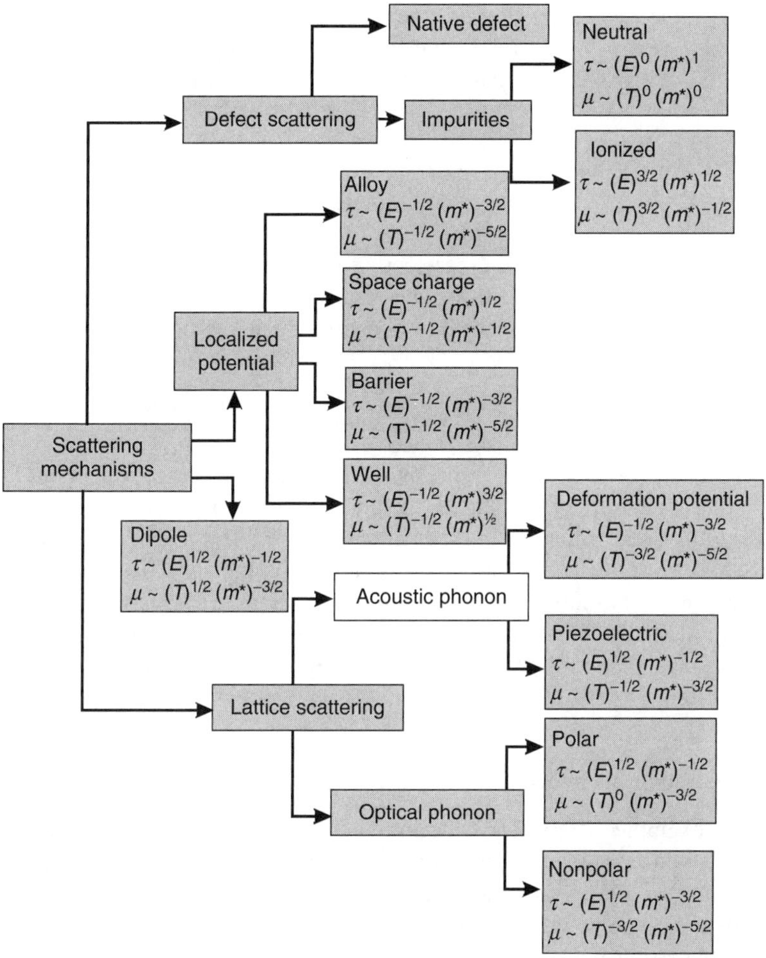

FIGURE 5.21 Summary of the major scattering mechanisms that influence the mobility and relaxation time of electron transport in semiconductors. The mobility temperature dependence for the nonpolar scattering is for $k_B T \gg \hbar\omega_0$, where ω_0 is the frequency of the nonpolar optical phonon.

are significant. The theoretical analyses of scattering processes are usually treated using quantum mechanics. The Fermi golden rule is employed to calculate the scattering rate or the transition matrix element when the charge carrier is scattered from one state to another.

The general procedure of calculation is to identify the scattering potential and then use the first-order perturbation theory to calculate the transition matrix element. The scattering relaxation time can be obtained from the following relation:

$$\tau^{-1} = N_t \sigma_t \upsilon_{th}, \tag{5.149}$$

where N_t is the density of the total scattering centers, σ_t is the total scattering cross section, and υ_{th} is the average thermal velocity [$\upsilon_{th} = (3k_B T/m^*)^{1/2}$]. The total scattering cross section can be obtained from the differential cross section using

$$\sigma_t = 2\pi \int_0^{\pi} \sigma(\theta')(1 - \cos\theta') \sin\theta' d\theta', \tag{5.150}$$

where $\sigma(\theta')$ is the differential cross section defined as the total number of particles that makes the transition from one state to another per unit solid angle per unit time divided by the incident flux density, which can be written as

$$\sigma(\theta') = \frac{V^2 k'^2 |W_{k,k'}|^2}{(2\pi\hbar\upsilon_{k'})^2}, \tag{5.151}$$

where V is the volume of the crystal, $W_{k,k'}$ is the transition matrix element for the particle that is scattered from state k to state k', and $\upsilon_{k'}$ is the particle velocity in state k'. For elastic collisions, both momentum and energy are conserved, which means that $\upsilon_k = \upsilon_{k'}$ and $k = k'$. Derivations of the mobility and scattering relaxation time are thus straightforward, even though extensive mathematical manipulation is unavoidable. In this section, we provide the final results of the mobility and relaxation time for several scattering centers without going through the derivation.

5.7.1 Scattering from an Ionized Impurity

A typical example of elastic scattering in semiconductors is the scattering of electrons by an ionized shallow donor impurity. This is due to the fact that the mass of the electron is much smaller than the mass of the ionized impurity, so the change in the electron energy is negligible. The relaxation time for ionized impurity scattering with an ionic charge of e is given by

$$\frac{1}{\tau_i} = \frac{e^4 N_i \ln\left[\frac{8m^* E\epsilon\epsilon_0 k_B T}{\hbar^2 e^2 n'}\right]}{16\pi(2m^*)^{1/2}\epsilon^2\epsilon_0^2 E^{3/2}}, \tag{5.152}$$

and

$$\mu_i = \frac{e\langle\tau_i\rangle}{m^*} = \frac{64\sqrt{\pi}\epsilon^2\epsilon_0^2(2k_B T)^{3/2}}{N_i e^3 (m^*)^{1/2} \ln\left[\frac{12m^*(k_B T)^2\epsilon\epsilon_0}{\hbar^2 e^2 n'}\right]}, \tag{5.153}$$

where N is the density of the ionized impurity, n′ is the density of the screening electrons surrounding the ionized donor impurity, and the subscript “i” is introduced to indicate ionized impurity scattering. The natural log of the quantity in the brackets is a slowly varying function with temperature and n'. A slight variation is obtained for the quantity in the bracket when the bare Coulomb potential is used as the perturbing Hamiltonian (Conwell and Weisskopf).

5.7.2 Scattering from a Neutral Impurity

The scattering of electrons by a neutral impurity was first derived by Erginsoy, who assumed that the scattering is elastic scattering. The relaxation time and mobility for this type of scattering are given by

$$\frac{1}{\tau_{\mathrm{ni}}} = \frac{80\pi\epsilon\epsilon_0 N_{\mathrm{n}}\hbar^3}{m^{*2}e^2} \tag{5.154}$$

and

$$\mu_{\mathrm{ni}} = \frac{e\tau_{ni}}{m^*} = \frac{m^* e^3}{80\pi\epsilon\epsilon_0 N_{\mathrm{n}}\hbar^3}, \tag{5.155}$$

where the subscript "ni" stands for neutral impurity and N_{n} is the density of the neutral impurity. The mobility and the scattering time are independent of both temperature and the energy of the electron. The charge carrier scattering from neutral impurities is more significant at low temperatures since carrier freeze-out may occur at shallow-level impurity centers in extrinsic semiconductors. However, the low temperature mobility in many semiconductor materials does not always agree with the theoretical behavior of the mobility given by Equation 5.155.

5.7.3 Scattering from Acoustic Phonons: Deformation Potential

Electron scattering from longitudinal acoustic phonons is very significant in intrinsic semiconductors. It is usually considered elastic scattering since the change in the electron energy is proportional to the ratio of the sound velocity in a solid ($\sim 3 \times 10^5$ cm/s) and the average thermal velocity of the electron ($\sim 10^7$ cm/s). This ratio is smaller than unity for temperatures higher than 100 K. The acoustic-mode lattice vibrations induce changes in lattice spacing, which induces a local fluctuation in the band gap. The potential resulting from this fluctuation is called the *deformation potential*. This potential may be estimated as the band gap is changed per unit strain. Electron scattering from the deformation potential is important in undoped silicon and germanium at room temperature. The relaxation time and mobility due to the deformation potential were derived by Bardeen and Shockley and given as

$$\tau_{\mathrm{dp}} = \frac{\pi\hbar^4 C_1}{\sqrt{2}(m^*)^{3/2}k_{\mathrm{B}}T E_1^2 E^{1/2}} \tag{5.156}$$

and

$$\mu_{\mathrm{dp}} = \frac{e\langle\tau_{\mathrm{dp}}\rangle}{m^*} = \frac{2\sqrt{2\pi}e\hbar^4 C_1}{3(m^*)^{5/2}(k_{\mathrm{B}}T)^{3/2}E_1^2}, \tag{5.157}$$

where the subscript "dp" stands for deformation potential, C_1 is the longitudinal elastic constant given by $C_1 = (C_{11} + C_{12} + C_{44})/2 = \rho {v_{\mathrm{s}}}^2$ for wave propagation along the (110) direction and $C_1 = C_{11}$ for wave propagation along the (100) direction, where ρ is the crystal density, v_s is the sound velocity in the crystal, and E_1 is the deformation potential constant. Values reported for E_1 include $|E_1| = 16$ eV for Si, $|E_1| = 9.5$ eV for Ge, and $|E_1| = 9.3$ eV for GaAs.

The theoretical expression for mobility indicates that mobility is proportional to $T^{-3/2}$, which is in good agreement with the experimental measurements obtained for silicon and germanium at $T \ll 200$ K. At higher temperatures, intervalley optical phonon scattering contributes substantially to electron mobility. This leads to a different relation between mobility and temperature such as $\mu_{\text{pd}} \sim T^{-n}$, where n lies between 1.5 and 2.7.

5.7.4 Scattering from Acoustic Phonons: Piezoelectric Potential

For polar semiconductors, such as compound semiconductors with zinc blende and wurtzite structures, the bonds are partially ionic and the unit cell does not possess inversion symmetry. A strain-induced electric field can be generated because of the piezoelectric effect. The piezoelectric potential is thus generated by the acoustic-mode lattice vibrations. A relaxation time due to carrier-piezoelectric potential scattering can be defined as

$$\tau_{\text{pz}} = \frac{2\sqrt{2\pi}\hbar^2 \epsilon\epsilon_0}{e^2(m^*)^{1/2}k_{\text{B}}\text{T}P^2E^{1/2}}, \tag{5.158}$$

and the corresponding mobility is

$$\mu_{\text{pz}} = \frac{e\langle\tau_{\text{pz}}\rangle}{m^*} = \frac{2\sqrt{2\pi}\hbar^2 \epsilon\epsilon_0}{3(m^*)^{3/2}eP^2(k_{\text{B}}T)^{1/2}}, \tag{5.159}$$

where the subscript "pz" stands for piezoelectric and P is the piezoelectric coupling coefficient, which is of the order of $\sim 5 \times 10^{-2}$ for many zinc blende semiconductors and about an order of magnitude higher for many wurtzite compound semiconductors. The piezoelectric scattering rate is several orders of magnitude smaller than that for deformation potential scattering. Thus, piezoelectric scattering is not that important in compound semiconductors, but it can be significant at very low temperatures.

5.7.5 Optical Phonon Scattering: Polar and Nonpolar

Scattering from dipole moments formed by the interaction of the ionic charges on atoms with optical-mode lattice vibrations is called the *polar optical-mode scattering process*. This scattering mechanism is the dominant process in semiconductors at high temperature or high electric field. Solving the Boltzmann Equation 5.111 is very difficult in this case since the relaxation time becomes a function of the perturbation strength in addition to the charge particle energy. The relaxation time and mobility are obtained for this scattering process using specific temperature conditions. Further details are discussed by Look. For example, the relaxation time and mobility for $0 \leq (T_{po}/T) \leq 5$, where T_{po} is the energy of the longitudinal optical phonon divided by Boltzmann constant, are given by

$$\tau_{po} = \frac{2\sqrt{2}\pi\hbar^2(e^{T_{po}/T} - 1)\chi(T_{po}/T)E^{1/2}}{e^2(m^*)^{1/2}(k_B T_{po})(\epsilon_\infty^{-1} - \epsilon^{-1})\epsilon_0} \tag{5.160}$$

and

$$\mu_{pz} = \frac{e\langle\tau_{po}\rangle}{m^*} = \frac{2\sqrt{2}\pi\hbar^2(e^{T_{po}/T} - 1)}{e(m^*)^{3/2}(k_B T_{po})^{1/2}(\epsilon_\infty^{-1} - \epsilon^{-1})\epsilon_0}, \tag{5.161}$$

where $\chi(T_{po}/T)$ is a slowly varying function of temperature and ϵ_∞ is the high frequency dielectric constant. The optical phonon modes produce fluctuations in the band gap similar to those produced by the acoustic phonon modes. Electrons are scattered by the deformation potential produced by the optical phonon modes. This type of scattering is called *nonpolar scattering*. The relaxation time for nonpolar scattering is given by

$$\tau_{npo} = \frac{2\sqrt{2}\pi\rho\hbar^{3\sim}\omega_0}{D_0(m^*)^{3/2}n_0[\sqrt{E + \hbar\omega_0} + \mathcal{H}(E - \hbar\omega_0)(n_0 + 1)n_0^{-1}\sqrt{E - \hbar\omega_0}]}, \tag{5.162}$$

where $\hbar\omega_0$ is the optical phonon energy, ρ is the density of the crystal, D_0 is the deformation potential (energy per unit strain), $\mathcal{H}$ is Heaviside step function, and $n_0 = 1/[\exp(\hbar\omega_0/k_B T) - l]$. The corresponding mobilities are

$$\mu_{npo} = \frac{2\sqrt{2\pi}\rho\hbar^{4\sim}\omega_0^2 e}{3D_0^2(m^*)^{5/2}(k_B T)^{3/2}} \quad \text{for } k_B T \gg \hbar\omega_0$$

$$\mu_{npo} = \frac{\pi\rho\hbar^4 e\sqrt{2\hbar\omega_0}}{D_0^2(m^*)^{5/2}n_0} \quad \text{for } k_B T \ll \hbar\omega_0. \tag{5.163}$$

It is very difficult to estimate D_0, but this scattering process is believed to be much weaker than other lattice scattering mechanisms, at least for electron scattering.

5.7.6 Scattering from Short-Range Potentials

Charge carriers can be scattered from a variety of potentials, which can be approximated as short-range potentials that have constant strength over a small volume and zero strength elsewhere. Brief discussions about several short-range potentials are presented in the following paragraphs.

5.7.6.1 Scattering from Dislocations Charge carriers (both electrons and holes) can be scattered from dislocations in semiconductors. Dislocations may be considered as a line charge, and scattering can be viewed as scattering from ionized impurity centers. On the other hand, dislocations create strain fields, which produce deformation potential-like scattering. The scattering from dislocations is significant for dislocation densities larger than 10^8 cm^{-2}, and the relaxation time can be estimated by assuming that the dislocation line is cylindrical with a radius R and length L. This leads to an electron scattering time as follows:

$$\tau_{\mathrm{dis}} = \frac{3}{8N_{\mathrm{d}}R\upsilon}, \tag{5.164}$$

where υ is the electron velocity and N_{d} is the dislocation density. The electron mobility is obtained directly from Equation 5.164 and is given by

$$\mu_{\mathrm{dis}} = \frac{e\langle\tau_{\mathrm{dis}}\rangle}{m^*} = \frac{3e}{8N_{\mathrm{d}}R}\frac{1}{\sqrt{3m^*k_{\mathrm{B}}T}}\cdot\frac{4\sqrt{2}}{3\sqrt{\pi}} \approx \frac{3e}{8N_{\mathrm{d}}R}\frac{1}{\sqrt{3m^*k_BT}}. \tag{5.165}$$

As shown in this equation, the mobility is proportional to $T^{-1/2}$. Dislocation scattering can be significant in materials in which dislocation densities are high, such as GaN thin films grown on sapphire.

5.7.6.2 Scattering from δ-Function and Alloy Potentials A typical example of a localized potential is the δ-function potential of the form

$$V = V_\delta E_\delta \delta(\boldsymbol{r} - \boldsymbol{r}_0), \tag{5.166}$$

where V_δ is a small volume and E_δ is the strength of the potential. An expression for the relaxation time for a density of scattering centers, N, is obtained as

$$\tau_\delta = \frac{\pi}{\sqrt{2}}\frac{\hbar^4}{NV_\delta^2E_\delta^2(m^*)^{3/2}E^{1/2}}. \tag{5.167}$$

The corresponding mobility is derived (Anselm and Askerov) and given by

$$\mu_\delta = \frac{2\sqrt{2\pi}}{3}\frac{e\hbar^4}{NV_\delta^2E_\delta^2(m^*)^{5/2}(k_{\mathrm{B}}T)^{1/2}}. \tag{5.168}$$

The alloy scattering time can be obtained (Mott and Jones) by setting $NV_\delta^2E_\delta^2$ in Equation 5.167 to

$$NV_\delta^2E_\delta^2 = V_{\mathrm{c}}x(1-x)E_{\mathrm{AB}}^2 \tag{5.169}$$

and assuming that the alloy is composed of two binary compounds, A and B. The quantity E_{AB} is the difference between the band gaps of A and B, x is the fraction of compound A, and V_{c} is the volume of the unit cell. The mobility, thus, can be obtained as

$$\mu_{\mathrm{al}} = \frac{2\sqrt{2\pi}}{3}\frac{e\hbar^4}{V_{\mathrm{c}}E_{\mathrm{AB}}^2x(1-x)(m^*)^{5/2}(k_{\mathrm{B}}T)^{1/2}}. \tag{5.170}$$

Notice that E_{AB} defines the difference in the scattering potential between compounds A and B.

5.7.7 Scattering from Dipoles

When acceptor and donor atoms in semiconductors are close together, they may scatter electrons as a dipole instead of as individual monopoles. The scattering relation time for the unscreened case was first derived by Dimitrov and is given as

$$\tau_{\text{dipole}} = \frac{2\sqrt{2}\pi 3\hbar^2 \epsilon_0^2 \epsilon^2 E^{1/2}}{(m^*)^{1/2} e^2 N q_{\text{d}}^2}, \tag{5.171}$$

where q_{d} is the dipole moment given by $e^* l$ (where l is the distance between the charges and e^* is the ionic charge) and N is the density of the dipoles. The mobility for the nondegenerate electrons is derived straightforward from Equation 5.171 and is given by

$$\mu_{\text{dipole}} = \frac{2^{9/2}\sqrt{\pi}\hbar^2 \epsilon_0^2 \epsilon^2 (k_{\text{B}} T)^{1/2}}{(m^*)^{3/2} e N q_{\text{d}}^2}. \tag{5.172}$$

If the charge of the ion is assumed to be e and the distance between the two ions is averaged to l, then the mobility can be written as

$$\mu_{\text{dipole}} = 3.57 \times 10^7 \frac{T^{1/2}}{\{l(\text{cm})\}^2 N(\text{cm}^{-3})} \ \text{cm}^2/\text{V/s}. \tag{5.173}$$

If l is 10 Å and the dipole density is 10^{17} cm^{-3}, the electron mobility due to dipole scattering is 6.18×10^5 cm^2/V/s at 300 K. This high value for mobility indicates that dipole scattering is insignificant. However, if l is 100 Å, then the electron mobility drops to 6.18×10^3 cm^2/V/s, which means that dipoles contribute strong scattering. In reality, dipole scattering does not play a major role in semiconductors.

5.8 SCATTERING IN A TWO-DIMENSIONAL ELECTRON GAS

Electrons generated at the interfaces of heterojunctions, such as GaAs/AlGaAs structures, are confined along the growth directions and are free to move in the perpendicular plane. A typical example of this type of heterostructure is shown in Fig. 5.22, where the barrier (AlGaAs) is doped instead of the GaAs layer. The portion of the AlGaAs layer near the interface is not doped and is called the *spacer*. A typical doping level in the AlGaAs layer is of the order of 10^{18} cm^{-3}. Experimentally, it is found that the mobility increases as the spacer thickness increases, while the two-dimensional electron gas density is decreased as the spacer thickness increases. Thus, there is a trade-off between the electron concentration, mobility, and the spacer thickness. The electron mobility for GaAs structures can reach values higher than 1×10^6 cm^2/V/s. The reason for the high mobility is that electron scattering from ionized impurities is negligible since scattering from ionized impurities decreases as the spacer thickness increases.

Doping in the barrier is called *modulation doping*, and the structure shown in Fig. 5.22 is the basis for the modulation-doped field-effect transistor (MODFET). This is also known as a high electron mobility transistor (HEMT). A comparison between the mobility for bulk GaAs and a GaAs MODFET structure is shown in Fig. 5.23 as a function of temperature. The ionized impurity concentration in bulk GaAs material is typically of the order of $10^{17}-10^{18}/\text{cm}^3$. Scattering from ionized impurities dominates the mobility behavior at low temperatures for bulk material. Essentially, impurity scattering is almost eliminated in the MODFET structure, as shown in Fig. 5.23, in which the mobility continues to increase as the temperature is reduced and reaches a plateau at temperatures less than 10 K. This figure shows the improvement of mobility over the years, which accompanied the advancement of epitaxial growth techniques, in particular, molecular beam epitaxy.

The theory of electron scattering in low dimensional systems such as quantum wells, wires, and dots can be extensive and lengthy. In this section, we present the most important scattering mechanisms in quantum structures without going through the theoretical derivation, bearing in mind that the theory of charge carrier scattering in semiconductors and their heterojunctions contains many approximations and assumptions. As mentioned previously, the scattering mechanisms are described in terms of the transition probability, $W_{k,k'}$, when the charge carrier is scattering from the initial state, k, to the final state, k'. This transition probability is derived using Fermi's golden rule. Any scattering mechanism can be described by the relaxation time, which is defined as the average time between scattering events. The inverse of the average relaxation time is called the *scattering rate*. Moreover, the relaxation time can further be classified as the population, momentum, or energy relaxation time. For example, a system at equilibrium can be described by a physical parameter, such as population, momentum, or energy. When a scattering event occurs, the system is no longer at equilibrium and its

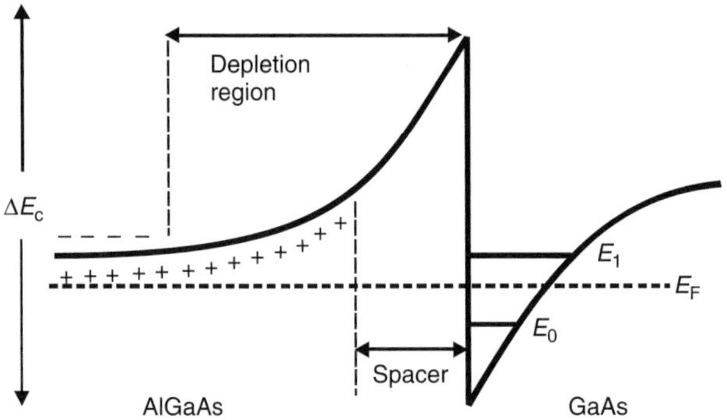

FIGURE 5.22 A sketch of the conduction band structure of a GaAs/AlGaAs heterojunction showing the confined states at the interfaces, the Fermi energy, the spacer, the depletion region, and the conduction band offset.

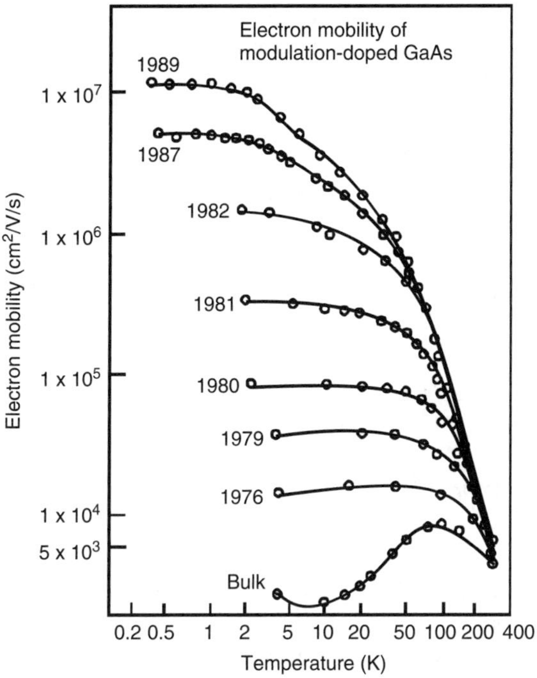

FIGURE 5.23 The electron mobility of n-type doped bulk GaAs and modulation-doped GaAs/AlGaAs heterostructure plotted as a function of temperature. The results show how the electron mobility in modulation-doped GaAs/AlGaAs heterojunction has risen over the years (after Pfeiffer *et al.*).

physical parameters have different values than at equilibrium. The time required for the physical parameters of the system to relax back to the equilibrium state is called the *relaxation time* of that particular physical parameter.

Electrons in low dimensional quantum systems scatter from imperfections, interface roughness, alloys, surface charges, dislocations, phonons, and from other charge particles. The following subsections briefly describe the most important scattering processes in low dimensional systems.

5.8.1 Scattering by Remote Ionized Impurities

The density of background impurities in the conduction channels of quantum wells and wires is usually small. For modulation-doped structures such as the structure shown in Fig. 5.22, scattering from remote ionized impurities can be significant, especially if the spacer thickness is too small. The perturbation part of the Hamiltonian of the electrostatic potential of the remote ionized impurities is mainly due to the fluctuation of the potential that breaks the translational symmetry parallel to the heterojunctions of the quantum well (Mitin *et al.*) This scattering, which causes a change of the momentum and average relaxation time, is called the *average momentum relaxation time*, $\tau_{\mathrm{p,i}}$, where subscripts p and i stand for momentum and ionized impurities. This relaxation time is derived (Mitin *et al.*) as follows:

$$\begin{aligned}
\langle \tau_{\mathrm{p,i}} \rangle^{-1} &= \frac{\pi m^* N_D^{2D}}{\hbar^3}\left(\frac{2e^2}{\epsilon_0\epsilon}\right)^2 \int_0^{\pi} \frac{(1-\cos\theta)}{(k_{\mathrm{sc}}+2k_{\mathrm{F}}\sin\frac{\theta}{2})} exp[-4k_{\mathrm{F}}|Z_0|\sin\tfrac{\theta}{2}]\mathrm{d}\theta \\
&\approx \frac{\pi m^* N_D^{2D}}{\hbar^3}\left(\frac{2e^2}{\epsilon_0\epsilon}\right)^2 \int_0^{\pi} \frac{\theta^2/2}{k_{\mathrm{F}}^2(1+\theta)} exp[-2k_{\mathrm{F}}|Z_0|\theta]\mathrm{d}\theta \\
&\approx \frac{\pi m^* N_D^{2D}}{2k_{\mathrm{F}}^2\hbar^3(2k_{\mathrm{F}}|Z_0|)^3}\left(\frac{2e^2}{\epsilon_0\epsilon}\right)^2 \int_0^{\infty} x^2 exp[-x]\mathrm{d}x \\
&\approx \frac{2\pi m^* N_D^{2D}}{2k_{\mathrm{F}}^2\hbar^3(2k_{\mathrm{F}}|Z_0|)^3}\left(\frac{2e^2}{\epsilon_0\epsilon}\right)^2, \qquad (5.174)
\end{aligned}$$

where the integration is made over the impurity coordinate, N_D^{2D} is the sheet concentration of the ionized impurity and is taken to be equal to the electron concentration in the well (n_{s}), $\boldsymbol{k}_{\mathrm{F}}$ is the Fermi wave vector and is given as $\boldsymbol{k}_{\mathrm{F}} = \sqrt{2\pi n_{\mathrm{s}}}$ (where n_{s} is the density of the two-dimensional electron gas), the angle θ is defined as $\boldsymbol{k} = 2\boldsymbol{k}_{\mathrm{F}}\sin\frac{\theta}{2}$ (where $\boldsymbol{k}$ is the electron wave vector before scattering), $\boldsymbol{k}_{\mathrm{sc}}$ is a characteristic wave vector that determines the range of wave vectors in which the electrons effectively screen the electric field, and Z_0 is defined as the distance at which a thin layer of impurities is present. The doping is assumed to be δ-doping. The integral in Equation 5.174 is evaluated after assuming that $\theta < 1$, which leads to $\sin(\theta/2) \sim \theta/2$ and $(1-\cos\theta) \sim \theta^2/2$, $\boldsymbol{k}_{\mathrm{F}}|Z_0| > 1$, and $\boldsymbol{k}_{\mathrm{sc}} \sim \boldsymbol{k}_{\mathrm{F}}$. Moreover, we assume that $x = 2\boldsymbol{k}_{\mathrm{F}}|Z_0|\theta$ and the upper limit of the integral is approximated to ∞, which leads to a value of 2 for the integral. Numerical calculations of the integral may lead to values other than 2. By putting all these assumptions together, the electron mobility associated with scattering by remote ionized impurities takes the following form:

$$\begin{aligned}
\mu_{\mathrm{p,i}} = \frac{e\langle \tau_{\mathrm{p,i}} \rangle}{m^*} &= 16\frac{(\epsilon_0\epsilon)^2\boldsymbol{k}_{\mathrm{F}}^5\hbar^3|Z_0|^3}{4e^3 2\pi n_{\mathrm{s}}(m^*)^2} \\
&= 16\frac{(\epsilon_0\epsilon)^2(2\pi n_{\mathrm{s}})^{5/2}\hbar^3|Z_0|^3}{4e^3 2\pi n_{\mathrm{s}}(m^*)^2} \\
&= 16\frac{(\epsilon_0\epsilon)^2(\pi n_{\mathrm{s}})^{3/2}\hbar^3|Z_0|^3}{\sqrt{2}e^3(m^*)^2}. \qquad (5.175)
\end{aligned}$$

Notice that the mobility increases as the third power of the distance between the 2DEG in the well and the doped layer. For a GaAs/AlGaAs quantum well, in which $Z_0 = 150$ Å, $n_{\mathrm{s}} = 10^{12}/\mathrm{cm}^2$, and a dielectric constant of 12.91, the mobility is estimated to be $\mu_{p,i} \approx 2.13 \times 10^5$ cm²/V/s. It is obvious that electron scattering by remote ionized impurities is negligible when the spacer in Fig. 5.22 is made thicker than 150 Å. Notice that the electron mobility described in Equation 5.175 does not depend explicitly on the temperature.

5.8.2 Scattering by Interface Roughness

Since semiconductor heterojunctions and quantum wells are composed of materials of different band gaps, interfaces between these materials possess various degree of roughness. Ideally, the interfaces should be abrupt. In reality, there is a fluctuation in the material thicknesses at the interfaces, which causes charge carriers to scatter. In Fig. 5.24, the interface roughness between GaAs and AlGaAs is presented as a variation of the well thickness. The ideal interfaces are shown as dotted lines in the figure. If the average fluctuation of the GaAs well is taken as D and the spatial correlation of the roughness is described by a correlation length, λ, the electron momentum relaxation time for an infinite quantum well is derived (Mitin *et al.*) as

$$(\tau_{p,n})^{-1} = \frac{4\pi m^* D^2 \lambda^2 E_n^2}{\hbar^3 L^2} \cdot \int_0^{2\pi} \frac{1}{2\pi} \frac{(1-\cos\theta)}{\left[1 + \frac{k_{sc}}{2k_F}\sin\frac{\theta}{2}\right]^2} \exp\left\{-\lambda^2 k_F^2 \sin^2\frac{\theta}{2}\right\} d\theta, \tag{5.176}$$

where L is the width of the quantum well, $\boldsymbol{k}_F$ is the Fermi wave vector, and $\boldsymbol{k}_{sc}$ has been defined previously as the screening length given as $\boldsymbol{k}_{sc} = 2/a^*$, (where

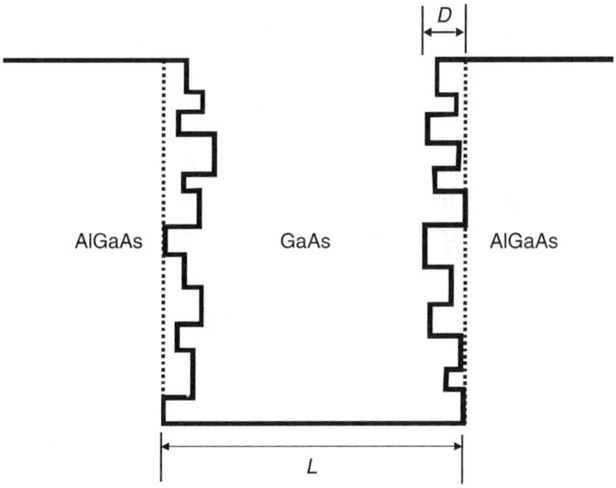

FIGURE 5.24 A sketch of a GaAs/AlGaAs quantum well showing the variation of the well thickness, which is the source of the interface roughness. The dotted lines represent prompt interfaces.

a^* is the bulk effective Bohr radius, which is ~100 Å for GaAs), and E_n is the quantized energy levels, which are given as $E_n = \hbar^2\pi^2 n^2/(2m^* L^2)$ for $n = 1,2,3,\ldots$ The relaxation time for the first subband ($n = 1$) can be written as

$$(\tau_{\mathrm{p},1})^{-1} = \frac{2\pi^3 D^2 \lambda^2 E_1}{\hbar L^4} \cdot \int_0^{2\pi} \frac{1}{2\pi} \frac{(1-\cos\theta)}{\left[1 + \frac{k_{\mathrm{sc}}}{2k_{\mathrm{F}}} \sin\frac{\theta}{2}\right]^2} \exp\{-\lambda^2 k_{\mathrm{F}}^2 \sin^2\frac{\theta}{2}\} d\theta. \quad (5.177)$$

For a GaAs/AlGaAs quantum well numerical calculation, let us set $L = 125$ Å, $m* = 0.067m_0$, $D/L = 0.03$, and $\lambda/L = 0.2$. For a carrier concentration of $10^{12}/\mathrm{cm}^2$, the Fermi wave vector is $k_{\mathrm{F}} = \sqrt{2\pi n_s} = 2.5 \times 10^{6}\,\mathrm{cm}^{-1}$ and $\lambda k_{\mathrm{F}} = 0.62$. For a screening length of $k_{\mathrm{sc}} = 2/a* = 2/100\mathrm{Å}$, we obtain $k_{\mathrm{sc}}/k_{\mathrm{F}} = 0.798$. Assuming that $\theta < 1$, we obtain for the integral an absolute value of 0.205. E_1 is obtained for the first subband as 35.99 meV. Substituting these values in Equation 5.177, we obtain $\tau_{p,1} = 3.97 \times 10^{-11}$s. The mobility associated with this scattering time is seen to be $\mu_{p,1} \approx 1.04 \times 10^6$ cm^2/V/s. If the 2π in the integral is dropped, the mobility reduces to $\mu_{\mathrm{p},1} \approx 1.7 \times 10^5/\mathrm{cm}^2$/V/s. Scattering from the interface roughness in quantum structures can dominate many scattering mechanisms at low temperatures.

5.8.3 Electron–Electron Scattering

Electron–electron interaction is usually considered as a part of the many-body effects. Electron–electron interaction is classified into two: long-range and short-range interactions. The long-range nature of the Coulomb interaction due to the collective response of the electrons comes from the collective oscillation of the electron gas. The resulting oscillations are called *plasma oscillations*, which have a range greater than the characteristic screening length of the system. Plasmon is the quanta of the plasma oscillations. On short-range scales, the electron gas behaves more as a collection of individual charged particles. The short-range electron–electron interaction is considered to be elastic scattering, in which the momentum and energy of the entire system are conserved. For short-range electron–electron scattering, the relation time is given (Brennan) by

$$(\tau_{\text{e-e}})^{-1} = \frac{m^* e^4 k_{12}}{4\pi\hbar^3\epsilon^2 k_{\mathrm{sc}}^{-2}(k_{\mathrm{sc}}^{-2} + k_{12}^2)}, \quad (5.178)$$

where electron spin is neglected, k_{sc} is the screening length, and k_{12} is the magnitude of the difference between wave vectors of two electrons in their initial states. For an electron gas in a GaAs quantum well, $m* = 0.067m_0$, $k_{\mathrm{sc}} = 2/a^* = 2 \times 10^{-8}$ m (where a^* is the effective Bohr radius taken as 10^{-8} m), $\epsilon = 12.91$, and it is assumed that $k_{12} = 10^{-2}$ m. The assumed value for k_{12} is quite reasonable knowing that the Fermi wave vector is 2.5×10^8 m for a carrier concentration of $10^{12}/\text{cm}^2$. This means that the two electrons have almost identical wave vectors in their initial states. These values give a relation time of $\tau_{e-e} \approx 1.54 \times 10^{-11}$s. The electron mobility associated with the relaxation time is obtained as $\mu_{e-e} \approx 4.03 \times 10^5$ cm²/V/s. This value is the same order of magnitude as obtained for remote ionized impurity and interface roughness mobilities. Electron–electron scattering is usually negligible for carrier concentration of the order of 10^{17} cm^{-3}. For detailed discussions regarding the relaxation time due to long-range electron–electron interaction, see Brennan and references therein.

5.9 COHERENCE AND MESOSCOPIC SYSTEMS

Electron or hole transport in commercially available electronic devices is governed by various scattering mechanisms. On the other hand, devices based on coherent transport (transport without scattering) are still at the developmental stage. To understand the coherent length or dephasing length, let us first consider an electron that undergoes an elastic collision, where the initial, $\psi_i(\boldsymbol{r}, t)$, and final, $\psi_f(\boldsymbol{r}, t)$, wave functions are (Mitin *et al.*)

$$\psi_i(\boldsymbol{r}, t) = e^{-i\omega t} e^{i\boldsymbol{k}.\boldsymbol{r}} \text{and} \psi_f(\boldsymbol{r}, t) = e^{-i\omega t} \sum_{\boldsymbol{k}', \boldsymbol{k}'=\boldsymbol{k}} A_{\boldsymbol{k}'} e^{i\boldsymbol{k}'.\boldsymbol{r}} = e^{-i\omega t} \psi(\boldsymbol{r}), \quad (5.179)$$

where $\boldsymbol{k}$ and $\boldsymbol{k}'$ are the wave vectors before and after the scattering event. For elastic scattering, we have $\boldsymbol{k} = \boldsymbol{k}'$, which means that the momentum is conserved, and $|A_{\boldsymbol{k}'}|^2$ is the probability of finding the electron with a wave vector $\boldsymbol{k}'$ after scattering. From Equation 5.179, one can obtain $|\psi_f(\boldsymbol{r}, t)|^2 = |\psi(\boldsymbol{r})|^2$, which means that the spatial distribution remains independent of time after scattering. The incident and scattered wave functions can produce complex wave patterns, but one of the most important properties of elastic scattering is that the phase of the electron is not destroyed, or elastic scattering does not destroy the coherence of the electron motion even for a distance larger than the elastic mean free path.

For inelastic scattering, the electron wave function after scattering has different energies and time dependencies according to the following:

$$\psi_i(\boldsymbol{r}, t) = e^{-i\omega(\boldsymbol{k})t} e^{i\boldsymbol{k}.\boldsymbol{r}} \quad \text{and} \psi_f(\boldsymbol{r}, t) = \sum_{\boldsymbol{k}', \boldsymbol{k}' \neq \boldsymbol{k}} A_{\boldsymbol{k}'} e^{-i\omega(\boldsymbol{k}')t} e^{i\boldsymbol{k}'.\boldsymbol{r}}. \quad (5.180)$$

The time-dependent component of the wave function of the scattered electron cannot be factored out of the sum, and $|\psi_f(\boldsymbol{r}, t)|^2$ is now a function of time. For inelastic scattering, the electron preserves its quantum coherence for a distance equal to or less than the inelastic scattering length, l_i. In general, l_i is larger than the de Broglie wavelength. A comparison between various characteristic lengths as compared to the de Broglie wavelength is shown in Fig. 5.25.

The electrons' dephasing length or coherent length, l_φ, is the distance that the electrons travel before losing their quantum mechanical coherence, which is a

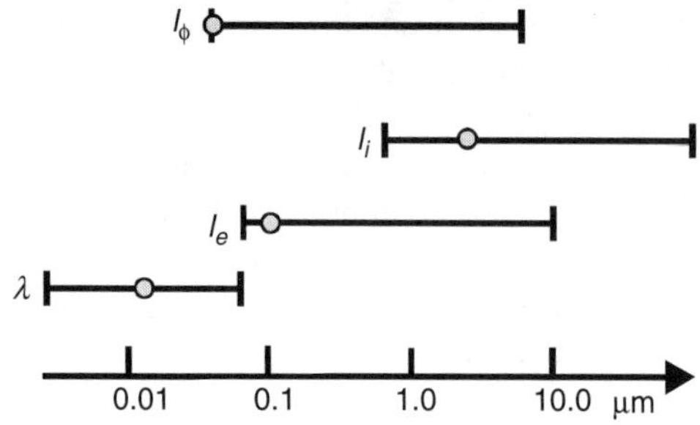

FIGURE 5.25 Intervals for the characteristic lengths:λ, de Broglie wavelength; l_e, mean free path; l_i, inelastic scattering length; and l_φ, coherence length in semiconductor materials. As an example, the lengths are marked by circles for Si at $T = 77$ K, assuming that the electron mobility equals 10^4 cm^2/V/s(after Mitinet *et al.*).

result of the large spreading of the wave function phases. The dephasing effect is caused by inelastic collisions, temperature spreading of phases, or both, which leads to the assumption that the dephasing length is determined by the smaller value of the inelastic length or the thermal diffusion length. The dephasing length, l_φ, is thus the distance that the electron transport has quantum characteristics. Systems in which electrons maintain coherence and remain in phase during transport are called *mesoscopic systems*, which have properties strongly depending on the geometry of the sample, contacts, and quantum structures.

The theoretical analysis of various quantum transport regimes is too complex and outside the scope of this book. One of the simplest examples of quantum transport is electron transport in the absence of any scattering. A system with no scattering is a perfect crystalline solid in which the equation of motion of the electron is

$$\frac{d\boldsymbol{p}}{dt} = \hbar\frac{d\boldsymbol{k}}{dt} = e\boldsymbol{\mathcal{E}}. \tag{5.181}$$

The electron starts at the bottom of the energy band and moves along the E versus k curve until it reaches the Brillouin zone edge. Since we have a perfect crystal, the energy bands are periodic in the k-space. Thus, when the electron reaches the zone edge, it is reflected and starts to lose its energy and then continues the cycle under the influence of the electric field. The momentum of the electron changes direction as the electron passes through the zone edges, leading to oscillations in k-space (and consequently in the real space). These oscillations are called *Zener–Bolch oscillations*, and their frequency is given by

$$\omega = \frac{ae\boldsymbol{\mathcal{E}}}{\hbar} = \frac{ae\boldsymbol{\mathcal{E}}}{\hbar}, \tag{5.182}$$

where a is the lattice constant. For an electric field of the order of 10^7 V/m, the frequency of the Zener–Bloch oscillations is $\sim 1.21 \times 10^{12}$ Hz. This frequency range is very important for high speed devices. Experimentally, it is difficult to observe these oscillations because of various scattering mechanisms that prevent the coherent transport of electrons.

An expression for the conductance of a system in which the phase coherence is maintained can be derived for different contacts and sample geometries. Consider a one-dimensional simple barrier under bias voltage, as shown in Fig. 5.26. Under a small bias voltage, as shown in Fig. 5.26a, the electrons tunnel from both left to right and right to left. Under a bias voltage, each side of the barrier has its own Fermi energy level with a difference of $E_{\mathrm{F}}^{\mathrm{L}} - E_{\mathrm{F}}^{\mathrm{R}} = eV$. As the bias voltage is increased, as shown in Fig. 5.26b, the electron tunneling from right to left becomes negligible.

The electric current depends on the tunneling transmission coefficient and is given by

$$I_{\mathrm{L}} = \frac{2e}{2\pi} \int_{E_{\mathrm{L}}}^{\infty} f(E, E_{\mathrm{F}}^{\mathrm{L}}) \upsilon(k) T(k) \mathrm{d}k = \frac{2e}{h} \int_{E_L}^{\infty} f(E, E_{\mathrm{F}}^{\mathrm{L}}) T(E) \mathrm{d}E, \qquad (5.183)$$

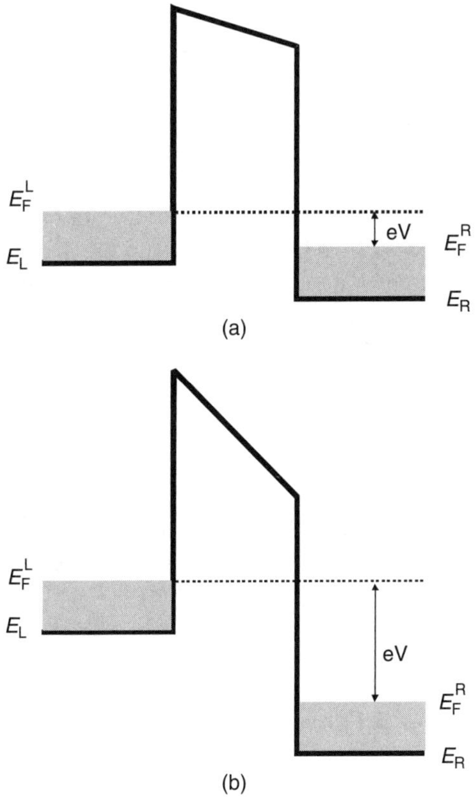

FIGURE 5.26 A single barrier with Fermi electron sea on both sides is shown for (a) small bias and (b) large bias.

where the factor 2 is for spin degeneracy, $\upsilon(k)$ is the electron group velocity, $f(E, E_F^L)$ is the Fermi–Dirac distribution function, and $dk/2\pi$ is introduced to account for the k-states. In this equation, we used $dE = (\hbar^2 k/m)dk = \hbar\upsilon dk$ so that the velocity is cancelled in the current expression. Similarly, the current from right to left under a small bias voltage can be approximated as

$$I_R = -\frac{2e}{2\pi}\int_{E_R}^{\infty} f(E, E_F^R)\upsilon(k)T(k)dk = -\frac{2e}{h}\int_{E_R}^{\infty} f(E, E_F^R)T(E)dE \qquad (5.184)$$

assuming that the transmission coefficient is the same from both sides of the barrier. The total current is the sum of I_L and I_R as follows:

$$I = I_L + L_R = \frac{2e}{h}\int_{E_L}^{\infty} \{f(E, E_F^L) - f(E, E_F^R)\}T(E)dE, \qquad (5.185)$$

where the lower limit of Equation 5.184 is changed to E_L since the electrons in the range E_R to E_L do not contribute to the current. For a small bias voltage, the Fermi–Dirac distribution functions can be expanded to first order in a Taylor expansion to give

$$f(E, E_F^L) - f(E, E_F^R) = -eV\frac{\partial f(E, E_F)}{\partial E}, \qquad (5.186)$$

where E_F is the distribution function at equilibrium. In this equation, the Fermi energy levels on both sides of the barriers were approximated as $E_F^L = E_F + \frac{1}{2}eV$ and $E_R^L = E_F - \frac{1}{2}eV$. Substitute Equation 5.186 into Equation 5.185 to obtain

$$I = \frac{2e^2V}{h}\int_{E_L}^{\infty}\left\{-\frac{\partial f(E, E_F)}{\partial E}\right\}T(E)dE. \qquad (5.187)$$

From Equation 5.188, one can obtain the conductance, G, as

$$G = \frac{I}{V} = \frac{2e^2}{h}\int_{E_L}^{\infty}\left\{-\frac{\partial f(E, E_F)}{\partial E}\right\}T(E)\,dE. \qquad (5.188)$$

At low temperatures, the Fermi–Dirac distribution function can be approximated as a step function and its derivative is simply a δ-function. Thus, $-\frac{\partial f(E,E_F)}{\partial E} \approx \delta(E - E_F)$, and Equation 5.188 becomes

$$G = \frac{2e^2}{h}\int_{E_L}^{\infty} \delta(E - E_F)T(E)dE$$

$$= \frac{2e^2}{h} T(E_F). \tag{5.189}$$

The factor e^2/h is known as the *quantum unit of conductance* and the corresponding resistance is $R = h/e^2 \approx 25.829$ kΩ. Equation 5.189 shows that the conductance is independent of the length of the sample, and depends solely on the transmission coefficient. For $T(E_F)1$, the conductance is $2e^2/h$, which is independent of the sample geometry. For higher temperatures, the above δ-function approximation is no longer valid and the integration of Equation 5.188 should be performed.

The conductance result expressed in Equation 5.189 is very simplistic since it is derived for only one mode or one path that the electron will take when traveling from one contact to another through the sample. In reality, one has to sum the electron contributions from all different paths the electron can take as it moves from one contact to another. For many different paths or propagating states, Equation 5.189 can be written as

$$G = \frac{2e^2}{h} \sum_{n,m} T(E_F, m, n) = 2G_0 \sum_{m,n} T(E_F, m, n), \tag{5.190}$$

where $G_0 = e^2/h$ and the sum is over all electron states, m and n, with energy $E < E_F$. Equation 5.190 is called the *Landauer formula*. Each channel or mode has two quantum numbers, m and n, where, for example, m represents the mode or state of the electron when leaving the left contact and n represents the mode or state of the electron when arriving at the right contact, as illustrated in Fig. 5.27.

Landauer formalism provides a means to understand the transport in terms of scattering process, as illustrated in Fig. 5.27, where, for mesoscopic structures, the electron waves can flow from one contact to maintain phase coherence. The phase coherence is maintained at low temperatures at which scattering processes due to phonons are suppressed. Thus, Landauer formalism is valid only at low temperatures and small bias voltages. An important property of phase coherence transport is the fluctuation observed in the conductivity (resistivity) as a function of magnetic field. In Equation 5.190, the sum is over the electron contribution

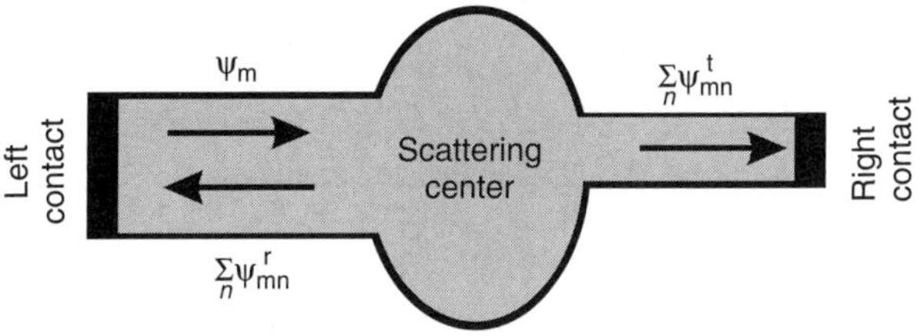

FIGURE 5.27 Illustration of coherent transport through a device with two leads. Each contact has many propagating states.

from all modes (channels) that the electron takes when traveling from the left contact to the right contact in Fig. 5.27. A fluctuation in the conductance is observed in a large number of experiments and systems, and it is found to be independent of the sample size. An example of the conductance fluctuation, or more precisely quantization, is shown in Fig. 5.28, obtained by van Wees *et al*., where the conductance was measured as a function of the gate voltage for a GaAs/AlGaAs high electron mobility transistor. In the van Wees *et al*. experiment, the HEMT or MODFET structure was fabricated with a pair of contacts to produce a short channel of a one-dimensional electron gas with high mobility, as shown in the inset of Fig. 5.28. The Fermi level and the electron wave functions are altered by controlling the gate voltage.

The conductance is discussed briefly for a mesoscopic system of two leads and one electron path and for a system with two contacts and many electron paths. For a mesoscopic system with four contacts or probes, as shown in Fig. 7.29, the conductance can be derived as (Singh 2003 and Davies 1998)

$$G_{4\text{-probe}} = \frac{2e^2}{h}\frac{T}{R} = \frac{2e^2}{h}\frac{T}{1-T}, \tag{5.191}$$

where T is the transmission coefficient and R is the reflection coefficient. Recall that $T+R = 1$. It appears that there is a difference in the conductance obtained from two probes and four probes. For a weak transmitting barrier, there is a small difference between the conductance obtained from two probes and that obtained from four probes. But when the barrier is transparent or the transmission coefficient is approaching unity, the conductance expressed in Equation 5.191 approaches infinity, while the conductance obtained for two probes and expressed in Equation 5.189 takes the value $2e^2/h$. This behavior may be explained as

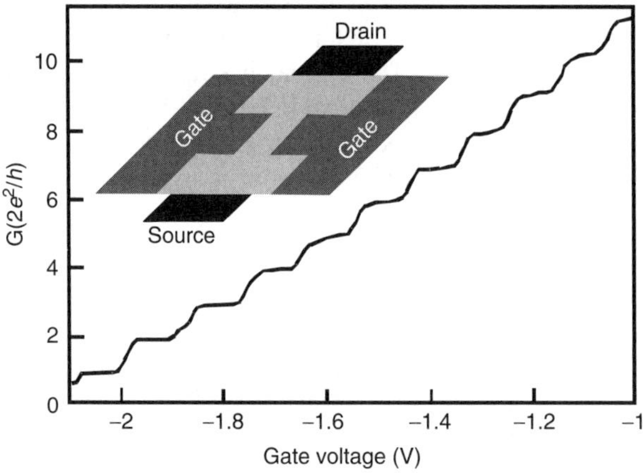

FIGURE 5.28 Conductance as a function of the gate voltage is plotted for GaAs/AlGaAs high electron mobility transistor (after van Wees *et al*.). The inset is a sketch of the MODFET showing the slit gate, drain, and source.

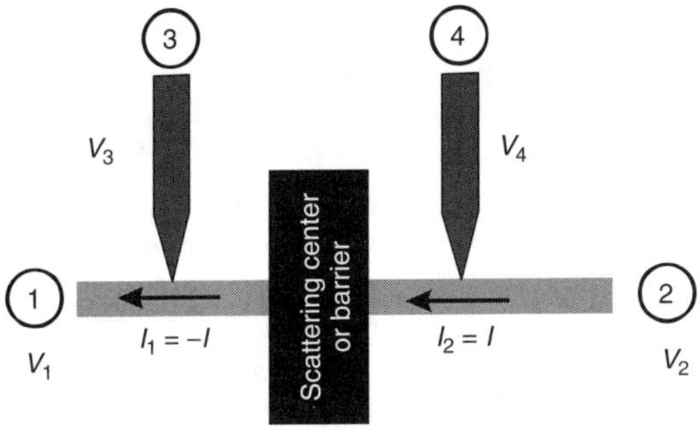

FIGURE 5.29 Four-probe measurements of the conductance of a scattering center (tunneling barrier) showing the four terminals at which the current and voltages can be measured.

follows: for a system in which the scattering center or the barrier is absent, the distribution of the electrons should be the same everywhere within the channel such that the voltage probe, in the case of a four-probe experiment, should read the same value at any point. Thus, the voltage difference between any two points is zero, giving rise to an infinite value for the conductance. When a bias voltage is applied to the two-probe configuration, an extra voltage appears because of an extra contact resistance of $h/(2e^2)$ in series with the sample. The extra resistance exists, even though the electrons are transmitted without any scattering.

5.10 SUMMARY

The general formalisms of the charge carrier transport properties in semiconductors are presented with emphasis on the basic concepts that allow one to investigate nanoscale materials and devices. The aim of the formalisms presented here is to show the reader in a broader sense how the bulk materials are treated and to present the limitations of classical treatment of the subject. Various characteristic lengths, including de Broglie wavelength, and time scales of several physical processes are presented. These characteristic lengths show the transport regimes in bulk, mesoscopic, and low dimensional semiconductor systems.

Hall effect, QHE, and SdH measurements are widely used to investigate the transport properties in bulk and heterojunction semiconductors. These experimental techniques were discussed, and the basic formalisms to interpret the data are presented.

The discussion in this chapter started with the charge carrier transport in bulk materials, in which the drift and diffusion current densities are dominant. The phenomenon of hot electron is briefly discussed, and the Gunn diode is presented as an example of the hot electron transport in bulk semiconductors. Gunn diode operates in the gigahertz region where a negative differential resistance is observed at high electric fields in compound semiconductors, such as GaAs and InP. The negative differential resistance is the primary cause of the instability in the device, which leads Gunn oscillations with frequencies in the microwave region. In addition to drift and diffusion current densities, the generation and recombination processes in semiconductor materials were discussed. The continuity equation, which combines the drift, diffusion, recombination, and generation is derived for both n-type and p-type semiconductors. From the continuity equation, it is realized that the transport properties of devices that are based on carrier injections, such as bipolar transistors and p–n junction diodes, are governed by the minority carrier transport.

When a bias voltage is applied to semiconductor materials and devices, the carriers are no longer in equilibrium. To derive the transport parameters at the nonequilibrium case, the Boltzmann transport equation is introduced. This equation is very complicated to solve without extensive approximations. One of these approximations is called *relaxation time approximation*, in which the integral part of the Boltzmann transport equation is replaced by the difference between the nonequilibrium and equilibrium distribution functions divided by the relaxation time required for the system to relax from the nonequilibrium case to the equilibrium case. Examples of how to derive the transport coefficients using the Boltzmann transport equation are presented.

Charge carriers suffer many scatterings during transport in semiconductors. Expressions for the electron mobility and relaxation time were presented for various scattering mechanisms ranging from defect scattering to lattice scattering. Scattering mechanisms in two-dimensional systems were discussed. In the absence of scattering, the electron transport is called *ballistic transport*, and the electrons maintain their phase during transport. Coherent transport in the mesoscopic system is briefly discussed, and the conductance in two and four terminal-mesoscopic structures with many channels is derived using the Landauer formalism.

PROBLEMS

5.1. Calculate the coherence length, l_{phi}, for an electron traveling with a velocity of 3×10^4 m/s in a bulk GaAs semiconductor material. Assume that the lifetime (time between two successive inelastic collisions) is 1.0 ps. What would be the coherence length in GaAs quantum wells and quantum wires?

5.2. A Hall effect device is fabricated from GaAs bulk material with $d =$ 0.5 mm, $W = 3$ mm, and $L = 10$ mm. The electrical current is $I_x = 5$ mA, the bias voltage is $V_x = 2.5$ V, and the magnetic field is $B_z = 0.1$ T. The Hall voltage was measured to be $V_{\mathrm{H}} = -3.0$ mV. Calculate the majority carriers, the mobility, and the resistivity. What is the conductivity type?

5.3. The filling factor, ν, is defined as the number of Landau levels lying below the Fermi energy level. Show that this factor is given by $\nu = n_s h/(eB)$. Assume that Landau levels are spin and valley degenerate.

5.4. Derive Equation 5.34, and show that for the steady-state case, the magnetoconductivity tensor is give by Equation 5.35.

5.5. Show that the magnetoresistivity tensor is given by Equation 5.38.

5.6. The Shubnikov-de Haas experiment was performed on GaAs/AlGaAs single quantum well. Two consecutive minima in ρ_{xx} were observed at $B_z =$ 1.56 and 2.34 T. Calculate the density of the 2DEG in the GaAs well in unit of cm-^{2}.

5.7. An n-type GaAs sample was subject to light illumination with photon energy larger than the band gap energy. Assume that the light illumination was turned off at time t_0 and the minority carriers and the excess minority carriers are much smaller than the majority carriers. Derive an expression for the minority carrier concentration as a function of time.

5.8. Derive an expression for the indirect net recombination rate for an indirect band gap semiconductor. Use Fig. P5.8, where the Fermi energy (E_F) is pinned at the recombination center, $\mathcal{E}_\mathrm{e}$ and $\mathcal{E}_\mathrm{h}$ are the emission rates of electrons and holes, and C_e and C_h are the capture rates of electrons and holes. The recombination center has electron and hole cross sections of σ_n and σ_p, respectively.

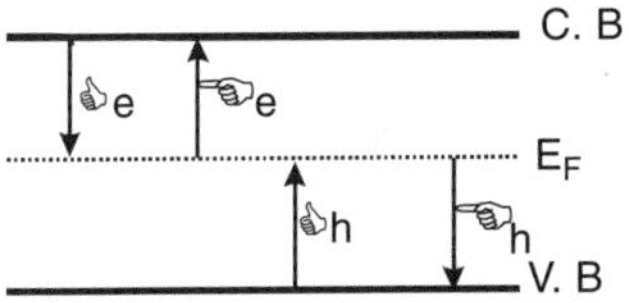

Fig. P5.8

5.9. Derive Equation 5.95, and then show that $D = \frac{D_\mathrm{p} D_\mathrm{n}(n+p)}{D_\mathrm{n} n + D_\mathrm{p} p}$.

5.10. Calculate the minority excess carrier concentration as a function of time (t) for an n-type semiconductor sample that is subject to generation rate of $t \geq 0$. Assume that the sample is in equilibrium with zero applied electric field at $t > 0$.

5.11. A p-type semiconductor sample is subject to a process in which the excess minority carriers are generated only at $x = 0$ and then begin to diffuse in both $-x$ and $+x$ directions. Derive an expression for the steady state excess minority carrier concentration as a function of x. Plot the concentration as a function of x.

5.12. The maximum amplitude of the minority carrier pulse measured at $t_1 = 50 \mu$s in Haynes–Shockley experiment for an n-type semiconductor sample is 6 times larger that amplitude of the pulse measured at $t_2 = 150 \mu$s. Calculate the minority carrier lifetime.

5.13. Derive an expression for the current in an n-type GaAs sample of length 1 mm and a cross section of 10^{-6} mm^2. Assume that the electron -concentration is 5×10^{16}/cm^3, the hole concentration is negligible, a bias voltage of 5 V is applied across the sample, the sample was illuminated uniformly with light at $t<0$ where the generation rate Γ is $5 \times 10^{21}/\text{cm}^3/\text{s}$, and the minority lifetime is 0.3 μs. The light was turned off at $t = 0$. The electron and hole mobilities are given as 1350 and 480 $\text{cm}^2/\text{V/s}$, respectively.

5.14. Excess carriers are generated at $x = 0$ in an n-type GaAs sample, which was subject to a constant electric field, $\mathcal{E}$. Derive an expression for the carrier concentration in the steady-state case.

5.15. Show that the average kinetic energy of electrons in a bulk semiconductor sample is given by $\langle E \rangle = \frac{3}{2} k_B T$. Assume that the electrons obey the Maxwell–Boltzmann distribution function.

5.16. Derive the electron Hall coefficient expression shown in Equation 5.143. Notice that this expression is derived for small magnetic field.

5.17. Assume that $\tau(E) = \tau_0 E^{\beta}$, where β is a positive real number. Show that the electron Hall coefficient is given by $R_H = -\frac{1}{ne} \frac{\Gamma(\frac{5}{2} + 2\beta)\Gamma(\frac{5}{2})}{[\Gamma(\frac{5}{2} + \beta)]^2}$, where $\Gamma(n) = \int_0^{\infty} x^{n-1} e^{-x} dx$ and the distribution function is assumed to follow the Maxwell–Boltzmann statistics.

5.18. Use the relaxation time expression shown in Equation 5.152, which is derived for the scatting of electrons from ionized impurities. Derive the mobility as shown in Equation 5.153. Calculate the electron mobility in GaAs due to the ionized impurity scattering at 300 K and 77 K. Assume that the electron concentration is $10^{16}/\text{cm}^3$, the impurity concentration is $10^{17}/\text{cm}^3$, dielectric constant is 12.91, and the effective mass is $0.067m_0$.

5.19. Consider the modulation-doped GaAs/AlGaAs structure shown in Fig. 5.22. The spacer is chosen as 5 nm, and doped AlGaAs layer is replaced by δ doping located at the far edge of the spacer, that is, the δ-doped layer is located at 5 nm from the interface. Assume that the two-dimensional dopant density is $5 \times 10^{12}/\text{cm}^2$. Calculate the momentum relaxation time due to the electron scattering from remote ionized impurity. Obtain the mobility, and compare your results to the data reported in Fig. 5.23. What would be the relaxation time and mobility for a spacer thickness of 20 nm?

5.20. A GaAs sample has an electron mobility of 10^5 $\text{cm}^2/\text{V/s}$ and an effective mass of $0.067m_0$. Calculate the thermal diffusion length, elastic scattering length, inelastic scattering length, dephasing length, and de Broglie wavelength. Assume that the electron group velocity is 5×10^7 cm/s.

5.21. A GaN thin film grown on sapphire was found to contain considerable amount of dislocations, where the inelastic scattering dominates the electron transport. The room temperature electron mobility of this thin film was measured by the Hall effect and found to be 35 $\text{cm}^2/\text{V/s}$. Calculate the phase coherence length. Compare your result to the interatomic distance.

5.22. Show that the nonequilibrium distribution, when an electric field is applied along the x-direction, can be written as follows: $f_k(k_x, k_y, k_z) = f_k^0((k_x - e\mathcal{E}_x\tau/\hbar), k_y, k_z)$, where f_k^0 is the equilibrium distribution function.

5.23. Assume that only a temperature gradient is applied along the x-direction in a semiconductor sample. Show that the steady-state nonequilibrium function can be written as $f_k(k_x, k_y, k_z) = f_k^0((k_x + \Delta k_x), k_y, k_z)$, where $\Delta k_x = \frac{\tau\hbar k_F}{m^*T}(k - k_F)\frac{dT}{dx}$ and k_F is the Fermi wave vector.

BIBLIOGRAPHY

Ando T, Fowlr AB, Stern F. Rev Modern Phys 1982;54:437.

Anselm AI, Askerov BP. Fiz Tverd Tela (Leningrad) 1961;3:3668.[Sov Phys Solid State 1962;3:2665].

Conwell E, Weisskopf VF. Phys Rev 1950;77:388.

Davies JH, Wilkins JW. Phys Rev B 1988;38:1667.

Dimitrov HD. J Phys Chem Solids 1976;37:825.

Erginsoy C. Phys Rev 1956;79:1013.

Gunn JB. Solid State Commun 1963;1:88.

Haynes JR, Shockley W. Phys Rev 1951;81:835.

Klitzing KV, Dorda G, Pepper M. Phys Rev Lett 1980;45:494.

Lambe J, Jaklevic RC. Phys Rev Lett 1969;22:1371.

Laughlin RB. Phys Rev Lett 1983;50:1395.

Look DC. Electrical characterization of gaas materials and devices. New York: Wiley;1989.

Mitin VV, Kochelap VA, Stroscio MA. Quantum heterostructures microelectronics and optoelectronics. Cambridge: Cambridge University Press; 1999.

Mott NF, Jones H. The theory of the properties of metals and alloys. New York: Dover; 1958.

Palik ED, Picus GS, Teitlee S, Wallis RF. Phys Rev 1961;122:475.

Pfeiffer L, West KW, Stromer HL, Baldwin KW. Appl Phys Lett 1989;55:1888.

Singh J. Electronic and optoelectronic properties of semiconductor structures. Cambrideg: Cambridege University press; 2003.

Smith RA. Semiconductors. 2nd ed. London: Cambridge; 1978.

Tsui DC, Störmer HL, Huang JCM, Brooks JS, Naughton MJ. Phys Rev B 1983;28:2274.

Sze SM. Physics of semiconductor devices. 2nd ed.. New York: Wiley; 1981.

Sze SM. Semiconductor devices physics and technology. 2nd ed. New York: Wiley; 2002.

van Wees BJ, van Houten H, Beenakkar CW, Williamson JG, Kouwenhoven LP, van der Marel D, Foxon CT. Phys Rev Lett 1988;60:848.

6

ELECTRONIC DEVICES

6.1 INTRODUCTION

The electronic and transport properties of electronic devices were reviewed in Chapter 5. These devices are usually microelectronic devices based on homojunction and heterojunction structures. Electronic devices are divided into two classes depending on their operational mode. The first class is called potential-effect devices, in which the transport properties are due to carrier injections. Bipolar transistors, which include heterojunction bipolar transistors (HBTs), and hot electron transistors (HETs) are examples of this class of devices. HETs include both ballistic injection devices and real-space transfer devices. The second class is called field-effect or voltage-controlled devices. Metal oxide–semiconductor field-effect transistors (MOSFETs), homogeneous field-effect transistors, and heterostructure field-effect devices all belong to the second class. There are several varieties of MOSFETs, such as semiconductor on insulator, complimentary MOSFETs, n-type MOSFETs, and p-type MOSFETs. Metal–semiconductor field-effect transistors (MESFETs) and junction field-effect transistors (JFETs) are examples of homogeneous field-effect devices. An example of heterojunctions field-effect devices is modulation-doped field-effect transistors (MODFETs), which are also called high electron mobility transistors (HEMTs). This chapter focuses on heterojunction devices, and thus, bipolar transistors and MOSFETs are not discussed, since they are the subjects of many textbooks.

The simplest electronic device is the ohmic contact that is based on metal–semiconductor interfaces. For any electronic device to be functional,

Introduction to Nanomaterials and Devices, First Edition. Omar Manasreh.

ohmic contacts are required to allow the charge carrier to move with ease in and out of the device. In other words, the current–voltage ($I-V$) curve must be linear (nonrectifying) for both positive and negative voltage with a very large slope. When a metal is brought in contact with a semiconductor, an energy barrier is formed. This barrier is referred to as the work function, which restricts the flow of charge carriers. One method to form an ohmic contact is to choose a metal that has a work function smaller than that of the semiconductor. But since the Fermi energy level lies within the band gap of nondegenerate semiconductors, the formation of an ohmic contact is difficult to obtain. This is due to the fact that the Fermi energy level is pinned at the metal–semiconductor interface, causing the formation of an energy barrier. Essentially, this is a Schottky contact. A practical solution to this problem is to heavily dope a small thickness of the semiconductor material near the surface before depositing the metal at the surface of the semiconductor. The metal–semiconductor interface is illustrated in Fig. 6.1 for an n-type semiconductor, where a small thickness near the surface of the semiconductor is heavily doped (n^+ region).

The heavily doped portion of the semiconductor reduces the depletion region such that the electron can easily tunnel through the barrier. The $I-V$ curve for the ohmic contact is linear as shown in the figure (the larger the slope, the better is the ohmic contact). In the absence of applied bias voltage, the depletion width in Fig. 6.1 can be written as

$$W = \left(\frac{2\epsilon V_{\mathrm{bi}}}{eN_{\mathrm{d}}} \right)^{1/2}, \tag{6.1}$$

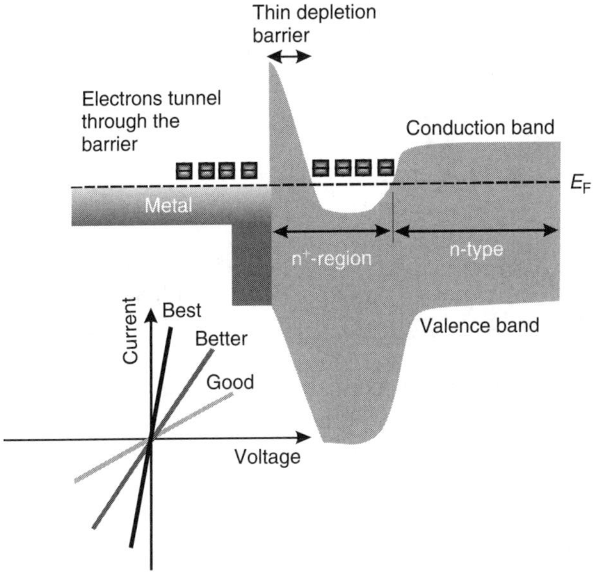

FIGURE 6.1 A band bending of metal-n^+-n-type semiconductor contact.

where V_{bi} is the built-in voltage, ϵ is the permittivity of the semiconductor material given as the product of the dielectric constant (ϵ_r) and the permittivity of space (ϵ_0), and N_d is the donor concentration in the semiconductor. The depletion width is inversely proportional to the square root of the dopant concentration. Assuming that each dopant atom contributes an electron to the conduction band, the electron concentration is N_d. However, in many cases, the carrier concentration is smaller than N_d. For a built-in voltage of 0.3V and for $N_d = 5 \times 10^{18}$ cm^{-3}, the depletion width for GaAs with a refractive index of 3.5 is ~90 Å.

The specific contact resistance in units of Ω-cm^2, R_c, of an ohmic contact is defined as the product of the contact resistance, R, and the area of the contact, A, or

$$R_c = RA = A\left(\frac{\partial I}{\partial V}\right)^{-1} = \left(\frac{\partial J}{\partial V}\right)^{-1}, \tag{6.2}$$

where J is the tunneling current density. The tunneling current density is proportional to the tunneling probability of a triangular barrier as described in Appendix H. The built-in electric field, $\mathcal{E}$, can be replaced by the built-in voltage divided by the depletion width. The tunneling current density becomes

$$J \sim \exp\left(-\frac{4}{3}\frac{W\sqrt{2m^*}}{\hbar e V_{bi}}(\Delta E_c - E)^{3/2}\right). \tag{6.3}$$

Substituting Equation 6.1 into Equation 6.3 and assuming that $\Delta E_c \approx eV_{bi}$, we have

$$J \sim \exp\left(-\frac{4}{3}\frac{\sqrt{2m^*}}{\hbar e V_{bi}}\left(\frac{2\epsilon V_{bi}}{eN_d}\right)^{1/2}(eV_{bi} - eV)^{3/2}\right). \tag{6.4}$$

Thus,

$$R_c \propto \exp\left(\frac{V_{bi}}{\sqrt{N_d}}\right) \quad \text{or} \quad \ln(R_c) \propto \frac{V_{bi}}{\sqrt{N_d}}, \tag{6.5}$$

where the electron energy E is taken as eV. This equation tells us that the specific contact resistance is minimized by using a metal with a small work function and by choosing N_d as high as possible. Table 6.1 lists the work functions of a few metals.

A Schottky diode is formed between a metal and a semiconductor. Its $I-V$ characteristic is similar to that of a homojunction p–n diode, except that it has a much faster response. The forward current of the Schottky diode is due to the majority carrier injection from the semiconductor side. Both ohmic and Schottky contacts are the building blocks of metal–semiconductor field-effect devices. For example, the drain and source are ohmic contacts, while the gate is a Schottky contact (rectifying contact). Thus, it is very useful to discuss Schottky diodes in more detail.

TABLE 6.1 A List of the Work Function, ϕ_m, for Several Metals that are Common in the Metallization Used for Semiconductor Devices, and the Electron Affinity, χ, for Four Semiconductors

Metal	ϕ_m (V)	Semiconductor	χ (V)
Aluminum (Al)	4.28	AlAs	3.5
Chromium (Cr)	4.50	GaAs	4.07
Cobalt (Co)	5.00	Ge	4.13
Gold (Au)	5.10	Si	4.01
Molybdenum (Mo)	4.60	—	—
Nickel (Ni)	5.15	—	—
Osmium Os)	5.93	—	—
Palladium (Pd)	5.65	—	—
Platinum (Pt)	5.65	—	—
Rhenium (Re)	4.72	—	—
Silver (Ag)	4.26	—	—
Tantalum (TA)	4.25	—	—
Titanium (Ti)	4.33	—	—
Tungsten (W)	5.55	—	

6.2 SCHOTTKY DIODE

The metal–semiconductor rectifying contact is called a Schottky barrier diode. The ideal energy band diagrams for both n-type and p-type Schottky barrier diodes are shown in Fig. 6.2. The electron affinity, $e\chi$, is defined as the energy required to remove an electron from the surface of the semiconductor to the vacuum level. The electron affinities are approximately 4.01 and 4.07eV for silicon and GaAs, respectively. At thermal equilibrium, the Fermi energy levels in both the metal and the semiconductor materials are equal. From this figure, one can find that the barrier height for the n-type semiconductor material, eV_{Bn}, can be written as

$$eV_{Bn} = e(\Phi_m - \chi). \tag{6.6}$$

The barrier height in case of a p-type semiconductor shown in Fig. 6.2d can be written as

$$eV_{Bp} = E_g - e(\Phi_m - \chi), \tag{6.7}$$

where E_g is the band gap of the semiconductor. From the above two equations, one can realize that the sum of the barrier heights of n-type and p-type of any semiconductor is equal to the band gap of the semiconductor regardless of the type of metal used for the Schottky barrier. Experimentally, however, the Schottky barrier height is usually smaller than the predicted values obtained from Equations 6.6 and 6.7. The discrepancy may be attributed to the presence of surface states.

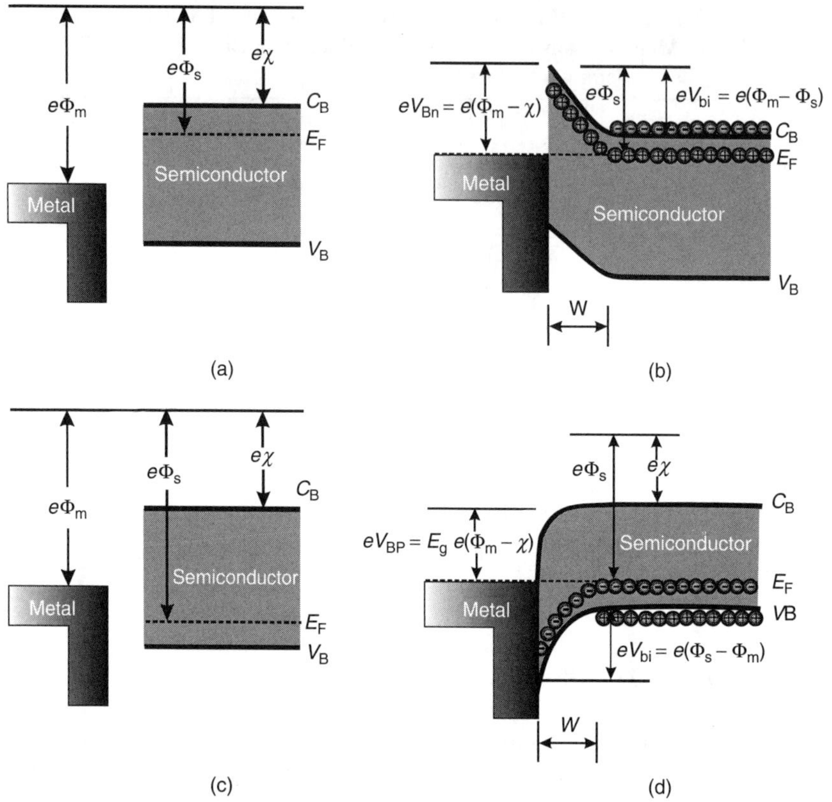

FIGURE 6.2 Band diagram of a Schottky barrier diode for an n-type semiconductor (a) before and (b) after contact. Similarly, the Schottky barrier diode for p-type semiconductor is shown (c) before and (d) after contact.

The quantity V_{bi} shown in Fig. 6.2 is called the built-in potential, which, according to Fig. 6.2b, can be written for an n-type semiconductor as

$$eV_{\mathrm{bi}} = e(\Phi_{\mathrm{m}} - \Phi_{\mathrm{s}}) = e(V_{\mathrm{Bn}} - V_{\mathrm{n}}), \tag{6.8}$$

where Φ_{s} is the work function of the semiconductor and V_{n} is the energy difference between the conduction band minimum and the Fermi energy level. Similarly, the built-in potential in p-type semiconductors is the same as Equation 6.8, except that V_{n} is replaced by V_{p}, where V_{p} is the energy difference between the Fermi energy level and the valence band maximum.

The energy band diagram of a Schottky barrier on an n-type semiconductor is shown in Fig. 6.3a for three different conditions. The corresponding $I-V$ curve is shown in Fig. 6.3b. When a large reverse bias voltage is applied (top panel in the figure), the barrier height on the semiconductor side increases, making it difficult for the electrons to flow from the semiconductor to the metal. Hence,

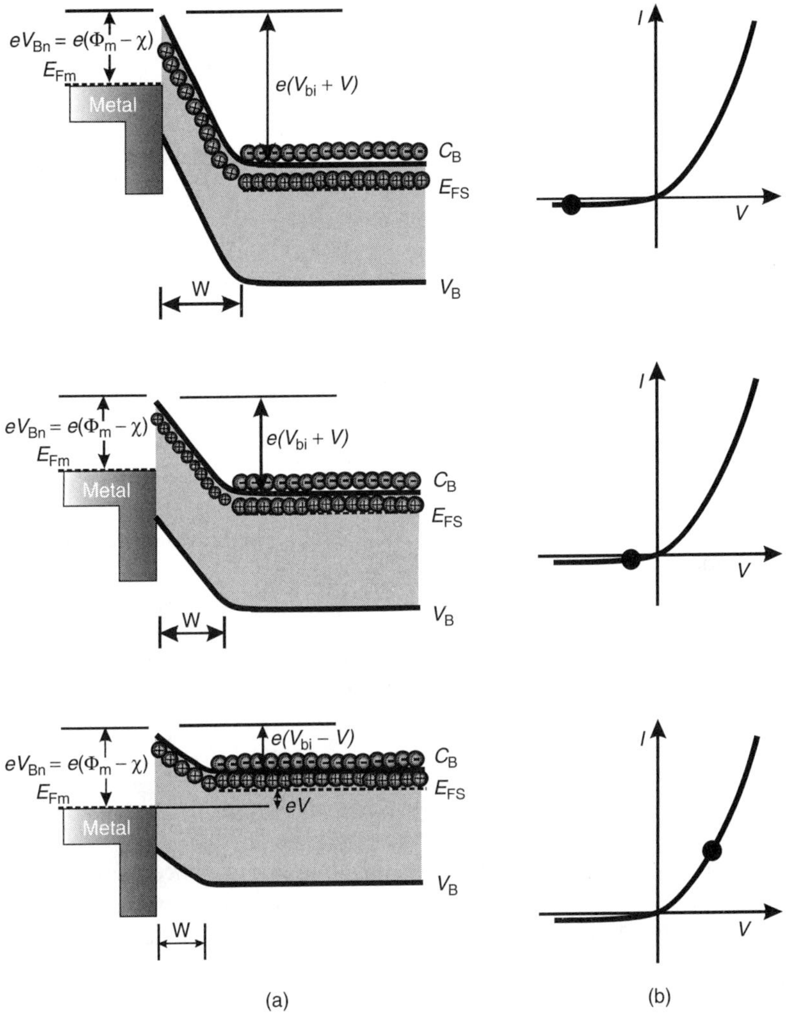

FIGURE 6.3 (a) Energy band diagram of a Schottky barrier diode for three different bias voltages. (b) The $I-V$ curves correspond to the energy band diagrams.

the reverse current is very small. In the case of forward bias, the barrier height decreases as the forward bias increases, which permits electrons to flow as illustrated in the bottom panel of this figure. The dot in the $I-V$ curve represents the operational modes (reverse or forward) of the Schottky barrier diode. Notice that when the bias voltage is applied, the metal–semiconductor junction is no longer in equilibrium and the Fermi energy levels in the semiconductor (E_{FS}) and in the metal (E_{FM}) are necessarily the same.

For an ideal metal–semiconductor junction, the electric field, $\mathcal{E}$, in the space charge region (depletion region) can be obtained from Poisson's equation, which relates the electric field to the space charge volume density, $\rho(x)$, as follows:

$$dE = \frac{\rho(x)}{\epsilon_s} \mathrm{d}x, \tag{6.9}$$

where ϵ_s is the dielectric permittivity of the semiconductor material. The electric field is zero at $x = W$, which yields after integrating Equation 6.9 the following relation:

$$E(x) = -\frac{eN_\mathrm{d}}{\epsilon_\mathrm{s}}(W - x), \tag{6.10}$$

where W is the depletion width, N_d is the electron concentration in the depletion region, and $\rho(x) = eN_d$. For a uniformly doped semiconductor, $\mathcal{E}$ is linear as a function of distance. The depletion width can be obtained as

$$W = \left(\frac{2\epsilon_\mathrm{s}(V_\mathrm{bi} - V)}{eN_\mathrm{d}}\right)^{1/2}, \tag{6.11}$$

and the space charge density in the semiconductor region can be expressed as

$$\rho_\mathrm{SC}(x) = eN_\mathrm{d}W = (2\epsilon_s eN_\mathrm{d}\,(V_\mathrm{bi} - V))^{1/2}\ {}_{\mathrm{C/cm^2}}. \tag{6.12}$$

Notice that the voltage V is positive for forward bias and negative for reverse bias. Differentiate Equation 6.12 with respect to V to obtain the capacitance of the depletion region

$$C = \left|\frac{\partial\rho_\mathrm{SC}(x)}{\partial V}\right| = \left(\frac{\epsilon_\mathrm{s}eN_\mathrm{d}}{2(V_\mathrm{bi} - V)}\right)^{1/2} = \frac{\epsilon_s}{W}\ \mathrm{F/cm^2}. \tag{6.13}$$

This equation is very useful in calculating the carrier concentration from the capacitance measurement as a function of bias voltage. Recently, C–V profiling has been used to calculate the carrier concentration as a function of the depth in heterojunctions and multiple quantum wells.

The Schottky barrier height is usually much larger than the thermal energy $k_\mathrm{B}T$. This barrier is usually obtained when the doping level in the semiconductor is smaller than the density of states in the conduction or valence band. As mentioned earlier, the transport in a Schottky diode is due to the majority carriers, and the thermionic mechanism is dominant at room temperature. For a Schottky diode with an n-type semiconductor under a forward bias voltage, V_F, the electron concentration emitted by the thermionic mechanism can be expressed as

$$N_\mathrm{th} = N_\mathrm{c}\mathrm{e}^{\left(-\frac{e(\Phi_\mathrm{Bn} - V_\mathrm{F})}{k_\mathrm{B}T}\right)}. \tag{6.14}$$

Thus, the current density from the semiconductor to the metal can be expressed as

$$J_{\mathrm{s}\to\mathrm{m}} = C_1 N_\mathrm{c} e^{\left(-\frac{e(\Phi_\mathrm{Bn} - V_\mathrm{F})}{k_\mathrm{B}T}\right)}. \tag{6.15}$$

Similarly, the current density from the metal to the semiconductor is given as follows:

$$J_{\mathrm{m}\to\mathrm{s}} = C_1 N_\mathrm{c} \mathrm{e}^{\left(-\frac{e\Phi_{\mathrm{Bn}}}{k_\mathrm{B}T}\right)}. \tag{6.16}$$

Notice that the forward bias voltage is not included in Equation 6.16, since the barrier height seen from the metal side is not affected by bias voltage. Under the influence of bias voltage, the current densities shown in Equations 6.15 and 6.16 are no longer the same. The net current density is the difference between the expressions in the above two equations and is given by

$$\begin{aligned} J = J_{\mathrm{s}\to\mathrm{m}} - J_{\mathrm{m}\to\mathrm{s}} &= C_1 N_\mathrm{c} \mathrm{e}^{\left(-\frac{e(\Phi_{\mathrm{Bn}}-V_\mathrm{F})}{k_\mathrm{B}T}\right)} - C_1 N_\mathrm{c} \mathrm{e}^{\left(-\frac{e\Phi_{\mathrm{Bn}}}{k_\mathrm{B}T}\right)} \\ &= C_1 N_\mathrm{c} \mathrm{e}^{\left(-\frac{e\Phi_{\mathrm{Bn}}}{k_\mathrm{B}T}\right)} \left\{ \mathrm{e}^{\left(\frac{eV_\mathrm{F}}{k_\mathrm{B}T}\right)} - 1 \right\}, \end{aligned} \tag{6.17}$$

where C_1 is a constant. The product $C_1 N_\mathrm{c}$ is found to be equal to $A^* T^2$, where T is the temperature and A^* is called the *effective Richardson constant* taken in units of $\mathrm{A}/(\mathrm{K}^2\ \mathrm{cm}^2)$ and is expressed as

$$A^* = \frac{4\pi m^* k_\mathrm{B}^2}{h^3}, \tag{6.18}$$

where m^* is the electron- or hole-effective mass and h is the Planck constant. The values of A^* are 110 and 32 for n-type and p-type silicon, respectively, and 8 and 74 for n-type and p-type GaAs, respectively. A more familiar form of the current density is

$$J = J_s \left(\mathrm{e}^{\left(-\frac{eV_\mathrm{F}}{k_\mathrm{B}T}\right)} - 1 \right), \tag{6.19}$$

where

$$J_\mathrm{s} = A^* T^2 \mathrm{e}^{\left(-\frac{e\Phi_{\mathrm{Bn}}}{k_\mathrm{B}T}\right)}. \tag{6.20}$$

The parameter J_s is called the *saturation current density*.

6.3 METAL–SEMICONDUCTOR FIELD-EFFECT TRANSISTORS (MESFETs)

Recent advances in materials growth using molecular beam epitaxy (MBE) and metal–organic chemical vapor-phase deposition (MOCVD) techniques enable the growth of high purity thin films. With these growth techniques, one can precisely control the doping level in semiconductor materials, which allows the growth of semiconductor structures for device processing and fabrication. The

MESFET was first proposed by Mead. MESFETs are usually associated with GaAs. A cross section of an n-channel MESFET is shown in Fig. 6.4. The drain and source are ohmic contacts, while the gate metal is a Schottky contact. Ion plantation is usually used to produce the n^+ regions under the drain and source. The active region is an epitaxially grown n-type thin film, and the substrate is semi-insulating. The semi-insulating GaAs substrates are usually grown by LEC under high pressure arsenic conditions or by doping the materials with chromium. The substrate resistivity is usually very high ($\rho > 10^7$ Ω-cm), preventing the electric current from flowing through the substrate.

From Fig. 6.4, one can derive the resistance of the n-channel according to the following relation:

$$R = \rho \frac{L}{A}, \tag{6.21}$$

where L is the length of the gate and A is the depletion region area given by $x(a\text{-}W)$, where x and a are shown in Fig. 6.4 and W is the width of the depletion region under the Schottky contact. The resistivity, ρ, is given by $(e\mu_n N_d)^{-1}$, where μ_n is the electron mobility in the n-channel and N_d is the electron concentration. Thus, the magnitude of the drain current, I_D, is given by V_D/R in the nonsaturation region.

A reverse bias gate-source voltage induces a space charge region (depletion region) under the metal gate, which modulates the channel conductance. The operation of the MESFET is depicted in Fig. 6.5, in which only the gate section is sketched. The source is usually grounded, while the gate and drain voltages are measured with respect to the source. Under normal operation, the gate voltage, V_G, is either zero or reverse biased ($V_G \leq 0$) and the drain voltage, V_D, is either zero or forward biased ($V_D \geq 0$). For $V_G = 0$ and small V_D, a small drain current, I_D, flows in the channel, as shown in Fig. 6.5a. The dot on the curve represents the $I_D - V_D$ coordinates associated with the sketch on the left-hand side. Notice that

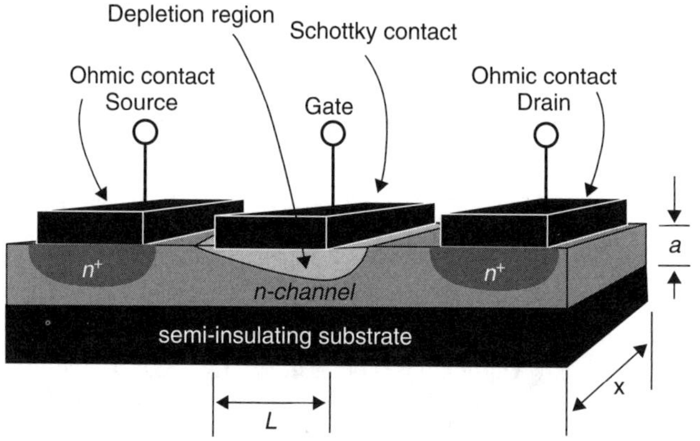

FIGURE 6.4 A sketch of a cross section of an n-channel MESFET.

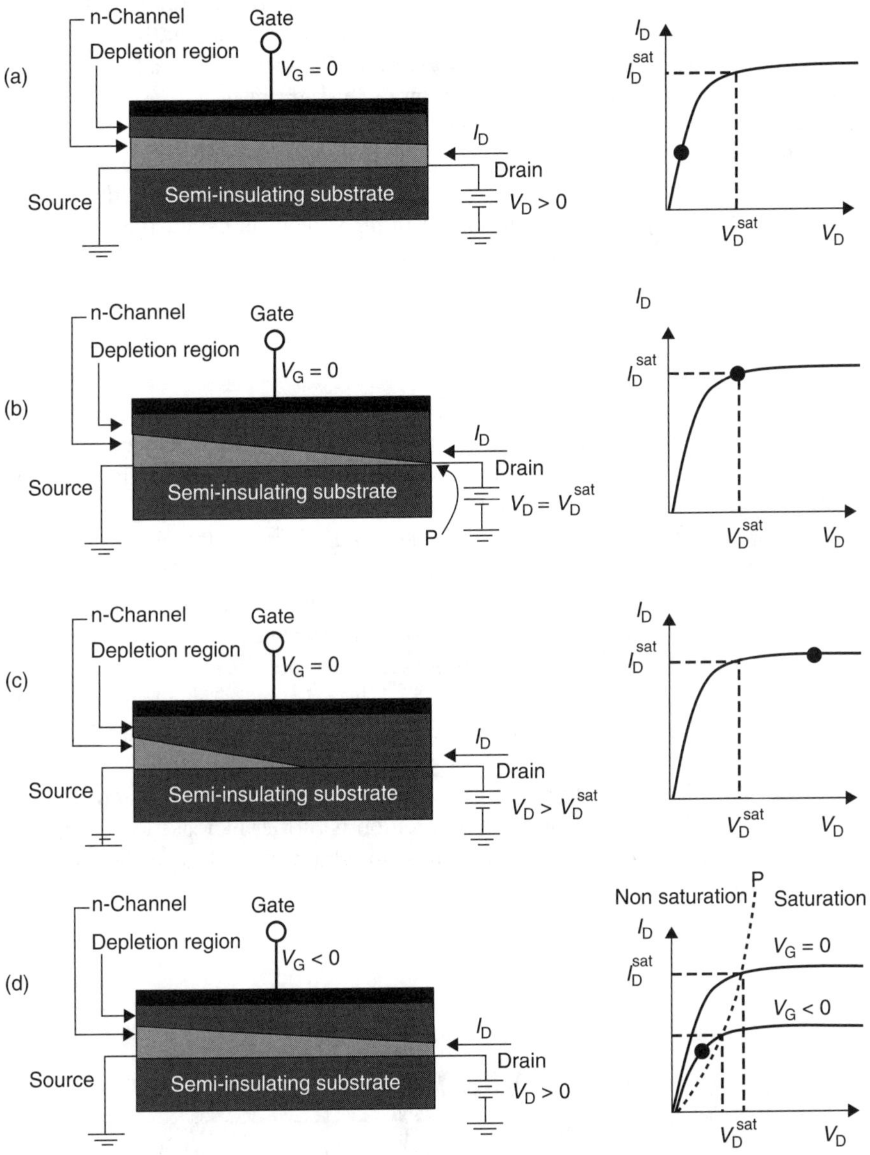

FIGURE 6.5 A schematic presentation of the gate cross-sectional area of a MESFET and the output I_D–V_D characteristic curves. (a) $V_G = 0, 0 < V < V_D^{sat}$. (b) $V_G = 0, V = V_D^{sat}$. (c) $V_G = 0, V > V_D^{sat}$. (d) $V_G < 0, 0 < V < V_D^{sat}$.

the channel is reverse biased and the voltage is zero at the source and increases to V_D at the drain terminal. As the drain voltage increases, the width of the space charge region (depletion region) increases. Thus, the cross-sectional area for the current flow decreases, causing a reduction of the current flow rate.

When $V_G = 0$ and the drain voltage is large enough such that the depletion region touches the substrate, as shown in Fig. 6.5b, the depletion region width is equal to the thickness of the n-type epitaxial layer at the drain side, that is, $W = a$. The corresponding value of the drain voltage is called the *saturation voltage*, which can be obtained from Equation 6.11 as

$$V_D^{sat} = a^2 \frac{eN_d}{2\epsilon_s} - V_{bi}, \tag{6.22}$$

where V_{bi} is the built-in voltage. Notice that V_D^{sat} is larger than V_{bi}. The built-in voltage for an ideal MESFET is shown in Fig. 6.6 in which the energy band diagram is sketched. The condition at which the depletion region touches the substrate at the drain side is called *pinch-off*, which is "P" in Fig. 6.5b. The current at point "P" is called the *saturation current*, I_D^{sat}, and the electrons tunnel through depletion region from the source to the drain. The voltage at point P remains the same and the current flowing in the channel remains independent of V_D. This is because the potential drop from the source to point P does not change. Figure 6.5c illustrates this process where the current is constant for $V > V_D^{sat}$. Again the processes described in Fig. 6.5 are for an ideal MESFET. In real devices, there are many parasitic effects that cause the drain current to depend on the drain voltage for $V_D > V_D^{sat}$.

Figure 6.5d illustrates the condition when the gate is reverse biased, that is, $V_G < 0$ and $0 < V < V_D^{sat}$. The depletion region is further increased as compared to the conditions shown in Fig. 6.5a. Thus, the drain current is further reduced

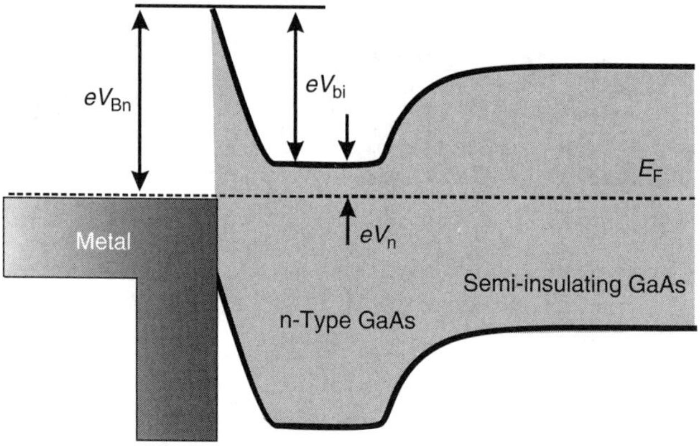

FIGURE 6.6 Sketch of a band gap diagram of an ideal n-channel MESFET showing the barrier height, built-in voltage, and the degenerate voltage, V_n.

as shown in the output characteristic curve. When the drain voltage is equal to V_D^{sat}, where the depletion region just touches the substrate, we have

$$V_D^{sat} = a^2 \frac{eN_d}{2\epsilon_s} - V_{bi} - |V_G|, \tag{6.23}$$

where V_G values are taken as the absolute values. This equation tells us that V_D^{sat} is reduced by applying a reverse bias voltage to the gate.

The current–voltage relation for a MESFET has been derived in many textbooks (Sze, 2002; Neamen; and Mitin *et al.*). The analyses by different authors are identical. The geometry of the depletion region under the gate contact is shown in Fig. 6.7. The voltage drop across an elemental section dy of the channel can be written as

$$\mathrm{d}V = I_D \,\mathrm{d}R = I_D \rho \frac{\mathrm{d}y}{A} = \frac{I_D \mathrm{d}y}{e\mu_n N_d x(a - W(y))}, \tag{6.24}$$

where x is the length of the gate. By using Equation 6.23, the depletion region width can be expressed as

$$W^2(y) = \frac{2\epsilon_s\{V(y) + V_{bi} + |V_G|\}}{eN_d}. \tag{6.25}$$

Take the first derivative of the depletion region width with respect to the voltage and substitute the results into Equation 6.24 to obtain the following expression for the drain current:

$$I_D \,\mathrm{d}y = e\mu_n N_d x\{a - W\} \frac{eN_d}{\epsilon_s} W \mathrm{d}W. \tag{6.26}$$

Integrating this equation over the limits of $y = 0$ to $y = L$ and $W = W_1$ to $W = W_2$, we get

$$I_D = \frac{e^2 \mu_n N_d^2 x}{2\epsilon_s L} \left[a(W_2^2 - W_1^2) - \frac{2}{3}(W_2^3 - W_1^3) \right]. \tag{6.27}$$

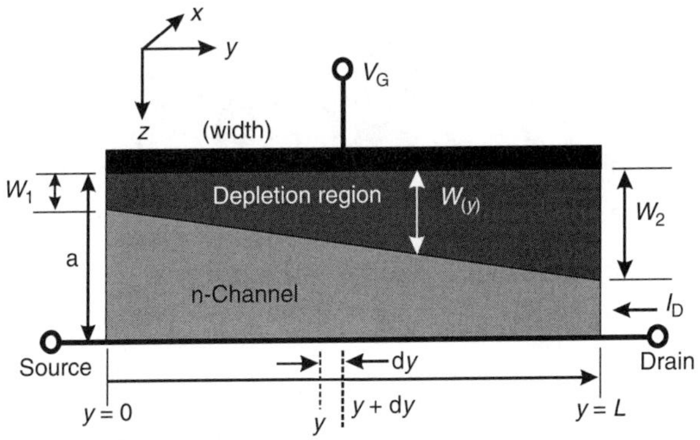

FIGURE 6.7 Sketch of the n-channel MESFET region under the gate contact.

This equation can be rewritten as

$$I_D = I_p\left[\frac{V_D}{V_p} - \frac{2}{3}\left(\frac{V_D + |V_G| + V_{bi}}{V_p}\right)^{3/2} + \frac{2}{3}\left(\frac{|V_G| + V_{bi}}{V_p}\right)^{3/2}\right], \tag{6.28}$$

where

$$I_p = \frac{e^2\mu_n N_d^2 x a^3}{2\epsilon_s L} \quad \text{and} \quad V_p = \frac{eN_d a^2}{2\epsilon_s}. \tag{6.29}$$

The quantities I_p and V_p are the pinch-off current and voltage, respectively. The saturation current can be obtained by setting $V_p = V_D + |V_G| + V_{bi}$. Thus, the saturation drain voltage is given as

$$V_D^{sat} = V_p - |V_G| - V_{bi}. \tag{6.30}$$

Substituting Equation 6.30 into Equation 6.28, we obtain the saturation current as

$$I_D^{sat} = I_p\left[\frac{1}{3} - \left(\frac{|V_G| + V_{bi}}{V_p}\right) + \frac{2}{3}\left(\frac{|V_G| + V_{bi}}{V_p}\right)^{3/2}\right]. \tag{6.31}$$

The I_D–V_D curves of a MESFET can be obtained by plotting Equation 6.28 or by using PSPICE. The results are shown in Fig. 6.8 for different gate voltages ranging between −3.0 and 0.0V. The saturation voltage is shown as the dashed curve, which can be considered as the boundary between saturation and nonsaturation regions of the transistor. For small V_D, the transistor is in the nonsaturation mode or linear mode. In this case, the channel behaves like an ohmic contact. When V_D is too large as compared to the saturation drain voltage, the current may substantially increase, causing an avalanche breakdown.

The transconductance, g_m, of the transistor is a parameter V_t that relates the output current to the input voltage and is a measure of the transistor gain. It is defined as

$$g_m = \left.\frac{\partial I_D}{\partial V_G}\right|_{V_D=\text{constant}}. \tag{6.32}$$

For small V_D, the first derivative of the drain current given by Equation 6.28 with respect to V_G is

$$g_m = \frac{\partial I_D}{\partial V_G} \approx \frac{1}{2}\frac{I_P}{V_P}\frac{V_D}{\sqrt{V_P(|V_G| + V_{bi})}}. \tag{6.33}$$

It is possible that the depletion region is large enough such that the device is normally off for $V_G = 0$. In this case, a forward biased gate ($V_G > 0$) is needed to turn on the device. The voltage needed to turn on the MESFET is called

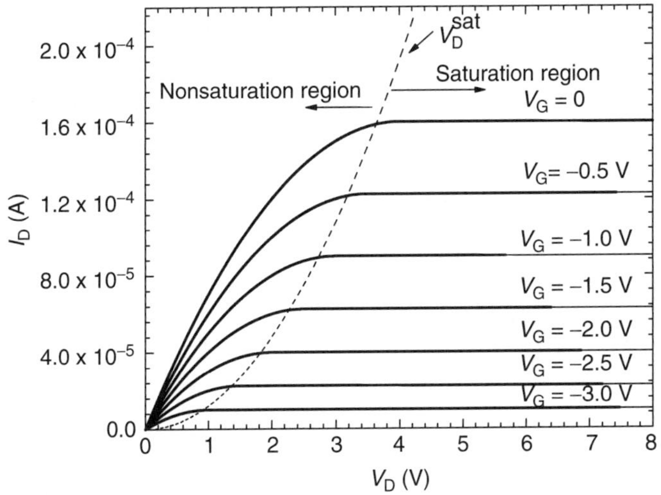

FIGURE 6.8 Drain current as a function of the drain voltage for an n-channel MESFET with different values of the gate voltage obtained using PSPICE. The saturation and nonsaturation regions are indicated.

the threshold voltage (V_t), which is given by $V_t = V_{bi} - V_p$ or $V_{bi} = V_t + Vp$. Substituting this expression into Equation 6.31 yields

$$I_D^{sat} = I_p \left[\frac{1}{3} - \left(1 - \frac{V_G - V_t}{V_p}\right) + \frac{2}{3}\left(1 - \frac{V_G - V_t}{V_p}\right)^{3/2} \right]. \tag{6.34}$$

Notice that $|V_G|$ in Equation 6.31 is replaced by $-V_G$. For $(V_G - V_t) \ll Vp$, the transconductance, when the device in saturation mode, can be obtained as

$$g_m \approx \frac{x\mu_n \epsilon_s}{aL}(V_G - V_t). \tag{6.35}$$

The MESFET simplistic and ideal model presented above describes the main characteristics of the transistor, such as the current–voltage and transconductance. These Field-Effect Transistor (FET) devices are usually used as amplifiers and converters. They are also used in microwave and radio frequency (RF) systems. For AC operation, the cut-off frequency, f_T, is a very important figure of merit. It is defined as the frequency at which the MESFET can no longer amplify the input signal. The generic equivalent small AC signal circuit of an FET is shown in Fig. 6.9a. The AC equivalent circuit shows the parasitic resistances and capacitances related to the drain–gate and gate–source, such as the drain–gate capacitance, the gate resistance, source resistance, etc. The gate–source input resistance represents the leakage current through the input junction. For an ideal MESFET, the input small-signal current through the gate at a frequency $\omega = 2\pi f$

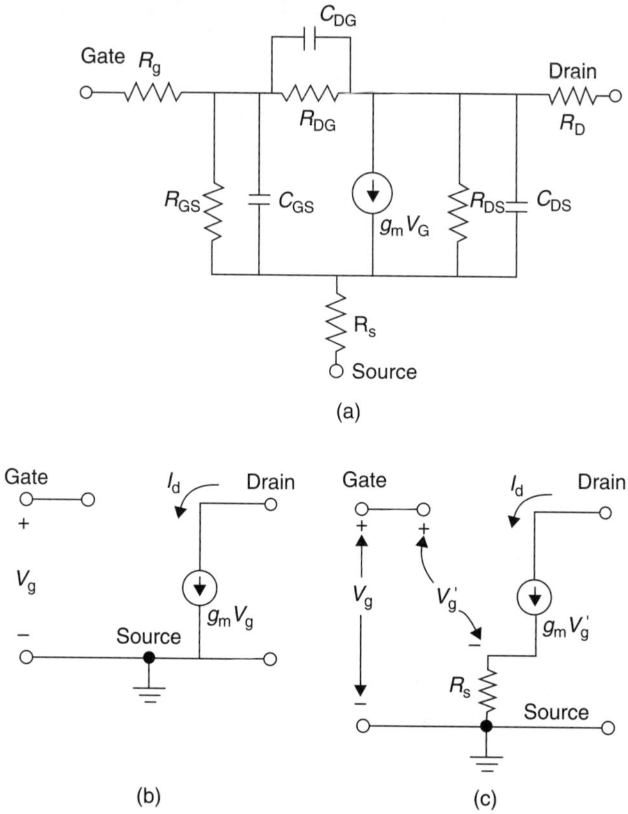

FIGURE 6.9 (a) An AC small-signal equivalent circuit model for an n-channel MESFET. (b) An ideal small-signal circuit. (c) An ideal small-signal circuit with R_s.

is mainly the displacement current, i_{in}, given by

$$i_{in} = 2\pi f C_G v_g, \tag{6.36}$$

where C_G is the gate to channel capacitance, $C_G = C_{gs} + C_{gd}$, given by

$$C_G = \frac{xL\epsilon_s}{\overline{W}}, \tag{6.37}$$

where $\overline{W}$ is the depletion region width at the gate Schottky contact. The small-signal output current is obtained from the definition of the transconductance given by Equation 6.32 such that

$$g_m = \frac{\partial I_D}{\partial V_G} = \frac{i_{out}}{v_g} \quad \text{or} \quad i_{out} = g_m v_g. \tag{6.38}$$

From the definition of the cut-off frequency, that is, the MESFET does not amplify, the output current must be equal to the input current. Thus, an expression for the cut-off frequency, f_T, can be obtained by equating Equations 6.36 and 6.38, which gives

$$2\pi f_T C_G v_g = g_m v_g \Rightarrow f_T = \frac{g_m}{2\pi C_G}. \tag{6.39}$$

By using the maximum value of $g_m = I_p/V_p$ obtained from the drain current in the saturation mode of the normally on MESFET (Eq. 6.31) and the definition of C_G given by Equation 6.37, one can rewrite Equation 6.39 as

$$f_T = \frac{g_m}{2\pi C_G} = \frac{I_p}{V_p}\cdot\frac{1}{2\pi C_G} = \frac{e^2\mu_n N_d^2 x a^3}{2\epsilon_s L}\cdot\frac{2\epsilon_s}{eN_d a^2}\cdot\frac{\overline{W}}{2\pi x L\epsilon_s}$$
$$= \frac{e\mu_n N_d a\overline{W}}{2\pi L^2\epsilon_s} = \frac{e\mu_n N_d a^2}{2\pi L^2\epsilon_s}. \tag{6.40}$$

Notice that in the saturation mode, the depletion region width is equal to a, where a is defined in Fig. 6.7 as the total thickness of the epitaxial layer. Equation 6.40 implies that to obtain a high cut-off frequency, the MESFET should be designed such that the electron mobility is as high as possible and the gate length is as short as possible.

The cut-off frequency can be written in terms of the transient time, τ_{tr}, required for the electron to travel through the channel as

$$f_T = \frac{1}{2\pi\tau_{tr}} = \frac{v_s}{2\pi L}, \tag{6.41}$$

where v_s is the carrier saturation velocity, which is essentially the drift velocity when the MESFET is operating in the saturation region. Again, the cut-off frequency can be increased by reducing the gate length and by choosing material with a high drift velocity. For example, if GaAs has a drift velocity of the order of 2.0×10^7 cm/s and the transistor gate length is 10 μm, the cut-off frequency is ~3.20GHz. As a comparison, the electron drift velocity in silicon is about an order of magnitude smaller than that of GaAs, which leads to a cut-off frequency of the order of ~0.32GHz for a transistor made of silicon.

The ideal AC small-signal circuit of the MESFET is shown in Fig. 6.9b, in which all the diffusion parasitic resistances are infinite and the series resistances are zero. All the capacitances are open circuit for low frequency. In the small-signal analysis, the drain current becomes

$$I_d = g_m V_g. \tag{6.42}$$

The effect of the source resistance on the drain current can be understood by examining Fig. 6.9c where R_s is added to the ideal small-signal equivalent circuit.

From this figure, the drain current is $I_d = g_m V_g'$, where V_g' and V_g are related through the following relation

$$V_g = V_g' + g_m V_g' R_s = (1 + g_m R_s) V_g'. \tag{6.43}$$

Thus, Equation 6.42 can be rewritten as

$$I_d = g_m V_g' = \frac{g_m}{(1 + g_m R_s)} V_g = g_m' V_g, \tag{6.44}$$

where g_m' is called the extrinsic transconductance and is smaller than g_m. The transconductance is decreased when the source resistance is added.

For high power applications, it is desired to have V_D^{sat} as small as possible. This requirement can be achieved by choosing a material with high electron mobility and very small source and drain resistances. It is also required that the breakdown voltage be as high as possible. The breakdown voltage, V_B, occurs at the drain end of the channel, where the reverse voltage is highest. Thus, $V_B = V_D + |V_G|$. Breakdown voltage is usually higher for larger band gap materials.

6.4 JUNCTION FIELD-EFFECT TRANSISTOR (JFET)

The basic structure of the JFET is shown in Fig. 6.10a for an n-channel device. The n-region between the two p+ -regions is called the channel. The majority carriers in the channel flow from the source to the drain. The gate is the control terminal. A more realistic sketch of a cross section of a JFET is shown in Fig. 6.10b. A p-channel JFET can be fabricated in which the p and n regions are reversed. Usually the p-channel JFET is slower than the n-channel JFET because of the fact that the hole mobility is smaller than the electron mobility.

If the gate is kept at zero bias and a small voltage is applied to the drain terminal, a drain current is formed between the drain and source. The source is usually grounded. The JFET under these conditions behaves as a resistance and the $I-V$ curve is linear so long as the drain voltage is small enough. If a reverse bias voltage (negative voltage) is applied to the gate, the depletion region widens and the channel region becomes narrower leading to an increase in the channel resistance. Thus, the slope of the $I-V$ linear curve is reduced. For sufficiently large reverse bias applied to the gate, the two depletion regions touch each other causing the channel to be completely depleted. This condition is called the *pinch-off*.

If the drain voltage is increased, the JFET behaves differently. The operation of the JFET under increasingly forward bias voltage applied to the drain and increasingly reverse bias voltage applied to the gate is shown in Fig. 6.11a. As seen from this figure, the $I-V$ behavior of the JFET is similar to that of the MESFET. Starting from zero biased gate and $V_{D1} < V_D^{sat}$, the JFET acts like a resistor and the drain current is linear with the drain voltage. Again, the saturation voltage, V_D^{sat}, is the drain voltage required to produce a pinch-off condition. The

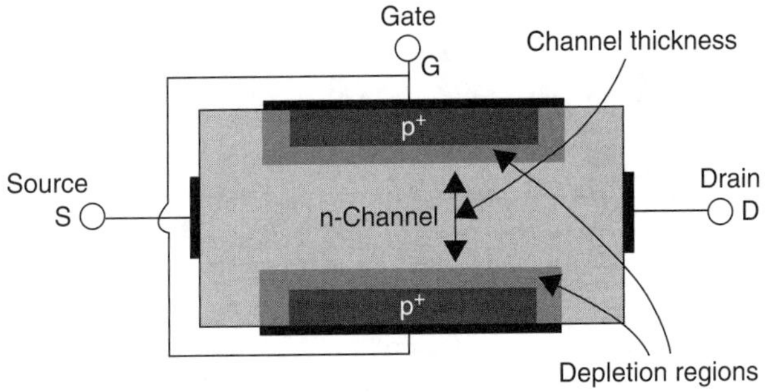

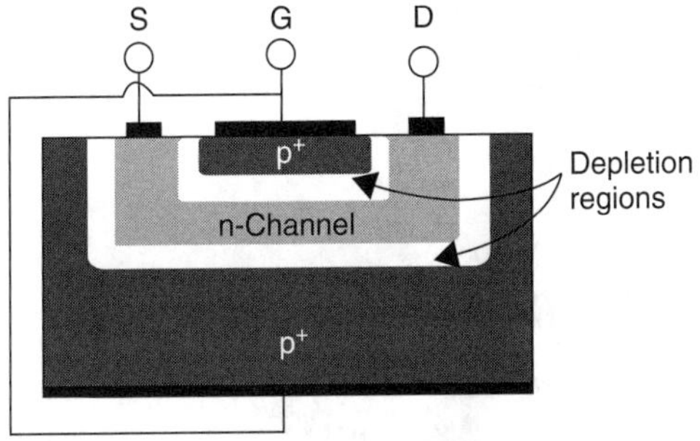

FIGURE 6.10 (a) The basic structure of the JFET. The drain and source metals are ohmic contacts, and the gate metal is a Schottky contact. (b) A sketch of a more realistic n-channel JFET. Notice that the gate metal is deposited on the p^+-type semiconductor and the drain and source contacts are made on the n-type semiconductor.

dot on the curve represents the approximate coordinates of the drain current and voltage. The JFET remains in the linear region so long as $V_D < V_D^{sat}$. For example, when $V_G = 0$ and $V_{D2} < V_D^{sat}$, where $V_{D1} < V_{D2}$, the two depletion regions are still separated as shown in Fig. 6.11b. When $V_G = 0$ and $V_D = V_D^{sat}$, the two depletion regions meet at the drain terminal. This is the pinch-off condition, which is shown in Fig. 6.11c. For $V_G = 0$ and $V_D > V_D^{sat}$, the device reaches the saturation mode and the drain current becomes independent of the drain voltage as shown in Fig. 6.11d. In the saturation mode, the electrons are injected into the depletion region and swept to the drain region under the influence of the electric field force. Ideally, the drain current remains constant in the saturation region until the avalanche breakdown voltage is reached, at which the device acts as if it shorted.

When a reverse bias voltage is applied to the gate, the depletion region will be increased even for a small drain forward bias voltage. Thus, the drain current is reduced as shown in Fig. 6.11e. For a fixed gate voltage, the drain current behavior as a function of the drain voltage is identical to the behavior observed

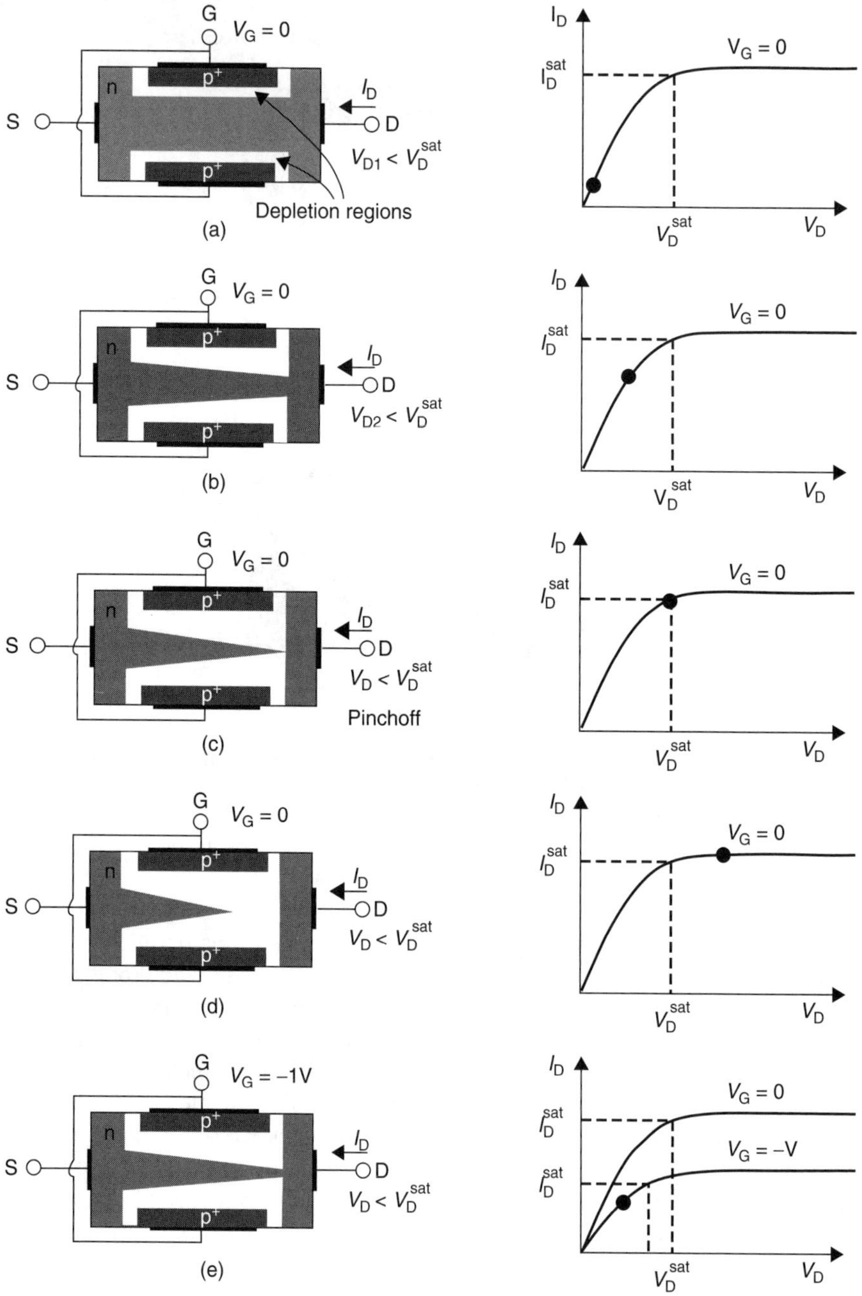

FIGURE 6.11 The basic operation of an n-channel JFET. The depletion regions and the $I-V$ curves are shown for the following conditions: (a) $V_G = 0$, $V_{D1} < V_D^{sat}$, linear region; (b) $V_G = 0$, $V_{D2} < V_D^{sat}$, where $V_{D1} < V_{D2}$, linear region; (c) $V_G = 0$, $V_D = V_D^{sat}$, pinch-off condition; (d) $V_G = 0$, $V_D > V_D^{sat}$, saturation region; and (e) $V_G = -1.0\text{V}$, $V_D < V_D^{sat}$, linear region.

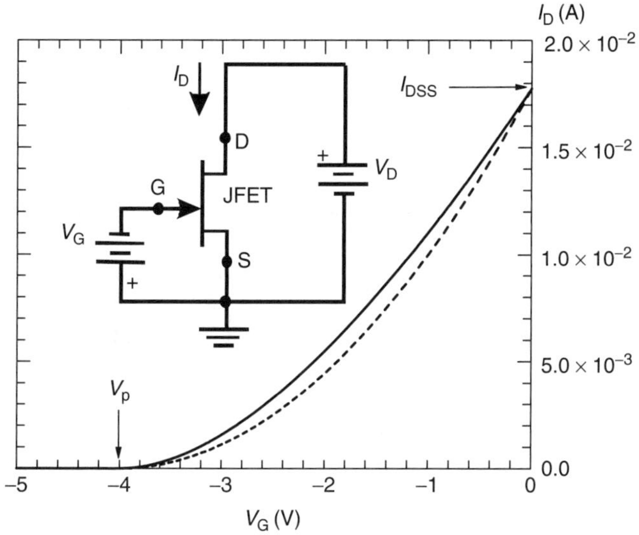

FIGURE 6.12 The drain current as a function of the gate voltage in the saturation region obtained by using PSPICE (solid line). The pinch-off voltage was chosen as −4 V. The inset is the DC circuit of the JFET. The dashed line is obtained using Equation 6.45.

in Fig.6.11a–d, except that the current value is lower. Using PSPICE, one can obtain the typical $I-V$ characteristic curves of the JFET, which are similar to those obtained for the MESFET as shown in Fig. 6.8.

Figure 6.12 shows the drain current as a function of the gate voltage at a constant drain voltage (solid line) obtained using PSPICE. The actual circuit used for PSPICE simulation is shown in the inset of the figure. The I_D versus V_G curve is obtained for a pinch-off voltage of $Vp = -4.0$V and the drain voltage was fixed at $V_D = 15$ V. The source resistance was chosen as 50 Ω, which is not shown in the figure. The derivation of the drain current as a function of the gate voltage is quite similar to that of the MESFET. The drain current can be approximated as

$$I_D \approx I_{DSS}\left(1 - \frac{V_G}{V_p}\right)^2, \qquad (6.45)$$

where I_{DSS} is the drain–source current obtained when the drain voltage is in saturation. It is obvious from Equation 6.45 that the drain current is a parabolic function of V_G. The current I_{DSS} is taken as the drain current at $V_G = 0$. Equation 6.45 is plotted in Fig. 6.12 as the dashed line, which is in good agreement with the result obtained from PSPICE.

6.5 HETEROJUNCTION FIELD-EFFECT TRANSISTORS (HFETs)

The MESFETs and JFETs have their own limitations because of the fact that the high level doping degrades the carrier mobilities and drift velocities. The next generation of FETs is based on epitaxially grown heterojunction structures, such as GaAs/AlGaAs, and GaN/AlGaN. The epitaxial growth techniques, such as MBE and MOCVD, allow one to deposit stacks of different band gap semiconductor layers. Heterojunction semiconductors are simple to grow with a precise control on the doping level, layer thickness, and alloy compositions. There are several types of heterojunction field-effect transistors (HFETs). The simplest structure is called the HEMT. This class of electronic devices is also called MODFETs. The conventional GaAs/AlGaAs MODFET structure is shown in Fig. 6.13a. The basic structure is composed of an undoped GaAs layer grown on a semi-insulating GaAs substrate. Buffer layers are usually introduced between the substrate and the undoped GaAs layer to reduce the dislocation density as well as to prevent the propagation of the dislocation into the device structure. An AlGaAs layer (larger band gap) is then grown on top of the undoped GaAs layer (smaller band gap). The AlGaAs is usually n-type doped, leaving a small portion of it undoped at the GaAs/AlGaAs interface. The undoped AlGaAs portion is called the *spacer layer*. The term *modulation doping* comes from the fact that the AlGaAs barrier is doped instead of the GaAs layer. The n^+-layer under the drain and source is usually produced by ion implantation. Silicon is usually used as a donor dopant in both the AlGaAs barrier and in the n^+-layers.

The thickness of the spacer layer is usually between 20 and 200 Å, and it plays a major role in the determination of the carrier density and mobility. For example, the carrier density decreases, while the mobility increases as the spacer thickness is increased. The effect of the spacer layer thickness can be understood in terms of electron–electron many-body interactions. When the barrier is doped, a triangular quantum well is formed at the GaAs/AlGaAs interfaces. This quantum well is populated by electrons forming what is called the two-dimensional electron gas (2DEG). The 2DEG is generated from the thermally ionized dopant atoms in the AlGaAs barrier. If the spacer thickness is larger, the probability of the electron being transferred to the quantum well is decreased. On the other hand, the electron–electron interactions (mainly exchange interactions) increase as the 2DEG density is increased. These interactions become significant when the carrier concentration exceeds 10^{18} cm^{-3}. The carrier mobility decreases as the magnitude of these interactions is increased as discussed in Chapter 7.

The 2DEG in the triangular well forms the transistor channel. More complicated MODFET structures can be fabricated such that several channels can be formed by repeating the epitaxially grown structure in Fig. 6.13a; each period produces a channel. The conduction band diagram of the MODFET is shown in Fig. 6.13b–d under different gate voltages. In Fig. 6.13b, we sketched the conduction band diagram at thermal equilibrium. The thickness of the doped AlGaAs layer is d and the spacer thickness is d_0. The conduction band offset is designated as ΔE_c, while the pinch-off voltage is shown as V_p. The gate metal is shown

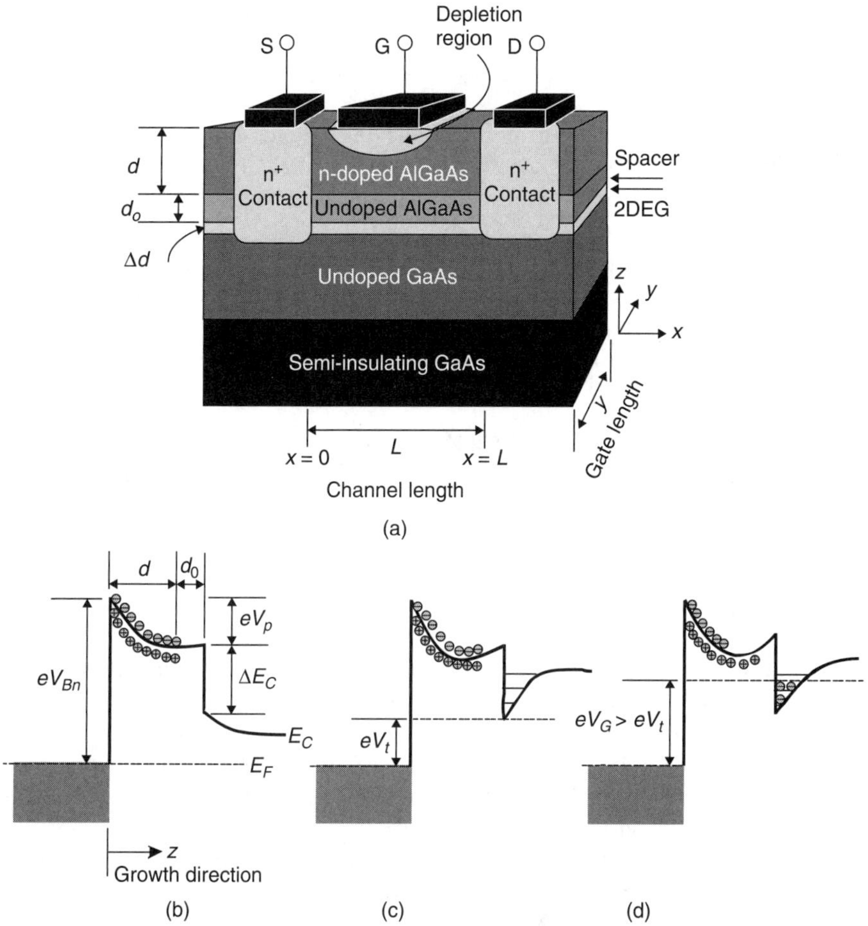

FIGURE 6.13 (a) Sketch of a MODFET based on GaAs/AlGaAsheterojunction. The conduction band diagram for (a) $V_G = 0$, thermal equilibrium; (b) $V_G = V_T$, where V_T is the threshold voltage; and (c) $V_G > V_T$.

on the left-hand side of the diagram. The pinch-off voltage can be obtained by integrating the electric field over the total thickness of the doped AlGaAs barrier. Assuming that the dopant density, N_d, is uniform across the barrier, V_p can be written as

$$V_p = \frac{e}{\epsilon_s} \int_0^d N_d z \mathrm{d}z = \frac{e N_d d^2}{2\epsilon_s}, \tag{6.46}$$

where ϵ_s is the permittivity of the AlGaAs layer.

The threshold voltage, V_t, shown in Fig. 6.13c is defined as the voltage applied to the gate such that the Fermi energy level is touching the bottom of the GaAs conduction band. This corresponds to the beginning of the formation of the channel at the GaAs/AlGaAs interface. The threshold voltage can be written as

$$eV_t = eV_{Bn} - \Delta E_c - eV_p, \tag{6.47}$$

where eV_{Bn} is the Schottky barrier height. The threshold voltage can be adjusted for a MODFET by choosing a specific Schottky metal and by adjusting the pinch-off voltage. The latter can be adjusted by varying the dopant concentration in the barrier and the spacer thickness.

When V_t is positive, the MODFET is normally off. This is called the enhancement mode. On the other hand, the normally on MODFET corresponds to the depletion mode or negative V_t. When the gate voltage is larger than V_t, as shown in Fig. 6.13d, a charge sheet or 2DEG density, $\mathcal{N}_s$, is induced by the gate at the heterojunction interface, which can be given as

$$\mathcal{N}_s = \frac{C_i(V_G - V_t - V_x)}{e}, \tag{6.48}$$

where the capacitance per unit area, C_i, is given as

$$C_i = \frac{\epsilon_s}{d + d_0 + \Delta d}, \tag{6.49}$$

where Δd is the thickness of the channel or the width of the triangular quantum well. The potential V_x is the channel potential with respect to the ground (source). Thus, V_x varies through the channel from zero at the source terminal to V_D at the drain terminal. According to Equation 6.48, the 2DEG density increases as the positive gate voltage, V_G, (forward bias) increases. For negative $V_G, \mathcal{N}_s$ is reduced drastically and the device is turned off.

The drain current in MODFET can be obtained by assuming that the channel width varies along the x-direction (Fig. 6.13a) similar to the behavior of the channel in the MESFET and MOSFET. The only difference in the MODFET is that the channel is a two-dimensional system in which the electrons are confined along the growth axis (z-axis), but not confined in the x–y plane. Thus, the drain current, I_D, at any point along the channel of the MODFET sketched in Fig. 6.13a can be written as

$$I_D = ye\mu_n\mathcal{N}_s\mathcal{E}_x, \tag{6.50}$$

where $\mathcal{E}_x$ is the electric field at any point along the channel. Substituting Equations 6.48 and 6.49 into Equation 6.50, we obtain

$$I_D = y\mu_n\epsilon_s\frac{(V_G - V_t - V_x)}{(d + d_0 + \Delta d)}\frac{dV_x}{dx}. \tag{6.51}$$

Integrating Equation 6.51 from source ($x = 0$ and $V_x = 0$) to the drain ($x = L$ and $V_x = V_D$) yields

$$I_D = \frac{y}{2L} \frac{\mu_n \epsilon_s}{(d + d_0 + \Delta d)} [2(V_G - V_t)V_D - V_D^2]. \tag{6.52}$$

For the linear region or nonsaturation region $V_D \ll V_G - V_t$, which reduces Equation 6.52 to the following

$$I_D = \frac{y}{L} \frac{\mu_n \epsilon_s}{(d + d_0 + \Delta d)} (V_G - V_t)V_D. \tag{6.53}$$

For a large drain voltage, the depletion region reaches the channel and causes the pinch-off, where the density of the 2DEG is reduced to zero at $x = L$. Under this condition, V_D^{sat} can be obtained from Equation 6.48 as

$$V_D^{sat} = V_G - V_t. \tag{6.54}$$

The saturation current can now be obtained by replacing V_D in Equation 6.52 with V_D^{sat} and by using Equation 6.54 as follows:

$$I_D^{sat} = \frac{y}{2L} \frac{\mu_n \epsilon_s}{(d + d_0 + \Delta d)} (V_G - V_t)^2. \tag{6.55}$$

The transconductance is obtained by using the definition given by Equation 6.32, which yields

$$g_m = \frac{y}{L} \frac{\mu_n \epsilon_s}{(d + d_0 + \Delta d)} (V_G - V_t). \tag{6.56}$$

Notice that this transconductance expression is similar to that of the MESFET given by Equation 6.35.

The electric field along the channel reaches high values in high speed devices, such as MODFETs, causing carrier velocity saturation. The drain current in the high speed operation mode can be written as

$$I_D^{hs} = ye\mu_n \mathcal{N}_s \epsilon_x = yev_s \mathcal{N}_s = yv_s C_i (V_G - V_t), \tag{6.57}$$

where the superscript "hs" is introduced to indicate "high speed" and v_s is the carrier saturation velocity. Using the definition of the cut-off frequency given by Equation 6.39 and knowing that the capacitance C_i defined in Equation 6.49 is the capacitance per unit area, one can express the cut-off frequency as

$$f_T = \frac{v_s}{2\pi \left(L + \dfrac{C_p}{yC_i}\right)}, \tag{6.58}$$

where C_p is the total parasitic capacitance. For designing a high speed MODFET, one needs to choose a material with a high carrier saturation velocity, small channel length, and very small parasitic capacitances.

6.6 GaN/AlGaN HETEROJUNCTION FIELD-EFFECT TRANSISTORS (HFETs)

Unique materials properties of GaN-based semiconductors have stimulated a great deal of research and development in materials growth and optoelectronic and electronic devices using this semiconductor system. Gate current collapse and device stability are among many other issues of interest in the research community. Recently, electronic devices based on GaN/AlGaN have been the subject of various investigations by many groups throughout the world. The interest in GaN HFETs stems from high power, high temperature, and high frequency applications. While the reported results are very encouraging thus far, there are several problems associated with GaN-related electronic devices. For example, the high dislocation densities can have a detrimental effect on the performance of the device. Gate leakage current or gate current collapse is another problem of the nitride HFETs.

The GaN/AlGaN heterostructures are normally grown by MBE or MOCVD techniques. The MOCVD reactors require triethylgallium and ammonia as the precursor gases to deposit a GaN layer in the conventional deposition regime in which precursors enter the growth chamber simultaneously. $Al_xGa_{1-x}N$ layers are deposited in the atomic layer regime when precursors enter the chamber in a cyclic manner using triethylgallium, triethylaluminum, and ammonia as precursors. The precursors are introduced into the chamber using hydrogen or nitrogen as a carrier gas. The common substrates used for HFETs are either sapphire or SiC substrates placed on a graphite susceptor, which is heated to the growth temperature by RF induction. Other substrates such as Si(111) have also been used for III-nitride HFETs.

Several device structures have been reported in the literature. However, the two most common structures are shown in Fig. 6.14. The 2DEG is formed at the GaN/AlGaN interface. The buffer layer varies. The most common buffer layer is the low temperature grown AlN layer. More elaborate buffer layers are composed of low temperature AlN and GaN/AlGaN superlattices. One of the main functions of the buffer layer is to prevent the dislocations formed at the substrate surfaces from propagating in the HFET structure. It also acts as an insulator between the device and the substrate. The main difference between the two structures is

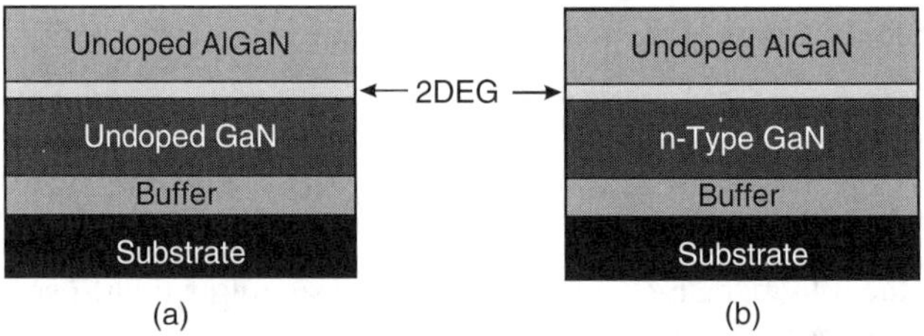

FIGURE 6.14 Two sketches of the most common GaN/AlGaN HFET structures.

that the GaN layer in structure (a) is undoped, while it is doped in structure (b). Silicon is usually used as the n-type dopant in GaN. The formation of the 2DEG in structure (a) relies on the spontaneous polarization-induced charge sheet. This requires that the polarity of the GaN surface should be Ga rich. The sheet carrier densities in nominally undoped GaN/AlGaN structures can, in fact, be comparable to those achievable in extrinsic doped structures, but without the degradation in mobility that can result from the presence of ionized impurities. A simple electrostatic analysis shows that the sheet carrier concentration, $\mathcal{N}_s$, of the 2DEG at the GaN/$Al_xGa_{1-x}N$ heterojunction interface should be given approximately by (Morkoç and Yu)

$$\mathcal{N}_s = \sigma_{pol}/e - (\epsilon_{AlGaN}/de^2)(e\phi_b + E_F - \Delta E_c) + \tfrac{1}{2}N_d d, \qquad (6.59)$$

where ϵ_{AlGaN} is the dielectric constant of $Al_xGa_{1-x}N$, ϕ_b is the Schottky metal work function, and E_F is the Fermi energy at the heterojunction interface. The thickness of the intentionally doped layer is d and N_d is the donor concentration, assumed to be uniformly distributed throughout the layer. For structure (a) in Fig. 6.14, N_d is zero. The polarization-induced sheet charge density σ_{pol}, shown in Equation 6.59 at the GaN/$Al_xGa_{1-x}N$ heterojunction interface is given, in the linear interpolation realm, approximately by (Yu)

$$\sigma_{pol}/e \approx -2[e_{31} - (c_{13}/c_{33})e_{33}](a_{GaN}/a_{AlN} - 1)x + P_{sp,z}^{GaN} - P_{sp,z}^{AlGaN}, \qquad (6.60)$$

where e_{31}, e_{33}, c_{13}, and c_{33} are the relevant piezoelectric and elastic constants for $Al_xGa_{1-x}N$, a_{GaN} and a_{AlN} are the lattice constants of GaN and AlN, respectively, and $P_{sp,z}^{GaN}$ and $P_{sp,z}^{AlGaN}$ are the spontaneous polarizations of GaN and $Al_xGa_{1-x}N$, respectively. Using averages of the values (Morkoç and Simin) of the spontaneous polarization fields, piezoelectric coefficients, and elastic constants, one can estimate that $\sigma_{pol} \approx (5\text{–}6.5) \times 10^{13} xe/\text{cm}^2$, where x is the Al mole fraction. Nonlinear models can be used for a more accurate estimation of the sheet charge in this manner.

There are two methods of fabricating GaN/AlGaN HFETs. The first method relies on depositing the ohmic contacts for the drain and the source and the Schottky contact for the gate directly on the AlGaN layer as shown in Fig. 6.15a. Ion implantation is not necessary to produce the n^+ region as it is in the case of GaAs/AlGaAs HFETs. This is because the metallic atoms from the ohmic contacts diffuse down to the channel. An example of this method is described by Gaska *et al*. The second method utilizes an oxide layer, such as SiO_2, Al_2O_3, and silicon oxynitrides, deposited underneath the gate metal, as shown in Fig. 6.15b. The acronym MOSHFET is given to this type of HFET, which indicates a metal oxide–semiconductor HFET. The advantage of the second method is that the gate current is reduced because of the presence of the oxide underneath the gate. The top view of an HFET is shown in Fig. 6.15c.

There are several steps involved in the fabrication of the GaN/AlGaN HFETs. The main fabrication steps include photolithography for ohmic contact openings;

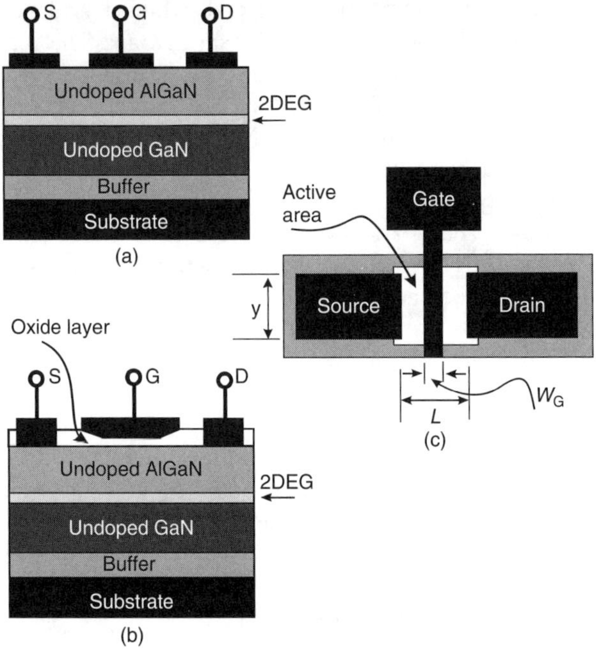

FIGURE 6.15 (a) A schematic structure of a GaN/AlGaN HFET showing the gate, drain, and source metals deposited directly on the surface of the AlGaN layer. (b) A schematic structure of GaN/AlGaN HEFT with an additional oxide layer deposited underneath the gate metal. (c) A sketch of the top view of an HFET.

ohmic contact metallization, which requires the deposition of four metal composition Ti/Al/Ti/Au; rapid thermal annealing of ohmic contacts; photolithography for device isolation level; reactive ion etching or ion implantation for electric device isolation; contact or electron beam lithography for gate openings; and gate metallization, which requires the deposition of Ni/Au, Pd/Au, or Pt/Au layers. Generally speaking, the source to drain spacing, or channel length, is $L \approx 2\mu$m and the gate length is $y \approx 5\mu$m. However, the total gate length could be as large as 50μm–200 µm. The width of the gate metal is $W_G \approx 0.2\,\mu$m

The drain current–voltage (I_D–V_D) characteristics of the GaN/AlGaN HFET are identical to those of the GaAs/AlGaAs HFET discussed in Section 6.5. A typical example is shown in Fig. 6.16 for $V_t = -5$ V and $0 \leq V_G \leq -4$ V. Notice that the gate voltage, channel current, and drain voltage were taken with respect to the source, which is usually grounded. In integrated circuits, the following notations are usually used: V_{DD} is the drain bias voltage, V_{SS} is the source voltage, V_{DS} is the drain–source voltage, I_{DS} is the drain–source current, and V_{GS} is the gate–source voltage.

There are several structural variations to the GaN/AlGaN HFETs. For example, Fig. 6.17 shows three different variations. The first structure, Fig. 6.17a, is already

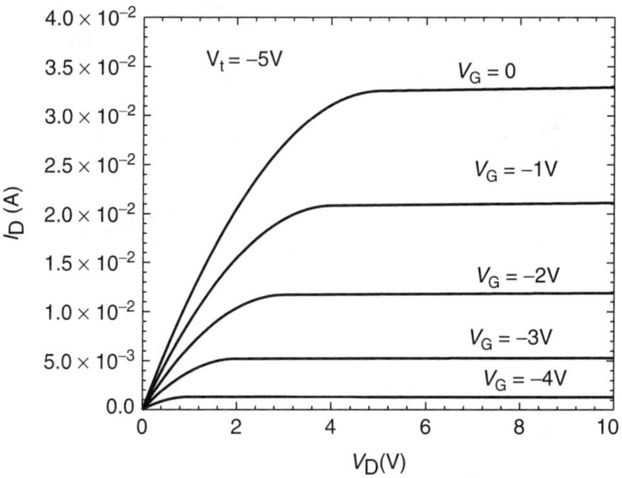

FIGURE 6.16 $I_D - V_D$ characteristic curves of a GaN/AlGaN single-channel HFET with a threshold voltage of −5 V, channel length of 0.5 μm, and gate length of 20 μm.

discussed above, and it has only one channel formed at the GaN/AlGaN interface. It is characterized by a single heterojunction as shown in the conduction energy band diagram. The double heterojunction HFET is shown in Fig. 6.17b. It is simply a single quantum well. The conduction energy band diagram shows the positive and negative charge carriers generated from the spontaneous polarization effect. A third type is shown in Fig. 6.17c, which is composed of two 2DEG channels. The 2DEG is almost doubled in this structure. Additional channels can be added to even further increase the 2DEG density.

The buffer layers in these structures vary from structure to structure. The most common buffer layer schemes are low temperature grown AlN, a thick undoped GaN layer, and GaN/AlGaN superlattices. More complicated buffers are composed of more than one scheme, such as GaN/AlGaN superlattices sandwiched between undoped GaN layers. The superlattice-GaN layer could be repeated several times to ensure the presence of high quality surfaces on which the device structure can be deposited. The analysis for multichannel HFETs is usually more complicated. For example, each channel has its own resistance.

6.7 HETEROJUNCTION BIPOLAR TRANSISTORS (HBTs)

A vertical device that has been used in many applications is called the HBT. The p-type base creates a barrier for electron diffusion from the n-type emitter to the n-type collector. One of the advantages of this electronic device is that the base thickness can be made too short and therefore the transient time for the electrons to tunnel through the base is very short. A typical example of an HBT is shown

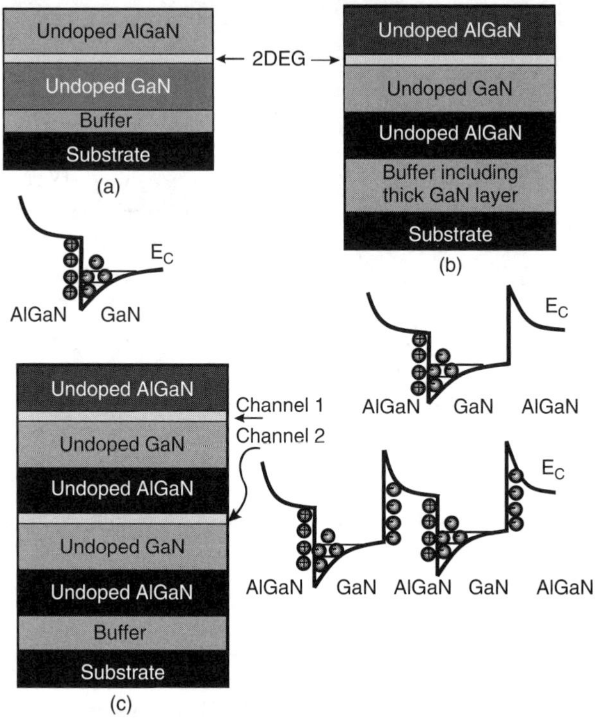

FIGURE 6.17 (a) A basic single heterojunction structure of a GaN/AlGaN HFET. (b) A double heterojunction structure. (c) Two channel structures. The conduction energy band diagram is sketched for the three structures showing the spontaneous polarization-induced charge distribution.

in Fig. 6.18, which is based on GaAs/AlGaAs heterojunctions. A sketch of the energy band diagram is also shown. The device isolation is achieved by either deep ion implantation or by making mesa structures.

Another advantage of HBTs over the regular bipolar transistors is that the junction between the emitter and base is abrupt on an atomic scale as compared to the macroscopic junction in the case of bipolar transistors. Since the base current is due to the carrier injection from the base into the emitter, the heterojunction barrier for holes reduces the base current and therefore enhances the current gain and allows a higher base doping. Higher doping in the base reduces the base resistance.

The formalism of the transport properties of HBTs is similar to that encountered in the bipolar transistor. The collector current density is given by

$$J_c = \frac{D_n e n_{po}}{W_B}\left(e^{eV_{BE}/k_B T} - 1\right), \tag{6.61}$$

where V_{BE} is the base–emitter voltage, n_{po} is the equilibrium electron density in the base (minority-carrier density), D_n is the electron diffusion coefficient, and

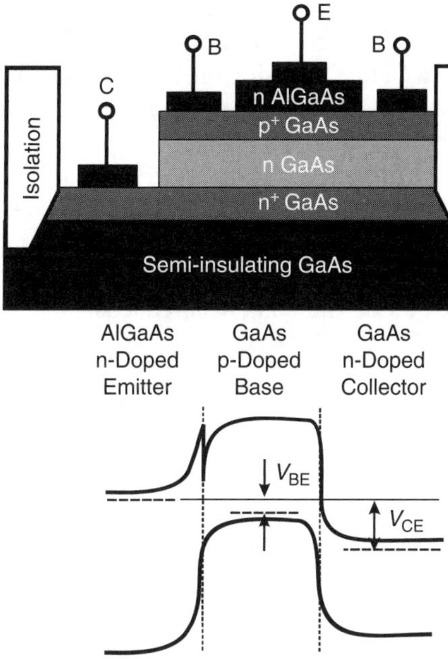

FIGURE 6.18 A schematic structure for a GaAs/AlGaAs HBT is shown with the emitter (E), base (B), and collector (C). The energy band diagram is also sketched showing the base–emitter voltage (V_{BE}) and collector–emitter voltage (V_{CE}).

W_B is the base width. For $eV_{BE} \gg k_B T$, the transconductance can be obtained as

$$g_m = \frac{\partial I_C}{\partial V_{BE}} = \frac{e}{k_B T} I_C. \tag{6.62}$$

The exponential dependence of g_m on V_{BE} produces large transconductance values, which are advantageous for digital applications.

The capacitance due to the minority charge stored in the base is given by

$$C = \frac{\partial Q_B}{\partial V_{BE}} = \frac{e}{k_B T} \frac{e n_p W_B}{2}, \tag{6.63}$$

where n_p is the minority-carrier density at the base entrance. The time required for the electron to travel through the base,τ_{tr}, can be obtained as

$$\tau_{tr} = \frac{W_B}{v_d} = \frac{W_B}{2D_n/W_B} = \frac{W_B^2}{2D_n}, \tag{6.64}$$

where the average diffusion velocity is taken as $v_d = 2D_n/W_B$. The time defined in Equation 6.64 can be drastically reduced by minimizing the base width leading to a very high speed device.

In the GaAs/AlGaAs HBTs, the aluminum mole fraction in the emitter material is usually kept to >30%. Carbon is usually used as the dopant in the p-type GaAs base material. It has a low diffusion coefficient compared to other dopants, such as zinc and magnesium. The dopant diffusion from the base to the emitter causes a significant problem in HBTs. Thus, there is a trade-off between the doping level and the performance of the device. HBTs have been fabricated from other semiconductor systems, such as Si/SiGe and InGaP/GaAs. The InGaP/GaAs system has a larger valence band offset as compared to the conduction band offset. This is an advantage since the holes see a larger barrier, and hence the hole injection from the base to the emitter is reduced.

6.8 TUNNELING ELECTRON TRANSISTORS

As we have seen above, the performance of the electronic devices is improved by the introduction of heterojunctions. The presence of potential barriers allows one to fabricate unipolar transistors in which the structure is composed of only n-type layers. The simplest device of this class is called the HET. The tunneling hot electron transistor (THET) is another variation of the HET. The conduction energy band diagrams of both HET and THET are shown in Fig. 6.19. The HET structure utilizes a graded AlGaAs barrier between the emitter and the base as shown in Fig. 6.19a. When a high base–emitter voltage is applied, it causes the electrons in the emitter to gain enough kinetic energy and be swept across the base to the collector. These are hot electrons and can ballistically transverse the base region very rapidly, allowing high frequency operation. By varying the base–collector voltage, it is possible to use the second AlGaAs barrier as an analyzer of the energy distribution of the electrons that arrive at the collector terminal.

The THET basically operates in a manner similar to HET, except that the hot electrons tunnel through the AlGaAs barrier placed between the emitter and the base. This barrier serves to inject electrons from the emitter to the base. Most of these devices operate at temperatures lower than room temperature.

A resonant tunneling hot electron transistor (RHET) was first proposed by Yokoyama *et al*. The principle of operation of the RHET is depicted in the conduction energy band diagram shown in Fig. 6.20. The structure is composed of a resonant tunneling double-barrier structure inserted between the base and the emitter. The structural design is made such that there is a single bound state in the GaAs well. The base and collector are separated by a thick AlGaAs layer. In Fig. 6.20a, the collector–emitter bias voltage, V_{CE}, is kept constant, while the base–emitter voltage, V_{BE}, is varied. This is a common emitter configuration. When V_{BE} is zero, there is no current flowing from the emitter to the base and the emitter current, I_E, is zero as illustrated in the $I-V$ curve on the right-hand side panel in the figure. The dot on the figure indicates the approximate current–voltage coordinates. A peak in the emitter current can be observed when eV_{BE} is equal to $2E$, where E is the bound energy level in the resonant tunneling structure as shown in Fig. 6.20b. For $eV_{BE} > 2E$, the emitter current

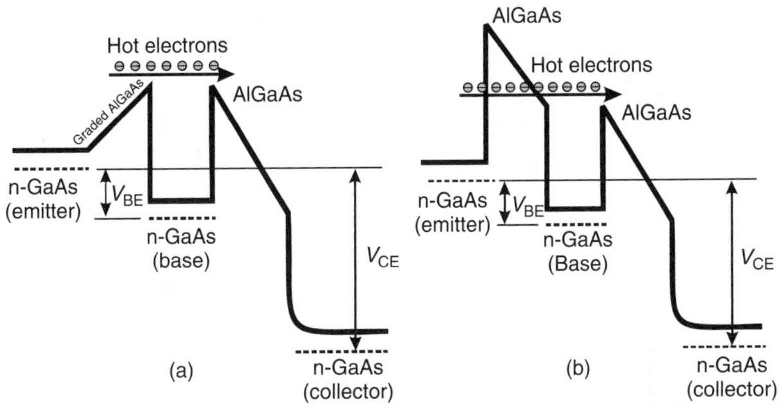

FIGURE 6.19 The conduction energy band diagram is shown for unipolar transistors. (a) Hot electron transistor with graded AlGaAs barrier and (b) a tunneling hot electron transistor.

starts to decrease as V_{BE} increases, as shown in Fig. 6.20c. The $I-V$ characteristic shown in this figure is similar to the negative differential resistance behavior encountered in the resonant tunneling diode discussed in Chapter 4.

Although the first report on RHET indicates that the device operates at cryogenic temperature, it is possible to fabricate this class of device using different semiconductor systems, such as InAs/AlSbAs, that can operate at room temperature. The application of RHETs can be realized in digital electronics and logic circuits. For example, if two inputs, A and B, are connected to the base, the output will be high (transistor is off) if both A and B are low or high. Otherwise, the output is low (transistor is on). This is an exclusive NOR logic circuit.

6.9 THE p–n JUNCTION TUNNELING DIODE

At present, tunneling diodes are used in many applications including locking circuits, low power microwave devices, and local oscillators. A typical structure of a p–n junction tunneling diode is shown in Fig. 6.21. One of the basic requirements for this homojunction is that both the n- and p-type junctions should be degenerate, which means they are heavily doped, such as the Fermi energy is pinned above (below) the conduction (valence) band minimum (maximum) as shown in the figure. The depletion region acts as the potential barrier. Carrier tunneling requires that occupied energy states exist on the side from which the electron tunnels and an empty state exists on the side to which the electron can tunnel. Both states should be at the same energy level, and tunneling can occur from n- to p-junction or vice versa.

The current of the tunneling diode is composed of three components (Sze, 1981): the tunneling, excess, and thermal currents. At thermal equilibrium, the tunneling current from the valence band to the conduction band ($I_{v \to c}$) and the

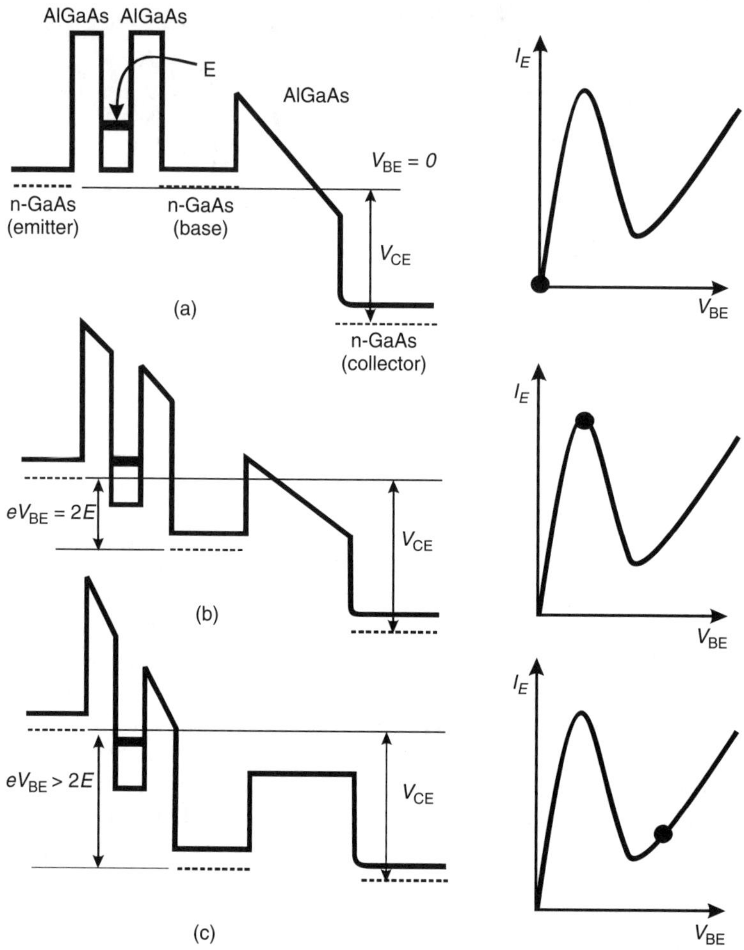

FIGURE 6.20 Conduction energy band diagram of a unipolar resonant tunneling hot electron transistor (RHET) for a constant V_{CE}. The band diagram and the $I-V$ curves were plotted for (a) $V_{BE} = 0$, (b) $eV_{BE} = 2E$, and (c) $eV_{BE} > 2E$.

current from valence conduction band to the valence band ($I_{c\to v}$) are balanced and they can be expressed as

$$I_{c\to v} = A \int_{E_c}^{E_v} f_c(E) n_c(E) T_t (1 - f_v(E)) n_v(E) \mathrm{d}E \tag{6.65a}$$

$$I_{v\to c} = A \int_{E_c}^{E_v} f_v(E) n_v(E) T_t (1 - f_c(E)) n_c(E) \mathrm{d}E, \quad \text{(b)} \tag{6.65b}$$

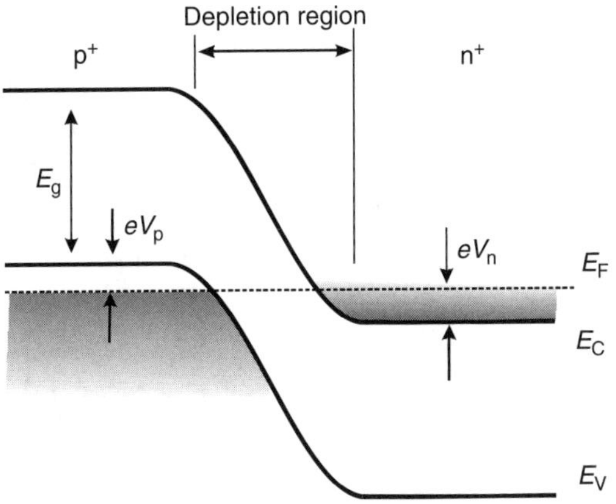

FIGURE 6.21 A sketch of a typical band diagram of a p–n junction tunneling diode.

where A is a constant, $f_c(E)$ and $f_v(E)$ are Fermi–Dirac distribution functions, $n_c(E)$ and $n_v(E)$ are the density of states in the conduction and valence bands, and T_t is the tunneling probability (transmission coefficient), which is taken as

$$T_t = \exp\left(-\frac{\pi\sqrt{m^*}E_g^{3/2}}{2\sqrt{2}e\hbar\mathcal{E}}\right)\exp\left(-\frac{2E_\perp}{\overline{E}}\right), \tag{6.66}$$

where E_g is the band gap of the semiconductor, $\mathcal{E}$ is the built-in electric field, $E_\perp$ the energy associated with the momentum perpendicular to the direction of the tunneling, and $\overline{E} = 4\sqrt{2}e\hbar\mathcal{E}/(3\pi m^* E_g^{1/2})$, which is a measure of the significant range of transverse momentum. $\overline{E}$ is usually small, which means that only electrons with small transverse momentum can tunnel. When a bias voltage is applied to the p–n junction, the observed tunneling current (I_t) is

$$I_t = I_{c\to v} - I_{v\to c} = A\int_{E_c}^{E_v}[f_c(E) - f_v(E)]T_t n_c(E)n_v(E)\mathrm{d}E. \tag{6.67}$$

The tunneling current in the above equation is derived by Demassa *et al*. and is simplified according the following expression:

$$I_t = I_0^p\frac{V}{V_0^p}\exp\left(1 - \frac{V}{V_0^p}\right), \tag{6.68}$$

where I_0^p and V_0^p are the peak current and peak voltage defined in Fig. 6.22. V_0^p was determined by Demassa *et al*. to be $V_0^p = (V_n + V_p)/3$ where V_n and V_p are the degeneracy voltages defined in Fig. 6.21.

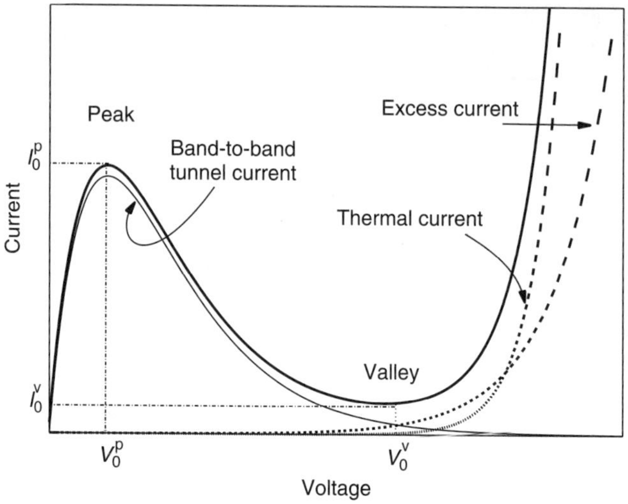

FIGURE 6.22 Static current–voltage characteristics of a typical tunnel diode. The current is composed of three components: band-to-band tunnel current, excess current, and thermal current. The sum of the three currents is shown as the thick curve with the well-known peak and valley.

The excess current is mainly due to defect-assisted tunneling, in which the carriers tunnel through defect states in the band gap. This component of the total current is usually present when the bias voltage is higher than the normal operational conditions. The access current can be understood by inspecting Fig. 6.23, where the bias voltage, V is high. Notice that V is multiplied by the electron charge (e) in order to project the voltage on the energy scale. The main process of the excess current is that an electron can drop from the conduction band (point A) to the defect state (point B) and then tunnel to the valence band (point C). Other routes are possible, but the above process is the most common route for an electron to tunnel from the conduction band to the valence band. For a bias voltage, V, the energy (E_x) that the electron must have to tunnel is given by

$$E_x \approx E_g - eV + e(V_n + V_p) \approx e(V_{bi} - V), \tag{6.69}$$

where V_{bi} is the built-in potential. The tunneling probability is essentially identical to the expression given in Appendix H, Equation H.8, except that $(\Delta E_c - E)$ is replaced by E_g

$$T_x \simeq \exp\left(-\frac{4}{3}\frac{\sqrt{2m^*}}{\hbar e \mathcal{E}} E_x^{3/2}\right) \simeq \exp\left(-\frac{\alpha_x}{\mathcal{E}} E_x^{3/2}\right), \tag{6.70}$$

where $\alpha_x \simeq 4\sqrt{2m^*}/(3\hbar e)$ and $\mathcal{E}$ is the electric field. The electric field across a step function can be written as $\mathcal{E} = 2(V_{bi} - V)/W$, where W is the depletion

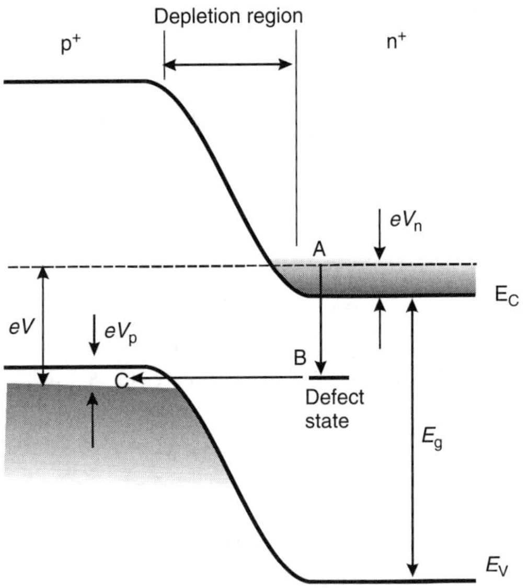

FIGURE 6.23 A sketch of the p–n junction band gap showing the defect-assisted tunneling, which causes the excess current under high bias voltage.

region width given by

$$W = \left[\frac{2\mathcal{E}_0}{e} \left(\frac{N_a + N_d}{N_a N_d} \right) (V_{bi} - V) \right]^{1/2}, \tag{6.71}$$

where ϵ_0 is the permittivity, N_a is the concentration of acceptors, and N_d is the concentration of donors.

The current density, J_x, associated with the excess current process can be written as

$$J_x \simeq A D_x T_x, \tag{6.72}$$

where D_x is the volume density of the occupied levels at energy E_x above the top of the valence band and A is a constant. Substituting Equation 6.69 through Equation 6.71 into Equation 6.72, we have (Chynoweth *et al.*)

$$J_x \simeq A_1 D_x \exp[-\alpha'_x (E_g - eV + 0.6(V_n - V_p))], \tag{6.73}$$

where A_1 is a constant. This relation shows that the excess current depends on the density of states and the applied voltage. Equation 6.73 can be rewritten as (Roy)

$$\begin{aligned} J_x &\simeq J_V \exp\left[-\frac{4}{3}\sqrt{\frac{m^* \epsilon_0}{N^*}} (V - V_0^V) \right] \\ &\simeq J_V \exp(A_2(V - V_0^V)), \end{aligned} \tag{6.74}$$

where J_V is the valley current density, V_0^V is the valley voltage, A_2 is a constant and $N^* = N_a N_d/(N_a + N_d)$. Equation 6.74 is plotted as the long dashed curve in Fig. 6.22.

Finally, the third component of the tunneling diode current density is the minority-carrier injection current given by

$$J_{th} = J_0 \left[\exp\left(\frac{eV}{k_B T} \right) - 1 \right], \tag{6.75}$$

where J_0 is the reverse saturation current density, k_B is the Boltzmann constant, and T is the temperature. The thermal current density is plotted as the short dashes in Fig. 6.22. The sum of the three currents is plotted in this figure as the thick curve which shows the characteristic peak and valley encountered in tunneling diodes.

6.10 RESONANT TUNNELING DIODES

The band gap alignment of the resonant band structure consists of two doped layers (bottom- and top-contact layers), two barriers, and one well with at least one bound energy state. When the structure is biased, the electrons tunnel from the bottom contact through the barriers, with a tunneling probability approaching unity, when the bound state in the well is resonant with the Fermi energy levels of the bottom layer. A typical structure consists of thick degenerate GaAs:Si bottom layer, 50-Å $Al_{0.3}Ga_{0.7}As$ barrier, 50-Å GaAs well, 50-Å $Al_{0.3}Ga_{0.7}As$ barrier and thick degenerate GaAs:Si top-contact layer. The structure is shown in Fig. 6.24 under zero bias. The $I-V$ characteristic of the resonant tunneling diode can be understood by examining Fig. 6.25. For small bias voltage, the bound state is assumed to be above the Fermi sea of electrons that is present in the bottom contact layer, and the tunneling through the bound state is minimum,

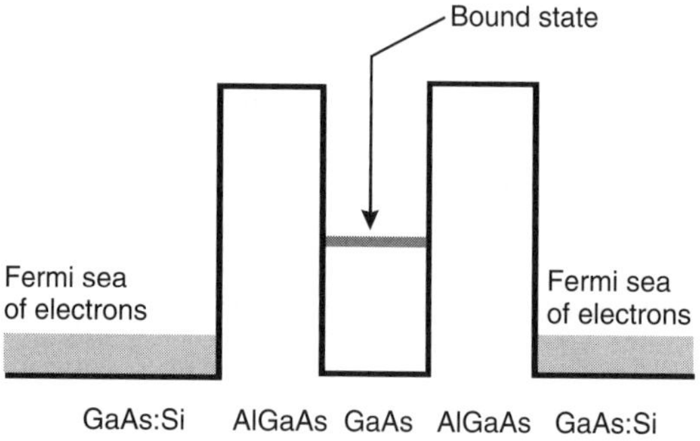

FIGURE 6.24 A sketch of a typical conduction band structure of a resonant tunneling diode is shown with the degenerate contact layers.

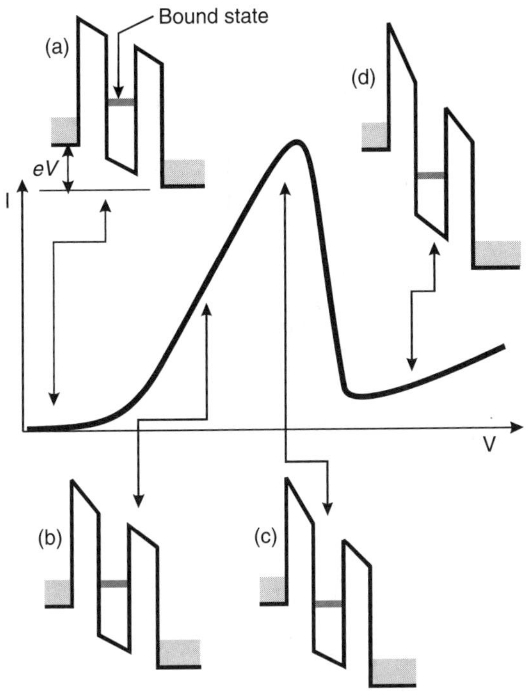

FIGURE 6.25 The $I-V$ characteristic of the resonant tunneling diode with a single bound state is sketched as a function of bias voltage (V).

case (a). As the bias voltage is increased, the bound state becomes resonant with the Fermi energy level of the contact layer and the electrons start to tunnel giving rise to current. The current continues to rise and peaks when the bound state is aligned with the bottom of the electron band as shown in case (c). The current drops abruptly as the bound state moves further down from the Fermi electron sea by increasing the applied bias voltage. If the GaAs well contains more than one bound state, the number of peaks in the $I-V$ curve will increase accordingly. The negative differential conductivity exhibited in the $I-V$ curve is very useful in amplifiers and oscillators.

Another example of resonant tunneling diodes is the InAs/AlSb double-barrier structure shown in Fig. 6.26a. The band offset is ~1.0eV. This band offset can accommodate more energy levels as compared to the GaAs/AlGaAs system. In addition, the electron effective mass in InAs is about three times smaller than that of GaAs, which leads to higher mobility and better transport properties. Four confined energy levels are shown in this figure. The tunneling of electrons occurs mostly through the ground state (E_1), but additional tunneling can occur through the excited states, giving rise to additional peaks in the $I-V$ characteristic curve as shown in Fig. 6.26b. The tunneling occurs when E_i (i = 1, 2, 3, and 4) is aligned, under bias voltage, with the Fermi sea electrons in the InAs to the left

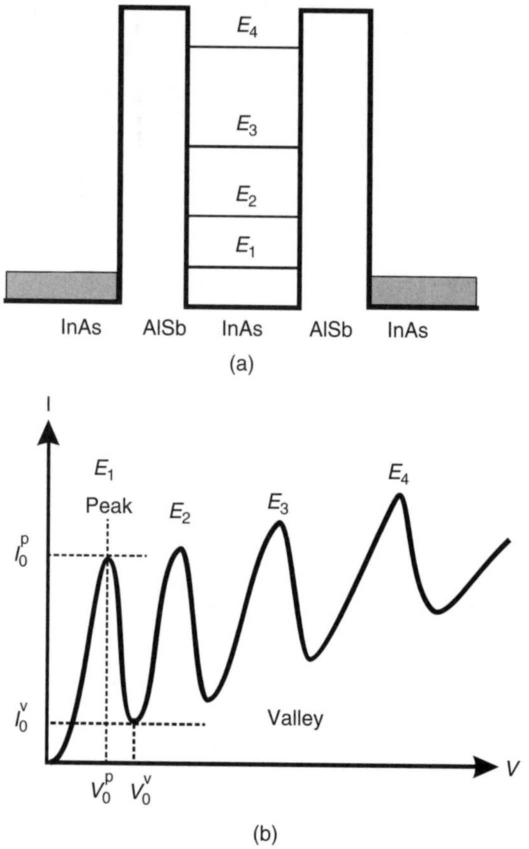

FIGURE 6.26 (a) A sketch of the InAs/AlSb double-barrier resonant tunneling diode showing four bound states. (b) A typical $I-V$ characteristic of the tunneling resonant diode showing the peak and valley voltages.

of the structure. The typical $I-V$ characteristic of the resonant tunneling diode is sketched in Fig. 6.24b, which exhibits four peaks corresponding to four bound energy levels in the well. In general, any peak voltage (V_0^{P}) should be larger than $2E_i/e$ because of voltage drops in accumulation and depletion regions.

For zero temperature and δ-function lineshape, the current density of the resonant tunneling diode is derived by Ferry as

$$J_z = \frac{e^2 V m^* \mathcal{E}_{\mathrm{w}}}{2\pi^2 \hbar^3} \mathrm{for} 2(\mathcal{E}_0 - \mathcal{E}_{\mathrm{F}}) \leq eV \leq 2\mathcal{E}_0, \tag{6.76}$$

where $\mathcal{E}_{\mathrm{w}}$ is the width of the transmission, V is the applied bias voltage, $\mathcal{E}_0$ is the energy of the ground state in the well, and $\mathcal{E}_{\mathrm{F}}$ is the Fermi energy level. Equation 6.75 is sketched in Fig. 6.27. For temperatures other than 0 K, one would expect to observe broadening in the $I-V$ curve and the current density requires to be

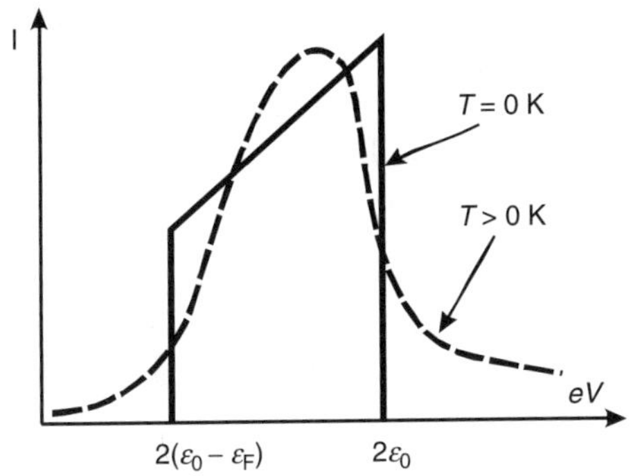

FIGURE 6.27 The $I-V$ curves obtained from the simple model derived by Ferry is plotted for both δ-function ($T = 0$ K) and Lorentzian ($T > 0$ K) lineshapes.

convoluted with a lineshape such as a Lorentzian. A sketch of the $I-V$ curve at $T > 0$ K is shown as the dashed line in Fig. 6.27.

The limitation of the resonant tunneling diode is the "valley" current as shown in Fig. 6.22. For device application, particularly in digital circuits, it is desired to have low valley current. In reality, it is difficult to achieve zero valley current, but with creative designs, one can reduce the valley current to an acceptable value. One possible design is proposed by Kitabayashi *et al*., which is based on InAs/AlSb/GaSb structure as shown in Fig. 6.28a. The structure in this figure is called *resonant interband tunneling diode*.

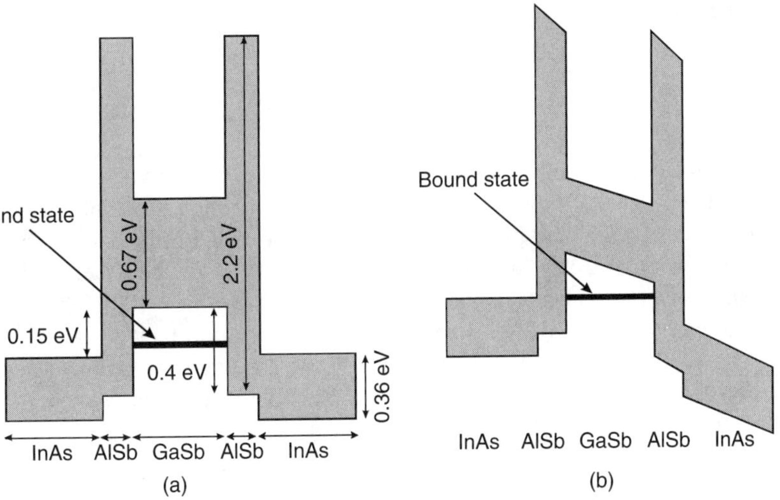

FIGURE 6.28 A schematic band-edge diagram of InAs/AlSb/GaSb,AlSb/InAs double-barrier resonant interband tunneling diode under zero bias voltage (a) and with nonzero bias voltage (b).

The bound state in this case is in the valence band of the GaSb well. The tunneling current will flow when the bound state is lined up with conduction band of the n-type InAs contact layer as illustrated in Fig. 6.28b. The electrons tunnel from the conduction band of the n-type InAs (left layer) through the bound state and then tunnel from the bound state to the conduction band of the n-type InAs on the right-hand side of the structure. With the large AlSb barriers, the peak current can be maintained at higher values by reducing the barrier thicknesses. The valley current in this structure is significantly reduced when the bound state becomes resonant with the band gap of the InAs layer on the left of Fig. 6.28b.

6.11 COULOMB BLOCKADE

For many electronic devices such as MOSFETS and bipolar transistors, the number of electrons involved in the transport is very large such that the energy quantization is irrelevant. However, the role of energy discreteness becomes increasingly very important in nanoscale devices in which the capacitance in the structure is extremely small. The capacitance of parallel plates of area A and separated by a distance d is given by

$$C = \frac{\epsilon\epsilon_0 A}{d}, \tag{6.77}$$

where ϵ is the dielectric constant of the material between the two plates and ϵ_0 is the permittivity of space. Let us consider a p–n junction made of GaAs having a cross section of $0.1 \times 0.1\ \mathrm{mm}^2$ and a thickness of 10 nm. The energy associated with one electron process is (Aleiner *et al.*)

$$E_s = \frac{e^2}{2C}. \tag{6.78}$$

Substituting Equation 6.77 into Equation 6.78, one can obtain a value of E_s to be much smaller than $k_B T$ even for temperature as low as liquid helium temperature. On the other hand, if the junction cross section is of the order $10 \times 10\mathrm{nm}^2$ and has a thickness of 20 Å, the single-electron energy is of the order of 15 meV. This single-electron energy is larger than $k_B T$ even for T = 100 K. This implies that the energy of a single-electron process is very important in nanostructures. One can think of the single-electron energy as the energy required to add one electron to the capacitor. The energy E_s is thus defined as the *charging energy*.

For any structure with a very small capacitance of the order of *atto* Farad (aF), the electrostatic potential caused by a single electron has a profound effect on the tunneling process. Generally speaking, this happened in quantum dot systems in which the transport property is regulated by the quantization of the charge in unit of the elementary charge, e, inside the nanostructure. This effect is called *Coulomb blockade*. The conditions of observing Coulomb blockade are such that the capacitance and conductance (G) of the device satisfy the following inequalities:

$$C \ll \frac{e^2}{k_B T} \quad \text{and} \quad G \ll \frac{e^2}{2\pi\hbar}, \tag{6.79}$$

where k_B is the Boltzmann constant and $e2/(2\pi\hbar)$ is called the quantum conductance (inverse of the quantum resistance). The main feature of Coulomb blockade is the total suppression of the current in a finite interval of external bias voltage such that

$$-\frac{e}{2C} < V_b < +\frac{e}{2C}, \tag{6.80}$$

where V_b is the applied bias voltage. To illustrate this process, let us examine the current (I)–bias voltage (V_b) characteristic of a thin tunnel junction as shown in Fig. 6.29. For the Coulomb blockade effect to be observed, the thermal energy should be smaller than the charging energy. When the bias voltage is zero, there is no electron tunneling through the barrier. When the bias voltage is increased, electron flow through the barrier remains zero as long as the bias voltage energy

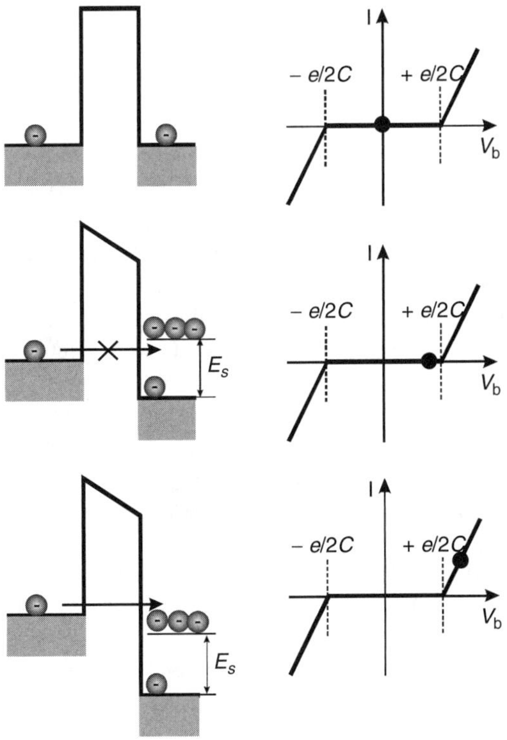

FIGURE 6.29 Illustration of Coulomb blockade in a thin junction with a small capacitance (~aF).

(eV_b) is smaller than the charging energy (E_s). Electron flow occurs when eV_b is larger than E_s as illustrated in the last panel. The current–voltage profile is shown with a solid dot indicating the values of the bias voltage. The charge quantization and Coulomb blockade effect are the basis of electronic nanostructure devices such as single-electron transistors.

6.12 SINGLE-ELECTRON TRANSISTOR

Investigating low dimensional systems has been a dominant force in recent years. The advances in semiconductor growth technology enable scientists and engineers to fabricate electronic and optoelectronic structures on a size scale in which quantum effects, such as quantization, are the dominant factors in describing the structure properties. Semiconductor quantum dots imply that they have dimensions much larger than the atomic scale and a single dot may contain a few million atoms. Electrons in these dots are tightly bound and behave like waves with de Broglie wavelengths of the order of the dot size. The number of electrons that occupy a quantum dot ranges between one and several hundreds depending on the size of the dot. The energy required to add or remove an electron from the dot is called the *charging energy*. It is analogous to the ionization energy of an atom. The atoms are investigated through their interaction with light, whereas quantum dots are investigated both by using light and by measuring their current–voltage characteristics.

The effect of a single electron was first observed by Millikan in his famous oil drop experiment in 1911, whereas electron tunneling was investigated by Guaver *et al*. (1968), and Lambe *et al*. Most of the theoretical aspects of electron charging effects and Coulomb oscillations were developed in the 1970s and 1980s. The invention of scanning tunneling microscopy (STM) had renewed the interest in Coulomb blockade since STM can both image the surface and measure the current–voltage characteristics of a single grain of size <10 nm. The lateral quantum dot defined by metallic surface gates has been widely investigated. The single-electron transistors operate at low temperature ($T<1$ K), but there are indications that these devices may operate at higher temperatures. According to Korotov *et al*., the voltage gain is less than unity for temperatures as low as $0.015e^2/(k_B C) = 27.9$ K/C (aF). Thus, the unity gain condition places a stringent requirement on either the operating temperature or the size of the capacitance, C. For $T = 300$ K, the capacitance is $0.015e^2/(300k_B) \sim 0.09$ aF. On the other hand, the lowest capacitance obtained experimentally is of the order of 0.1 fF.

The single-electron transistor describes a single electron transport through a quantum dot. There are many variations to the structure of the single-electron transistor. The main components of the single-electron structure are shown in Fig. 6.30. The island is the quantum dot, which is connected to the drain and source terminals. Electron exchange occurs only with the drain and source terminals, which are connected to current and voltage meters. The gate terminal provides electrostatic or capacitive coupling. When there is no coupling to

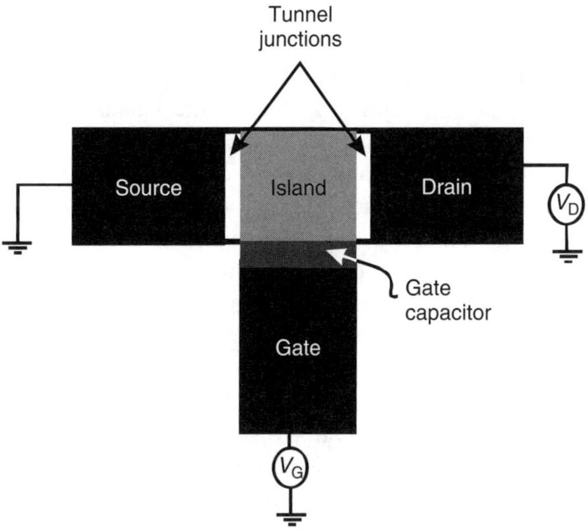

FIGURE 6.30 A sketch of a single-electron transistor showing the tunneling junctions, gate capacitor, island, and the three terminals (source, drain, and gate).

source and drain, there is an integer number, N, of electrons in the quantum dot (island). The total charge on the island is quantized and equal to eN. If tunneling is allowed between the island and drain and source terminals, the number of electrons N adjusts itself until the energy of the total system is minimized. The tunneling junctions (barriers) are made thick enough so that the electrons exist in the island, source, or drain, such that the quantum fluctuation in the number N due to tunneling through the barrier is much smaller than unity.

The electrostatically influenced electrons traveling between the source and the drain terminals need to tunnel through two junctions (barriers). The island is charged and discharged as the electrons cross it, and the relative energies of the island containing 0 or 1 extra electron depends on the gate voltage. Thus, the charge of the island changes by a quantized amount e. The change in the Coulomb energy associated with adding or removing an electron from the island is usually expressed in terms of the island capacitance, C. The charging energy, E_c, can be expressed as $E_c = e^2/C$. As mentioned in Chapter 4, this charging energy becomes important when it exceeds the thermal energy $k_B T$. The time, Δt, needed to charge or discharge the island can be expressed as $\Delta t = R_t C$, where R_t is the lower bound tunnel resistance. From the Heisenberg uncertainty principle, we have $\Delta E \Delta t = (e^2/C)(R_t C) > h$, or $R_t > h/e^2$. The quantity $h/e^2 = 25.813\text{k}\ \Omega$ is called the *quantum resistance* or *quantum conductance* ($G = 38.74\ \mu\text{S}$). Thus, two conditions must be met to observe the charge quantization:

$$R_t \gg \frac{h}{e^2} \text{ and } \frac{e^2}{C} \gg k_B T. \tag{6.81}$$

The capacitance can be made small by reducing the quantum dot size since $C = 4\pi\epsilon_s R$ for a sphere and $C = 8\pi\epsilon_s R$ for a flat disc, where R is the radius and ϵ_s is the permittivity of the material.

The gate voltage, V_g, is applied to change the island electrostatic energy in a continuous manner. The total gate voltage-induced charge on the island is expressed as $q = C_g V_g$. This charge is considered continuous. By sweeping the gate voltage, the buildup of induced charge will be compensated for in periodic interval because of the tunneling of discrete charges. The competition between the induced charges and the discrete compensation leads to so-called Coulomb oscillations. A typical example of these oscillations is shown in Fig. 6.31 where the conductance is measured as a function of the gate voltage for a fixed drain–source voltage. The sequence of the number of electrons residing in the dot (island) is $N \rightarrow N+1 \rightarrow N \rightarrow N+1 \rightarrow N \ldots$.

Consider that the gate voltage is fixed for the single-electron transistor, while the drain–source voltage is varied. The current–voltage results exhibit a staircase-like behavior known as a *Coulomb staircase*. Figure 6.32 illustrates the Coulomb staircase behavior. A simple capacitor–resistor circuit is shown in Fig. 6.32a, where the current as a function of applied voltage is a straight line and lacks the charge quantization. In Fig. 6.32b, we sketched a circuit that presents a single-electron transistor. The drain and source were presented as resistors and capacitors. In this circuit, the gate voltage is fixed, while the drain–source voltage is varied. The drain current–source characteristic curve is shown as the steplike curve (gray line) in Fig. 6.32c. The black thin line in this figure is plotted for higher-resistive tunnel junctions and lower temperature. Notice that the drain voltage interval is e^2/C, where C is the sum of gate, drain, and source capacitances. The steplike behavior in Fig. 6.32c is called the *Coulomb staircase*.

The basic operation of the single-electron transistor is shown in Fig. 6.33, where the tunneling junctions are presented as barriers. The island is represented as the well between the two barriers. The structure is symmetrical, where the source and the drain can be exchanged without losing the transistor characteristic.

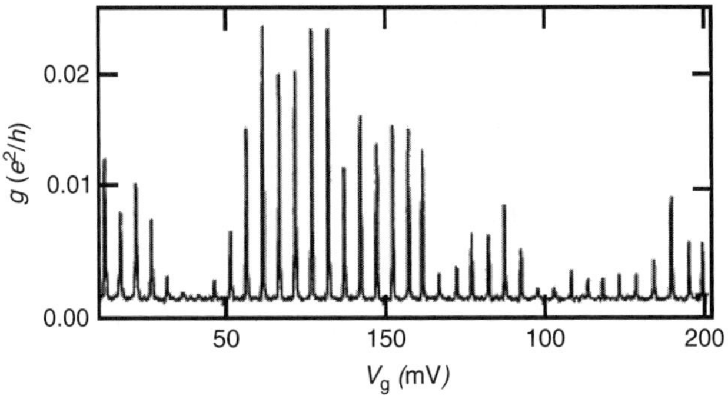

FIGURE 6.31 The conductance, g, as a function of the gate voltage, V_g, measured for a GaAs quantum dot showing Coulomb oscillations (after Folk et al.).

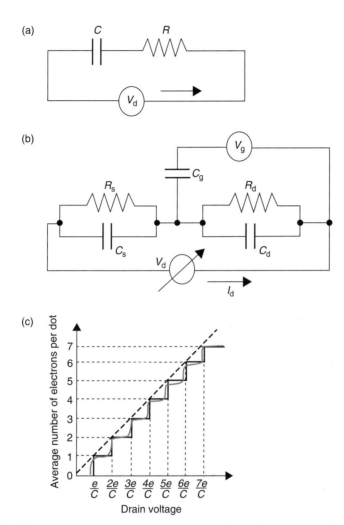

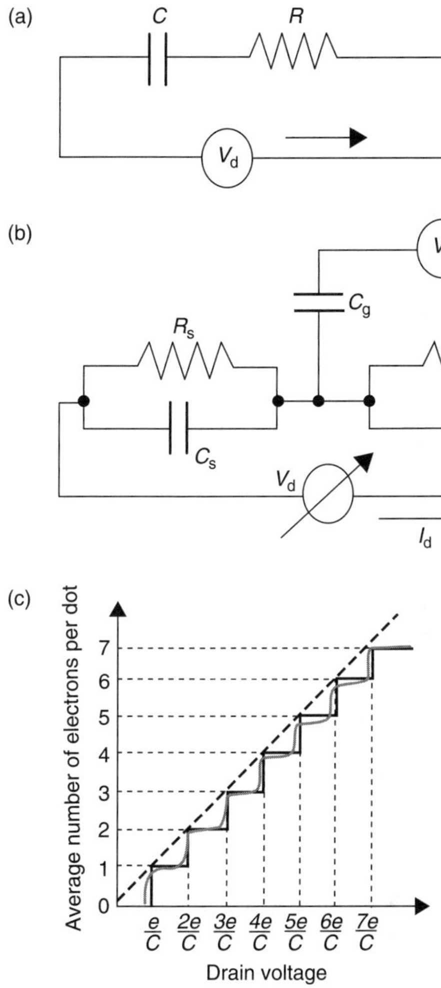

FIGURE 6.32 (a) A resistor–capacitor circuit with a bias voltage V_d. The DC current–voltage characteristic is shown as the dashed line in (c), which lacks charge quantization. (b) A representation of a single-electron transistor showing the resistances and capacitances of the drain and source. A fixed gate voltage is applied. (c) The drain current–voltage characteristic (gray line) of the circuit is shown in (b). The steplike line is for more resistive tunnel junctions and at lower temperature.

A small drain–source voltage is applied as shown in Fig. 6.33a. The solid lines in the island represent the occupied energy levels, while the dashed lines represent the empty energy levels. The energy required to move one electron from the full top energy level to the bottom empty level in the island can be derived as follows. The Fermi Energy levels (chemical potentials) of the island (dot), the drain, and the source are shown in the figure as E_F^{Dot}, E_F^{D}, and E_F^{S}, respectively.

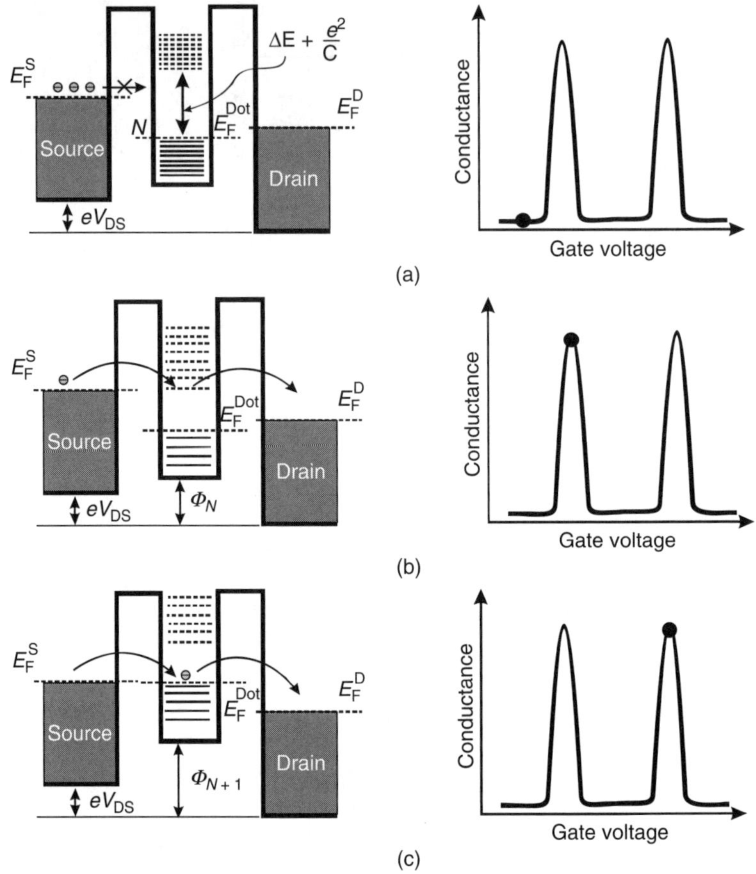

FIGURE 6.33 (a) A sketch of the single-electron transistor when Coulomb blockade exists. (b) An electron tunnel when the bottom energy level is aligned with E_F^S. (c) An electron tunnels when E_F^{Dot} is aligned with E_F^S. Thus, the sequence of electron tunneling is $N \rightarrow N+1 \rightarrow N \rightarrow N+1 \ldots$.

When a drain–source bias voltage is applied, these Fermi energy levels are no longer aligned as shown in the figure. At zero temperature (zero thermal energy), the current is zero when the gap between the bottom empty state and the top full state is aligned with E_F^D and E_F^S. This is called *Coulomb blockade*. The minimum energy required to add an electron to the dot is $E_F^{Dot} = E_N - E_{N-1}$, where N is the total number of electrons in the dot (island). For the linear response time, E_F^{Dot} can be written as (Kouwenhoven *et al.*).

$$E_F^{Dot}(N) = E_N + \frac{e^2(N - N_0 - 1/2)}{C} - e\frac{C_g}{C}V_G, \tag{6.82}$$

where N_0 is the number of electrons in the island at $V_G = 0$, N is the number of electrons in the dot at gate voltage V_G, and $C = C_g + C_s + C_d$. The energy E_N

is the single-particle energy for the Nth electron measured from the bottom of the conduction band. When an electron is added to the island at a constant gate voltage, the Fermi energy becomes

$$E_{\text{F}}^{\text{Dot}}(N+1) = E_{N+1} + \frac{e^2(N - N_0 + 1/2)}{C} - e\frac{C_{\text{g}}}{C}V_{\text{G}}. \tag{6.83}$$

By taking the difference between Equations 6.66 and 6.67, one can find that

$$E_{\text{F}}^{\text{Dot}}(N+1) - E_{\text{F}}^{\text{Dot}}(N) = E_{N+1} - E_N + \frac{e^2}{C} = \Delta E + \frac{e^2}{C}, \tag{6.84}$$

where $\Delta E = E_{N+1} - E_N$ is the energy separation between the energy levels in the quantum dot. Thus, the energy gap between the filled energy levels (solid lines in Fig. 6.33) and empty energy levels (dashed lines in Fig. 6.33) in the dot is composed of the energy separation between the energy levels that exist below $E_{\text{F}}^{\text{Dot}}(N)$ and the e^2/C. Notice that the e^2/C is a many-body contribution, and it exists only at the Fermi energy level. Below the Fermi energy levels, the separation between the energy levels is only ΔE. If ΔE is too small, the energy gap between the filled (solid lines) and empty (dashed lines) states is approximately e^2/C. The energy levels in the dot can be adjusted by applying a gate voltage as shown in Fig. 6.33b,c such that the electron can tunnel from the source to the drain. The conductance versus gate voltage is shown in the curve on the right in Fig. 6.33, where the dot on the curve represents the conductance and gate voltage coordinates. The minima of the conductance correspond to the presence of the Coulomb blockade.

Another useful example is shown in Fig. 6.34 in which a silicon-based single-electron transistor is sketched. The configuration of the single-electron transistor consists of a tellurium (Te)-doped silicon material with SiO_2 between the aluminum and Si:Te. The aluminum layer forms the gate, source, and drain terminals. The gaps between the gate and drain and gate and source form the tunneling barriers. The island is formed under the gate metal. Contrary to the above model in which the energy levels in the island are assumed full and empty with an energy gap of $\Delta E + e^2/C$, the energy levels in the island in Fig. 6.34 are similar to the regular confined energy levels in a quantum well.

An electron tunnels from the source to the drain via the island when any of the energy levels in the island is aligned with the Fermi energy levels in the source. A small drain–source voltage (V_{DS}) is applied such that the Fermi energy levels in both the drain and the source are almost the same. Thus, Coulomb blockade exists only when the Fermi energy level in the source is aligned between the energy levels in the island. The bottom of the potential in the island is adjusted by applying a gate voltage. When the gate voltage is increased, it forces one of the electrons on the Te atom to come close to the gate metal, pushing the bottom of the island upward as illustrated in Fig. 6.34. The conductance is increased when electrons tunnel from the source to the gate as shown in the curves plotted on

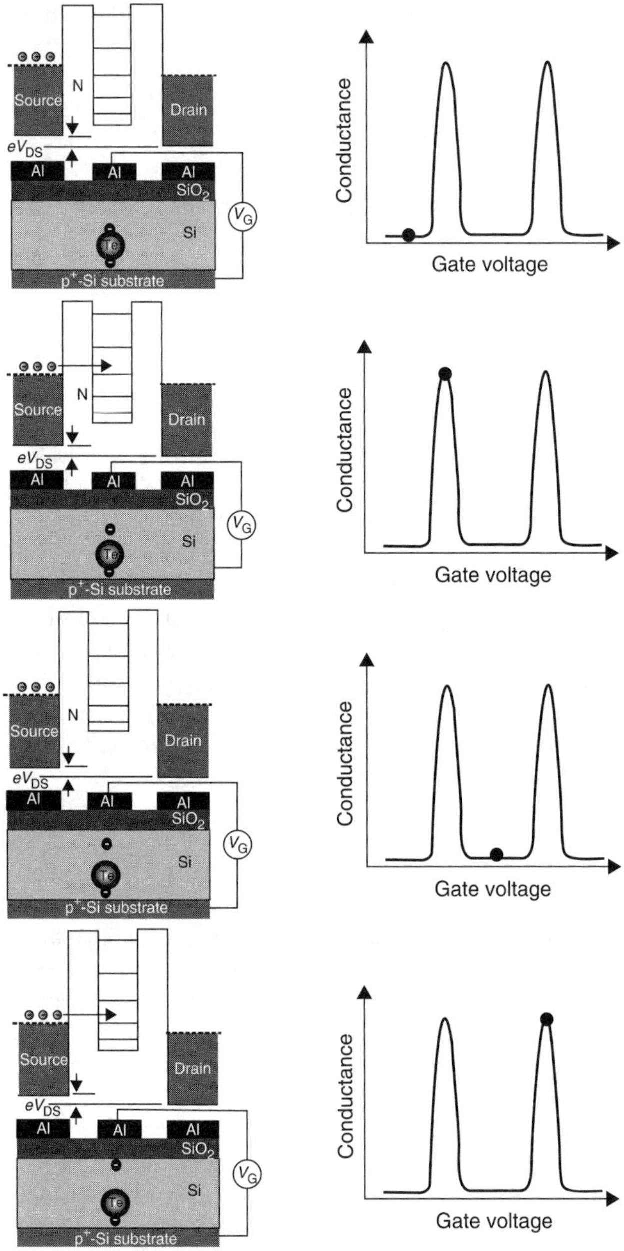

FIGURE 6.34 A schematic of single-electron transistor based on Si:Te. The conductance as a function of the gate voltage at a fixed drain–source voltage is shown on the right-hand side where the dot represents the conductance–gate voltage coordinates.

the right-hand side of the figure. To emphasize, the above description of single-electron tunneling is valid when the two conditions stated in Equation 6.81 are satisfied (Averin and Likharev).

Inverter and complimentary circuits based on single-electron transistors have been reported in the literature (Korotkov *et al.*). A typical example of an inverter is shown in Fig. 6.35. The V_{out} as a function of V_{in} is also shown in the figure with $C = C_g + C_s + C_d$, $C_s = C_d$, $C_g = 8C_d$; $R_s = R_d = R_t$; and $V_{DD} \sim e/2C$. The maximum current gain (Timp) is approximately $1/(\omega C_G R_t)$, which could be very large depending on the frequency. The maximum theoretical voltage gain, A_v, is the slope of the falling portion of the sawtooth of the V_{out} curve as a function of V_{in}, which can be written as $A_v = -C_G/C_s = -8$. In practice, this gain is much smaller.

The charge transport in a single-electron transistor is based on the quantum ballistic electron transport, which is briefly discussed in Chapter 7. This means that the single-electron transistor is a mesoscopic device in which the coherence length is larger than the device dimension. The theory of ballistic transport is

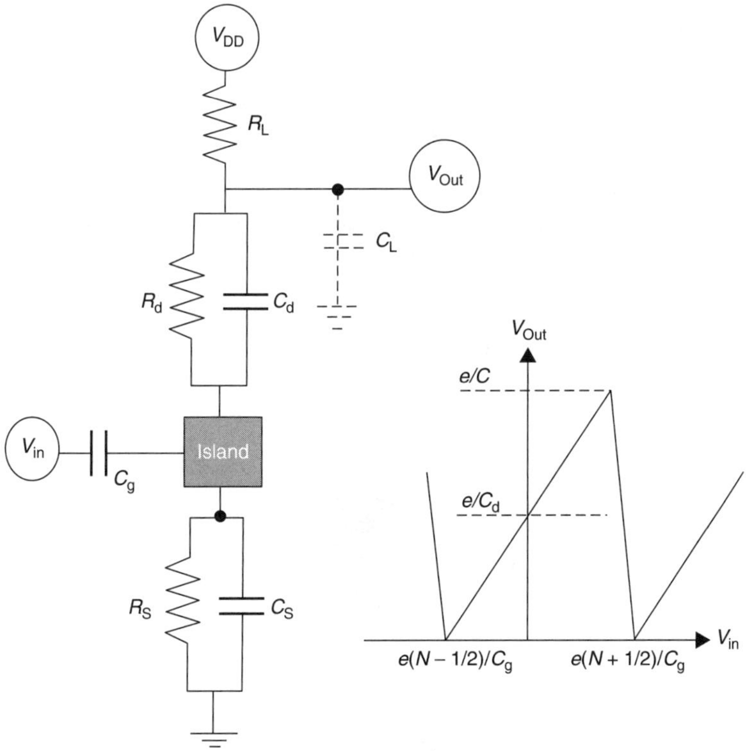

FIGURE 6.35 An illustration of a single-electron transistor as an inverter. The load is chosen as a resistance, but it can be an enhanced mode transistor. The V_{out} as a function of V_{in} is plotted as sawtooth curve on the right-hand side.

too complicated to be presented in this book. For a single-channel transport, the Landauer formula for the device conductance is derived in Chapter 7 and is given as

$$G = \frac{I}{V} = \frac{2e^2}{h} T(E_F), \tag{6.85}$$

where $T(E_F)$ is the electron transmission probability, E_F is the Fermi energy level, and the factor 2 is due to the spin degeneracy. This formula relates the quantum transmission probability directly to the conductance of the device. For maximum conductance, where $T(E_F) = 1$, we have G = 77.44 μS per spin.

The single-electron transistor can be considered as an electron waveguide which is schematically shown in Fig. 6.36. The gate voltage modulates the depletion region such that the penetration length, L^*, of the electron wave in the gate arm is changed. Thus, the electron transmission between the source and the drain is affected by changing the length L^*. The drain current–voltage (Sols *et al.*) can be written as

$$I_D = \frac{2e^2}{h} T(E_F) V_D. \tag{6.86}$$

The transconductance of the device is obtained by taking the first derivative of the drain current with respect to the gate voltage such that

$$g_m = \frac{\partial I_D}{\partial V_G} = \frac{2e^2}{h} V_D \frac{\partial T(E_F)}{\partial V_G}. \tag{6.87}$$

The electron transmission probability is affected by the length L^*, which is modulated by the gate voltage. Thus, the transmission probability is also a function of L^* or $T(E_F, L^*)$, which means that Equation 6.86 can be rewritten as

$$g_m = \frac{\partial I_D}{\partial V_G} = \frac{2e^2}{h} V_D \frac{\partial T(E_F, L^*)}{\partial L^*} \frac{\partial L^*}{\partial V_G}. \tag{6.88}$$

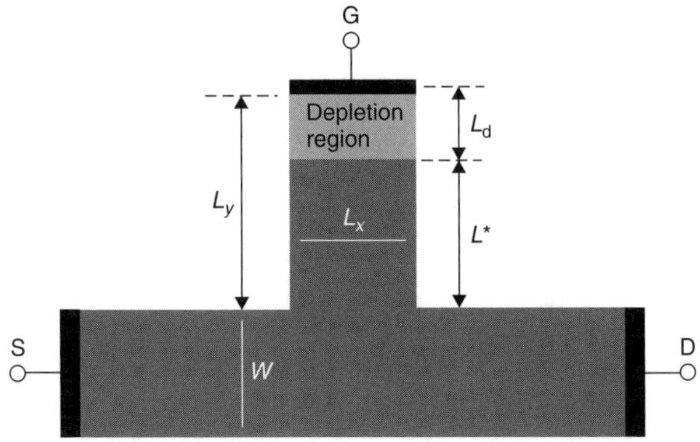

FIGURE 6.36 A planar view of a single-electron transistor as an electron waveguide.

The task now is to evaluate the terms $\frac{\partial T(E_F,L^*)}{\partial L^*}$ and $\frac{\partial L^*}{\partial V_G}$. The first term can be obtained by utilizing the tunneling probability reported by Glazman *et al*. for adiabatic constriction

$$T(E_F, L^*) = \frac{1}{1 + \exp\left\{-\left(\frac{k_F L^*}{\pi} - n\right)\pi^2\sqrt{2R/L^*}\right\}}, \tag{6.89}$$

where k_F is the Fermi wave vector, n is the energy level quantum number, and R is the curvature of the restriction. Using Equation 6.89, Timp *et al*. obtained the following expression

$$\frac{\partial T(E_F, L^*)}{\partial L^*} = \frac{0.35}{\sqrt{\left(\frac{\mathcal{N}_s}{10^{12}\,cm^{-2}}\right)}}\text{nm}^{-1}, \tag{6.90}$$

where $\mathcal{N}_S$ is the density of the 2DEG in cm^{-2}. The second term, $\frac{\partial L^*}{\partial V_G}$, can be evaluated according to the following procedure (Sols *et al*.). The gate capacitance can be defined as

$$C_G = \left|\frac{\partial Q}{\partial V_G}\right| = e\mathcal{N}_s L_x \left|\frac{\partial L_d}{\partial V_G}\right|, \tag{6.91}$$

where the lengths L_x and L_d are defined in Fig. 6.36 and $\mathcal{N}_s$ is the 2DEG density. From Fig. 6.36, one can write $L_y = L_d + L^* = \text{constant}$, which yields

$$\frac{\partial L^*}{\partial V_G} = -\frac{\partial L_d}{\partial V_G}\text{or}\left|\frac{\partial L^*}{\partial V_G}\right| = \frac{C_G}{en_s L_x}. \tag{6.92}$$

Substituting Equation 6.76 into the transconductance Equation (6.72) yields

$$g_m = \frac{2e^2}{h}V_D\frac{\partial T(E_F, L^*)}{\partial L^*}\frac{\partial L^*}{\partial V_G} = I_{Do}\frac{C_G}{e\mathcal{N}_s L_x}\frac{\partial T(E_F, L^*)}{\partial L^*}, \tag{6.93}$$

where $I_{Do} = 2e^2 V_D/h$ is the drain current at $T(E_F, L^*) = 1$.

Using the definition of the cut-off frequency, f_T, given by Equation 6.39, one can write the cut-off frequency of the single-electron transistor as

$$f_T = \frac{|g_m|}{2\pi C_G} = I_{Do}\frac{1}{2\pi e\mathcal{N}_s L_x}\left|\frac{\partial T(E_F, L^*)}{\partial L^*}\right|. \tag{6.94}$$

The carrier density, $\mathcal{N}_S$, can be written as

$$\mathcal{N}_s = \frac{\alpha}{W^2}, \tag{6.95}$$

where W is given in Fig. 6.36 and α is a scaling constant that relates the Fermi energy levels to the zero-point energy, E_0. According to Sols et al. ., $\alpha = E_F/E_0$

and $E_0 = \hbar^2\pi^2/(2m * W^2)$, which is related to higher subband energies, E_n, such that $E_n = (n^2-1)E_0$. Substituting Equation 6.80) into Equation 6.79 yields

$$f_\text{T} = 7.693 \times 10^{13}\frac{V_\text{D}}{\alpha\beta}W\left|\frac{\partial T(E_\text{F}, L^*)}{\partial L^*}\right| \text{Hz}, \tag{6.96}$$

where $\beta = L_x/W$. For $V_\text{D} = 30$ mV, $\alpha = 1.5, \beta = 2$, and $W\left|\frac{\partial T(E_\text{F},L^*)}{\partial L^*}\right| = 1$; the cut-off frequency is 1.15 THz. This frequency is very high as compared with those of HFETs and HBTs.

The term $\frac{\partial L^*}{\partial V_\text{G}} = -\frac{\partial L_\text{d}}{\partial V_\text{G}}$ can be estimated by taking the first derivative of the depletion width (L_d) with respect to the gate voltage. The depletion length can be written for the single-electron transistor in a manner similar to Equation 6.25 by neglecting the drain voltage, that is, $V(y) \approx 0$, or

$$L_\text{d} = \sqrt{\frac{2\epsilon_s(V_\text{bi} + V_\text{G})}{eN_\text{d}}}, \tag{6.97}$$

where the stub in Fig. 6.36 is terminated by a Schottky barrier and V_bi is the built-in voltage. Taking the derivative of Equation 6.82 with respect to V_G yields

$$\left|\frac{\partial L^*}{\partial V_\text{G}}\right| = \left|\frac{\partial L_\text{d}}{\partial V_\text{G}}\right| = \frac{1}{2}\sqrt{\frac{2\epsilon_s}{eN_\text{d}}}\frac{1}{\sqrt{(V_\text{bi} + V_\text{G})}}. \tag{6.98}$$

If one assumes that $N_\text{d} = 10^{18}$ cm^{-3}, this formula can be rewritten as

$$\left|\frac{\partial L^*}{\partial V_\text{G}}\right| = \frac{189.6}{\sqrt{(V_\text{bi} + V_\text{G})}}\text{Å/V}, \tag{6.99}$$

where V_G and V_bi are given in volts. Thus, the stub length, L^*, can be modulated or controlled by the gate voltage. This implies that the basic operation of the single-electron transistor is determined by the control of the interference patterns of conducting electrons using an external gate voltage. The interference patterns are generated from traveling electrons in two or more channels.

To illustrate the formation of the interference patterns, consider the device shown in Fig. 6.37a. This device is simply an HFET with a channel short enough to permit ballistic transport without scattering, that is, the coherence length is larger than the channel length. An embedded barrier is introduced parallel to the current, which splits the main channel into two. The symmetry of channels 1 and 2 is distorted when a gate voltage is applied. The widths of the channel and the split channels are small enough that the electron energy levels are quantized in the z-direction. The first two subband energy levels are shown in Fig. 6.37b. Notice that the energy levels in the split channels are displaced from each other to indicate the broken symmetry between channels 1 and 2.

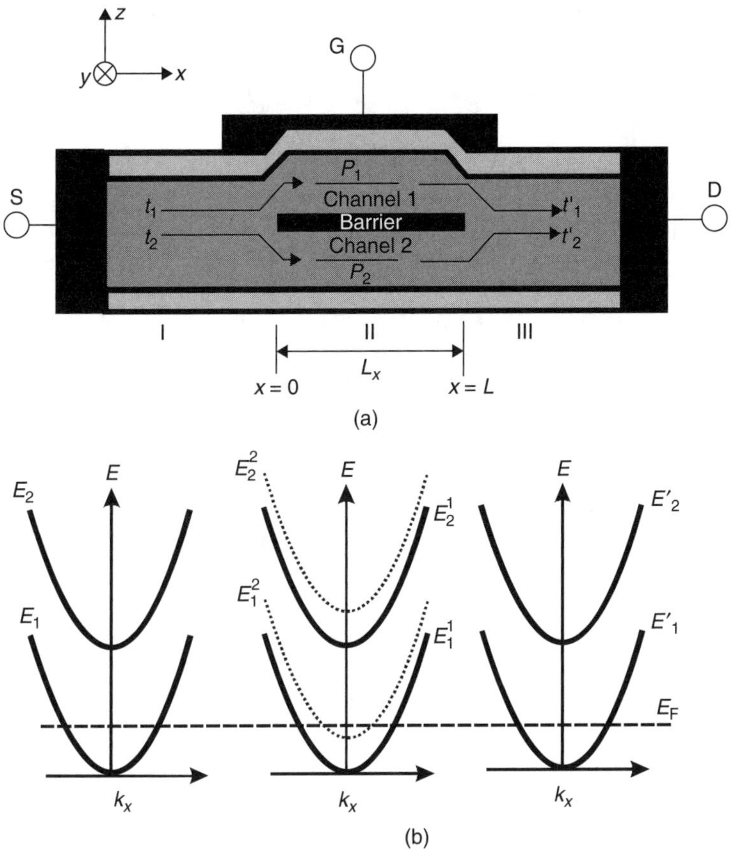

FIGURE 6.37 (a) A sketch of a single-electron transistor is shown with a barrier forming two electron channels. The parameters are defined in the text. (b) The electron subband energy levels are sketched for three different regions of the channel. The energy levels of the two channels in region II are displaced.

If an electron is injected from the source to the gate, its wave functions in the three regions (regions I, II, and III defined in Fig. 6.37a) of the channel can be written according to Mitin *et al*. as

$$\psi = \begin{cases} u_{\mathrm{I}}(z)\mathrm{e}^{i(k_x x + k_y y))} & \text{for region I} \\ t_1 u_{\mathrm{II}}^1(z)\mathrm{e}^{i(k_x(x-L)+k_y y))} & \text{for region II (channel1)} \\ t_2 u_{\mathrm{II}}^2(z)\mathrm{e}^{i(k_x(x-L)+k_y y))} & \text{for region II (channel2)} \\ (t_1 P_1 t_1' + t_2 P_2 t_2')u_{\mathrm{III}}(z)\mathrm{e}^{i(k_x(L-x)+k_y y))} & \text{for region III,} \end{cases} \tag{6.100}$$

where $u_{\mathrm{I}}(z)$, $u_{\mathrm{II}}^1(z)$, $u_{\mathrm{II}}^2(z)$, and $u_{\mathrm{III}}(z)$ are the wave functions of the subbands in their perspective regions. The quantities $\mathrm{t}_1, \mathrm{t}_2, t_1'$, and t_2' are the amplitude of the split waves in region II and interfering waves in region III, and the quantities P_1

and P_2 are the factors of the different phase shifts in the split channels (region II), and they can be taken as

$$P_1 = e^{ik_{x1}L} \quad \text{and} \quad P_2 = e^{ik_{x2}L}., \tag{6.101}$$

where k_{x1} and k_{x2} are the wave vectors for channel 1 and channel 2, respectively. If multiple reflections between the source and drain are neglected and $u_{\text{I}}(z), u_{\text{II}}^1(z), , u_{\text{II}}^2(z),$, and $u_{\text{III}}(z)$ functions in Equation 6.100 are normalized to unity, the transmission probability coefficient, $T(E)$, can be written as

$$T(E) = 2(t_1 t_1')^2 (1 + \cos\theta), \tag{6.102}$$

where θ is the relative phase shift in both channels and is given as

$$\theta = (k_{x1} - k_{x2})L. \tag{6.103}$$

Notice that the channel is assumed to be symmetrical around $z = 0$ where the wave amplitudes satisfy the conditions $t_1 = t_2$ and $t_1' = t_2'$ for zero applied gate voltage.

If the average electron velocity, υ_x, in the split channels (region II in Fig. 6.37a) is expressed as

$$\upsilon_x = \frac{\hbar(k_{x1} + k_{x2})}{2m^*}, \tag{6.104}$$

the relative phase shift can be written as

$$\theta = \frac{L}{\upsilon_x} \frac{(E_1^1 - E_1^2)}{\hbar}, \tag{6.105}$$

where L is the length of the channel as shown in Fig. 6.37a, and E_1^1 and E_1^2 are the ground energy levels for channels 1 and 2, respectively. Notice that the superscripts 1 and 2 are introduced to indicate channels 1 and 2. It is clear from Equation 6.105 that if the ground states of the two channels are different, that is, $E_1^1 \neq E_1^2$, then different phases of waves traveling from the source to the gate exist. The difference in the phases causes the quantum interference.

For a zero gate voltage, the ground states of the two channels in region II of Fig. 6.37a are equal ($E_1^1 = E_1^2 = E_0$), which leads to a zero phase shift. When a nonzero gate voltage is applied, the potential energy, $U(z)$, changes according to the following relation:

$$U(z) = U_0(z) - e\varphi(z), \tag{6.106}$$

where $U_0(z)$ is the potential energy at zero gate voltage and $\varphi(z)$ is the potential induced by the applied gate voltage. This induced potential causes a shift to the subband energy levels in the two channels according to the following relations:

$$E_1^1 = E_0 - e\langle u_{\text{II}}^1|\varphi(z)|u_{\text{II}}^1\rangle$$
$$E_1^2 = E_0 - e\langle u_{\text{II}}^2|\varphi(z)|u_{\text{II}}^2\rangle. \tag{6.107}$$

Substitute Equation (6.91) into Equation (6.89) to obtain

$$\theta = \frac{L}{\upsilon_x}\frac{e\varphi_{12}}{\hbar}, \tag{6.108}$$

where φ_{12} is given by

$$\varphi_{12} = \langle u_{\text{II}}^2|\varphi(z)|u_{\text{II}}^2\rangle - \langle u_{\text{II}}^1|\varphi(z)|u_{\text{II}}^1\rangle. \tag{6.109}$$

Thus, the electron transmission in the device is determined by the value of φ_{12}.

For the case of the device described in Fig. 6.37, the Landauer formula can be obtained by combining Equations 6.85 and 6.102 to obtain the following expression for the transconductance:

$$G = \frac{4e^2}{h}(t_1 t_1')^2(1 + \cos\theta).. \tag{6.110}$$

The conductance is an oscillatory function of the relative phase shift. The transconductance, g_{m}, is maximum when $\theta = n\pi$, where $n = 0, 2, 4, \ldots$ and zero when $\theta = \text{m}\pi$ where $m = 1, 3, 5, \ldots$ When θ is zero, Equation 6.108 implies that φ_{12} is also zero. But when $\theta = \pi$, then $e\varphi_{12} = \hbar\pi\upsilon_x/L$, which corresponds to destructive interference.

The transit time, τ_{tr}, required for the electron to cross the channel of length L can be defined as $\tau_{\text{tr}} = L/\upsilon_x$, which leads to a cut-off frequency similar to the frequency form given by Equation 6.41 and is given by

$$f_{\text{T}} = \frac{1}{2\pi\tau_{\text{tr}}} = \frac{\upsilon_x}{2\pi L}. \tag{6.111}$$

For a device with a gate length of $L = 100$ Å and an average velocity of $\upsilon_x = 10^7$ cm/s, the cut-off frequency is ~1.6 THz. This frequency is easily achieved in the single-electron transistor, provided that the channel is undoped, the transport is ballistic, and the channel is made too short (shorter than the coherence length).

6.13 SUMMARY

In this chapter, we have mainly discussed the electronic devices (transistors) based on heterojunctions, quantum wells, and quantum dots. From the discussion in this chapter, one can conclude that there are three generations of electronic devices. The first generation includes the MESFET and JFETs. These devices are based on doped semiconductors, and therefore, they are the slowest of the three generations. The presence of dopants degrades the mobility and transport properties, and therefore the cut-off frequency is low. The second generation of field-effect transistors is based on quantum wells, in which the channel of the device is the two-dimensional electron gas formed at the interfaces of the heterojunctions or in quantum wells. An example of this class of devices is the heterojunction field-effect devices, such as GaAs/AlGaAs and GaN/AlGaN HFETs. The frequency of these devices can reach the GHz range. This high cut-off frequency is due to the high mobility of the two-dimensional electron gas as compared to bulk mobility. Most of the HFETs contain a two-dimensional electron gas with mobility larger than 10^5 $cm^2/V/s$. The GaN/AlGaN HFETs were discussed in more detail because of their potential use in high frequency, high temperature, and high power applications.

The HBTs were also discussed since they can operate in the GHz frequency range. Another class of devices based on tunneling effects, such as tunneling electron transistors, resonant tunneling transistors, and HETs, were discussed briefly. The latter class of devices is unipolar, which means they are based only on n-type materials. The disadvantage of this class of transistors is that most of them operate at a temperature lower than room temperature, but their high cut-off frequency and potential for digital applications have made them attractive devices to investigate.

Tunneling electron transistors were briefly discussed including HETs and THETs. These transistors are unipolar and operate at low bias voltage. A p–n junction tunneling diode was discussed in this chapter as an example of the tunneling process. A discussion was presented for the resonant tunneling diodes, which are important devices in microwave applications. A variation of the resonant tunneling diodes was discussed.

The third generation of the electronic devices discussed in this chapter is the single-electron transistor, which consists of an island or quantum dot, drain, source, gate, gate capacitance, and two barriers. The single-electron transistors operate in the THz frequency range, provided that the temperature is <1.0 K. The limiting factor of this operational temperature is the gate capacitance. Significant research is currently being conducted on the single-electron transistors for their applications in quantum and molecular computing.

While the theory of single-electron transistors is very complicated, we presented simplistic ideas on how these devices operate. The operation of single-electron transistors is based on charge quantization. To observe this charge quantization, two important conditions must be met. First, the lower bound tunneling resistance should be much larger than the quantum resistance (h/e^2) and second, the charging energy (e^2/C) must be much larger than the thermal energy $(k_B T)$.

PROBLEMS

6.1. Calculate the depletion width at the interface of a nickel film deposited at the top of an n-type doped GaAs film with a dopant concentration of

$5 \times 10^{18}/\text{cm}^3$. The electron affinity in GaAs is $\chi = 4.07$ V. What would be the depletion width if silver is used instead of nickel? Does silver form a better ohmic contact as compared to Nickel?

6.2. Show that the width of the depletion region in the Schottky barrier diode is given by Equation 6.11. A Schottky barrier diode is made of chromium metal deposited on a GaAs surface. Calculate the depletion width in GaAs at room temperature for the following bias voltage values: V $= -1.0$, 0.0, and $+0.5$V. Assume that the GaAs material is uniformly doped with $N_\text{d} = 10^{16}/\text{cm}^3$ and the dielectric constant of GaAs is 12.25.

6.3. Consider Fig. P6.3 in which $1/C^2$ is plotted as a function of the bias voltage for a Schottky barrier diode made of tungsten and GaAs. Calculate the carrier concentration from the graph. Estimate the Schottky barrier height.

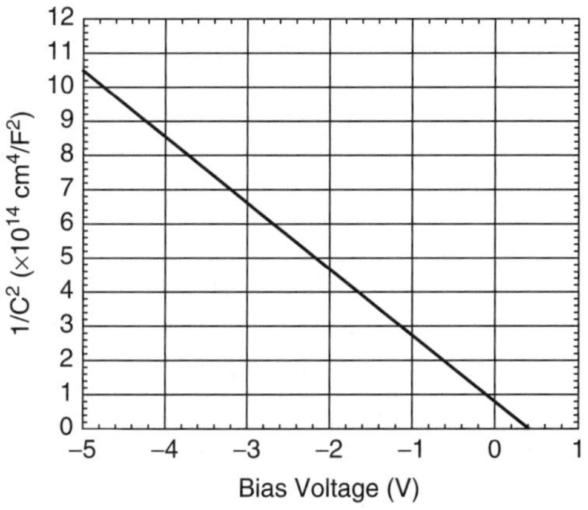

Fig. P6.3

6.4. Show that the effective Richardson constant, A^*, is given by Equation 6.18.

6.5. Show that the drain current of an n-channel MESFET is given by Equation 6.28.

6.6. Show that the transconductance, g_m, for an n-channel MESFET is given by Equation 6.33 for $V_\text{D} < V_\text{D}^\text{sat}$.

6.7. Derive the expression of the transconductance of an n-channel MESFET for the normally off condition, as shown in Equation 6.35.

6.8. What is the cut-off frequency of an n-channel MESFET that is composed of 1-μm thick n-type epitaxial layer of GaAs grown on a semi-insulating GaAs substrate? The gate length is 20 μm, the dopant concentration is $7 \times 10^{16}/\ \text{cm}^3$, and the electron mobility is 2500 cm²/V/s.

6.9. High power MESFETs usually operate under high drain–source voltages. Assume that the desired drain–source breakdown voltage is 75V. Calculate the gate length and the corresponding cut-off frequency for the following materials: silicon (drift velocity is 5×10^6 cm/s and breakdown field 5×10^5 V/cm), GaAs (drift velocity is 1×10^7 cm/s and breakdown field 1×10^6 V/cm), GaN (drift velocity is 5×10^6 cm/s and breakdown field 4×10^6 V/cm), and SiC (drift velocity is 2×10^7 cm/s and breakdown field 5×10^6 V/cm).

6.10. Consider a GaAs/$Al_{0.3}$ $Ga_{0.7}$As MODFET, in which the AlGaAs barrier is silicon doped, producing a carrier concentration of $N_d = 1 \times 10^{18}/cm^3$. Assume that the doped AlGaAs layer is 400 Å, the spacer is 100 Å, the Schottky barrier height is $V_{Bn} = 0.7$ V, the triangular quantum well width is 60 Å, the conduction band offset is 0.3eV, and the AlGaAs refractive index is 3.5. Calculate the 2DEG density at zero gate voltage and zero channel potential.

6.11. Derive an expression for the transconductance, g_m, of a MODFET in the nonsaturation region.

6.12. Show that the cut-off frequency of a single-channel MODFET is given by Equation 6.58. Consider a GaAs/AlGaAs MODFET in which d = 40nm, $d_0 = 10$ nm, and $\Delta d = 6$ nm. Assume that the parasitic capacitance is small (1×10^{-15} F) and the carrier saturation velocity is 1×10^7 cm/sec. Calculate the cut-off frequency for a channel of length 0.2 μm and width of 50 μm.

6.13. Calculate the density of the 2DEG in an undoped 100-Å GaN/$Al_{0.3}$ $Ga_{0.7}$N HFET. Assume that the band offset is 0.4eV, the Fermi energy is 0.1eV above the bottom of the conduction band, the Schottky barrier height is 0.75eV, and the dielectric constant of $Al_{0.3}$ $Ga_{0.7}$N is 4.84.

6.14. Assume that the drift velocity in the base of an HBT is given by $\upsilon(x) = D_n/x$ for $0<x<W$, where D_n is the electron diffusion coefficient and W is the width of the base. Show that the transient time required for an electron to cross the base is given by Equation 6.64.

6.15. A single-electron transistor is fabricated from GaAs/AlGaAs heterojunction. The island is assumed to be spherical with a radius of 0.1 μm. Calculate the charging energy assuming that the refractive index of GaAs is 3.5. What is the charging energy if the dot is a flat disc with the same radius? What is the radius of the spherical island needed for the transistor to operate at $T \leq 125$ K?

6.16. Consider a single-electron transistor with a gate capacitance that can be represented as two parallel plates with a material of a dielectric constant of 13 and an area of $100nm^2$. Calculate the separation between the plates that can yield a capacitance of 1.0 aF.

6.17. Consider a single-electron transistor in which the condition to produce oscillations in the conductance as a function of the gate voltage, V_G, is given as $E_F^{Dot}(N, V_G) = E_F^{Dot}(N + 1, V_G + \Delta V_G)$. (a) Derive an expression for the change in gate voltage between oscillations. (b) Use the initial conditions $E_F^{Dot}(N) = -\frac{N_0 e^2}{C}$ and $E_{N0} = 0$ when $V_G = 0$ to derive an expression for the gate voltage for the Nth conductance peak.

6.18. Calculate the cut-off frequency of a single-electron transistor using the following parameters: $a = 1.5$, $b = 1$, $W\left|\frac{\partial T(E_F, L^*)}{\partial L^*}\right| = 4$, and $V_D = 45$ mV. Design a single-electron transistor that possesses a cut-off frequency of 2.0 THz for a drain voltage of 50mV.

6.19. Show that the electron transmission coefficient for a single-electron transistor with a split channel, as shown in Fig. 6.37, is given by Equation 6.102. Plot $T(E)$ as a function of the relative phase shift, θ. Explain the results.

6.20. Show that when the ground energy levels of the split channels in region II of Fig. 6.37a are different, the relative phase shift, θ, is given by Equation 6.105.

6.21. The difference between the average potential in the two channels (region II) in the single-electron transistor shown in Fig. 6.37a is given by $e\varphi_{12} = \hbar\pi\upsilon_x/L$ for a zero transconductance, where υ_x is the electron velocity and is given by $\upsilon_x = \hbar k_F/m*$, where k_F is the Fermi wave vector. The voltage φ_{12} can be considered as the threshold voltage needed to destruct the interference pattern in the single-electron transistor. Show that $e\varphi_{12} = eV_t^{FET}\lambda_F/L$, where V_t^{FET} is the threshold voltage of a typical field-effect transistor and λ_F is the de Broglie wavelength associated with Fermi energy. Consider that $eV_t^{FET} = E_F$, where E_F is the Fermi energy level.

6.22. The empirical tunneling current in the p–n junction is given by Equation 6.68. Derive an expression for the negative differential resistance. Find the largest negative differential resistance and the corresponding voltage. Assume that $V_0^p = 0.4$ V, and $I_0^p = 30$ mA.

6.23. Consider a thin GaAs tunnel junction with a thickness of 7nm and area of 100 Å × 100 Å. What is the temperature needed to generate a thermal energy equivalent to the charging energy? Assume that the dielectric constant of GaAs is 11.56. Derive expressions for the transmission and reflection coefficients of a particle of mass m traveling from right to left and tunneling through a δ-function potential barrier of a form $V(x) = \lambda\delta(x)$.

6.24. Derive expressions for the transmission and reflection coefficients of a particle of mass m traveling from right to left and tunneling through a δ-function potential barrier of a form $V(x) = \lambda\delta(x)$.

BIBLIOGRAPHY

Aleiner IL, PW Brouwer, Glazman LI. Phys Rep 2002;358:309.

Averin DV, Likharev KK. In: Altshuler BL, Lee PA, Webb RA, editors. Mesoscopic phenomena in solids. Amsterdam: Elsevier Science; 1991. p.173.

Chynoweth AG, Feldmann WL, Logan RA. Excess tunell current in silicon esaki junctions. Phys Rev 1961;121:684.

Demassa TA, Knott DP. The prediction of tunnel diode voltage–current characteristics. Solid State Electron 1970;13:131.

Ferry DK. Quantum mechanics: an introduction for device physicists and electrical engineers. Bristol: IOP; 2001.

Folk JA, Patel SR, Godijn SF, Huibers AG, Cronewett AM, Marcus CM, Campman K, Gossard AC. Phys Rev Lett 1996;76:1699.

Gaska R, Shur MS, Khan A. In: Yu ET, Manasreh MO, editors. Volume 16, III-V Nitride semiconductors: applications and devices. New York: Taylor and Francis; 2003.

Giaever I, Zeller HR. Phys Rev Lett 1968;20:1504.

Glazman LI, *et al*. JETP Lett 1988;48:238 [Pis'ma Zh Teor Fiz 1988;48:218].

Kitabayashi H, Waho T, Yamamoto M. Appl Phys Lett 1997;71:512.

Korotkok AN, Chen RH, Likharev KK. J Appl Phys 1995;78:2520.

Lambe J, Jaklevic RC. Phys Rev Lett 1969;22:1371.

Morkoç H, Cingolani R, Gil B. Mater Res Innov 1999;3:97.

Mitin VV, Kochelap VA, Stroscio MA. Quantum heterostructures microelectronics and optoelectronics. Cambridge: Cambridge University Press; 1999.

Neamen DA. Semiconductor physics and devices: basic principles. 4th ed. New York: McGraw-Hill; 2011.

Roy DK. On the prediction of tunnel diode I-V characteristics. Solid State Electron 1971;14:520.

Simin G, Koudymov A, Fatima H, Zhang J, Yang J, Asif Khan M, Hu X, Tarakji A, Gaska R, Shur MS. IEEE EDL 2002;23:458–460.

Simin G, Hu X, Tarakji A, Zhang J, Koudymov A, Saygi S, Yang J, Asif Khan M, Shur MS, Gaska R Jpn J Appl Phys 2001;400:L921–L924.

Sols F, Macucci M, Ravaioli U, Hess K. J Appl Phys 1989;66:3892.

Timp G, editor. Nanotechnology. New York: Springer; 1999.

Timp G, Howard RE, Mankiewich PM. In: Timp G, editors. Nanotechnology. New York: Springer; 1999. Chapter 2.

Wu WY, Schulman JN, Hsu TY, Efron U. Effect of size nonuniformity on the absorption spectrum of a semiconductor quantum dot system. Appl Phys Lett 1987;51:710.

Sze SM. Semiconductor devices physics and technology. 2nd ed. New York: Wiley; 2002.

Sze SM. Physics of semiconductor devices. 2nd ed. New York: Wiley; 1981.

7

OPTOELECTRONIC DEVICES

7.1 INTRODUCTION

Semiconductor heterojunctions and nanostructures have been investigated for their applications in electronic and optoelectronic devices. This chapter is directed toward optoelectronic devices such as detectors and emitters. There is a myriad of applications for the optoelectronic devices including 1.31- and 1.55-μm optical communications where the silica fibers exhibit the lowest losses, terahertz application, infrared and long wavelength infrared detectors, multijunctions solar cells, and so on. One of the mechanisms used to generate light from a semiconductor is the radiative recombination of electrons and holes across the fundamental band gap, which gives rise to photon emission. Soon after the invention of laser, p–n junction GaAs lasers were demonstrated with an emission in the 0.827–0.886-μm spectral range, which is basically limited by the band gap of the GaAs material. This spectral range is outside the energy spectrum that is visible to the human eye. Substantial research has been performed since then toward the development and production of emitters in the visible, ultraviolet, and infrared spectral regions (see Fig. 7.1 for various spectral region limits). For example, recent research efforts in III-nitride semiconductor materials led to the production of blue and green light-emitting diodes (LEDs) and diode lasers. Further research has pushed the performance of the III-nitride materials to the ultraviolet and far ultraviolet LEDs.

The development of infrared emitters goes back to the 1960s and continues to develop as new applications and needs emerge. In theory, the wavelength of emitters based on interband transitions in semiconductors can be constructed to

Introduction to Nanomaterials and Devices, First Edition. Omar Manasreh.

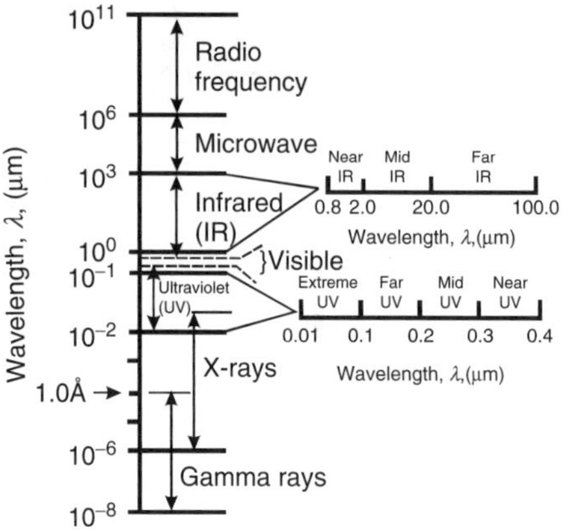

FIGURE 7.1 Wavelength of various spectral regions. Notice that the spectral region visible to human eye is very narrow.

cover long and very long wavelength spectral regions. This is due to the fact that one can grow ternary and quaternary alloys with a precise control on the composition using the latest growth technologies, such as MBE and MOCVD growth techniques. The limiting factors for III–V semiconductors, however, are the nonradiative recombination processes and the internal losses due to free carrier absorption. These factors become significant and detrimental to the device performance as the band gap of the semiconductor decreases (or increases in the wavelength). Advances in the long wavelength infrared diode lasers continue in the spectral range of 3–30 μm using IV–VI compound semiconductors, such as PbTe, which have their nondegenerate direct band gap located at the L-point of the first Brillouin zone. The nonradiative recombination processes at this point, such as Auger recombination processes, are less likely to happen.

There are many types of light sources that operate using different operational principles. Gunn diode operates on the principle of fast charge oscillations, which acts as classical dipole source. This type of diodes emits electromagnetic waves with wavelength larger than 1000 μm (microwave source). Optically pumped gas lasers usually emit lights in the 40–1000 μm spectral range. Generally speaking, the lasers can be categorized into four major classes: (i) gas lasers such as HeNe, CO_2, and Argon-ion; (ii) dye (liquid) lasers such as oxazine and polyphenyl; (iii) solid-state lasers such as Ti:Sapphire, Nd:YAG, and ruby; and (iv) semiconductor lasers. This chapter, however, is focused on lasers based on interband and intersubband transitions in semiconductor quantum wells and quantum dots, such as edge-emitting lasers, vertical cavity surface-emitting lasers (VCSELs), and quantum cascade lasers.

Another class of optoelectronic devices, whose history goes back to the 1800s, is the infrared detectors. The first infrared detector is the thermometer, which was discovered by Hershel in 1800. Nowadays, there are many types of infrared detectors ranging from a single element to large focal plane arrays and from thermal detectors to multiple quantum well infrared detectors. Again, the discussion is focused on the quantum detectors based on interband and intersubband transitions in semiconductor quantum wells and dots.

7.2 INFRARED QUANTUM DETECTORS

The two major categories of infrared detectors are thermal and photonic (quantum) detectors. The principle of operation of the thermal detectors is based on the temperature change of the detector materials on absorbing the photons. Accompanying this change in temperature is a change in at least one physical property of the material, which leads to generation of an electrical signal. In general, the electrical output signal of thermal detectors is independent of the incident photon wavelength, but of course depends on the radiant power. An example of thermal detectors is the bolometer, which could be made of metal, semiconductor, superconductor, or ferroelectric material. The physical parameter that changes when photons are absorbed is the electrical conductivity or electrical resistivity. Other examples of thermal detectors are as follows: (i) thermocouples where the generation of voltage is caused by change in the temperature of the junction of two dissimilar materials; (ii) Golay cells, which is based on the change in the thermal expansion of the gas; (iii) pyroelectric detectors in which the spontaneous polarization is changed when absorbing photons; and (iv) pyromagnetic detectors where the name implies that the magnetic properties are changed.

Quantum detectors, on the other hand, operate on the principle of electron–photon interaction. Thus, these detectors are much faster than the thermal detectors. Two basic processes are involved in quantum detectors. First, conduction electrons or electrons bound to the lattice atoms or impurities absorb light and get excited to higher energy levels. Second, the excited electrons are swept by applied bias voltage and collected as an electrical signal. Ideally, all excited electrons can be collected, leading to 100% quantum efficiency. This means that each photon absorbed excites an electron and all the excited electrons are collected under bias voltage. In practice, 100% efficiency is very difficult to achieve because of the fact that many of the excited electrons recombine with the holes, trapped by positively charged ions, or lose their energy as phonons. Thermal generation of charge carriers (dark current) can be significantly reduced by cryogenic cooling.

7.2.1 Figures of Merit

There are too many variables, such as electrical, radiometric, and device design parameters, involved when measuring the photoresponse of a detector regardless of whether it is a thermal or a photonic detector. Thus, it is difficult to measure the performance of the device. Several figures of merit have evolved over the years and are used to characterize and quantify infrared detectors. While some figures of merit that were developed may not be of any use to quantify many of the quantum detectors, the currently accepted figures of merit are briefly discussed in the following section.

7.2.1.1 ***Responsivity*** Responsivity of a detector is difficult to quantify, especially when comparing various detectors. For example, the photoresponse of thermal detectors is independent of the photon wavelength, but the responsivity of a quantum detector is a linear function of the wavelength. For quantum detectors, the responsivity is defined as the ratio between the detector output electrical signal and the input radiant optical power. The detector output signal is either voltage or current. For the voltage output signal, the spectral responsivity is given as

$$\mathcal{R}_{\mathrm{v}}(\lambda, f) = \frac{V_{\mathrm{s}}}{P_{\mathrm{in}}(\lambda)}, \tag{7.1}$$

where $\mathcal{R}_{\mathrm{v}}(\lambda, f)$ is the voltage spectral responsivity, which is a function of the incident photon wavelength (λ) and the operating electrical chopping frequency (f), V_{s} is the output signal voltage, and $P_{\mathrm{in}}(\lambda)$ is the spectral radiant input power given by

$$P_{\mathrm{in}}(\lambda) = \frac{A_{\mathrm{d}}\Phi_{\mathrm{ph}}hc}{\lambda}, \tag{7.2}$$

where A_{d} is the area of the detector, Φ_{ph} is the incident photon flux density expressed in $photons/m^2/s^1$, and λ is the incident light wavelength. The current responsivity, $\mathcal{R}_i(\lambda, f)$, is similar to Equation 7.1 and it can be written as

$$\mathcal{R}_i(\lambda T, f) = \frac{I_{\mathrm{ph}}}{P_{\mathrm{in}}(\lambda)}, \tag{7.3}$$

where I_{ph} is the photocurrent. The spectral responsivity expressed in Equations 7.1 and 7.3 should be multiplied by the photoconductive gain, which is defined as the ratio between the recombination time and the transit time. For now, we assume that the photoconductive gain is maximum (unity).

The blackbody responsivity is a very useful parameter, and it is defined as the output of a detector produced in response to a watt of input optical radiation from a blackbody at temperature T modulated at an electrical frequency f. The blackbody source is usually calibrated and standardized at specific temperatures. For example, infrared blackbody sources are calibrated at 500 K, while the near infrared and visible sources are calibrated at 2856 K (see Dereniak and Boreman). When measuring the blackbody responsivity, the radiant power on the detector contains all wavelengths of radiation regardless of the spectral response curve of the detector. Thus, the current responsivity shown in Equation 7.3 can be modified as

$$\mathcal{R}_i(T, f) = \frac{I_{\mathrm{ph}}}{\int_0^{\infty} P_{\mathrm{in}}(\lambda)d\lambda}. \tag{7.4}$$

Notice that the blackbody responsivity depends on temperature (T) and electrical frequency (f). Notice that when Fourier-transform spectroscopy is used to measure the responsivity of the detector, the chopping frequency becomes irrelevant, since choppers are not used. Furthermore, the entire flux incident on the detector appears in the calculations of blackbody responsivity. The responsivity is usually a function of the bias voltage, f and λ.

*7.2.1.2 **Noise-Equivalent Power*** The responsivity is a good figure of merit used to estimate an expected signal level for a given radiant power on the detector. However, it does not provide useful information regarding the sensitivity of the detector. In addition to the signal level, the noise level is important. The question that one may ask is what is the minimum radiant flux level a detector can measure. The detector output photocurrent due to an input power must be larger than the noise current level. The signal ($\mathcal{S}$)-to-noise ($\mathcal{N}$) ratio can be defined as

$$\frac{\mathcal{S}}{\mathcal{N}} = \frac{\mathcal{R}_i \Phi_e}{i_\mathrm{n}}, \tag{7.5}$$

where Φ_e is the radiant flux (W), i_n is the noise current, and $\mathcal{R}_i$ is the current responsivity (A/W). The noise-equivalent power (NEP) is the radiant power incident on the detector that produces $\mathcal{S}/\mathcal{N} = 1$. Setting Equation 7.5 equal to unity yields the NEP

$$\Phi_e = \mathrm{NEP} = \frac{i_\mathrm{n}}{\mathcal{R}_i}. \tag{7.6}$$

Similarly, the NEP can be written in terms of voltage responsivity $\mathcal{R}_\mathrm{v}$ as

$$\Phi_e = \mathrm{NEP} = \frac{V_\mathrm{n}}{\mathcal{R}_\mathrm{v}}, \tag{7.7}$$

where V_n is the noise voltage. When the responsivity is a spectral responsivity, the NEP is called *spectral NEP*. The term *blackbody NEP* is used when the blackbody responsivity is used. The unit of NEP is watt and a more sensitive detector has a lower NEP. In this sense, the NEP is a defect function rather than a figure of merit. In addition to NEP, a detector performance is measured by other parameters such as the optimum bias voltage, operating temperature, and noise-equivalent bandwidth.

*7.2.1.3 **Detectivity*** The normalized spectral detectivity, known as D^*, is another important figure of merit, which is defined as

$$D^* = \frac{\sqrt{A_\mathrm{d}\Delta f}}{\mathrm{NEP}}, \tag{7.8}$$

where Δf is the noise-equivalent bandwidth and A_d is the area of the detector. The larger the D^* is, the better the detector. The D^* is usually normalized to $A_\mathrm{d} = 1\ \mathrm{cm}^2$ and $\Delta f = 1$ Hz, and it can be interpreted as the signal-to-noise ratio

(SNR) of the detector when 1 W of radiant power is incident on 1 cm^2 active area of the detector given a noise-equivalent bandwidth of 1.0 Hz. The units of D^* is expressed as Jones = $\mathrm{cm}\sqrt{\mathrm{Hz}}/\mathrm{W}$. The detectivity can be written in different forms such as

$$D^* = \frac{\sqrt{A_\mathrm{d}\Delta f}}{\mathrm{NEP}} = \frac{\sqrt{A_\mathrm{d}\Delta f}}{V_\mathrm{n}/\mathcal{R}_\mathrm{v}} = \frac{\sqrt{A_\mathrm{d}\Delta f}}{i_\mathrm{n}/\mathcal{R}_\mathrm{i}} = \frac{\sqrt{A_\mathrm{d}\Delta f}}{\Phi_e}\frac{\mathcal{S}}{\mathcal{N}}. \tag{7.9}$$

This equation implies the following definition of D^*: It is the root-mean-square (rms) SNR of a detector of 1.0 cm^2 area in 1-Hz bandwidth per unit rms incident radiant power.

The blackbody detectivity, $D^*(T)$, can be obtained from the spectral detectivity according to the following relation

$$D^*(T) = \frac{\int_0^\infty D^*\Phi_e(T,\lambda)\mathrm{d}\lambda}{\int_0^\infty \Phi_e(T,\lambda)\mathrm{d}\lambda} = \frac{\int_0^\infty D^* Q_\mathrm{B}(T,\lambda)\mathrm{d}\lambda}{\int_0^\infty Q_\mathrm{B}(T,\lambda)\mathrm{d}\lambda}, \tag{7.10}$$

where $\Phi_e(T,\lambda) = Q_\mathrm{B}(T,\lambda)A_\mathrm{d}$ is the incident blackbody radiant flux in watt and $Q_\mathrm{B}(T,\lambda)$ is the blackbody irradiance in watt per square centimeter.

When the detector noise is low as compared to the photon noise, the detector is said to reach its maximum performance. The photon noise arises from the detection process as a result of the discrete nature of the radiation field. For most infrared detectors, the practical operating limit is the background fluctuation limits and not from the signal fluctuation limits. Thus, when the background photon flux is much larger than the signal flux, the photon flux is the dominant noise source. This condition is called *background limited infrared performance* (BLIP).

The spectral detectivity can be derived under the BLIP conditions by considering the photon irradiance from the signal source (Q_s), and the background (Q_B). The rms noise current, i_n, can be established by assuming that the shot noise in the detector is due to the DC-photogenerated current I_ph flowing across a potential barrier as

$$i_\mathrm{n} = \sqrt{2e\overline{I}\Delta f}. \tag{7.11}$$

The shot noise was first shown by Schottky in 1918 that the random arrival of electrons on the collecting electrode of a vacuum tube was responsible for what is called *shot noise*. The total photocurrent, $\overline{I}$, caused by both background and signal photons is given as

$$\overline{I} = e\eta(Q_\mathrm{s} + Q_\mathrm{B})A_\mathrm{d}, \tag{7.12}$$

where η is the quantum efficiency defined as the ratio between the electron–hole pairs generated per incident photon. Substituting Equation 7.12 into Equation 7.11 and assuming that $Q_\mathrm{s} \ll Q_\mathrm{B}$ yields

$$i_\mathrm{n} = \sqrt{2e^2\eta Q_\mathrm{B}A_\mathrm{d}\Delta f}. \tag{7.13}$$

The assumption that $Q_s \ll Q_B$ is valid since we are trying to determine the minimum detectable signal. The SNR can now be written as

$$\frac{\mathcal{S}}{\mathcal{N}} = \frac{I_{ph}}{i_n} = \frac{e\eta Q_s A_d}{\sqrt{2e^2\eta Q_B A_d \Delta f}}, \tag{7.14}$$

where I_{ph} is the photocurrent due to the signal alone. In order to detect the minimum detectable signal irradiance, the SNR in Equation 7.14 is set to unity, which yields

$$Q_s = \sqrt{\frac{2Q_B \Delta f}{\eta A_d}}. \tag{7.15}$$

This equation can be considered as the noise-equivalent irradiance on the surface of the detector. Using the definition of the power given by Equation 7.2, one can write the signal power as

$$\Phi_{e,\text{signal}} = Q_e \frac{hc}{\lambda} A_d. \tag{7.16}$$

Since Q_s is determined by setting Equation 7.14 to unity, it is defined as the noise-equivalent irradiance. This implies that the power defined in Equation 7.16 is the NEP or

$$\text{NEP}(\lambda) = \sqrt{\frac{2Q_B \Delta f}{\eta A_d} \frac{hc}{\lambda}}. \tag{7.17}$$

Combining Equations 7.8 and 7.17 to obtain a definition of the detectivity at the BLIP conditions

$$D^*_{\text{BILP}} = \frac{\sqrt{A_d \Delta f}}{\text{NEP}(\lambda)} = \frac{\lambda}{hc}\sqrt{\frac{\eta}{2Q_B}}. \tag{7.18}$$

The background photon irradiance, Q_B, (or the background photon flux density) reaches the detector can be written as (Dereniak and Boreman)

$$Q_B = \sin^2(\theta/2) \int_0^{\lambda_c} Q(\lambda, T_B) d\lambda, \tag{7.19}$$

where $\sin^2(\theta/2)$ is the numerical aperture, λ_c is the cutoff frequency, T_B is the blackbody temperature, and $Q(\lambda, T_B)$ is the Planck's photon emittance (in units of photons/cm^2/s/μm) and is given by (see, e.g., Hudson)

$$Q(\lambda, T_B) = \frac{2\pi c}{\lambda^4 \left[e^{hc/(\lambda k_B T_B)} - 1\right]} = \frac{1.885 \times 10^{33}}{\lambda^4 \left[e^{14388/(\lambda T_B)} - 1\right]}, \tag{7.20}$$

where λ is given in microns. A plot of $Q(\lambda, T_B)$ is shown in Fig. 7.2 for different values of T_B. The integral in Equation 7.19 can be obtained numerically. For additional analysis of the detectivity at the BLIP conditions, see Rogalski and Dereniak *et al*.

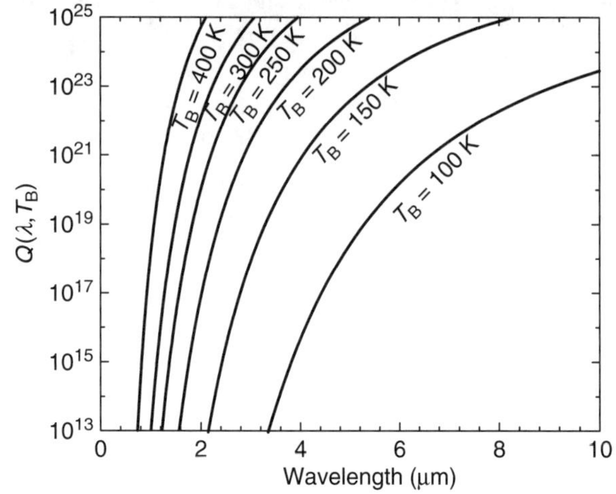

FIGURE 7.2 Planck's photon emittance is plotted as a function of the wavelength for different blackbody temperatures.

7.2.1.4 Noise-Equivalent Temperature Difference Many of the long wavelength infrared detectors are used as thermal imagers. One of the figures of merit used to describe the performance of the thermal imaging system is the noise-equivalent temperature difference (NETD). Thermal imager systems are used to map the temperature difference related to spatial flux and emissivity differences across an extended object. Thus, thermal sensitivity is concerned with the minimum temperature difference that can be distinguished above the noise level. The thermal imaging systems have been discussed and presented by Lloyd. The NETD is derived by many authors (Lloyd, Dereniak *et al*., and Rogalski) and is given as

$$\text{NETD} = \frac{4}{\pi}\left[\frac{(F/\#)^2\sqrt{\Delta f}}{D^*\dfrac{\partial L}{\partial T}\sqrt{A_d}}\right], \tag{7.21}$$

where $F/\#$ is known as the *f-number*, which is the distance between the pupil and the detector divided by the entrance pupil area. The parameter L is the radiance at the location instantaneous field of view.

Another useful parameter is the minimum resolvable temperature difference (MRTD), which is useful as a summary measure of performance and design criterion. It combines both thermal sensitivity and spatial resolution. Smaller is better for both NETD and MRTD.

7.2.2 Noise in Photodetectors

There are different sources of noise in semiconductor photodetectors. The most important noise sources are Johnson noise, $1/f$ noise, generation–recombination noise, and preamplifier noise. Johnson noise is also called *thermal noise*, which is associated with the finite resistance of the device. It is due to the random thermal motion of charge carriers in the detector material. This noise is present, in the absence of bias voltage, as a fluctuation in the current or voltage, which is due to the random arrival of the charge carriers at the device electrodes. The Johnson noise is generated in both the detector and the load resistance. The rms of the Johnson current noise can be expressed as

$$i_{\mathrm{J}} = \sqrt{\frac{4k_{\mathrm{B}}T_{\mathrm{d}}\Delta f}{R_{\mathrm{d}}} + \frac{4k_{\mathrm{B}}T_{\mathrm{L}}\Delta f}{R_{\mathrm{L}}}}, \tag{7.22}$$

where T_{d} and T_{L} are the temperatures of the detector and the load resistor, respectively, and R_{d} and R_{L} are the detector resistance and the load resistance, respectively. If the temperature of the detector and the load resistor is the same ($T_{\mathrm{d}} = T_{\mathrm{L}} = T$), then the thermal fluctuation current can be written as

$$i_{\mathrm{J}} = \sqrt{\frac{4k_{\mathrm{B}}T\Delta f}{R_{\mathrm{eq}}}} \quad \text{or} \quad V_{\mathrm{J}} = R_{\mathrm{eq}} i_{\mathrm{J}} = \sqrt{4k_{\mathrm{B}}T\Delta f R_{\mathrm{eq}}}, \tag{7.23}$$

where $R_{\mathrm{eq}} = R_{\mathrm{d}}R_{\mathrm{L}}/(R_{\mathrm{d}} + R_{\mathrm{L}})$ and V_{J} is the Johnson-noise voltage.

The intrinsic noise mechanism of a photodiode is called *shot noise*, which is the noise in the current passing through a diode. The general form of the noise in the current, i_{s}, in an ideal diode is expressed as

$$i_{\mathrm{s}} = \sqrt{[2e(I_{\mathrm{D}} + 2I_{\mathrm{s}}) + 4k_{\mathrm{B}}T(G_j - G_0)]\Delta f}, \tag{7.24}$$

where I_{D} is the diode current given by $I_{\mathrm{D}} = I_{\mathrm{s}}(\mathrm{e}^{eV/k_{\mathrm{B}}T} - 1)$, I_{s} is the saturation reverse current, G_j is the conductance of the junction, and G_0 is the value of G_j at low frequency. For zero bias voltage and low frequency, shot noise reduces to Johnson noise.

The origin of the $1/f$ noise is not well understood, but it appears to be associated with potential barriers. The general expression for the $1/f$ noise current is

$$i_{1/f} = \sqrt{\frac{A i_{\mathrm{b}}^{\alpha}\Delta f}{f^{\beta}}}, \tag{7.25}$$

where A is constant, $\alpha \sim 2$, $\beta \sim 1$, and i_{b} is the DC bias current. The $1/f$ noise is dominant at low frequency.

The current generation–recombination noise results from the random number of free carriers due to the background photons and thermal excitations in the detector. The general form of this noise is expressed as

$$i_{\mathrm{gr}} = 2eG\sqrt{\eta\Phi_{\mathrm{b}}A_{\mathrm{d}}\Delta f + \mathcal{G}_{\mathrm{th}}\Delta f}, \tag{7.26}$$

where G is the photoconductive gain, Φ_b is the background flux density, and $\mathcal{G}_{\text{th}}$ is the thermal generation rate of carriers. Many long wavelength infrared detectors operate at temperatures lower than room temperature where the thermal noise is negligible. In this case, the current recombination–generation noise can be expressed as the noise due to background photons, which is simply the first expression in Equation 7.26

$$i_{\text{gr}} = 2eG\sqrt{\eta \Phi_{\text{b}} A_{\text{d}} \Delta f}. \tag{7.27}$$

The expression given by Equation 7.27 is further modified when $h\nu/k_{\text{B}}T < 1$, giving the following relation

$$i_{\text{gr}} = 2eG\sqrt{\eta \Phi_{\text{b}} A_{\text{d}} \Delta f (1 + b_{\text{Bose}})}, \tag{7.28}$$

where b_{Bose} is Bose factor, which is not negligible at long wavelengths.

There are other sources of noise, such as the preamplifier noise, which will not be discussed here. Summing all the sources of noise, one can write the total noise as

$$i_{\text{noise}} = \sqrt{i_{\text{J}}^2 + i_{\text{s}}^2 + i_{1/f}^2 + i_{\text{gr}}^2}. \tag{7.29}$$

A sketch of Johnson noise, $1/f$ noise, generation–recombination noise, and the total noise as a function of frequency is shown in Fig. 7.3. The dominant noise at low frequencies is $1/f$ noise, while the generation–recombination noise is dominant at mid frequencies. Note that the recombination–generation noise given by Equation 7.28 is the low frequency noise.

For higher frequencies, the generation–recombination noise can be obtained as

$$i_{\text{gr}}(f) = \frac{i_{\text{gr}}(0)}{\sqrt{1 + (\omega\tau)^2}}, \tag{7.30}$$

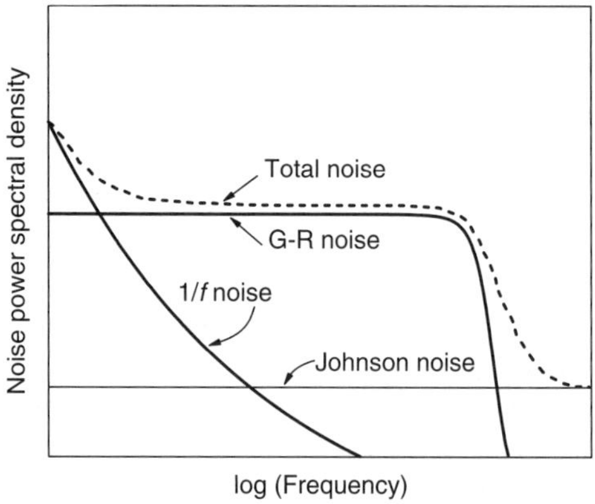

FIGURE 7.3 A sketch of $1/f$, Johnson, generation–recombination, and total noise are plotted as a function of frequency.

where $\omega = 2\pi f$, and τ is the recombination lifetime. At high frequencies, the generation–recombination noise is rolled off and the dominant noise is Johnson noise.

7.2.3 Multiple Quantum Well Infrared Photodetectors (QWIPs)

There are many types of quantum photodetectors, such as p–n junction photodiodes, Schottky barrier photodiodes, and metal-insulator-semiconductor photodiodes. This book is focused on heterojunctions and nanostructures, and, therefore, we focus our discussion on detectors based on quantum wells, superlattices, and quantum dots. Different quantum well designs were investigated for long wavelength infrared detectors. The basic principle of operation is based on the photon absorption by electrons that exist in the quantum well ground state, which then excited to a higher energy level. These excited electrons are collected under a bias voltage to give a photoconductive signal in the infrared regions. The quantum well infrared photodetectors (QWIPs) is based on the intersubband transition, which is shown in Fig. 7.4 for different quantum well designs. The first design shown in Fig. 7.4a is a typical quantum well structure with two bound states. The applied bias voltage causes the conduction band (CB) to bend where the excited state is now located near the edge of the barrier CB. Electrons in the ground state can be excited by illuminating the sample with infrared light and then collected under the influence of the bias voltage. The detector output signal is usually called a *photocurrent* or *photoresponsivity*.

The second quantum well design is shown in Fig. 7.4b, where the quantum well is sandwiched between two thin layers with a band gap larger then the barrier materials. A typical example is a GaAs quantum well and $Al_xGa_{1-x}As$ barriers. The thin layer between the well and barrier is $Al_yGa_{1-y}As$, where $y > x$. The addition of the $Al_yGa_{1-y}As$ thin layers reduces the dark current and produces narrower photoresponse as compared to the structure shown in Fig. 7.4a. The excited electrons in the latter structure tunnel through the thin barrier and are then collected at one of the electrodes.

An alternative structure, commonly used with low dark current devices, is shown in Fig. 7.4c, where the barrier is composed of superlattices rather than a simple bulk barrier. A miniband is formed in the superlattice barrier, which is extended into the quantum well itself. When the sample is irradiated with photons, the electrons are excited from the ground state to the miniband and then transported via the miniband when the sample is biased. The dark current is further reduced by designing a quantum well, such as InGaAs, embedded between the GaAs/AlGaAs superlattices as shown in Fig. 7.4d.

Since the quantum wells in these structures are required to be populated with electrons, the wells are usually doped with silicon. Modulation doping in the barriers yields similar results. To increase the optical length, the quantum well/barrier repetition is usually chosen between 20 and 50 periods. A typical example of the optical absorption of an intersubband transition in n-type 75 Å GaAs/

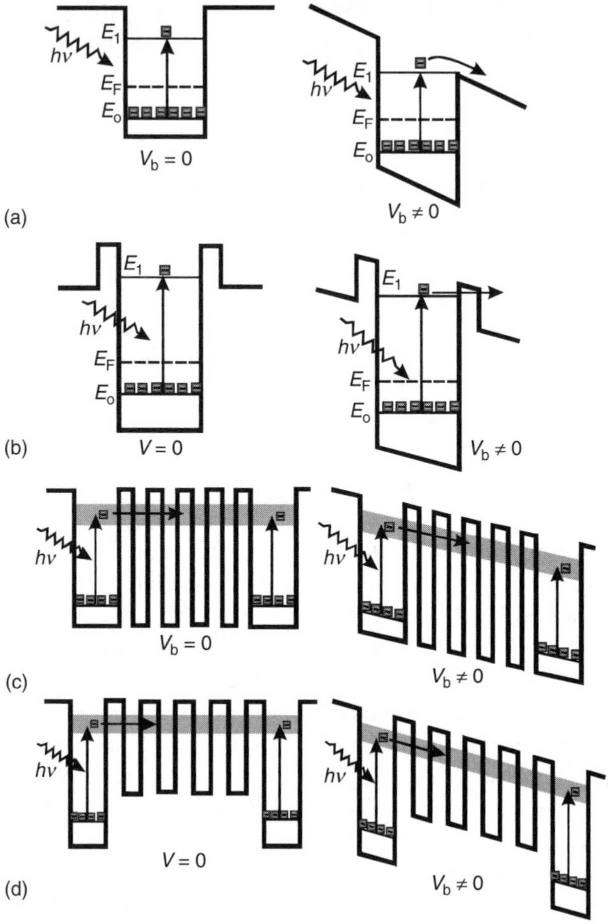

FIGURE 7.4 Intersubband transitions in multiple quantum wells with different design. (a) Typical quantum well design structure with two bound states, for example, GaAs/AlGaAs. (b) Quantum well structure with additional thin barrier, for example: GaAs/$Al_yGa_{1-y}As/Al_xGa_{1-x}As$, where $y < x$. (c) Quantum well structure where the barrier is composed of superlattice, for example: GaAs (well)/{GaAs/AlGaAs} (superlattice barrier). (d) Embedded quantum well structure with superlattice barrier, for example InGaAs (well)/{GaAs/AlGaAs} (superlattice barrier). The left panel is for zero bias voltage and the right panel is sketched for applied bias voltage.

100 Å $Al_{0.3}Ga_{0.7}As$ multiple quantum wells grown on semi-insulating GaAs wafers is shown in Fig. 7.5a. In this figure, we plotted the optical absorption spectra obtained at room temperature at 77 K for both the Brewster's angle and waveguide configurations. The intensity of the spectra collected using the waveguide configuration is much larger than those obtained using the Brewster's angle configuration because the incident light makes multiple passes in the waveguide

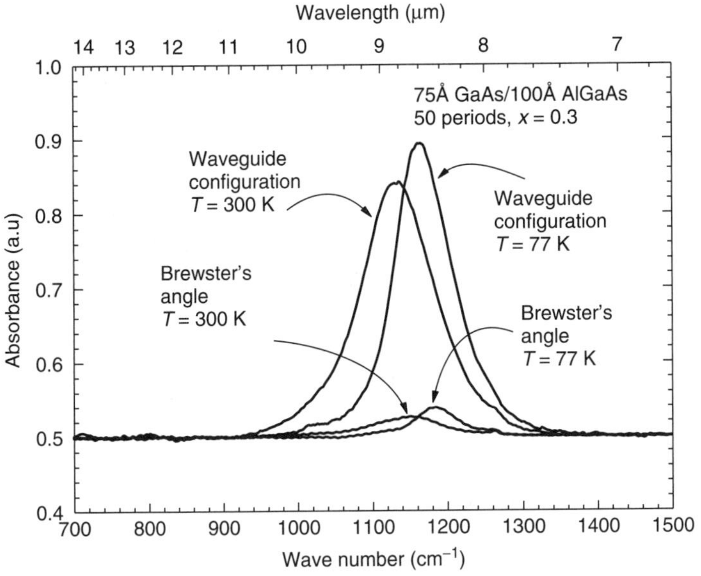

(a)

(b)

FIGURE 7.5 (a) Optical absorption spectra of the intersubband transition in GaAs/AlGaAs multiple quantum wells obtained at room temperature and 77 K using both Brewster's angle and waveguide configurations. (b) A scanning electron microscopy image of a waveguide made of GaAs/AlGaAs multiple quantum wells grown in a semi-insulating GaAs wafer of a thickness of ~0.450 mm.

configuration. A typical waveguide is shown in Fig. 7.5b, where the width of the sample is ~2.0 mm and the length is ~5 mm. The waveguide is usually cut from the wafer, and then the facets along the 5-mm edges were polished at an angle of 45°. The optical absorption coefficient and the selection rules have been discussed in more detail in Chapter 4.

Photoconductivity measurements of a photodetector require fabrication of a mesa, where the electrodes are attached to the mesa and a bias voltage is applied. A typical mesa structure is shown in the inset of Fig. 7.6a, where the quantum

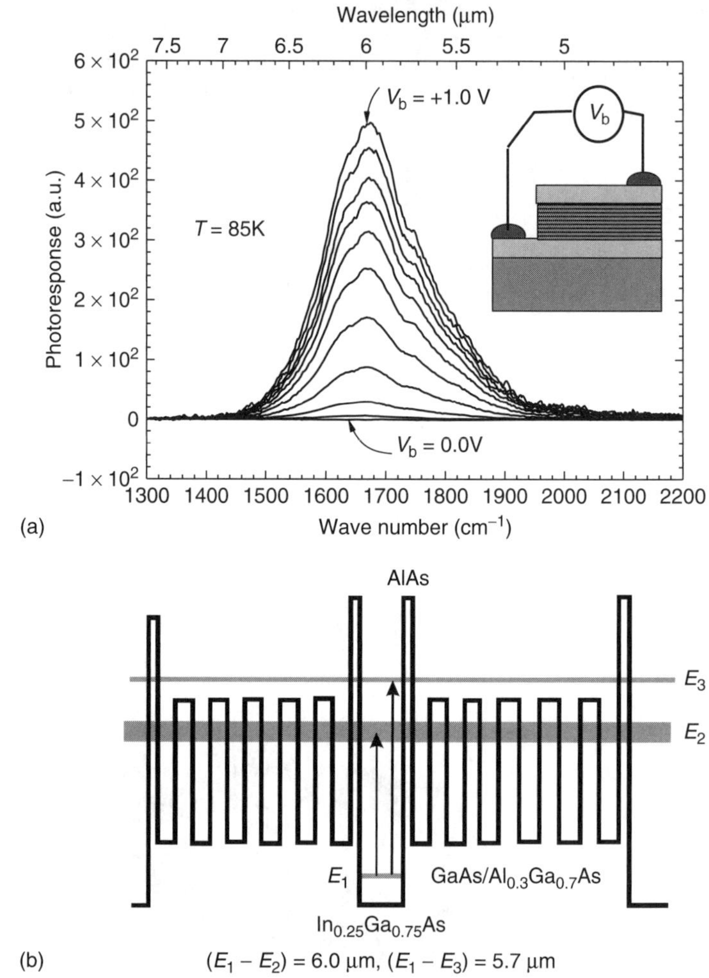

FIGURE 7.6 (a) The photoresponse spectra of the multiple quantum well infrared photodetector obtained for different bias voltages applied to the mesa (shown in the inset). (b) The structure of the multiple quantum wells used to fabricate a photodetector device.

well structure is sandwiched between contact layers. The photoresponse spectra shown in Fig. 7.6a were collected under different bias voltages in the range $0 \leq V_b \leq 1.0V$ with a step of 0.1 V. The structure of the multiple quantum wells is shown in Fig. 7.6b. It is composed of five periods of $In_{0.25}Ga_{0.75}As$ quantum wells and $GaAs/Al_{0.3}Ga_{0.7}A$ superlattice barriers. Thin layers of AlAs were also inserted as shown in Fig. 7.6b to reduce the dark current. The photoresponse spectra show the dominant transition between ground state, E_1, and the miniband, E_2, around 6.0 μm and the weaker transition between the ground state, E_1, and the excited state, E_3, at around 5.7 μm.

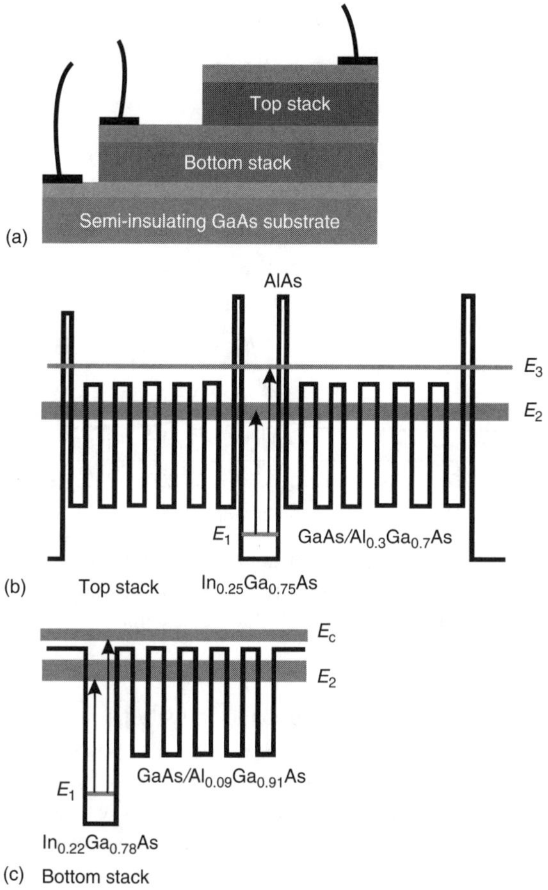

FIGURE 7.7 (a) A sketch of two stacks of multiple quantum wells separated by n-type doped GaAs layers, which act as contact layers. The contact pads are shown: (b) the multiple quantum well structure used for the top stack, (c) the multiple quantum well structure used for the bottom stack.

Multicolor detectors can be easily fabricated by growing different stacks of multiple quantum wells separated by contact layers. An example of a two-color detector is sketched in Fig. 7.7a, where n-type doped GaAs layers were grown as contact layers. The top stack multiple quantum well structure is shown in Fig. 7.7b, which is essentially the same structure as that shown in Fig. 7.6b. The structure of the bottom stack is shown in Fig. 7.7c, which is designed such that the photoresponse is at ~10.5 μm. From the design of the bottom stack, one can see that the dominant transition is between the ground state, E_1, and the miniband, E_2. The transition between the ground state and the continuum, E_c, is also allowed, but with a smaller oscillator strength.

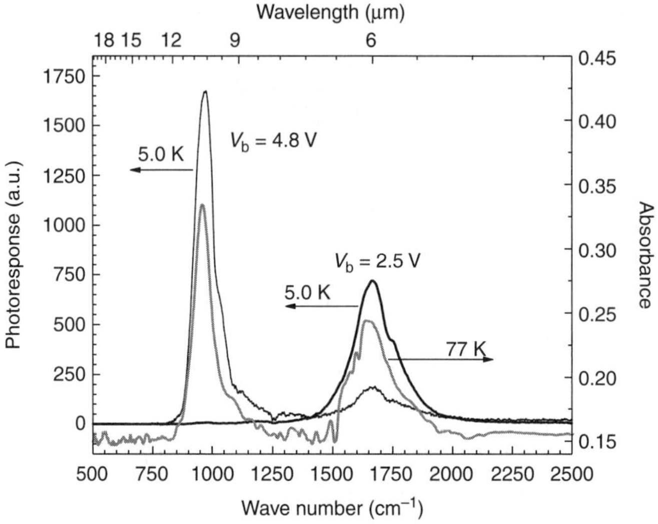

FIGURE 7.8 The optical absorption spectrum (gray line) obtained at 77 K for two-color infrared photodetector. The photoresponse spectra obtained at 5.0 K for two bias voltages of (a) 2.5 V, thick black line, and (b) 4.8 V, thick black line.

The optical absorption of the intersubband transitions in the two-stack sample described in Fig. 7.7 was obtained at 77 K using a 45° polished facets waveguide configuration. The incident light makes two to three passes inside the waveguide before exiting the sample. The result is shown as the gray spectrum in Fig. 7.8. A bias voltage is not needed for the optical absorption measurement. The two dominant transitions in both stacks are associated with the transition from the ground state to the miniband in each stack. The miniband is formed from the ground states in the superlattice barrier, which is extended into the quantum wells. Each period in the superlattice barrier contributes one energy level to the miniband. The optical absorption spectrum shown in Fig. 7.8 exhibits two peaks. The first peak is observed around 5.7 μm, which is originated from the top stack, and the second peak is observed around ~10.5 μm, which is originated from the bottom stack.

The device photoresponse is measured at 5.0 K under different bias voltages. For this test, the bias voltage is applied between the contact layer deposited on top of the top stack and the contact layer deposited underneath the bottom stack (Fig. 7.7a). When the voltage is low ($V_b = 2.5$ V), the peak from the top stack (~5.7 μm) is dominant, while the peak from the bottom stack is below the detection limit as shown in the photoresponse spectrum (thick black line). However, when the voltage is increased to $V_b = 4.8$ V, the photoresponse peak (~10.5 μm) from the bottom stack is dominant (thin black line). Both peaks coincide with the peaks observed in the optical absorption measurements.

The multiple quantum well detectors described in Figs. 7.7 and 7.8 are called *voltage tunable photodetectors*. It has the ability to respond to different photon

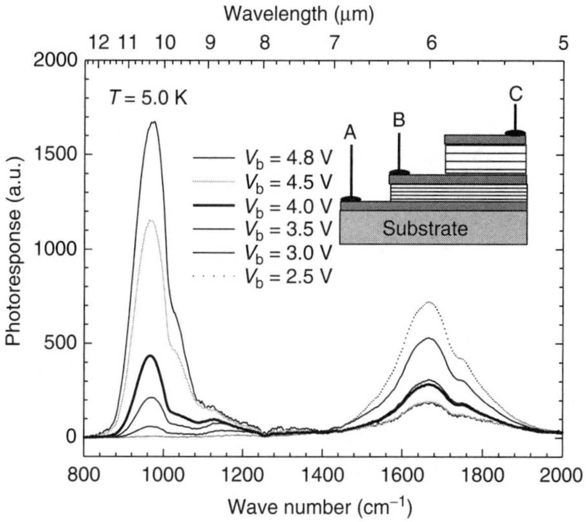

FIGURE 7.9 Several photoresponse spectra obtained as a function of the bias voltage for a two-stack photodetector. The inset is a sketch of the two stacks with A, B, and C electrodes.

wavelengths under different bias voltages. Further illustration of how the photoresponse depends on the bias voltage is shown in Fig. 7.9. Several photoresponse spectra were recorded for different bias voltages applied across points A and C in the figure inset. The measurements were obtained at 5.0 K. Similar results were obtained at 77 K, but with lower peak intensities. It is clear from this figure that different stacks can be turned on or off depending on the bias voltage. If a bias voltage is applied between points A and B, one can only observe the peak at ~10.5 μm. Similarly, the peak around 5.7 μm is observed when a bias voltage is applied between points B and C.

The internal quantum efficiency of the QWIPs can be obtained from the optical absorption coefficient. If the light incident intensity is I_{i}, then the quantum efficiency is the fraction of the light incident intensity that is absorbed by the electrons that undergo the intersubband transition. The light intensity absorbed by the quantum wells can be defined as

$$\begin{aligned} I_1 &= I_{\mathrm{i}}(1-\alpha_0) \text{ for a single quantum well} \\ I_2 &= I_{\mathrm{i}}(1-\alpha_0)^2 \text{ for two quantum wells} \\ &\vdots \\ I_N &= I_{\mathrm{i}}(1-\alpha_0)^N \text{ for} N \text{ quantum wells,} \end{aligned} \tag{7.31}$$

where α_0 is the fractional optical absorption coefficient due to a single quantum well. If a waveguide configuration is used instead of a single-pass (Brewster's angle) configuration, the above equation can be rewritten as

$$I_N = I_{\rm i}(1 - \alpha_0)^{N \cdot P}, \tag{7.32}$$

where P is the number of passes that the light makes inside the waveguide. With the aid of Equation 7.32, the internal quantum efficiency, η can be defined as

$$\eta = \frac{(I_{\rm i} - I_N)}{I_{\rm i}} = \frac{(I_{\rm i} - I_{\rm i}(1 - \alpha_0)^{N \cdot P})}{I_{\rm i}} = \{1 - (1 - \alpha_0)^{N \cdot P}\}. \tag{7.33}$$

The internal quantum efficiency is plotted as a function of the number of quantum wells in Fig. 7.10. In this plot, the fractional optical absorption due to a single quantum well is assumed to be 0.005 and the total number of passes in the waveguide is $P = 2$.

Theoretically, α_0 can be calculated using the formalism presented in Chapter 6. Experimentally, the fractional optical absorption due to a single quantum well can be obtained from the following expression

$$A_{\rm max} = -\log_{10}(1 - \alpha_0)^{N \cdot P}, \tag{7.34}$$

where $A_{\rm max}$ is the amplitude of the optical absorbance of the intersubband transition. For example, the two peaks of the optical absorbance shown in Fig. 7.8 yield $A_{\rm max} = 0.18$ and $A_{\rm max} = 0.1$ for the two peaks observed at 10.5 μm and 5.7 μm, respectively. The total number of quantum wells is $N = 10$ and the number of passes is $P \approx 2$. Substituting these values in Equation 7.34, one can obtain $\alpha_0 = 0.0205$ and 0.0114 for the two peaks. Substituting the values of α_0 into Equation 7.33 yields internal quantum efficiencies of ~34.0% and ~20.0% for the two peaks observed at 10.5 μm and 5.7 μm, respectively.

Another important aspect of the QWIPs is the dark current or the measured electric current when the device is kept in the dark. There are different mechanisms that contribute to the dark currents, such as thermionic emission and

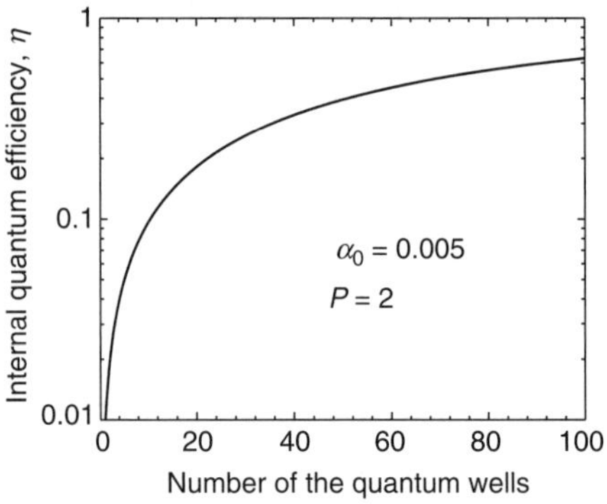

FIGURE 7.10 The internal quantum efficiency plotted as a function of the number of the quantum wells, assuming $\alpha_0 = 0.005$ and $P = 2$.

thermionic assisted tunneling. However, several methods can be used to reduce the dark current and enhance the QWIP performance. For example, the barrier width can be grown thick enough to reduce the tunneling of charger carriers. Reduction of the operating temperature reduces the thermionic emission significantly.

Several models were presented for the derivation of the dark current (Levine; Razeghi, 1996; Rogalski) can be obtained for QWIPs. The general form of the dark current, I_{dark}, can be written as (Levine)

$$I_{\text{dark}} = N^*(\mathcal{E}) e \upsilon_{\text{d}}(\mathcal{E}) A, \tag{7.35}$$

where $N^*(\mathcal{E})$ is the effective number of electrons that are thermally excited out of the well into the continuum transport state (which is a function of the applied electric field, $\mathcal{E}$), $\upsilon_{\text{d}}(\mathcal{E})$ is the average drift velocity (which is also a function of the applied electric filed, $\mathcal{E}$), and A is the area of the detector. The effective number of the thermally excited electrons, $N^*(\mathcal{E})$ can be written as

$$N^*(\mathcal{E}) = \left(\frac{m*}{\pi \hbar^2 L_{\text{p}}}\right) \int_{E_1}^{\infty} f(E) T_{\text{r}}(E, \mathcal{E}) \text{d}E, \tag{7.36}$$

where m^* is the electron effective mass, L_{p} is the total period length (the sum of the well and barrier thicknesses), $f(E)$ is the Fermi–Dirac distribution function given by

$$f(E) = (1 + \text{e}^{(E - E_1 - E_{\text{F}})/(k_{\text{B}} T)})^{-1}, \tag{7.37}$$

where E is the energy of the electron, E_1 is the bound ground state in the well, and E_F is the Fermi energy level. All these parameters are illustrated in Fig. 7.11. The quantity $T_r(E, \mathcal{E})$ is the tunneling transmission coefficient for a single barrier, which can by written according to the WKB approximation as (see appendix D)

$$T_{\text{r}}(E, \mathcal{E}) = \text{e}^{-2 \int_0^{Z_c} \sqrt{2 m_b^* (V - E - e\mathcal{E}z)/\hbar \text{d}z}}, \tag{7.38}$$

where m_{b}^* is the effective mass of the charge carrier in the barrier material, V is given by

$$V = V_0 - e\mathcal{E} L_{\text{w}}/2, \tag{7.39}$$

where V_0 is the barrier height at zero applied electric field and L_{w} is the well thickness, and $Z_{\text{c}} = (V - E)/(e\mathcal{E})$ defines the semiclassical turning point, as illustrated in Fig. 7.11. The average drift velocity in Equation 7.35 can be expressed as

$$\upsilon_{\text{d}}(\mathcal{E}) = \frac{\mu \mathcal{E}}{\sqrt{1 + (\mu \mathcal{E}/\upsilon_{\text{sat}})^2}}, \tag{7.40}$$

where μ is the charge carrier mobility and υ_{sat} is the saturation drift velocity.

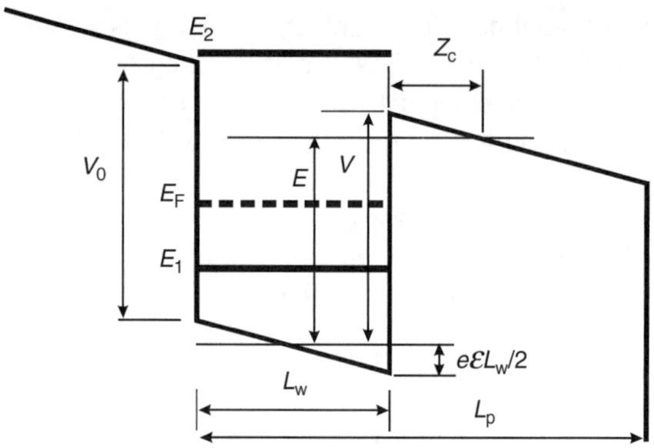

FIGURE 7.11 A sketch of a quantum well showing the energy levels, Fermi energy level, and other parameters used in the derivation of the dark current.

For low bias voltages, the tunneling transmission coefficient can be set as $T_r(E, \mathcal{E}) = 1$ for $E > V_0$ and $T_r(E, \mathcal{E}) = 0$ for $E < V_0$. In this case, Equation 7.36 can rewritten as

$$m_b^* N^*(\mathcal{E}) \approx \left(\frac{m^* k_B T}{\pi \hbar^2 L_p}\right) e^{-(E_{cut}-E_F)/(k_B T)}, \tag{7.41}$$

where $E_{cut} = V - E_1$ is the spectral cutoff energy and the Fermi energy can be obtained according the formalism in Chapter 5 as

$$N_d = \left(\frac{m^* k_B T}{\pi \hbar^2 L_w}\right) \ln(1 + e^{E_F/(k_B T)}), \tag{7.42}$$

where N_d is the doping density in the well. Thus, the dark current can be written as in terms of the temperature as

$$I_{dark} \propto T e^{-(E_{cut}-E_F)/(k_B T)}, \tag{7.43}$$

where E_{cut} is the cutoff energy taken as $E_{cut} = V_0 - E_1$.

The detectivity of QWIPs is derived in a similar manner as the detectivity described earlier in this chapter, which can be written as

$$D^* = R_i \frac{\sqrt{A \Delta f}}{I_n}, \tag{7.44}$$

where R_i is the spectral current responsivity, A is the area of the detector, and I_n is the shot noise introduced earlier in this chapter, which is given for a QWIP as (see Beck)

$$I_n = \sqrt{4 e I_{dark} G \Delta f}, \tag{7.45}$$

where G is the photoconductive gain and I_{dark} is the dark current derived above. This expression is obtained assuming that the capture probability of the electron by the quantum well is much smaller than unity. Substituting the expression of the dark current given by Equation 7.43 into Equation 7.45 gives the following general expression for the detectivity

$$D^* = D_0 e^{\{E_{\text{cut}}/(k_B T)\}}, \tag{7.46}$$

where D_0 is a constant. On the basis of the experimental measurements reported by Levine, the best of the results yields

$$D^* = 1.1 \times 10^6 e^{\left\{\frac{E_{\text{cut}}}{2k_B T}\right\}} \quad \text{for n-type QWIPs} \tag{7.47a}$$

$$D^* = 2.0 \times 10^5 e^{\left\{\frac{E_{\text{cut}}}{2k_B T}\right\}} \quad \text{for p-type QWIPs.} \tag{7.47b}$$

The cutoff energy can be rewritten as $E_{\text{cut}}(\text{eV}) = hc/\lambda_{\text{cut}} = 1.24/\lambda_{\text{cut}}$ (μm). A plot of the expression (a) in Equation 7.47a, 7.47b is shown in Fig. 7.12 for different temperatures.

The difference between the n-type and p-type QWIPs is that the quantum wells are doped with either donors or accepters. In case of n-type QWIPs, the wells are doped with silicon and the quantum wells are populated with electrons. For the p-type QWIPs, the quantum wells are doped with either beryllium or carbon. Thus, the charge carriers are holes. A typical example of the photoresponse of p-type QWIP is shown in Fig. 7.13, where the intersubband transition is between the ground state of the hole and the continuum in the valence band (VB).

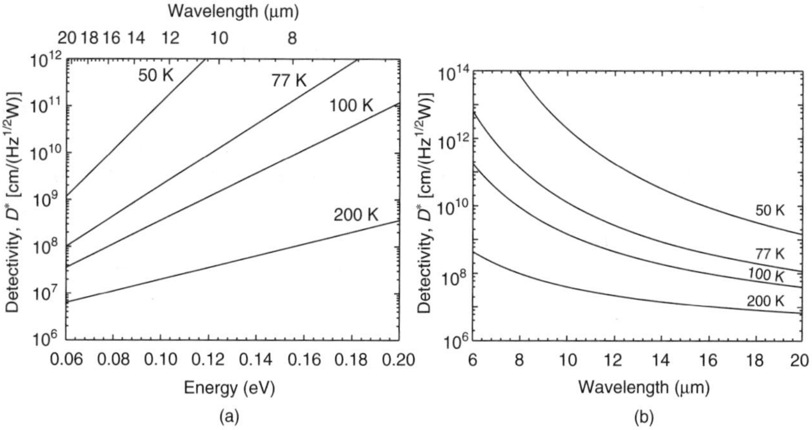

FIGURE 7.12 The detectivity plotted as a function of (a) energy and (b) wavelength for different temperatures.

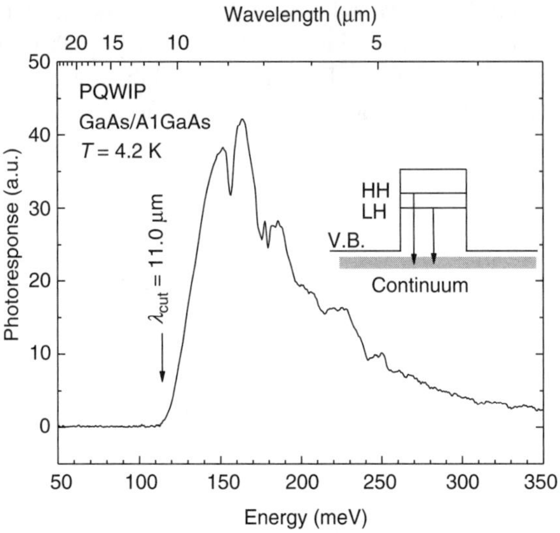

FIGURE 7.13 The photoresponse of a p-type QWIP plotted as a function of incident photon energy. The inset is a sketch of the intersubband transitions from the heavy (HH) and light (LH) hole to the continuum in the valence band (VB).

The advantage of the p-type QWIP is that normal incident photons can be absorbed by the charge carriers. The disadvantage is that the heavy hole effective mass is much larger than the electron effective mass causing a reduction in the detectivity as shown in Equation 7.47b. Further discussion about the p-type QWIPs is presented by Brown and Szmulowiczs.

7.2.4 Infrared Photodetectors Based on Multiple Quantum Dots

In case of n-type multiple quantum wells, the electron–photon coupling occurred when there was a polarization light component (TM) parallel to the growth axis. This required that the photons illuminate the sample at an incident angle different from the normal. The maximum photon–electron coupling occurs at the Brewster's angle. Grating layers are usually fabricated to scatter the light at an angle. Owing to the lack of energy dispersion in the zero dimensional electron systems such as quantum dots, the photon–electron selection rules are absent and the electron can absorb normal incident photons. This property made quantum dots attractive as potential alternative systems to the multiple quantum dots.

The quantum dot system can be designed to increase the optical path length by growing several layers of dots as shown in Fig. 7.14. In this figure, we sketched a structure depicting an infrared detector made of InAs quantum dots and the GaAs barrier. The wetting layer is also shown, which is the result of the layer-island growth mode known as the *Stranski–Krastanov* mode. Ohmic contacts are

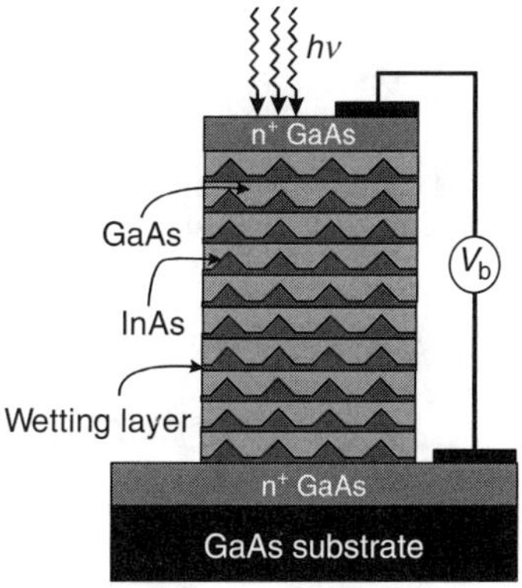

FIGURE 7.14 A simple sketch of an InAs/GaAs multiple quantum dot detectors showing the wetting layers and pyramidal-shaped quantum dots.

fabricated on top of the contact layers of n^+ GaAs material as shown in Fig. 7.14 to bias the device.

Several quantum dot systems have be reported as infrared detectors. The most common system is the InAs/GaAs multiple quantum dots. Different structures have been designed and grown to mainly reduce the dark current. Unlike the doped multiple quantum wells, where the charge carriers are needed to populate the quantum wells, the multiple quantum dots in many cases do not even need to be doped, leading to a significant reduction in the dark current. For undoped quantum dot detectors, the contact layers labeled n^+ GaAs provide the charge carriers under bias voltage. A few designs of the multiple quantum dot infrared detectors are shown in Fig. 7.15. The sketches in this figure do not show the wetting layer. There is a general consensus that the wetting layer is usually of the order of one monolayer InAs intermixed with the GaAs barrier. A simple InAs/GaAs quantum dot band structure is shown in Fig. 7.15a, where two bound states exist in the structure and the intersubband transition is indicated by the arrow.

A different structure composed of an InAs quantum dot embedded in an InGaAs quantum well is shown in Fig. 7.15b. This structure is usually referred to as *dot-well* structure. Since the charge carriers are often tunneled out of the dot, especially when a bias voltage is applied, the dark current can be reduced by adding a thin layer of AlAs as shown in Fig. 7.15c. The AlAs layer is usually referred to as the *blocking layer*. This layer can be chosen as AlGaAs with an Al mole fraction of $\sim 30\%$. The presence of the blocking layer has proven to reduce the dark current and increase the responsivity of the quantum dot infrared photodetectors. Short-period superlattices, such as one monolayer of AlAs and two monolayers of GaAs, can also be used as blocking layers. The fourth structure shown in Fig. 7.15d is referred to as *dot-graded well* structure. Several bound states are present in the dot-graded well structures.

The photoconductivity measurements of multiple quantum dot infrared detectors have been reported by many authors (Madhukar *et al.*) An example of

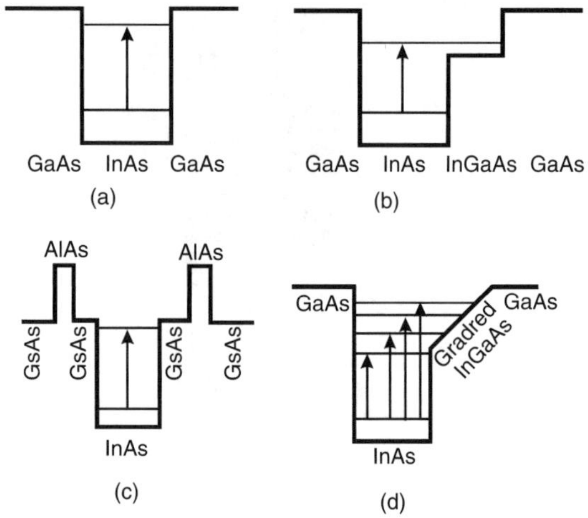

FIGURE 7.15 The conduction band structure of (a) InAs/GaAs quantum dot, (b) InAs/InGaAs/GaAs dot-well structure, (c) InAs/GaAs quantum dot with AlAs blocking layer, and (d) InAs/graded InGaAs well/GaAs barrier structure.

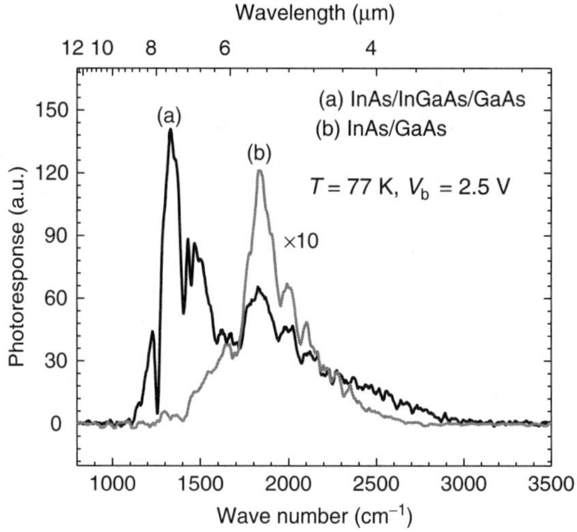

FIGURE 7.16 Photoconductivity measurements of (a) InAs/InGaAs/GaAs dot-well detector, and (b) InAs/GaAs multiple quantum dot detector. The bias voltage was 2.5 V.

the photoconductivity measurements obtained for multiple quantum dot infrared detectors is shown in Fig. 7.16. In this figure, the photoconductivity spectrum of InAs/InGaAs/GaAs (dot-well) structure is shown as trace (a). It has a cutoff wavelength around ~9.0 μm and it peaks around ~7.5 μm. As a comparison, the photoconductivity spectrum of a simple InAs/GaAs multiple quantum dots detector is shown as trace (b) in the figure. Both devices were biased under the same condition (bias voltage, $V_b = 2.5$ V). It peaks around ~5.5 μm. The growth conditions are identical for both devices. In addition, the number of periods of the undoped multiple quantum dots is the same for both structures. The number of InAs monolayers deposited by the MBE growth system is also the same for both structures, but a significant redshift in the peak position is observed in case of the InAs/InGaAs/GaAs dot-well device. This redshift is attributed to the lowering of the excited state, which exists in the quantum well. Thus, with a proper design, the intersubband transition in quantum dots can be tuned to cover a range of wavelengths.

Finally, a quantum dots system schematically presented in Fig. 7.15d exhibit a rich structure in the photoresponse as shown in Fig. 7.17. The three spectra were recorded as 77 K under three different reverse bias voltages. The structure is composed of 10 periods of InAs quantum dots embedded in InGaAs graded quantum wells with a GaAs barrier. There are several excited states in the InGaAs graded quantum well such that a few intersubband transitions are present. The maximum photoresponse is relatively high compared to other quantum dot systems examined here, and the structure in the spectra is repeatable. The fine structure observed in the spectra could be a combination of intersubband transitions in different sized quantum dots.

The dark current in multiple quantum dot infrared detectors is similar to that observed in the multiple quantum well infrared detectors, which is dominated by the thermionic emission and tunneling processes. The three major processes that

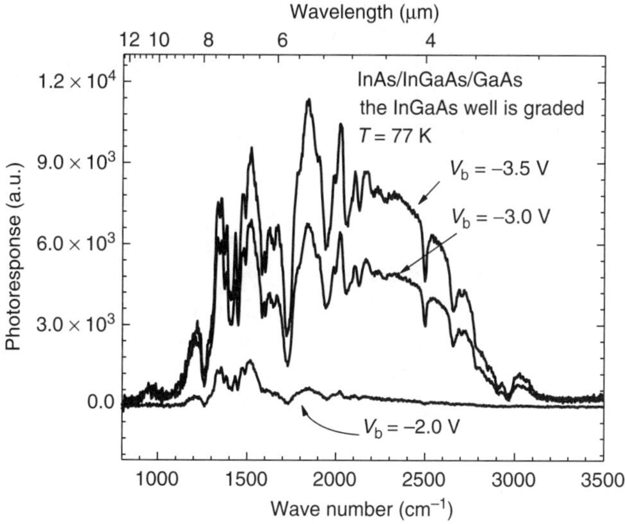

FIGURE 7.17 Photoresponse spectra measured at 77 K under three different reverse bias voltage for InAs multiple quantum dots embedded in InGaAs graded quantum well with GaAs barrier.

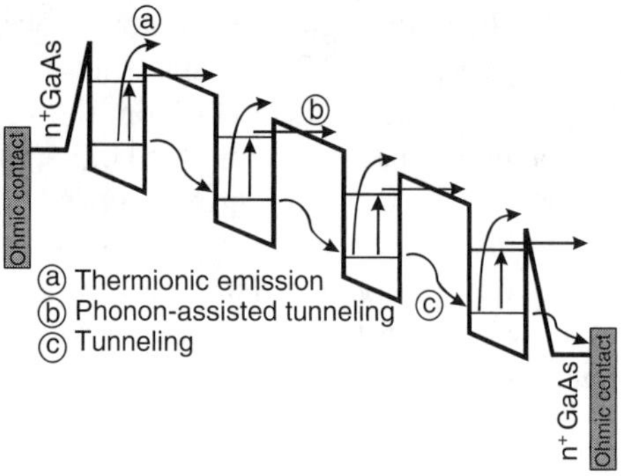

FIGURE 7.18 An illustration of the three major processes that contribute to the dark current in the multiple quantum dot infrared detector.

contribute to the dark current are labeled in Fig. 7.18 as (a) thermionic emission, where the electrons are thermally excited out of the quantum dot to the continuum and then swept by the bias voltage, and (b) phonon-assisted tunneling. This process indicates that the electrons are thermally excited from the ground states to the excited state, followed by tunneling through the trapezoidal barrier to the continuum. (d) Tunneling process by which the electrons in the ground states tunnel when these energy levels are aligned. Similarly, the current noise in quantum dot detectors is governed by the same processes discussed previously in this chapter, which are mainly the $1/f$ noise, Johnson noise, and generation–recombination noise. The sketch shown in Fig. 7.18 is for undoped multiple quantum dots. The quantum dots, however, are populated by the charge carriers originated from the n^+ GaAs contact layer when the bias voltage is applied.

The external quantum efficiency, η_{ex}, of the multiple quantum dot infrared detectors can be defined as the ratio of the number of collected output carriers to the number of incident photons, according to the following expression

$$\eta_{ex} = \frac{i_{ph}/e}{P_{in}/h\nu} = \mathcal{R}\frac{1.24}{\lambda}, \tag{7.48}$$

where $\mathcal{R}$ is the responsivity defined by Equation 7.3, i_{ph} is the photocurrent defined by Equation 7.22, P_{in} is the spectral radiant input power defined by Equation 7.2, and λ is the photon wavelength in micrometers. The external quantum efficiency is different from the internal quantum efficiency, η_{in}, defined by the optical absorption coefficient as shown in Equation 7.33. The internal and external quantum efficiencics can be related according to the following approximate relation

$$\eta_{in} = \frac{\eta_{ex}}{G(1-r)}, \tag{7.49}$$

where G is the photoconductive gain described in Chapter 4 and r is the reflectivity of the semiconductor surface (in this case, GaAs). If the photoconductive gain is unity, Equation 7.49 is reduced to $\eta_{ex} = \eta_{in}(1 - r)$. If the photoconductive gain is approximately equal to the noise gain, then one can write

$$G = \frac{i_n^2}{4eI_{dark}}, \tag{7.50}$$

where i_n is the current noise and I_{dark} is the dark current. These current components can be measured experimentally, from which the photoconductive gain can be estimated using the above relation.

The transport mechanisms in multiple quantum dot infrared detectors are similar to those in multiple quantum well detectors since the basic modes of operations are identical in both systems. Additional effects are presented in the quantum dot systems. For example, planar transports between quantum dots through tunneling exist. Furthermore, the space between the quantum dots is comparable to the space occupied by the dots. Thus, the fill factor is less than unity and the photoconductive gain of quantum dots detector can be written as

$$G = \frac{1 - p}{FNp}, \tag{7.51}$$

where F is the fill factor, N is the number of quantum dots periods, and p is the trapping or capture probability given by

$$p = \frac{\tau_{esc}}{\tau_{life} + \tau_{esc}}, \tag{7.52}$$

where τ_{esc} is the escape time for a photoexcited electron form the quantum dot and τ_{life} is the lifetime of the carrier.

The major transport mechanisms in a multiple quantum dot detector under a bias voltage are shown in Fig. 7.19. The photocurrent is usually a few orders

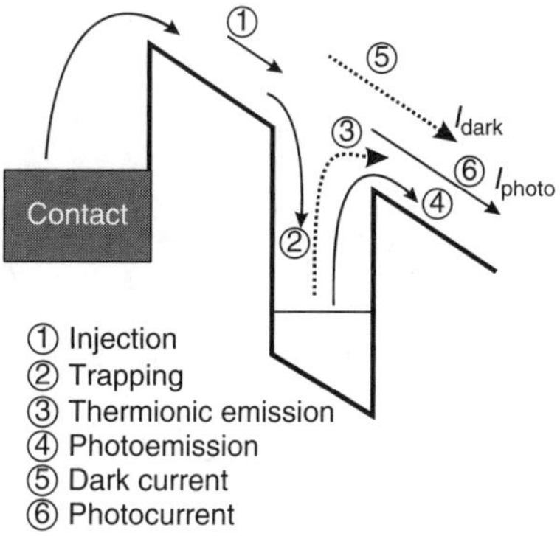

FIGURE 7.19 The major transport mechanisms in multiple quantum dot infrared detectors. The dashed arrows indicate the thermionic emission process that contributes to the dark current.

of magnitude larger than the dark current. The dark current components are the thermionic emission and the injected carriers that reach the collector contact without being trapped. The total photocurrent can be derived in a manner similar to that of the multiple quantum wells photodetectors presented by Liu (1996), which is given by

$$I_{\text{photo}} = i_{\text{photo}} + \delta i, \tag{7.53}$$

where i_{photo} is the total photocurrent due to only the photoexcited carriers. Taking into account the capture probability, i_{photo} can be written as

$$i_{\text{photo}} = i_{\text{photo}}^1 \sum_{n=1}^{N} (1-p)^{n-1} = i_{\text{photo}}^1 \frac{\{1-(1-p)^N\}}{p}, \tag{7.54}$$

where i_{photo}^1 is the photocurrent due to the photoexcited carriers from a single quantum dot, which is given as

$$i_{\text{photo}}^1 = e\Phi\eta \frac{1-p}{FN}, \tag{7.55}$$

where Φ is the incident number of photons per second, and η is the absorption quantum efficiency.

The second quantity, δi, in Equation 7.53 is the fraction of the extra injected current that reaches the collector contact, which can be written as

$$\delta i = (1-p)^N \delta I_{\text{inject}} = i_{\text{photo}}^1 \frac{(1-p)^N}{p}. \tag{7.56}$$

The quantity δI_{inject} is the fraction of the injected current that is needed to balance the net loss of electrons in the quantum dot because of photoemission, which is given as

$$\delta I_{\text{inject}} = \frac{i_{\text{photo}}^1}{p}. \tag{7.57}$$

Combining Equations 7.53, 7.54, and 7.56 yields

$$I_{\text{photo}} = i_{\text{photo}} + \delta i = \frac{i_{\text{photo}}^1}{p} = e\Phi\eta \frac{(1-p)}{FNp}. \tag{7.58}$$

This expression verifies the photoconductive gain given by Equation 7.51 or

$$G = \frac{I_{\text{photo}}}{e\Phi\eta} = \frac{(1-p)}{FNp}. \tag{7.59}$$

The relationship tells us that if the capture probability is unity, the photoconductive gain is zero. As the capture probability is decreased, the photoconductive gain increases.

7.3 LIGHT-EMITTING DIODES

LEDs were discovered early in the twentieth century. As the name implies, a semiconductor diode is subject to a small forward bias voltage, where the electrons are injected in a normally empty CB. The injected electrons recombine in the holes in the VB by emitting their energy as photons. This process is called *electroluminescence* or *spontaneous emission*. Since an optical cavity is not needed to provide photon feedback, the emitted photons have random phases and therefore the LEDs are incoherent light sources. Furthermore, the emitted photon energy is close to the band gap of the semiconductor material. A sketch of an LED p–n junction is shown in Fig. 7.20, where the injected electrons in the CB recombine with the holes in the VB by emitting photons. A small fraction of the injected electrons recombine with holes in the VB without emitting photons (nonradiative recombination process). Furthermore, the LEDs are usually made of direct band gap semiconductor materials, since the conversion efficiency of the electrical signal to the photon signal in indirect semiconductor materials is very low. This is due to the involvement of the phonon in the interband transitions.

For p–n junction LEDs, the wavelength of the emitted light depends on the band gap of the material. For narrow band gap materials, the wavelength is in the infrared region; for wide band gap materials, the wavelength is in the visible and ultraviolet spectral regions. In addition to the direct band gap requirements, the semiconductor material should be easily doped with both n- and p-type to form the junction. For example, GaAs is a direct semiconductor material and it can be easily doped with donors or acceptors. The wavelength of the LEDs made of GaAs is usually around 0.855 μm at room temperature. On the other hand, GaN is a direct semiconductor material with a band gap of ~3.40 eV at 0.365 μm). The wavelength of the LEDs can be easily tuned by choosing ternary materials such as AlGaAs, InGaN, and AlGaN.

In addition to the charge carrier injection and the radiative recombination processes, the generated photon should be able to exit the device and be used for the intended applications. The latter process is called the *extraction process*. Each of these processes has its own efficiency. The overall device efficiency, η_0,

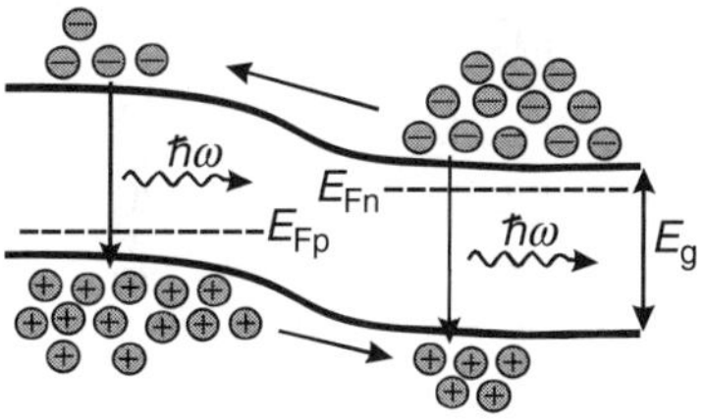

FIGURE 7.20 A sketch of a p–n junction under the influence of a small forward biased voltage. The spontaneous emission occurs when the electrons combine with the holes.

(also known as *external efficiency*) can be expressed as follows:

$$\eta_0 = \eta_{\text{inj}}\eta_{\text{r}}\eta_{\text{e}}, \tag{7.60}$$

where η_{inj} is the injection efficiency, η_{r} is the recombination efficiency, and η_{e} is the extraction efficiency. It is difficult to measure the individual efficiencies. But the external efficiency can be measured by taking the ratio of the output optical power to the input electrical power. For p–n junction LEDs, the external efficiency is usually in the range of 1–10%.

For parabolic electron–hole bands, the LED spontaneous emission rate, r_{sp}, can be written as

$$r_{\text{sp}} = P_{\text{em}} N_j(E) f(E_{\text{e}})\{1 - f(E_{\text{h}})\}, \tag{7.61}$$

where $f(E_{\text{e}})$ and $f(E_h)$ are the Fermi–Dirac distribution functions of the electrons in the CB and the holes in the VB, respectively, P_{em} is the emission probability, which is the inverse of the recombination lifetime, τ_{r}, and $N_j(E)$ is the joint density of state given by

$$N_j(E) = \frac{(2m_{\text{r}}^*)^{3/2}}{2\pi^2\hbar^3}\sqrt{E - E_{\text{g}}}, \tag{7.62}$$

where m_{r}^* is the reduced effective mass and E_{g} is the band gap. The unit of the emission rate is (/s/eV/cm^3). The total spontaneous emission rate, R_{sp}, per unit volume can be defined as

$$R_{\text{sp}} = \int r_{\text{sp}} \mathrm{d}E. \tag{7.63}$$

For $E \ll k_{\text{B}}T$, the Fermi–Dirac distribution function can be approximated as the Boltzmann distribution function or $f(E_{\text{e}})\{1 - f(E_{\text{h}})\} \approx \mathrm{e}^{-E/k_{\text{B}}T}$. With this approximation, the spontaneous emission rate can be written as

$$r_{\text{sp}} = \frac{(2m_{\text{r}}^*)^{3/2}}{2\pi^2\hbar^3\tau_{\text{r}}}\sqrt{E - E_{\text{g}}}\,\mathrm{e}^{\frac{-E_{\text{g}}}{k_{\text{B}}T}}\,\mathrm{e}^{\frac{-(E-E_{\text{g}})}{k_{\text{B}}T}}. \tag{7.64}$$

For a weak forward bias voltage (weak injection), the quasi-Fermi energy levels in the LED are still in the band gap (nondegenerate case). Thus, Equation 7.64 can be modified by including the quasi-Fermi energy levels as

$$r_{\text{sp}} = \frac{(2m_{\text{r}}^*)^{3/2}}{2\pi^2\hbar^3\tau_{\text{r}}}\sqrt{E - E_{\text{g}}}\,\mathrm{e}^{\frac{(E_{\text{Fn}}-E_{\text{Fp}}-E_{\text{g}})}{k_{\text{B}}T}}\,\mathrm{e}^{\frac{-(E-E_{\text{g}})}{k_{\text{B}}T}}. \tag{7.65}$$

The total photon flux, Φ_0, emitted from the LED is obtained by integrating over r_{sp}

$$\Phi_0 = V\int_0^\infty r_{sp}\mathrm{d}E = \frac{(2m_r^*)^{3/2}}{2\pi^2\hbar^3\tau_r}\,\mathrm{e}^{\frac{(E_{Fn}-E_{Fp}-E_g)}{k_BT}}\int_0^\infty \sqrt{E-E_g}\,\mathrm{e}^{\frac{-(E-E_g)}{k_BT}}\,\mathrm{d}E$$

$$= \frac{V}{\sqrt{2}\hbar^3\tau_r}\left(\frac{m_r^* k_BT}{\pi}\right)^{3/2}\mathrm{e}^{\frac{(E_{Fn}-E_{Fp}-E_g)}{k_BT}}, \qquad (7.66)$$

where V is the volume of the active region. In this expression, the recombination lifetime is assumed to be independent of the charge carrier energy, which may not be true. It is clear from Equation 7.66 that the photon flux depends on temperature and on the positions of the quasi-Fermi energy levels. By increasing the injection level, the quasi-Fermi energy levels are moved closer to the CB and the VB for the electrons and hole, respectively, and the separation between them ($E_{Fn} - E_{Fp}$) increases. Thus, the photon flux increases as the charge carrier injection level is increased.

Another important parameter is the responsivity, $\mathcal{R}_{LED}$ of the LED, which is defined as the ratio of the emitted optical power to the injection current

$$\mathcal{R}_{LED} = \frac{\Phi_0\hbar\omega}{I_{inj}} = \frac{hc}{e\lambda}\eta_0 = \frac{1.24}{\lambda(\mu m)}\eta_0, \qquad (7.67)$$

where the emitted optical power is defined as the product of the photon flux given by Equation 7.66 and the photon energy ($\hbar\omega$), and I_{inj} is the injection current. The units of responsivity is (W/A). For an external efficiency of 5% of a GaAs LED at room temperature, the responsivity is $\mathcal{R}_{LED} = (1.24)(0.05)/(0.855) =$ 72.50 mW/A = 72.5 μW/mA. The optical power can be defined as the product of the emitted optical power and the injected current or

$$\mathcal{P}_{LED} = \Phi_0\hbar\omega I_{inj} = \eta_0\frac{hc}{e\lambda}I_{inj} = \frac{1.24 I_{inj}\eta_0}{\lambda(\mu m)}. \qquad (7.68)$$

For an injected current of 5 mA and an external efficiency of 5%, an LED emits light at 0.532 μm and has power of $\mathcal{P}_{LED} = (1.24)(5\times10^{-3})(0.05)/(0.532) =$ 0.58 mW.

The surface nonradiative recombination and the carrier injections from the n-region to the p-region through the depletion region are two processes that reduce the external efficiency of the LED. These two disadvantages can be overcome by fabricating a heterojunction LED, where the active region can be made as thin as possible. Figure 7.21 shows a sketch of a GaAs/AlGaAs LED, where the p^- GaAs is the active region. The n^+ AlGaAs provides an interface with state density much lower than the free surface states, which reduces the nonradiative recombination significantly. It should be pointed out that the band gap of AlGaAs

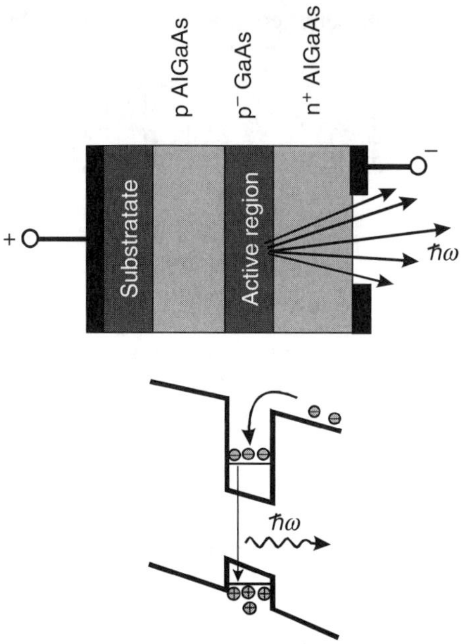

FIGURE 7.21 A sketch of GaAs/AlGaAs heterojunction LED is shown. The band gap of the LED is also sketched under forward bias voltage.

is larger than that of GaAs, and, therefore, the top AlGaAs layer acts as a window through which the emitted photons exit with minimum reabsorption.

The LED frequency response can be derived from the continuity equation that takes into account the excess electron loss due to the spontaneous recombination and diffusion. If the drift current is neglected, the continuity equation can be written as

$$\frac{\partial n(x)}{\partial t} = -\frac{n(x)}{\tau_r} + D_e \frac{\partial^2 n(x)}{\partial x^2}, \tag{7.69}$$

where τ_r is the recombination lifetime throughout the active region, D_e is the diffusion coefficient, and $n(x)$ is the excess carrier concentration. A possible solution of this equation is

$$n(x) = n_0(x) + n_1(x)e^{i\omega t}, \tag{7.70}$$

where the first term is due to the steady state due to the DC bias and the second term is due to the time-dependent small-signal modulation of the diode. By separating the DC and the frequency-dependent parts, Equation 7.68 can be rewritten as

$$D_e \frac{\partial^2 n_0(x)}{\partial x^2} - \frac{n_0(x)}{\tau_r} = 0 \text{ and } D_e \frac{\partial^2 n_1(x)}{\partial x^2} - \frac{n_1(x)\{1 + i\omega\tau_r\}}{\tau_r} = 0. \tag{7.71}$$

The diffusion length can be defined as

$$L_e^2 = \begin{cases} D_e\tau_r & \text{for the DC part} \\ \dfrac{D_e\tau_r}{\sqrt{1+\omega^2\tau_r^2}} & \text{for the AC part} \end{cases} \tag{7.72}$$

Substituting Equation 7.72 into Equation 7.71 yields

$$\frac{\partial^2 n_0(x)}{\partial x^2} - \frac{n_0(x)}{L_e^2} = 0 \text{ and } \frac{\partial^2 n_1(x)}{\partial x^2} - \frac{n_1(x)}{L_e^2(\omega)} = 0, \tag{7.73}$$

where $L_e(\omega)$ is taken as the second expression in Equation 7.72.

If the width of the active region, d, of the LED is much larger than the diffusion length, L_e, and the carrier concentration at the p AlGaAs/p$^-$ GaAs interface (Fig. 7.21) is zero, that is, $n_1(x = d) = 0$, then the solution to the frequency-dependent part of Equation 7.73 can be expressed as

$$n_1(x) = n_1^0 e^{-x/L_e(\omega)}, \tag{7.74}$$

where n_1^0 is the initial electron concentration injected in the active region. The frequency response of the LED, $r(\omega)$ can be defined as

$$r(\omega) = \frac{\Phi_1(\omega)}{J_1(\omega)/e}, \tag{7.75}$$

where $\Phi_1(\omega)$ is the AC photon flux given by

$$\Phi_1(\omega) = \frac{1}{\tau_r}\int_0^d n_1(x)\mathrm{d}x = \frac{n_1^0}{\tau_r}\int_0^d \mathrm{e}^{-x/L_e(\omega)}$$
$$= \frac{n_1^0 L_e(\omega)}{\tau_r} e^{-x/L_e(\omega)}\Big|_0^d = \frac{n_1^0 L_e(\omega)}{\tau_r}. \tag{7.76}$$

Notice that from the initial conditions, we assumed that $n_1(x = d) = 0$. The AC current density, $J_1(\omega)$ shown in Equation 7.75 can be obtained from the following equation:

$$J_1(\omega) = eD_e\frac{\partial n_1(x)}{\partial x} = -eD_e\frac{n_1^0}{L_e(\omega)}. \tag{7.77}$$

Substituting Equations 7.76 and 7.77 into Equation 7.75 gives

$$r(\omega) = \frac{L_e^2(\omega)}{\tau_r D_e} \quad n_1(x) = n_1^0 \mathrm{e}^{-x/L_e(\omega)}. \tag{7.78}$$

Substitute the expression in Equation 7.72 for the frequency-dependent diffusion length into Equation 7.78 to obtain

$$r(\omega) = \frac{1}{\sqrt{1+\omega^2\tau_r^2}}. \tag{7.79}$$

This expression implies that for high frequency response, the recombination lifetime should be very small.

7.4 SEMICONDUCTOR LASERS

The light amplification by stimulated emission of radiation (laser) is an optical waveguide terminated by mirrors or facets to form a resonant cavity. Einstein described the stimulated emission process in 1917 and Townes demonstrated this process at the microwave frequencies. In 1990, Maiman demonstrated the stimulated emission at the optical frequencies in a ruby crystal. In 1962, the semiconductor laser was introduced. Propagation of electromagnetic waves in waveguides has been the subject of many textbooks, and it is not discussed here.

7.4.1 Basic Principles

There are different types of lasers depending on the medium used. The semiconductor lasers depend on the design of the band structures and the confined energy levels. To illustrate the stimulated emission, consider two energy levels, E_1 and E_2, system populated with electrons of densities N_1 and N_2, respectively, as shown in Fig. 7.22. At thermal equilibrium, the electron populations in the energy levels follow the Maxwell–Boltzmann distribution function assuming that the two energy levels have the same degeneracy

$$\frac{N_2}{N_1} = e^{\frac{-(E_2-E_1)}{k_B T}} = e^{\frac{-\hbar\omega_{12}}{k_B T}}. \tag{7.80}$$

Three different processes can occur when a photon of energy $\hbar\omega_{12} = E_2 - E_1$ interacts with the system. The first process is called *absorption*, where the photon is absorbed by an electron causing the electron to jump from the ground state (E_1) to the excited state (E_2). This process is illustrated in Fig. 7.22a. The second process describes an electron relaxing from the excited state to the ground state releasing its energy as a photon. This process is called *spontaneous* emission, which is illustrated in Fig. 7.22b. The time that an electron spends in the excited state before relaxing to the ground state is called *spontaneous lifetime*. While the energy of the photons resulted from the spontaneous emission is the same, they have random propagation directions and random phases. The third process illustrated in Fig. 7.22c can be thought of as the inverse of absorption. When a

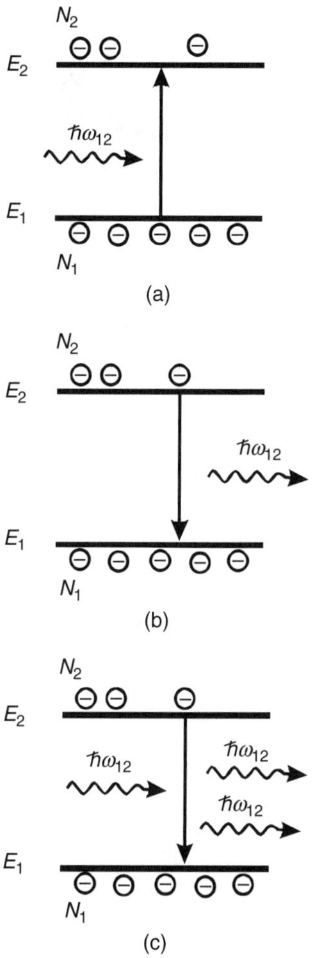

FIGURE 7.22 Interaction of a photon with an electron in two-energy level system showing the three different processes. (a) Absorption process, (b) spontaneous emission, and (c) stimulated emission.

photon of an energy $\hbar\omega_{12}$ interacts with the two-energy level system, it passes the system without being absorbed and forces an electron to relax from the excited state to the ground state emitting another photon with same energy $(\hbar/\omega_{12})$. This process is called *stimulated* emission. This process generates a photon in a time τ_{st}, called the *stimulated emission time*. The propagation direction and the phase of the stimulated photon are identical to those of the passing (stimulating) photon.

In 1917, Einstein showed that the three processes, such as absorption, spontaneous emission, and stimulated emission, described above are related. When the two energy levels in Fig. 7.23 have different degeneracies, Equation 7.80 takes

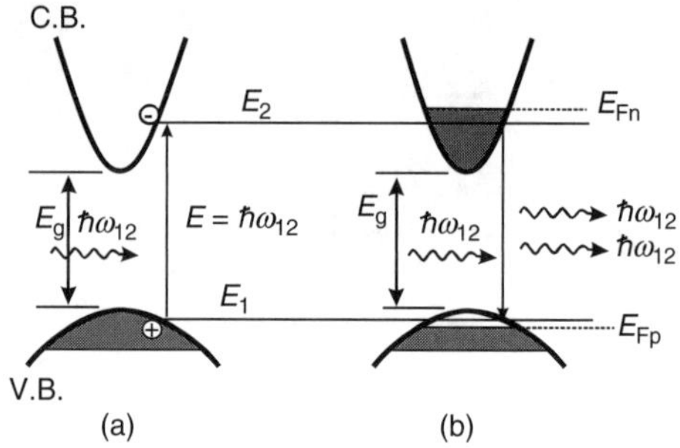

FIGURE 7.23 (a) Absorption and (b) stimulated emission in a direct semiconductor material at 0 K.

the following form:

$$\frac{N_2}{N_1} = \frac{g_2}{g_1} e^{\frac{-(E_2-E_1)}{k_B T}} = \frac{g_2}{g_1} e^{\frac{-\hbar\omega_{12}}{k_B T}}, \tag{7.81}$$

where g_1 and g_2 and the degeneracies of the two energy levels. The upward transition rate, R_{12}, of an electron from E_1 to E_2 can be expressed as

$$R_{12} = N_1 \Phi(\omega_{12}) B_{12}, \tag{7.82}$$

where B_{12} is the Einstein coefficient for absorption or stimulated upward transition and $\Phi(\omega_{12})$ the radiation density of the energy density of the radiation field at frequency ω_{12}.

The spontaneous emission rate, R_{21}^{sp}, is defined as

$$R_{21}^{\text{sp}} = N_2 \frac{1}{\tau_{21}} = N_2 A_{21}, \tag{7.83}$$

where τ_{21} is the spontaneous emission lifetime, which is equal to the spontaneous recombination lifetime, and A_{21} is the Einstein coefficient for spontaneous emission ($A_{21} = 1/\tau_{21}$). Finally, the stimulated emission rate, R_{21}^{st}, of electron from E_2 to E_1 is given by

$$R_{21}^{\text{st}} = N_2 \Phi(\omega_{12}) B_{21}, \tag{7.84}$$

where B_{21} is the Einstein coefficient for stimulated emission. Combining Equations 7.83 and 7.84 yields the total transition rate, R_{21}, from E_2 to E_1

$$R_{21} = N_2 A_{21} + N_2 \Phi(\omega_{12}) B_{21}. \tag{7.85}$$

At thermal equilibrium, the upward transition rate is equal to the downward transition rate, $R_{12} = R_{21}$, or

$$N_1 \Phi(\omega_{12}) B_{12} = N_2 A_{21} + N_2 \Phi(\omega_{12}) B_{21}. \tag{7.86}$$

It follows from this equation that

$$\Phi(\omega_{12}) = \frac{A_{21}}{B_{21}} \frac{1}{\left\{ \frac{N_1 B_{12}}{N_2 B_{21}} - 1 \right\}}. \tag{7.87}$$

By using Equation 7.81, the above expression can be rewritten as

$$\Phi(\omega_{12}) = \frac{A_{21}}{B_{21}} \frac{1}{\left\{ \frac{g_1 B_{12}}{g_2 B_{21}} e^{\frac{\hbar\omega_{12}}{k_B T}} - 1 \right\}}. \tag{7.88}$$

Without going through the detailed analysis, Einstein relations are given as

$$B_{21} = \frac{g_2}{g_1} B_{21} \tag{7.89a}$$

$$\frac{A_{21}}{B_{21}} = \frac{8\pi \nu^3 n_r^3}{c^3}, \tag{7.89b}$$

where c is the speed of light, n_r is the refractive index of the medium, and $\nu = \omega/(2\pi)$. When the spontaneous emission lifetime, τ_{21}, is equal to the recombination lifetime, τ_r. then $A_{21} = 1/\tau_r$ and $B_{21} = c^3/(8\pi \nu^3 n_r^3 \tau_r)$.

The ratio of the stimulated emission rate to the spontaneous emission rate is an important parameter for the laser action, which can be obtained by substituting Equation 7.89a into Equation 7.88 such as

$$\frac{B_{21}\Phi(\omega_{12})}{A_{21}} = \frac{1}{\left\{ e^{\frac{\hbar\omega_{12}}{k_B T}} - 1 \right\}}. \tag{7.90}$$

For stimulated emission to be dominant, the above ratio must be made large. This could be accomplished by letting $\Phi(\omega_{12})$ be very large and N_2 should be made larger than N_1. The latter condition is called the *population inversion*. To achieve the population inversion condition, a large amount of energy is required to excite the carrier from the ground state to the excited state. For two-energy level systems, the population ratio is unity at best ($N_1 = N_2$), which does not produce optically significant optical gain. Many of the laser systems, in particular, gas and solid-state systems, are based on three and four energy levels.

Lasing and optical gain in semiconductor systems is different from the two- or three-energy level systems. It is based on the creation of nonequilibrium condition of the charge carriers in both the CB and the VB. Figure 7.23 illustrates the (a) absorption and (b) lasing action of a semiconductor material. In the initial conditions at $T = 0$ K, the VB is completely full and the CB is completely empty. Incident photons with an energy $E = \hbar\omega_{12}$ generate electron–hole pairs such that the quasi-Fermi energies are above E_2 and below E_1 as shown in Fig. 7.23b. Under this condition, photons with energy between E_g and $(E_{Fn} - E_{Fp})$ cannot be absorbed. On the other hand, photons with this energy, that is, $E_g < \hbar\omega < (E_{Fn} - E_{Fp})$, can induce stimulated emission as shown in Fig. 7.23b. The absorption rate, R_{abs}, of photons with energy E lies between E_g and $(E_{Fn} - E_{Fp})$ and can be expressed as

$$R_{abs} = \mathcal{W}[1 - f_n(E_2)]f_p(E_1)N_p(E), \tag{7.91}$$

where $\mathcal{W}$ is the transition probability discussed in Chapter 4, $N_p(E)$ is the density of photons of energy E, and $f_n(E_2)$ and $f_p(E_1)$ are the Fermi–Dirac distribution functions of the electrons and holes, respectively. Similarly, the stimulated emission rate, R_{st}, can be written as

$$R_{st} = \mathcal{W}[1 - f_p(E_1)]f_n(E_2)N_p(E). \tag{7.92}$$

To achieve optical gain and population inversion, the following conditions must be satisfied

$$R_{st} > R_{abs} \text{ and } f_n(E_2) > f_p(E_1). \tag{7.93}$$

When the photon energy satisfies the condition $\hbar\omega = (E_{Fn} - E_{Fp})$, the photons are not absorbed by the charge carriers and they pass through without losing their energy. This is known as the *transparency point*. For a voltage-driven semiconductor laser, the applied bias voltage, V_b, must satisfy the following condition for the lasing to occur

$$E_g < eV_b = (E_{Fn} - E_{Fp}). \tag{7.94}$$

From the above formalism, the optical gain is positive when $\hbar\omega = E_g$ and continues to increase until it reaches the maximum, then starts to decrease, and approaches zero as $\hbar\omega$ approaches $(E_{Fn} - E_{Fp})$. The condition (Eq. 7.94) tells us that the lasing action occurs when the gain is equal to or larger than the loss. This can be expressed as

$$g(E) \geq -\alpha(E), \tag{7.95}$$

where $g(E)$ is the optical gain and $\alpha(E)$ is the absorption coefficient.

The recombination lifetime and the optical gain are two important parameters for lasing action. Another important parameter is the threshold conditions for lasing. If the semiconductor structure is made as a waveguide, and if the waveguide is cleaved at both ends, the two parallel facets form a resonant cavity that acts as a Fabry–Perot cavity, as shown in Fig. 7.24a. The length of the cavity, L, is given by

$$L = \frac{m\lambda}{2} = \frac{m\lambda_0}{2n_r}, \tag{7.96}$$

where m is the integer that represents the number of allowed modes in the cavity, n_r is the refractive index, λ is the photon wavelength within the cavity, and λ_0 is the photon wavelength in vacuum ($\lambda_0 = \lambda n_r$). The allowed wavelengths within the cavity are called *longitudinal optical modes*. Three longitudinal modes are

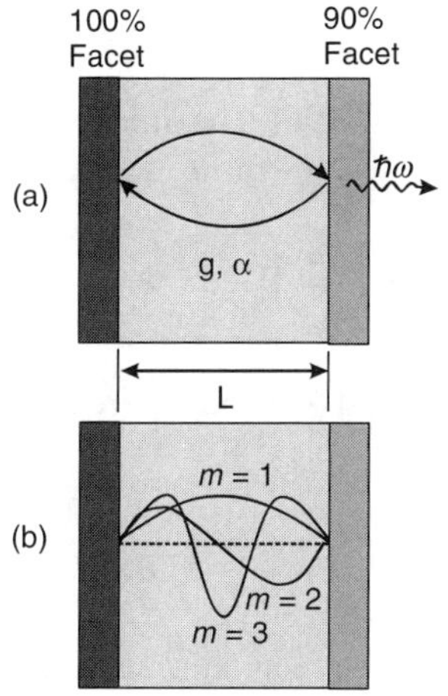

FIGURE 7.24 (a) A waveguide and two parallel facets form the laser cavity or the Fabry–Perot cavity. A round trip of the light in the cavity is shown. (b) Three longitudinal normal modes of the laser are sketched for parallel facets.

shown in Fig. 7.24b. The transverse optical modes can be observed when the facets or the mirrors at the end of the cavity are not quite parallel. This leads to a special intensity distribution at the exit facet (or mirror). Notice that once the reflectivity of one facet is made 100% while the reflectivity of the other facet is made <100%, the laser will exit the cavity.

The amplifying medium between the two mirrors or the two facets is characterized by a well-defined wavelength region in which the stimulated emission occurs. If the medium is characterized by an optical gain, g, and absorption (loss) coefficient, α, and if the reflectivities of the two facets are represented by R_1 and R_2, the intensity of the light, I, that made an around-trip ($2L$) within the cavity (Fig. 7.24a) can be expressed as

$$I = I_0 R_1 R_2 e^{2(g-\alpha)L}, \tag{7.97}$$

where I_0 is the initial intensity. When $g > \alpha$, the increase in light intensity causes amplification of light.

The threshold gain, g_t, can be obtained when the optical gain is equal to the optical loss. In this case $I = I_0$. and Equation 7.97 becomes

$$1 = R_1 R_2 e^{2(g_t-\alpha)L} \tag{7.98}$$

or

$$g_t = \alpha + \frac{1}{2L} \ln\left(\frac{1}{R_1 R_2}\right). \tag{7.99}$$

The second expression on the right-hand side of Equation 7.99 represents the useful laser output. In general, a few round trips of light between the two mirrors are needed before the lasing reaches a steady state and the gain reaches the threshold gain, g_t.

Let us reexamine Equation 7.96. The integer m represents the number of longitudinal modes. For large values of m, one can write Equation 7.96 as

$$\partial m = 2L\partial\left(\frac{n_r}{\lambda_0}\right) = \frac{2L}{\lambda_0}\partial n_r - \frac{2Ln_r}{\lambda_0^2}\partial\lambda_0. \tag{7.100}$$

For discrete changes in m and λ_0, one can rewrite the above equation as

$$\partial\lambda_0 = \left(\frac{2L}{\lambda_0}\frac{dn_r}{d\lambda_0} - \frac{2Ln_r}{\lambda_0^2}\right)^{-1}\delta m = \frac{\lambda_0^2}{2Ln_r}\left(\frac{\lambda_0}{n_r}\frac{dn_r}{d\lambda_0} - 1\right)^{-1}\delta m. \tag{7.101}$$

Thus, the wavelength change between adjacent longitudinal modes is

$$\partial\lambda_0 = \frac{\lambda_0^2}{2Ln_r}\left(1 - \frac{\lambda_0}{n_r}\frac{dn_r}{d\lambda_0}\right)^{-1}. \tag{7.102}$$

The frequency separation between adjacent longitudinal modes ($\Delta m = \pm 1$) is given by

$$\delta\nu = \frac{c}{2Ln_r}. \tag{7.103}$$

This separation is illustrated in Fig. 7.25a, where several longitudinal modes are sketched. The optical gain is sketched in Fig. 7.25b. A large number of laser longitudinal modes exist. However, the only modes amplified are those within the optical gain region as shown in Fig. 7.25c. In general, the optical gain profile resembles the spontaneous emission region. A semiconductor laser can

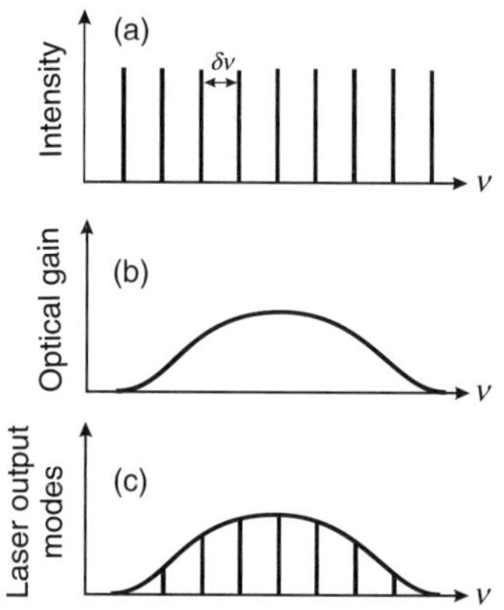

FIGURE 7.25 (a) Longitudinal modes of Fabry–Perot cavity, (b) the output optical gain profile, which is essentially the spontaneous emission profile of the laser. (c) Only the longitudinal modes that are within the optical gain can be amplified.

be designed to produce a few longitudinal modes. This can be accomplished by reducing the optical gain linewidth.

7.4.2 Semiconductor Heterojunction Lasers

The earliest semiconductor lasers are made of degenerate p–n junctions. A large forward bias voltage is applied to the junction, where a large density of electrons and holes are injected. The lasing action usually takes place in a narrow region in the p-type side, as shown in Fig. 7.26. Semiconductor lasers do not need external mirrors to form the cavity. The Fabry–Perot cavity is formed by cleaving the planes at the end of the waveguide. The small region in which the radiative electron–hole recombination occurs is called the *active region*. There are several disadvantages associated with the p–n junction laser, such as the active region is not well defined, the large cavity loss, and large threshold current density.

Semiconductor heterojunction lasers provide a better confinement for both the charge carriers and the longitudinal optical modes. The basic structure of a heterojunction laser is shown in Fig. 7.27a for GaAs/AlGaAs heterojunction at

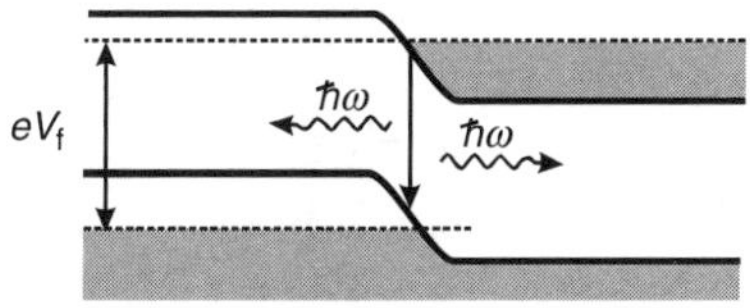

FIGURE 7.26 Degenerate p–n junction laser under forward bias voltage (V_f).

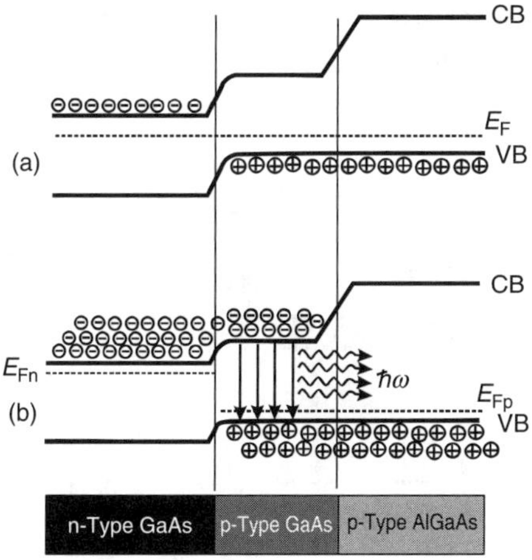

FIGURE 7.27 A GaAs/AlGaAs heterojunction used as diode laser (a) at equilibrium and (b) under the influence of a large forward bias voltage. Notice that the valence band offset between GaAs and AlGaAs is approximated to zero.

equilibrium. It consists of n-type GaAs, p-type GaAs, and p-type AlGaAs. The band diagram for a large forward bias is shown in Fig. 7.27b, where the carriers are injected at both sides of the junction. Population inversion is achieved, and lasing action is obtained. Notice that the VB offset between p-type GaAs and p-type AlGaAs is approximated to zero. This approximation is realistic only when a small aluminum mole fraction is used in the AlGaAs layer. For a more realistic GaAs/AlGaAs heterojunction diode laser, the total band offset is usually divided into 60%:40% ratio of the CB:VB offsets.

The current density needed to turn on the diode laser is called the *threshold current density*, J_t, which is due to the injection and diffusion processes. The p-type GaAs active region thickness is usually smaller than the diffusion length of the carrier, but in the following analysis, the diffusion current density is not neglected. Consider the heterojunction, shown in Fig. 7.28, where the injection level is assumed to be high. The effective confinement barrier, Φ, shown in the figure can be written as

$$\Phi = \Delta E_g - \Delta E_n - \Delta E_{p1} - \Delta E_{p2}, \tag{7.104}$$

where ΔE_g is the total band offset ($\Delta E_g = E_{g2} - E_{g1}$). For the steady state case, the threshold current density can be written as

$$\begin{aligned} J_t &= J_{inj} + J_{diff} \\ &= \frac{e n_t d}{\tau} + \frac{e D_{e2} n_2}{L_{e2}}, \end{aligned} \tag{7.105}$$

where, n_t is the injected carrier density in the active region (p-type GaAs), d is the thickness of the active region, D_{e2} is the electron diffusion coefficient in the barrier region (p-type AlGaAs), L_{e2} is the electron diffusion length in the barrier region, and n_2 is the injected minority carrier density in the barrier.

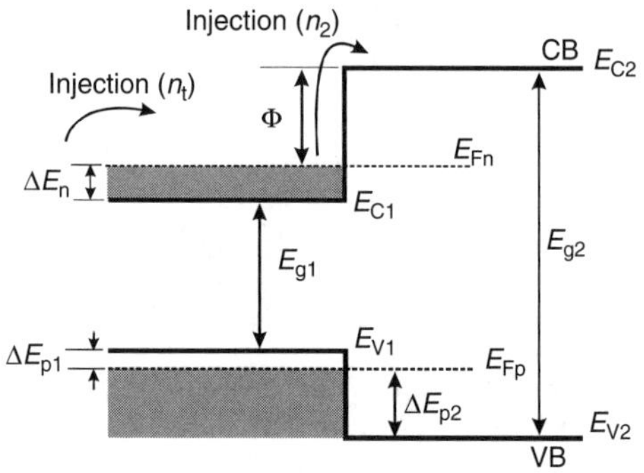

FIGURE 7.28 A sketch of the band diagram of GaAs/AlGaAs heterojunction diode laser used to derive the threshold current density.

Using the formalisms of the density of states presented in Chapter 5 and Fermi–Dirac distribution functions, the carrier densities, n_t, and n_2 can be written as

$$n_t = \frac{4\pi}{h^3}(2m_{e1}^*)^{3/2} \int_{E_{c1}}^{\infty} \frac{\sqrt{E - E_{c1}}}{1 + e^{\frac{(E-E_{Fn})}{k_B T}}} dE \tag{7.106a}$$

$$n_2 = \frac{4\pi}{h^3}(2m_{e2}^*)^{3/2} \int_{E_{c2}}^{\infty} \frac{\sqrt{E - E_{c2}}}{1 + e^{\frac{(E-E_{Fn})}{k_B T}}} dE, \tag{7.106b}$$

where E_{c1} and E_{c2} are the CB minima of the GaAs and AlGaAs, respectively, and m_{e1}^* and m_{e2}^* are the electron effective masses in GaAs and AlGaAs, respectively. For $(E_{c2} - E_{Fn}) \gg k_B T$, the Fermi distribution functions can be approximated as the Boltzmann distribution function. Thus, Equation 7.106b can be rewritten as

$$\begin{aligned} n_2 &= \frac{4\pi}{h^3}(2m_{e2}^*)^{3/2} \int_{E_{c2}}^{\infty} \frac{\sqrt{E - E_{c2}}}{1 + \mathrm{e}^{\frac{(E-E_{Fn})}{k_B T}}} dE \approx \frac{4\pi}{h^3}(2m_{e2}^*)^{3/2} \int_{E_{c2}}^{\infty} \frac{\sqrt{E - E_{c2}}}{\mathrm{e}^{\frac{(E-E_{Fn})}{k_B T}}} dE \\ &= \frac{2}{h^3}(2\pi m_{e2}^* k_B T)^{3/2} \mathrm{e}^{\frac{-(E_{c2}-E_{Fn})}{k_B T}} = N_{c2} \mathrm{e}^{\frac{-(E_{c2}-E_{Fn})}{k_B T}}, \end{aligned} \tag{7.107}$$

where N_{c2} is the effective density of state given by $N_{c2} = \frac{2}{h^3}(2\pi m_{e2}^* k_B T)^{3/2}$. The diffusion current density, which is also known as the *leakage current density*, can be written as

$$J_{\text{diff}} = \frac{eD_{e2}n_2}{L_{e2}} = \frac{eD_{e2}}{L_{e2}} N_{c2} e^{\frac{-(E_{c2}-E_{Fn})}{k_B T}} = \frac{eD_{e2}}{L_{e2}} N_{c2} e^{\frac{-(\Phi)}{k_B T}}. \tag{7.108}$$

Notice that $\Phi = (E_{c2} - E_{Fn})$ as illustrated in Fig. 7.28. Substituting Equation 7.104 into Equation 7.108 yields

$$J_{\text{diff}} = \frac{eD_{e2}}{L_{e2}} N_{c2} e^{\frac{-(\Delta E_g - \Delta E_n - \Delta E_{p1} - \Delta E_{p2})}{k_B T}}. \tag{7.109}$$

For high injection rate, $\Delta E_{p1} \approx 0$ and the majority carriers in the barrier can be written as

$$p_2 = N_{v2} \mathrm{e}^{\frac{-\Delta E_{p2}}{k_B T}}, \tag{7.110}$$

where N_{v2} is the effective density of states in the VB of the barrier (p-AlGaAs). Combine Equations 7.109 and 7.110 to obtain

$$J_{\text{diff}} \approx \frac{eD_{e2}N_{c2}}{L_{e2}} \frac{N_{v2}}{p_2} \mathrm{e}^{\frac{-(\Delta E_g - \Delta E_n)}{k_B T}}. \tag{7.111}$$

For high p-type doping, it can be assumed that $N_{v2} \approx p_2$, which yields

$$J_{\text{diff}} \approx \frac{eD_{e2}N_{c2}}{L_{e2}} e^{\frac{-(\Delta E_g - \Delta E_n)}{k_B T}}. \tag{7.112}$$

For a heterojunction such as GaAs/AlGaAs, ΔE_g can be made high enough such that the leakage current density described in the above equation is negligible. It is important to keep the aluminum mole fraction in the barrier material <40% so that the band gap of the barrier material remains direct. The effective density of states, N_{c2}, increases for indirect AlGaAs since the effective mass in the X and L minima are usually higher than of the Γ minimum.

By neglecting the diffusion current density, the threshold current density given by Equation 7.105 can now be approximated as

$$J_t \approx \frac{en_t d}{\tau}. \tag{7.113}$$

The laser output power, P_{out}, is related to the photon density, N_{ph}, according to the following expression

$$P_{\text{out}} = V\hbar\omega\upsilon_g\alpha_m N_{\text{ph}}, \tag{7.114}$$

where V is the volume of the active region, υ_g is the light group velocity, α_m is the optical loss due to the facets, and the photon density, N_{ph}, is given by

$$N_{\text{ph}} = \eta_i \frac{J - J_t}{eD}\tau_{\text{ph}}, \tag{7.115}$$

where η_i is the internal quantum efficiency, J is the current density ($J > J_t$), and τ_{ph} is the photon lifetime, which is given by

$$\tau_{\text{ph}} = \frac{1}{\upsilon_g(\alpha_m + \alpha_i)}, \tag{7.116}$$

where α_i is the optical loss due to the active region. Combining Equations 7.114–7.116 and converting the current density to current ($J = I/A$, where A is the area of the active region), we obtain

$$P_{\text{out}} = \frac{\hbar\omega}{e}\frac{\alpha_m}{(\alpha_m + \alpha_i)}\eta_i(I - I_t). \tag{7.117}$$

The typical laser power output as a function of applied bias current is shown in Fig. 7.29, where the P_{out} and Voltage are sketched as a function of the applied bias voltage. The threshold current, I_t, and the turn-on voltage (V_t) for the laser diode are indicated.

The laser diode resistance is obtained from the slope of the linear portion of the $I-V$ curve. The slope of the Pout curve in the stimulated emission region

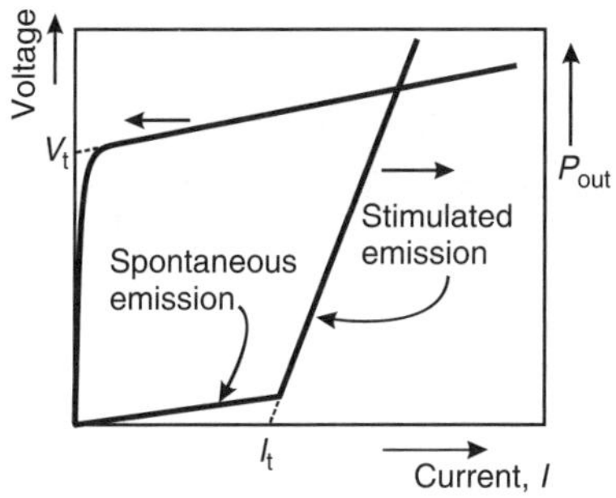

FIGURE 7.29 The characteristic behavior of the semiconductor laser power output as a function of the applied bias current. The spontaneous emission and stimulated emission regions are indicated.

is usually referred to as the *slope efficiency*. The external differential quantum efficiency is defined as the increase in the light output due to the increase in the applied bias current and is given as

$$\eta_d = \frac{dP_{out}/\hbar\omega}{d(I - I_t)/e} = \eta_i \frac{\alpha_m}{\alpha_i + \alpha_m}. \tag{7.118}$$

The facet loss, α_m, can be determined from the following relation:

$$1 = R_1 R_2 e^{-\alpha_m \cdot 2L} \text{ or } \alpha_m = \frac{1}{2L} \ln\left(\frac{1}{R_1 R_2}\right). \tag{7.119}$$

For $R_1 = R_2 = R$, the external differential efficiency can be written as

$$\eta_d = \frac{dP_{out}/\hbar\omega}{d(I - I_t)/e} = \eta_i \frac{\ln(1/R)}{\alpha_i L + \ln(1/R)}. \tag{7.120}$$

For most III–V semiconductor heterojunction laser diodes, the internal quantum efficiency is close to unity and the optical loss in the active region is of the order of 10–100 cm^{-1}. For a GaAs/AlGaAs heterojunction diode laser with a cavity length of 0.5 mm and if the reflectivity is of the order of $R \sim 0.3$, then $\eta_d \approx 0.7\eta_i$ for $\alpha_i = 10$ cm^{-1}.

7.4.3 Quantum Well Edge-Emitting Lasers

A single quantum well edge-emitting laser diode is sketched in Fig. 7.30. The structure of the laser consists of an undoped InGaAlAs quantum well, which is the active region, sandwiched between the barrier layers of n-type AlGaAs and p-type AlGaAs. These layers also act as the waveguide for the photons. The sample is cleaved and the facets act as the mirrors that form the laser cavity. The heat sink is a copper block. The laser output power is plotted as a function of the

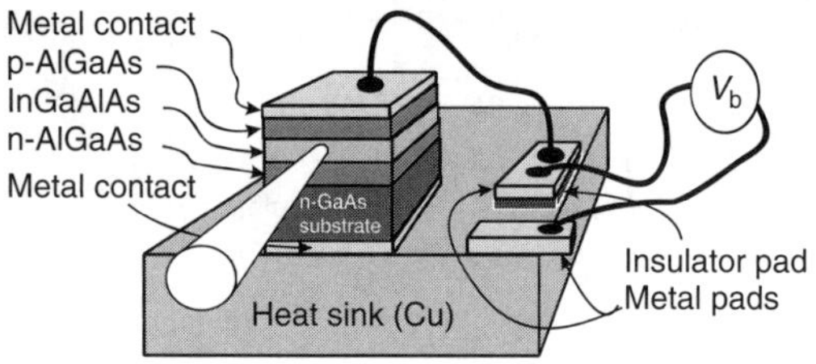

FIGURE 7.30 A sketch of an edge emitting laser based on InGaAlAs/AlGaAs single quantum well.

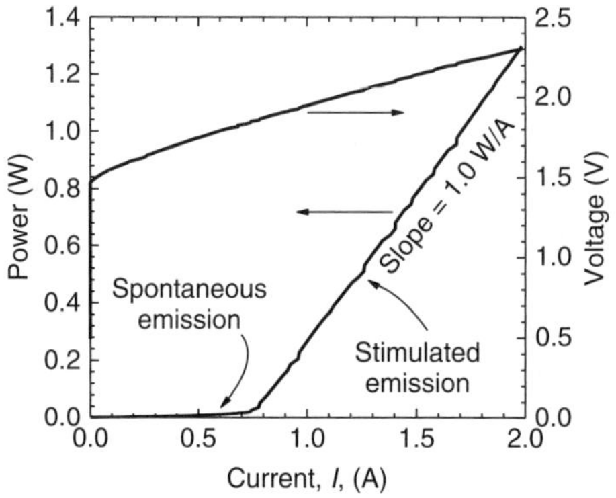

FIGURE 7.31 The output power of the edge emitting laser, shown in Fig. 7.30, plotted as a function of the applied current. The $I-V$ curve is also shown.

applied current as shown in Fig. 7.31. This figure shows the two characteristic regions: the spontaneous emission region below the threshold current and the stimulated emission above the threshold current.

When the current is below the threshold current, the laser operates in the spontaneous emission mode, which is identical to the LED operation mode. The spontaneous emission spectrum shown in Fig. 7.32a is obtained using the laser as the source for a Fourier-transform spectrometer. The applied current is approximately equal to the threshold current. The spectrum still resembles the LED spontaneously even when the applied current is just below the threshold current. The laser longitudinal modes are observed as shown in Fig. 7.32b when the applied current is increased above the threshold current. The longitudinal modes are observed at a spectral resolution as low as 0.1 cm^{-1}.

Ideally, the full width at half maximum of the longitudinal modes should be zero. The spectrum in Fig. 7.32a shows that these modes have a finite width, which is due to the attenuation arising from several factors, such as the finite facet transmission and the absorption in the cavity. Thus, instead of infinitely narrow longitudinal modes, we have modes with a finite full width at half maximum,

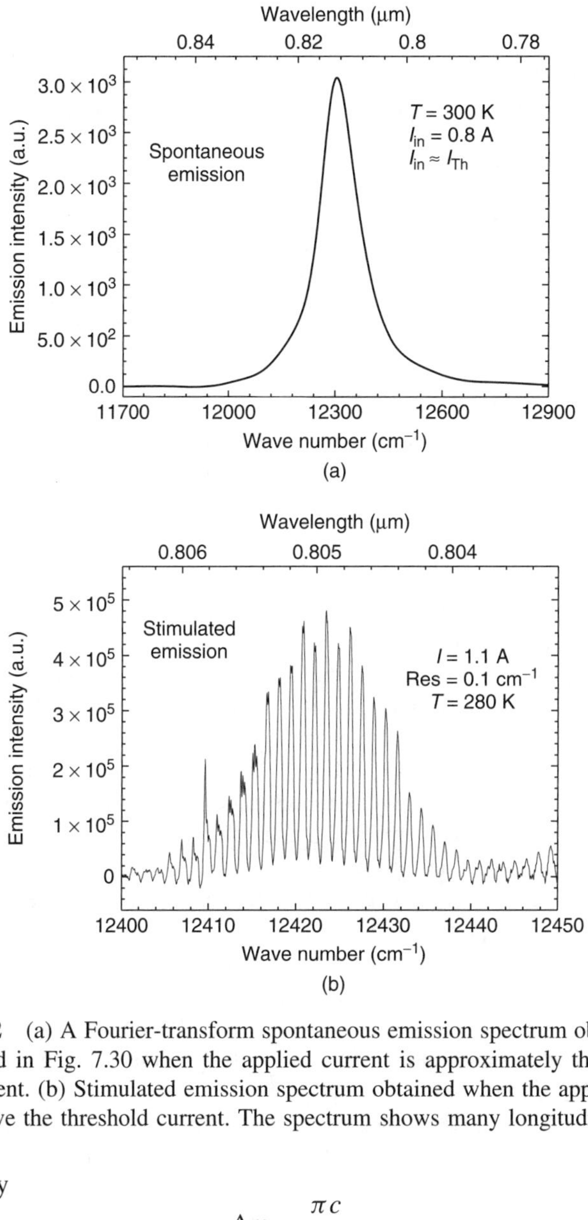

FIGURE 7.32 (a) A Fourier-transform spontaneous emission spectrum obtained for the laser described in Fig. 7.30 when the applied current is approximately the same as the threshold current. (b) Stimulated emission spectrum obtained when the applied current is increased above the threshold current. The spectrum shows many longitudinal modes.

$\Delta\omega$, given by

$$\Delta\omega = \frac{\pi c}{L\mathcal{F}}, \tag{7.121}$$

where L is the length of the cavity, c is the speed of light, and $\mathcal{F}$ is a factor called the *finesse* or the *contrast parameter* given by

$$\mathcal{F} = \frac{\pi\sqrt{\gamma}}{(1-\gamma)}, \tag{7.122}$$

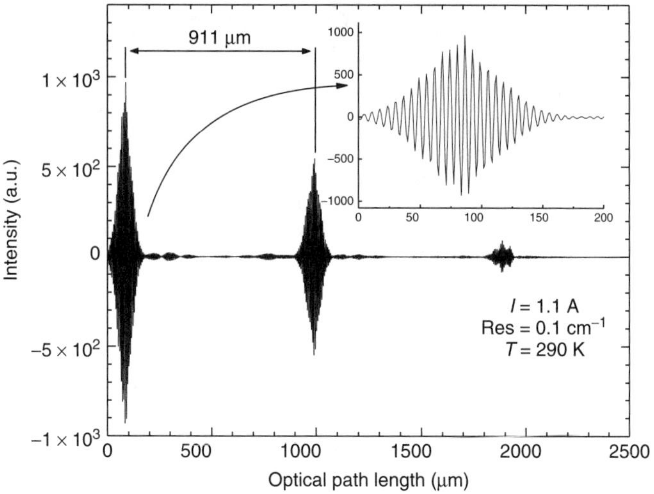

FIGURE 7.33 The interferogram of the edge-emitting laser, described in Fig. 7.30, operating in the stimulated emission mode. The inset is the enlarged first packet.

where γ is the attenuation or loss factor in the cavity. For known γ and known full width at half maximum, one can calculate the cavity length using Equation 7.121. The cavity length can also be calculated using the separation between the longitudinal modes given by Equation 7.103. A more accurate method of obtaining the cavity length of the laser is by obtaining the interferogram from the Fourier-transform spectrometer as shown in Fig. 7.33. The interferogram contains several interference packets separated by the cavity length. In this case, the packets are separated by 0.911 mm (911 μm). Each packet consists of an interference pattern, as shown in the inset of the figure.

The edge-emitting laser structure described above is considered a high power laser since the output power is of the order of watts rather than milliwatts. The quantum well is compressively strained because of the lattice mismatch. The actual structure is 70 Å $In_{0.06}Ga_{0.86}Al_{0.08}As$ quantum well (active region) and $Al_{0.3}Ga_{0.7}As$ barriers (waveguide). Owing to the difference in the refractive index of the quantum well (n_{rw}) and barriers (n_{rb}), the photons are confined in the well, as shown in Fig. 7.34.

In the edge-emitting laser, the feedback needed for lasing action is achieved by the cavity facets formed by cleaving. For higher mode purity, the stimulated emission can be obtained by introducing *distributed Bragg reflectors* (DBRs). These reflectors are usually corrugated structures or grading introduced in the waveguide.

7.4.4 Vertical Cavity Surface-Emitting Lasers

Optical interconnects between chips, large arrays of sources, narrow beam widths, and generation of high power are a few of the VCSEL applications.

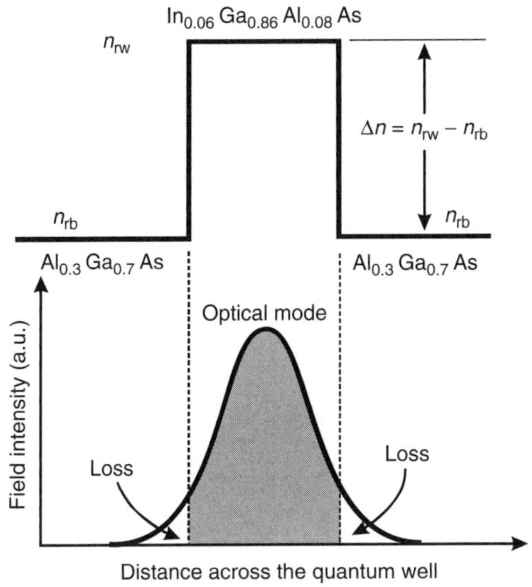

FIGURE 7.34 Optical mode distribution in the single quantum well laser due to the step discontinuity in the refractive index of the well (n_{rw}) and barriers (n_{rb}).

Optical interconnects based on VCSELs may play a major role at different levels of the interconnection hierarchy for data communication networks. Optical interconnect technology emerged in the late 1970s using AlGaAs/GaAs-based or InGaAsP/InP-based surface-emitting LEDs as the optical source.

One of the major research advances in VCSEL technology is the discovery of the selective wet oxidation of AlGaAs layers with a high aluminum content. This technique allows the fabrication of very small microcavity devices with low threshold current to be realized by eliminating the high carrier loss that occurs because of the defects created by ion implantation. Two-dimensional arrays of VCSELs are useful for implementing very dense optical interconnects spanning short distances that can potentially alleviate the peripheral pin-out limit of the very large scale integrated chips. These arrays can significantly reduce the power dissipation of parallel optical links. Parallel optical interconnects represent a compact and potentially low-cost approach for transmitting a large volume of data across longer distances, thereby replacing the large-profile shielded electrical cables that are needed to interconnect computer processors across local area networks.

Another important factor in the VCSEL technology is the development of the high reflectivity semiconductor DBRs, which are made possible with the advancement of the crystal growth using MBE and MOCVD growth techniques. A typical example of a VCSEL structure is shown in Fig. 7.36. It consists of GaAs/AlGaAs multiple quantum well active regions. This active region is referred to as *multiple quantum well graded index separate confinement heterostructure* (MQW-GRINCH), which is bounded by two DBR mirrors. The active region is composed of GaAs/$Al_{0.15}Ga_{0.85}As$ MQW, and the DBR consists of several pairs of n-type AlAs/$Al_{0.15}Ga_{0.85}As$ at the bottom and several pairs of p-type AlAs/$Al_{0.15}Ga_{0.85}As$. The proton-implanted isolation regions are used as isolation regions, which define the active region of the VCSEL. The active area is usually 15–20 μm in diameter. The aluminum profile of the structure is also shown in Fig. 7.35.

The threshold current of the proton implant isolated VCSEL is usually high ($I_t > 2$ mA). Thus, it is impractical to use such emitters in dense parallel arrays applications. Selective oxidation has been used recently for VCSEL isolation, which reduces the threshold current to the submilliampere range and with a typical power efficiency larger than 30%. An example of an oxide confinement VCSEL is shown in Fig. 7.36. The active region of the oxide-confined VCSEL is usually <10 μm in diameter. Owing to the small thickness of the active region in the VCSELs, the separation between the longitudinal modes is very large when compared to the separation between the longitudinal modes in the edge-emitting lasers.

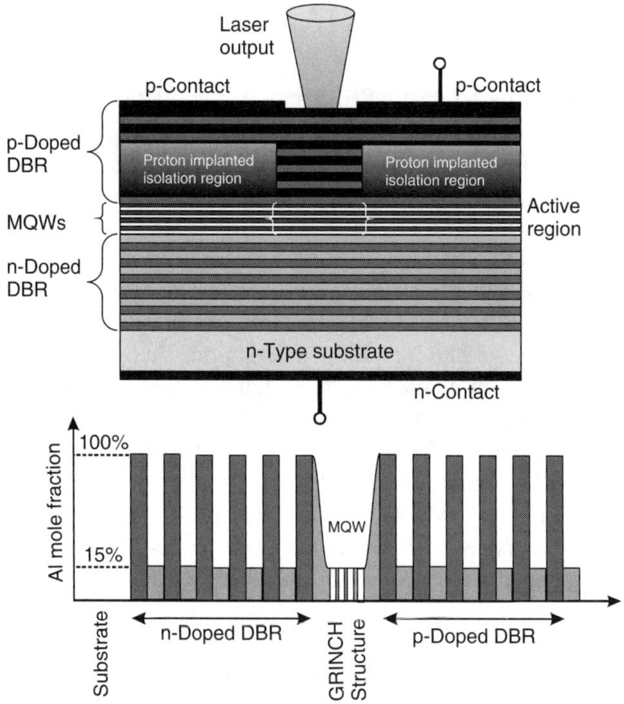

FIGURE 7.35 A schematic showing the VCSEL device structure, whose active region is defined by the proton implantation. The aluminum composition profile of the structure is shown in the lower panel.

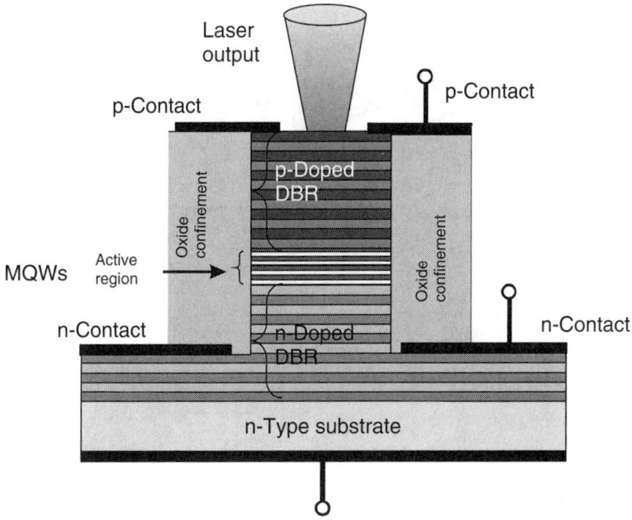

FIGURE 7.36 A sketch of an oxide-confined VCSEL structure.

The refractive index of AlGaAs is ~3.0, while the refractive index of the oxidized layer is of the order of ~1.6. The differential change in the refractive index provides an excellent index-guiding optical confinement. Unlike the ion-implanted VCSELs, the index guiding in the selectively oxidized VCSELs produces a monotonic decrease in the threshold currents. For additional discussion on VCSELs, see Cheng and Dutta.

7.4.5 Quantum Cascade Lasers

The advances in the epitaxial growth, in particular, the MBE and MOCVD growth methods, allow the deposition of layer-by-layer semiconductor materials with dissimilar band gaps and lattice constants with high accuracy. The precise control of the dopants and layer thicknesses permits one to engineer the band gap of complicated structures for specific applications. In many ways, the well-controlled growth conditions can be thought of as wave function engineering. An example of this sophisticated epitaxial growth is the elaborate structure of the quantum cascade lasers. Unlike quantum well laser diodes described in the earlier sections, the quantum cascade lasers are unipolar devices, where the charge carriers are either electrons or holes. Recent quantum cascade lasers were shown to emit coherent light in the far infrared regions. The basic operational principle of the quantum cascade laser is the downward intersubband transitions, where the injected electrons decay from an excited bound state to low-lying energy level emitting their energy as photons.

The structures and designs of the quantum cascade lasers are numerous. For example, three simple structures are sketched in Fig. 7.37. An interwell transition

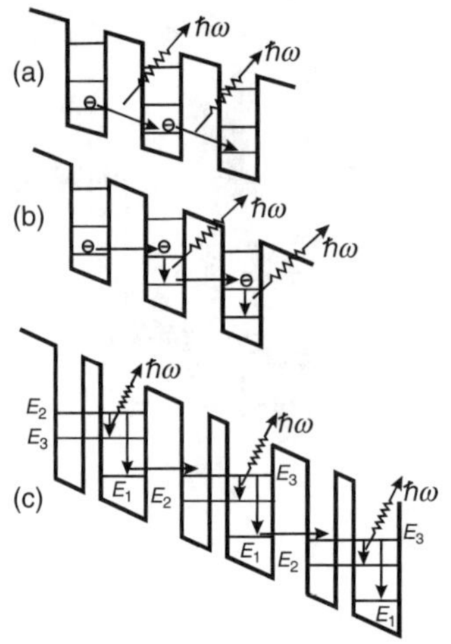

FIGURE 7.37 Schematics of three different structures of quantum cascade lasers. (a) A sketch of interwell radiative transition, (b) interwell radiative transition, and (c) coupled double well structures with three energy levels.

is shown in Fig. 7.37a, where the electrons decay under the influence of a forward bias from the ground state in one quantum well to an excited state in an adjacent quantum well emitting a photon in the process. The second structure shown in Fig. 7.37b is based on electron resonant tunneling followed by a radiative downward transition. The third structure is a coupled double quantum well with three bound energy levels. Population inversion is established in the excited state because of the spatial separation of the two states by the barrier layers. The cleaved facets of the structures act as the mirrors.

The lasing threshold gain, g_t, in quantum cascade laser can be written as

$$g_t = \frac{(\alpha_i + \alpha_m)}{\Gamma}, \tag{7.123}$$

where Γ is optical confinement factor, α_m is the mirror loss due to the finite facet reflectivities, and α_i is the internal loss for the optical wave, which results from various absorption mechanisms, such as scattering and free carrier absorption. The confinement factor can be approximated as

$$\Gamma \approx 2\pi^2 \left(n_1^2 - n_2^2\right) \frac{d^2}{\lambda^2}, \tag{7.124}$$

where n_1 and n_2 are the refractive index of the active region and the cladding regions, respectively, d is the well width, and λ is the wavelength.

The threshold gain depends on several factors and it is related to the threshold current density, J_t, according to the following relation:

$$g_t = \frac{J_t L T_s \eta_{pi} \eta_r \eta_{inj}}{e}, \tag{7.125}$$

where L is a parameter related to the emission wavelength and has the dimension length, T_s is a parameter with the dimension of time that depends on the inverse linewidth of the spontaneous emission, η_{pi} is the population inversion efficiency, η_r is the radiative efficiency, and η_{inj} is the injection efficiency. For more details on these parameters, see Helm.

More complicated quantum cascade structures have been reported by several groups (see, e.g., Helm). Two typical examples are shown in Fig. 7.38. The first structure shown in panel (a) consists of the active region of three coupled quantum wells, with the three energy levels separated by injection regions. The injection region is composed of several n-type short-period superlattice or digitally graded alloy, which serves as the collector for the proceeding active region and as an emitter for the following active region. The discrete tunneling scheme presented in this figure prevents the formation of the electric domains as observed in the continuous sequential tunneling in the superlattice structure. It also eases the rigorous requirement of position-dependent energy level alignment over the whole superlattice structure.

The second quantum cascade laser structure shown in Fig. 7.38b is basically two coupled quantum wells with three energy levels sandwiched between supelattices that act as DBRs. This scheme is less sensitive to the defects, interface

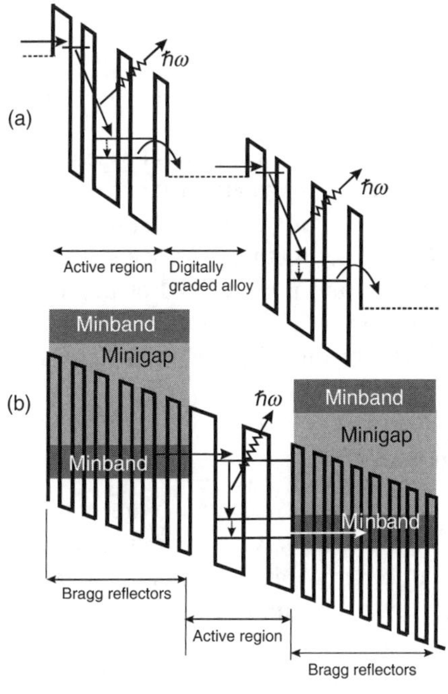

FIGURE 7.38 (a) A schematic presentation of the conduction band energy diagram of a portion of quantum cascade laser based on intersubband transition. (b) The conduction band energy diagraph of a portion of a vertical transition quantum cascade laser structure. The dashed arrows indicate electron transitions due to phonon scattering.

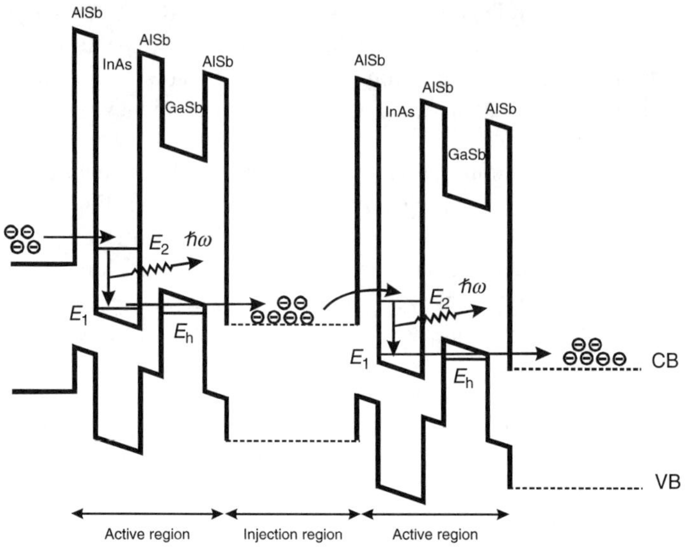

FIGURE 7.39 A portion of a quantum cascade laser structure based on type II superlattices.

roughness, and impurity fluctuations. In addition, the DBR possesses miniband and minigap as shown in Fig. 7.38b. The minigap prevents the electrons from escaping to the continuum.

The structure of the quantum cascade laser is based on type II superlattices or broken band gap superlattices. A schematic structure of a type II quantum cascade laser is shown in Fig. 7.39. Several variations in this structure have been reported by many research groups. The structure represents a discrete tunneling scheme, where the actual device structure is composed of many active regions separated by n-type doped injection region. These injection regions consist of either graded AlInAsSb quaternary layers or digitally graded InAs/AlSb superlattices.

7.4.6 Quantum Dots Lasers

The search for new structures for quantum wire and dot lasers is motivated by the low threshold current density, weaker temperature dependence of the threshold current density or temperature-insensitive threshold current, tuning the gain spectrum width and wavelength, and low shift of the laser wavelength with respect to the injection current (low chirp). The quantum well lasers are now used for most commercial applications. The new frontier, however, is the quantum wire and quantum dot lasers, where the density of states spans a small energy region. As shown in Fig. 7.40, the density of states and the optical gain of the laser are sketched in four confinement systems. The first system is the bulk semiconductor system, where the optical gain spectrum width is broad and the density of states

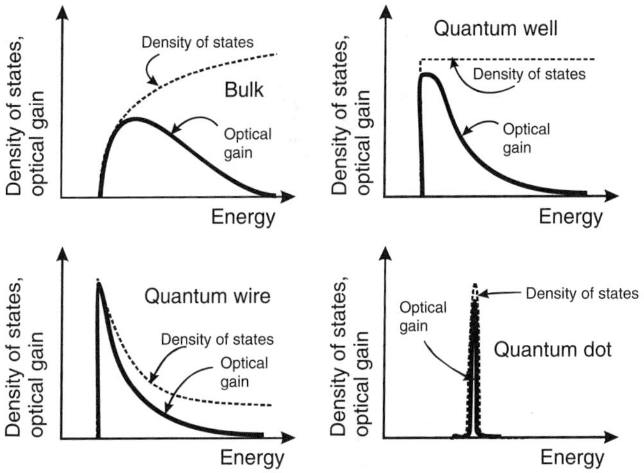

FIGURE 7.40 Schematic presentations of the density of states and the optical gain spectrum plotted for different confinement active region systems. Bulk semiconductor material is a zero confinement system, quantum well is a one-dimensional confinement system, quantum wire is a two-dimensional confinement system, and quantum dot is a three-dimensional confinement system.

is continuous. See Chapter 5 for a detailed discussion on the density of states. The density of states for a quantum well structure is merely a step and the optical gain is more localized when compared to the bulk system. For quantum wires, where the confinement is along two dimensions with one degree of freedom, and the density of states is proportional to the inverse of the square root of energy. This behavior provides a much narrower optical gain. In the case of quantum dots, the density of states and the optical gain are both spread over a very small region of energy. The quantum dot system thus provides a drastic change in the density of states and optical gain (discrete carrier distribution), which is highly ideal for lasing action with low threshold current and high temperature stability.

A typical quantum dot structure used for lasers is shown in Fig. 7.41a. The most commonly investigated quantum dots is the InAs matrixed in GaAs layer, which acts as a barrier. The cladding layers are added to form the optical cavity or the optical confinement layer. The quantum dots were depicted as semispherical shaped that is randomly distributed in the active region. The band structure diagram is shown in Fig. 7.41b. The downward interband radiative transition is shown as the sum of the band gap of the quantum dot material, ΔE_{n} and ΔE_{p}, where ΔE_{n} and ΔE_{p} are the confined energy level positions of the electron in the CB and the holes in the VB, respectively. The band structure shown in Fig. 7.41b is sketched for an ideal quantum dot structure with a cubic shape. In reality, the quantum dots are usually pyramidal in shape. The implication is that the band structure of the quantum dots resembles a graded structure, where the band gap decreases along the growth direction.

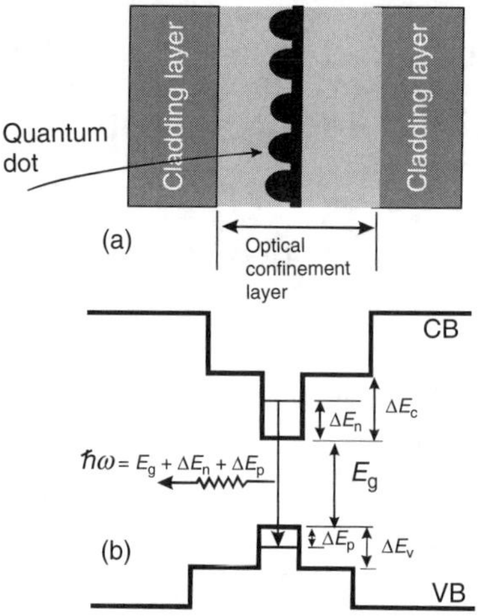

FIGURE 7.41 (a) An illustration of the quantum dot laser active layer defined by the cladding layers. (b) A sketch of the energy band diagram showing the downward radiative interband transition.

One drawback of the quantum dot structure is the nonuniformity of the self-assembled quantum dot size. For the ideal case, where all the quantum dots are identical in size, the optical gain should resemble the density of states in the quantum dots. This means that the optical gain spectrum should take the δ-function distribution form. In practice, the self-assembled quantum dots are nonuniform in size. The variation in the dot size produces an inhomogeneous line broadening in the optical gain as shown in Fig. 7.42. In this figure, three quantum dots with different sizes are sketched. The interband transitions are shown as E_1, E_2, and

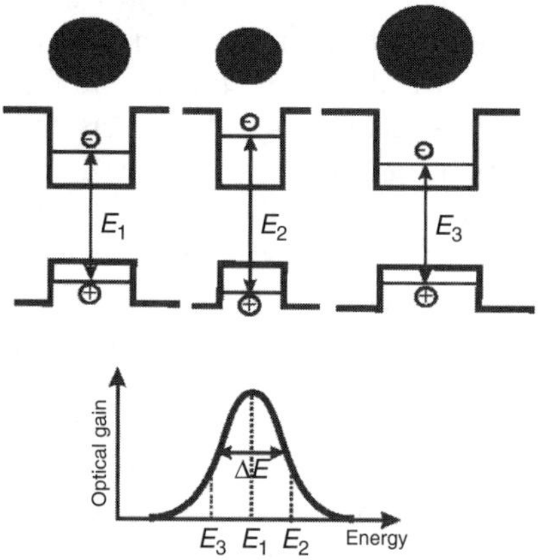

FIGURE 7.42 Illustration of the inhomogeneous line broadening due to the variation in the quantum dot size.

E_3. The variation in these energies causes an inhomogeneous broadening in the optical gain spectrum as shown in the figure.

The inhomogeneous line broadening or dispersion due to the quantum dot size affects the laser characteristic parameters. For example, the maximum gain decreases, the threshold current increases, the internal differential efficiency decreases, the output power decreases, and the threshold current density becomes more sensitive to temperature. The inhomogeneous line broadening, ΔE, can be expressed as (Asryan *et al.*)

$$\Delta E = \left(\Delta E_{\mathrm{n}} \frac{\partial \Delta E_{\mathrm{n}}}{\partial a} + \Delta E_{\mathrm{p}} \frac{\partial \Delta E_{\mathrm{p}}}{\partial a} \right) \delta, \tag{7.126}$$

where a is the quantum dot mean size and δ is the rms of relative quantum dot size fluctuation. The maximum optical gain is inversely proportion to ΔE. Thus, the larger the inhomogeneous line broadening, the smaller is the maximum optical gain.

Another parameter affected by the variation in the dot size is the chirp or the change in the laser emission wavelength during the current injection. The origin of this shift is related to the coupling of the real and imaginary parts of the complex susceptibility in the laser active region. The variation in the gain due to the current injection causes a variation in the refractive index. The variation in the refractive index modifies the phase of the optical mode in the laser cavity. The coupling strength is defined by the linewidth enhancement factor, α_{c}, which is defined as

$$\alpha_{\mathrm{c}} = \frac{\Delta n_{\mathrm{r}}}{\Delta k_{\mathrm{r}}}, \tag{7.127}$$

where Δn_{r} and Δk_{r} are the changes in the real part and imaginary part of the refractive index, respectively. The change in the imaginary part of the refractive index is related to the change in the net gain (Δg), which occurs during the laser relaxation oscillation, according to the following relation

$$\Delta g = -\frac{4\pi \nu}{c} \Delta k_{\mathrm{r}} = -\frac{4\pi}{\lambda} \Delta k_{\mathrm{r}}, \tag{7.128}$$

where ν is the laser frequency. Substituting Equation 7.128 into Equation 7.127 yields

$$\alpha_{\mathrm{c}} = -\frac{4\pi}{\lambda} \frac{\Delta n_{\mathrm{r}}}{\Delta g} = \frac{2E}{\hbar c} \frac{\Delta n_{\mathrm{r}}}{\Delta g}. \tag{7.129}$$

The change in the gain increases linearly with the injected carrier density, while the refractive index decreases linearly with the injected carrier density. Thus, one may rewrite Equation 7.129 as follows:

$$\alpha_{\mathrm{c}} = -\frac{4\pi}{\lambda} \frac{\partial n_{\mathrm{r}}/\partial N}{\partial g/\partial N}, \tag{7.130}$$

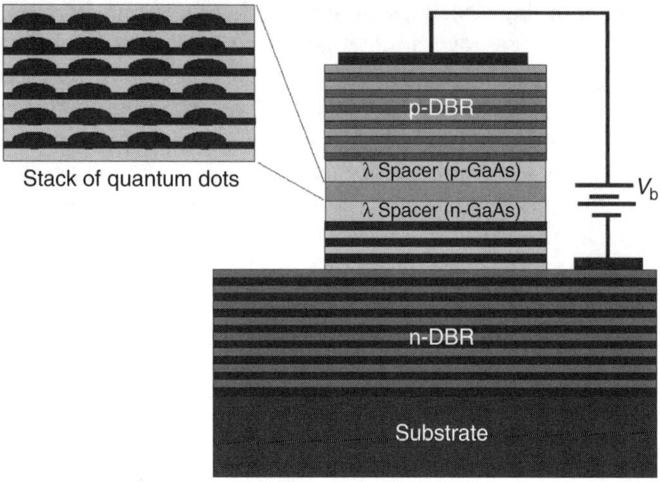

FIGURE 7.43 A sketch of the multiple quantum dot VSCEL with AlO/AlGaAs DBP.

where N is the carrier density. For an ideal quantum dot laser, the variation in the refractive index with respect to the carrier density is zero, which means $\alpha_c = 0$ and the laser is chirp-free.

A stack of multiple quantum dots can be used as the active region for VCSELs as shown in Fig. 7.43. To maintain a low loss, the mirrors need to be highly reflective, which is usually achieved by growing stacks of $\lambda/4$ DBRs. The mesa diameter is usually of the order of 10 μm with a vertical cavity determined by the total thickness of the multiple quantum dot stacks. Since the cavity length is very short, the separation between the laser modes (Fabry–Perot modes) is relatively very large. A typical structure of this laser is InGaAs quantum dots and GaAs barriers. The DBRs are usually made of AlGaAs/AlAs. Other DBRs, such as AlGaAs/AlO, are used for the quantum dot VCSELs. The spacer thicknesses are usually grown to match the wavelength of the laser cavity.

The difficulty in achieving multiple quantum dot VCSELs is the quantum dot size uniformity and precise epitaxial growth homogeneity, and control is required. For example, if the spacers and the DBR thicknesses do not match the laser mode wavelength, a larger threshold current may be needed to produce stimulated emission.

7.5 SUMMARY

This chapter presented detailed discussions on the optoelectronic devices fabricated from quantum wells and dots. As the name implies, optoelectronic devices are those that can either absorb light and then generate an electrical signal or emit light under the influence of an injected electrical current. The discussion in this chapter was first focused on the infrared quantum detectors, where the figures of merit, such as responsivity, NEP, detectivity, and NETD were derived.

The source of noise in infrared detectors was discussed. The most important noise sources are the Johnson noise, the shot noise, the generation–recombination noise, and the $1/f$ noise. Johnson noise is dominant at high frequency while the $1/f$ noise is dominant at low frequency.

Examples of quantum detectors were presented. The multiple quantum well infrared photodetectors, known as *QWIPs*, were discussed in more detail. Examples including miniband and embedded miniband quantum structures were presented. In principle, the basic operation of these detectors is based on the electrons that undergo the intersubband transitions in the quantum wells. This transition is excited in case of quantum wells by photons that are incident at certain angles or p-polarized light. For the intersubband transitions in quantum dot systems, the light is absorbed at any angle or any light polarization because of the fact that the energy dispersion is absent and the selection rules do not exist. This case is valid for an ensemble of quantum dots with high size uniformity.

The photoconductivity measurements were described for a single-color and two-color infrared detectors based on multiple quantum wells. The dark current or the current that is observed under bias voltage in the absence of light illumination is discussed for the quantum infrared detectors. The two major components of the dark currents are the thermionic and tunneling currents.

Different quantum dot structures used for infrared detectors were discussed with emphasis on the photoconductivity measurements. It should be pointed out that the quantum dot systems are inherently limited for certain spectral regions. This is because there is a limit on the lower number of monolayers that can be deposited to produce self-assembling quantum dots and there is an upper limit on the number of monolayers that can produce a quantum dots system with quantum confinement useful for infrared detectors.

LEDs based on p–n junctions and heterojunctions were briefly discussed. The responsivity and the frequency response of the LEDs were presented. These devices are based on the spontaneous emission of light, where the electrons are injected in the structure followed by downward transitions of the electrons from the CB to the VB. As a result of this downward transition, the electrons emit their energy as photons. The important parameters of these devices are the recombination lifetime and the optical gain.

Semiconductor lasers were discussed at the end of the chapter. The laser can be thought of as a waveguide with two mirrors to form a feedback cavity. The discussion covers the basic operational principles of the laser, in general. In particular, the spontaneous emission and stimulated emission processes were discussed. The discussion covers the population inversion and the transparency point, which are needed for the lasing action to occur.

Several laser systems were presented. In particular, the edge-emitting lasers, VCSELs, and quantum cascade lasers were discussed in reasonable detail without going into the full details of deriving their parameters. Finally, quantum dot laser systems were briefly discussed. The latter system is a fairly new subject and the full potential of the quantum dot systems has not been fully realized. In the next several years, we should witness the evolution of the quantum dots and their use as laser devices. Thus, an update on the subject will be presented in the future editions of this book.

PROBLEMS

7.1. Calculate the responsivity of a photoconductor of an area of 1.0 mm^2 when a blackbody source is used. Filters were used such that the light wavelength limits seen by the detectors are 1.0 and 100.0 μm. The measured photocurrent is 0.3 A and the photon flux was measured as 10^{17} photons/m^2/s.

7.2. Show that the spectral current responsivity in the frequency domain is given by Equation 7.26.

7.3. Show that the shot noise in a photodiode detector is reduced to Johnson noise when the modulation frequency is low and the bias voltage is zero.

7.4. Use the spectra that were measured at 77 K in Fig. 7.6 for both the waveguide and Brewster's angle configurations. Calculate the internal quantum efficiency for both spectra assuming that the light made three passes in the waveguide.

7.5. The photocurrent, $i_{\mathrm{ph}}(z)$ in a QWIP can be written as $i_{\mathrm{ph}}(z) = \frac{e}{h\nu}\alpha T_{\mathrm{r}} L_{\mathrm{t}}(z)\Phi_z$, where $h\nu$ is the photon energy, α is the absorption coefficient, T_{r} is the transmission tunneling probability, $L_{\mathrm{t}}(z)$ is the transport length given by $L_{\mathrm{t}}(z) = l(1 - \mathrm{e}^{-z/l})$. Here l is the mean free path of the electrons. The optical power,Φ_z, at a location z is given by $\Phi_z = \Phi_0 \mathrm{e}^{-\alpha z}$, where Φ_0 is the power at $z = 0$. Derive an expression for the quantum efficiency. (Hints: Integrate over the detector length to find an expression for the average photocurrent along the length of the detector (L). Then use the definition of the current spectral responsivity).

7.6. Derive Equation 7.65 and then show that Equation 7.66 is valid.

7.7. Calculate the number of modes of a GaAs laser that has a cavity length of 500 μm with an emission wavelength of 855 nm. The refractive index of GaAs is 3.5. What are the frequency and wavelength separations of the normal modes? How many longitudinal modes exist for an output gain bandwidth of 10 nm?

7.8. Two consecutive longitudinal modes in the spectrum shown in Fig. 7.32b were observed at 12424.833 cm^{-1} and 12423.422 cm^{-1}. Use Equation 7.103 to estimate the laser cavity length. Assume that the refractive index of the quantum well material is 4.0. Compare your results to the cavity length obtained from the interferogram shown in Fig. 7.33.

BIBLIOGRAPHY

Asryan LV, Luryi S. In: Steiner T, editor. Semiconductor nanostructures for optoelectronic applications. Boston: Artech House; 2004. Chapter 4.

Beck WA. Appl Phys Lett 1993; 63:3589.

Brown GJ, Szmulowicz F. In: Razeghi M, editor. Long wavelength infrared detectors. New York: Taylor and Francis; 1996.

Chuang SL. Physics of photonic devices. New York: Wiley; 2009.

Dereniak EL, Boreman GD. Infrared detectors and systems. New York: Wiley; 1996.

Helm M, editor. Volume 6. Long wavelength infrared emitters based on quantum wells and superlattices. New York: Taylor & Francis; 2000.

Hudson RD. Infrared system engineering. New York: Wiley; 1969.

Levine BF. J Appl Phys 1993; 74:R1.

Lloyd JM. Thermal imaging systems. New York: Plenum; 1975.

Liu HC. In: Razeghi M, editor. Long wavelength infrared detectors. New York: Taylor and Francis; 1996. Chapter 1.

Liu J-M. Photonic devices. New York: Cambridge; 2005.

Madhuker A, Kim ET, Chen Z, Campbell JC, Ye Z. In: Steiner T, editor. Semiconductor nanostrucutres for optoelectronic applications. Boston: Artech House; 2004. Chapter 3.

Razeghi M, editor. Volume 1. Long wavelength infrared detectors. New York: Taylor and Francis; 1996.

Rogalski A. Infrared detectors. Amsterdam: Gordon and Breach; 2000.

APPENDIX A

DERIVATION OF HEISENBERG UNCERTAINTY PRINCIPLE

Heisenberg uncertainty principles can be derived by different methods. This appendix uses the wave packet to derive the uncertainty principles. For the sake of simplicity, the case of the one-dimensional wave packet is presented here. Then the wave function only depends on x and t.

$$\psi(x,t) = \frac{1}{\sqrt{2\pi}} \int g(k) \mathrm{e}^{i(kx-\omega t)}\, \mathrm{d}k. \tag{A.1}$$

For t=0, we have

$$\psi(x,0) = \frac{1}{\sqrt{2\pi}} \int g(k) \mathrm{e}^{ikx}\, \mathrm{d}k \tag{A.2}$$

and

$$g(k) = \frac{1}{\sqrt{2\pi}} \int \psi(x,0) \mathrm{e}^{-ikx}\, \mathrm{d}x \tag{A.3}$$

Thus, $g(\boldsymbol{k})$ is the Fourier transform of $\psi(\boldsymbol{r},0)$. The wave packet is given by x-dependent wave function expressed in Equation A.2. If $|g(\boldsymbol{k})|$ has the shape depicted in Fig. A1 and $\psi(x)$, instead of having the form shown in Equation A.2, is composed of three plane waves with wave vectors of k_0, $k_0 + \Delta k/2$, and $k_0 - \Delta k/2$ and amplitudes proportional to 1, $\frac{1}{2}$, and $\frac{1}{2}$, then one can write the new wave packet as

$$\psi(x) = \frac{g(k_0)}{\sqrt{2\pi}} \left[e^{ik_0 x} + \frac{1}{2} e^{i(k_0 - \Delta k/2)x} + \frac{1}{2} e^{i(k_0 + \Delta k/2)x} \right]$$

Introduction to Nanomaterials and Devices, First Edition. Omar Manasreh.

$$= \frac{g(k_0)}{\sqrt{2\pi}} e^{ik_0 x} \left[1 + \cos\left(\frac{\Delta k}{2} x\right)\right]$$

$$\psi(x) = \frac{g(k_0)}{\sqrt{2\pi}} \left[e^{ik_0 x} + \frac{1}{2} e^{i(k_0 - \Delta k/2)x} + \frac{1}{2} e^{i(k_0 + \Delta k/2)x}\right]$$

$$= \frac{g(k_0)}{\sqrt{2\pi}} e^{ik_0 x} \left[1 + \cos\left(\frac{\Delta k}{2} x\right)\right]. \tag{A.4}$$

From Fig. A1, $|\psi(x)|$ is maximum at $x = 0$. This result is due to the fact that when x is zero, the three waves are in phase and interfere constructively as shown in the figure. As one moves away from the value $x = 0$, the waves become more and more out of phase and $|\psi(x)|$ decreases. The interference becomes completely destructive when the phase shift between $e^{ik_0 x}$ and $e^{i(k_0 \pm \Delta k/2)x}$ is equal to $\pm\pi$ and $|\psi(x)| = 0$ when $x = \pm\Delta x/2$. In other words, $|\psi(x)| = 0$ when $\cos(\Delta k.\Delta x/4) = -1$. This leads to the following equation:

$$\Delta k \cdot \Delta x = 4\pi. \tag{A.5}$$

This equation shows that the larger the width of $|\psi(x)|$, the smaller the width of $g(\boldsymbol{k})$. Equation A.4, however, shows that $|\psi(x)|$ is periodic in x and therefore has a series of maxima and minima. This arises from the fact that $\psi(x)$ is the superposition of a finite number of waves. For a continuous superposition of an

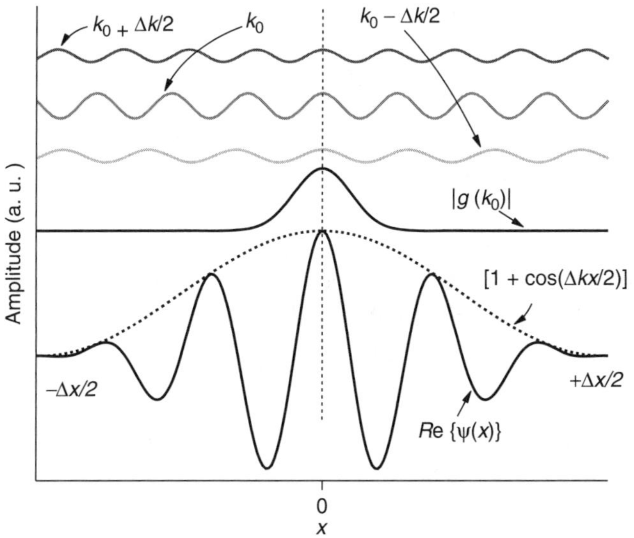

FIGURE A1 The shape of the function $|g(k)|$ is plotted along with the real parts of the three functions whose sum gives the function $\psi(x)$ of Equation A.3. The real part of $\psi(x)$ is also shown. The dashed-line curve corresponds to the function $[1 + \cos(\Delta k x/2)]$, which represents the form of the wave packet infinite number of waves as shown in Equation A.2, such a phenomenon does not exist and $|\psi(x,0)|$ can have only one maximum.

Let us return to the general wave packet formula shown in Equation A.2. Its form results from an interference phenomenon. Let $\alpha(\mathbf{k})$ be the argument of the function $g(\boldsymbol{k})$, that is,

$$g(k) = |g(k)|\ \mathrm{e}^{i\alpha(k)} \tag{A.6}$$

Assume that $\alpha(\boldsymbol{k})$ varies sufficiently smoothly within the interval $[k_0 - \Delta k/2, k_0 + \Delta k/2]$, where $|g(k)|$ is appreciable. Hence, when $\Delta\boldsymbol{k}$ is small enough, one can expand $\alpha(k)$ around $k \approx k_0$ such that $\alpha(k) \approx \alpha(k_0) + (k - k_0)\mathrm{d}\alpha/\mathrm{d}k|_{k=k0}$, which enables us to rewrite Equation A.2 in the following form:

$$\psi(x,0) \approx \frac{\mathrm{e}^{i[k_0x+\alpha(k_0)]}}{\sqrt{2\pi}} \int_{-\infty}^{+\infty} |g(k)|\mathrm{e}^{i(k-k_0)(x-x_0)}\mathrm{d}k, \tag{A.7}$$

where $x_0 = -[\mathrm{d}\alpha/\mathrm{d}k]_{k=k0}$. Equation A.7 is very useful for studying the variation of $|\psi(x)|$ in terms of x. When $|x - x_0|$ is large as compared to $1/(\Delta\boldsymbol{k})$, the wave function oscillates rapidly within the interval $\Delta\boldsymbol{k}$ as shown in Fig. A2. On the other hand, when $|x - x_0|$ is small as compared to $1/(\Delta\boldsymbol{k})$, the wave function oscillates only once as shown in the figure. Thus, when x moves away from x_0, $|\psi(x)|$ decreases. The decrease becomes appreciable if $\mathrm{e}^{i(k-k_0)(x-x_0)}$ oscillates approximately once. That is, when

$$\Delta k \cdot (x - x_0) \approx 1. \tag{A.8}$$

If Δx is the approximate width of the wave packet, one can write

$$\Delta \mathrm{k} \cdot \Delta x \geq 1 \tag{A.9}$$

This classic relation relates the widths of two functions that are Fourier transforms of each other. The important fact is that the product $\Delta k \cdot \Delta x$ has a lower bound, which depends on the precise definition of the each width.

With the help of the relation $\Delta\mathbf{p} = \hbar\Delta\mathbf{k}$, Equation A.9 can be rewritten as

$$\Delta \mathrm{p} \cdot \Delta x \geq \hbar \tag{A.10}$$

The relationship is called the *Heisenberg uncertainty principle*. The same procedure can be repeated by assuming

$$\psi(r,t) = A\mathrm{e}^{i\omega t} \tag{A.11}$$

to obtain a wave packet that is localized in time and frequency with their widths being related by

$$\Delta\omega \cdot \Delta t \approx 1. \tag{A.12}$$

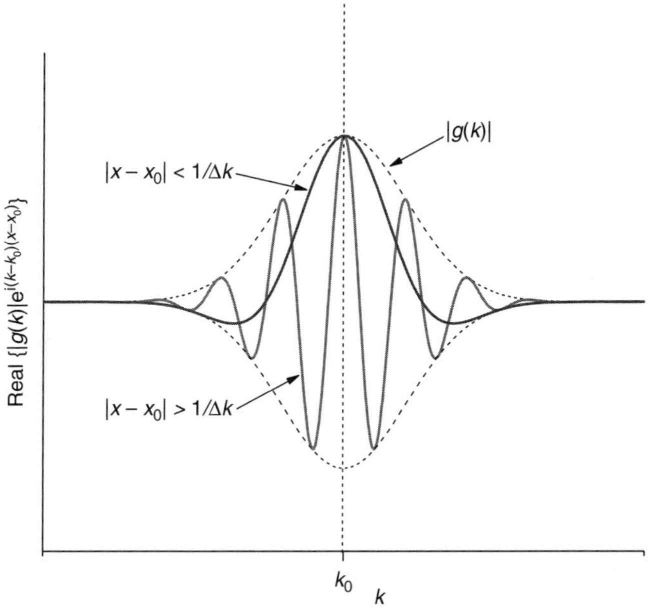

FIGURE A2 Variation of the wave function versus k. When $|x - x_0| > 1/\Delta \boldsymbol{k}$, we see several oscillations, but when $|x - x_0| < 1/\Delta \boldsymbol{k}$, we see only one oscillation

With the aid of the relation $\Delta E = \hbar\Delta\omega$, the uncertainty principle becomes

$$\Delta E \cdot \Delta t \geq \hbar. \tag{A.13}$$

These inequalities shown in Equations 1.34 and 1.37 are introduced to show that $\hbar$ is the lower limit. It is possible that one can construct wave packets for which the products of the quantities in these equations are larger than $\hbar$.

APPENDIX B

PERTURBATION

Simple physical systems such as simple harmonic oscillator and hydrogen atoms can be solved exactly where the Hamiltonian is simple enough to generate exact eigenvalues. In general, the Hamiltonian is very complicated for systems, such as many electron atoms, semiconductor nanostructures, multiple quantum wells, and quantum dots. Solving the Schrödinger equation for such complicated systems is difficult to handle, and, therefore, one needs to make several approximations to reach a reasonable answer. One of these approximations is the perturbation theory. In this appendix we treat the time-independent perturbation (stationary) approximation, which is widely used in many systems such as solid state physics. To understand this approximation, one needs to define a physical system and isolate the main effects that are responsible for the main features of the system. Once these features are understood, the finer details can be discussed by considering the less-important effects that were neglected in the first approximation. Treating these secondary effects can be performed using the perturbation theory. Thus, the Hamiltonian of the system can be presented in the following form:

$$H = H_0 + H_1, \tag{B.1}$$

where H_0 is the unperturbed Hamiltonian with known eigenvectors and eigenvalues, and H_1 is the perturbation that describes the secondary effects in the system. The problem now is to find the modification of the eigenvalues produced by adding H_1. The matrix elements of H_1 are assumed to be much smaller than

Introduction to Nanomaterials and Devices, First Edition. Omar Manasreh.

those of H_0. Let us now define a very small real number $\lambda \ll= 1$ such as

$$H_1 = \lambda \widehat{H}_1, \tag{B.2}$$

where $\widehat{H}_1$ is an operator whose matrix elements are comparable to those of H_0. The perturbation theory deals with expanding the eigenvalues and eigenvectors of H in terms of power λ with a finite number of terms.

Let us assume that the discrete eigenvalues (E_u^0) and eigenvectors ($|\varphi_{\mathrm{u}}^{\mathrm{i}}\rangle$) are known for H_0. The subscript "u" is to indicate the unperturbed terms and the superscript "i" is added in case we have degenerate states. For the first approximation (unperturbed system), we have

$$H_0|\varphi_{\mathrm{u}}^{\mathrm{i}}\rangle = E_{\mathrm{u}}^0|\varphi_{\mathrm{u}}^{\mathrm{i}}\rangle, \tag{B.3}$$

where the set of vectors $|\varphi_{\mathrm{u}}^{\mathrm{i}}\rangle$ forms an orthogonal basis such that $\langle\varphi_{\mathrm{u}}^{\mathrm{i}}|\varphi_{\mathrm{u}'}^{\mathrm{i}'}\rangle = \delta_{\mathrm{ii}'}\delta_{\mathrm{uu}'}$ and $\sum_{\mathrm{u}}\sum_i |\varphi_{\mathrm{u}}^{\mathrm{i}}\rangle\langle\varphi_{\mathrm{u}}^{\mathrm{i}}| = 1$. Now we can consider the system Hamiltonian that depends on the parameter λ by substituting Equation B.2 into Equation B.1

$$H(\lambda) = H_0 + \lambda\widehat{H}_1. \tag{B.4}$$

For $\lambda = 0$, we have only the unperturbed Hamiltonian and the eigenvalues of $H(\lambda)$ that depend on λ. To find the approximate solution of the Schrödinger equation, one needs to find $E(\lambda)$ and $|\psi(\lambda)\rangle$ of $H(\lambda)$:

$$H(\lambda)|\psi(\lambda)\rangle = E(\lambda)|\psi(\lambda)\rangle. \tag{B.5}$$

Let us assume that $E(\lambda)$ and $|\psi(\lambda)\rangle$ can be expanded in powers of λ as

$$E(\lambda) = E_0 + \lambda E_1 + \lambda^2 E_2 + \cdots + \lambda^n E_n \tag{B.6a}$$

and

$$|\psi(\lambda)\rangle = |0\rangle + \lambda|1\rangle + \lambda^2|2\rangle + \cdots + \lambda^n|n\rangle. \tag{B.6b}$$

Substituting these expansions into Equation B.5, we obtain

$$(H_0 + \lambda\widehat{H}_1)\sum_n \lambda^n|n\rangle = \sum_{n'} \lambda^{n'} E_{n'} \sum_n \lambda^n|n\rangle. \tag{B.7}$$

From this equation, we obtain the following relations: For the 0th order of λ, we have

$$H_0|0\rangle = E_0|0\rangle, \tag{B.8}$$

for the first order,

$$(H_0 - E_0)|1\rangle + (\widehat{H}_1 - E_1)|0\rangle = 0, \tag{B.9}$$

for the second order,

$$(H_0 - E_0)|2\rangle + (\widehat{H}_1 - E_1)|1\rangle - E_2|0\rangle = 0, \tag{B.10}$$

and so on.

From the orthogonal property, we have for the 0th order

$$\langle 0|0\rangle = 1. \tag{B.11}$$

For the first order, we have

$$\langle \psi(\lambda)|\psi(\lambda)\rangle = \langle 0|0\rangle + \lambda[\langle 1|0\rangle + \langle 0|1\rangle] + \mathrm{O}(\lambda^2), \tag{B.12}$$

where $\mathrm{O}(\lambda^2)$ is a term of the second order. Since λ is real number, $\langle 1|0\rangle = \langle 0|1\rangle = 0$.
Similarly, one can find from the second order in λ that

$$\langle 2|0\rangle = \langle 0|2\rangle = -\frac{1}{2}\langle 1|1\rangle. \tag{B.13}$$

Let us consider the modification to the unperturbed E_u^0 defined in Equation B.3. Consider first the 0th perturbation for $\lambda \to 0$. By comparing Equation B.3 and Equation B.8, we have $E_0 = E_\mathrm{u}^0$ and $|0\rangle = |\varphi_n\rangle$. This simple result demonstrates how to obtain the eigenvalues and eigenvectors of H_0. For the first-order correction, we need to determine E_1 and $|1\rangle$ from Equation B.9. Let us project Equation B.9 onto the eigenvector $|\varphi_n\rangle$ to obtain

$$\langle \varphi_n|(H_0 - E_0)|1\rangle + \langle \varphi_n|(\widehat{H}_1 - E_1)|0\rangle = 0. \tag{B.14}$$

By letting H_0 operate on the bra $\langle \varphi_n|$, we find that the first term in this equation is zero since $|0\rangle = |\varphi_n\rangle$. Hence, Equation B.14 is reduced to

$$E_1 = \langle \varphi_n|\widehat{H}_1|0\rangle. \tag{B.15}$$

Substituting Equation B.6a into Equation 2.6a, we have

$$\begin{aligned} E_n(\lambda) &= E_\mathrm{u}^0 + \lambda\langle \varphi_n|\widehat{H}_1|0\rangle + \mathrm{O}(\lambda^2) \\ &= E_\mathrm{u}^0 + \langle \varphi_n|H_1|0\rangle + \mathrm{O}(\lambda^2). \end{aligned} \tag{B.16}$$

Thus, the first-order correction to the unperturbed eigenvalue is the mean value of the perturbed term H_1.

For the eigenvector correction, let us project Equation B.9 onto the eigenvectors $|\varphi_\mathrm{p}^\mathrm{i}\rangle \neq |\varphi_\mathrm{n}\rangle$ to obtain

$$\langle \varphi_p^i|(H_0 - E_0)|1\rangle + \langle \varphi_p^i|(\widehat{H}_1 - E_1)|\varphi_n\rangle = 0. \tag{B.17}$$

Recall that $|0\rangle = |\varphi_n\rangle$, the index p is different from n, and i is the degeneracy index. Since the eigenvectors of H_0 associated with different eigenvalues are orthogonal, $E_1\langle\varphi_p^i|\varphi_n\rangle = 0$. Recall that $E_0 = E_\mathrm{u}^0$, and let H_0 act on the bra $\langle\varphi_p^i|$ to give E_p^0. Equation B.17 can then be written as

$$(E_p^0 - E_\mathrm{u}^0)\langle\varphi_p^i|1\rangle + \langle\varphi_p^i|\widehat{H}_1|\varphi_n\rangle = 0. \tag{B.18}$$

This equation gives the coefficients of the desired expansion of the eigenvector $|1\rangle$ on all the unperturbed basis states except $|\varphi_n\rangle$.

$$\langle\varphi_p^i|1\rangle = \frac{1}{(E_\mathrm{u}^0 - E_p^0)}\langle\varphi_p^i|\widehat{H}_1|\varphi_n\rangle, \quad p \neq n. \tag{B.19}$$

The last coefficient $\langle\varphi_n|1\rangle = 0$ according to Equation B.12. Finally, the eigenvector $|1\rangle$ can be written as

$$|1\rangle = \sum_{p\neq n}\sum_i \frac{\langle\varphi_p^i|\widehat{H}_1|\varphi_n\rangle}{(E_u^0 - E_p^0)}|\varphi_p^i\rangle. \tag{B.20}$$

Consequently, the eigenvector $|\psi_n(\lambda)\rangle$ has the following form:

$$|\psi_n(\lambda)\rangle = |\varphi_n\rangle + \sum_{p\neq n}\sum_i \frac{\langle\varphi_p^i|H_1|\varphi_n\rangle}{(E_u^0 - E_p^0)}|\varphi_p^i\rangle + \mathrm{O}(\lambda^2). \tag{B.21}$$

For the second-order perturbation theory, the energy correction is obtained by projecting Equation B.10 onto the vector $|\varphi_n\rangle$

$$\langle\varphi_n|(H_0 - E_0)|2\rangle + \langle\varphi_n|(\widehat{H}_1 - E_1)|1\rangle - \langle\varphi_n|E_2|0\rangle = 0. \tag{B.22}$$

By letting H_0 operate on the bra $\langle\varphi_n|$ and knowing that $E_0 = E_\mathrm{u}^0$, the first term is zero. The E_1 term is also zero since $\langle\varphi_n|1\rangle = 0$, and hence the above equation is reduced to

$$E_2 = \langle\varphi_n|\widehat{H}_1|1\rangle. \tag{B.23}$$

By substituting the expression of $|1\rangle$ as shown in Equation B.20 into Equation B.23, the second-order corrections to the eigenvalue can written as

$$E_2 = \sum_{p\neq n}\sum_i \frac{|\langle\varphi_p^i|\widehat{H}_1|\varphi_n\rangle|^2}{E_u^0 - E_p^0}. \tag{B.24}$$

The final expression for $E_n(\lambda)$ to the second-order perturbation takes the following form:

$$
\begin{aligned}
E_n(\lambda) &= E_u^0 + \lambda\langle\varphi_n|\widehat{H}_1|0\rangle + \lambda^2 \sum_{p\neq n}\sum_{i} \frac{|\langle\varphi_p^i|\widehat{H}_1|\varphi_n\rangle|^2}{E_{\mathrm{u}}^0 - E_{\mathrm{p}}^0} + \mathrm{O}(\lambda^3) \\
&= E_{\mathrm{u}}^0 + \langle\varphi_n|H_1|0\rangle + \sum_{p\neq n}\sum_{i} \frac{|\langle\varphi_p^i|H_1|\varphi_n\rangle|^2}{E_{\mathrm{u}}^0 - E_{\mathrm{p}}^0} + \mathrm{O}(\lambda^3).
\end{aligned}
\tag{B.25}
$$

For the eigenvector $|\psi_n(\lambda)\rangle$ corrections, one can project Equation B.22 onto $|\varphi_{\mathrm{u}}^i\rangle$ to obtain the following result:

$$
|\psi_n(\lambda)\rangle = |\varphi_n\rangle + \sum_{p\neq n}\sum_{i} \frac{\langle\varphi_p^i|H_1|\varphi_n\rangle}{(E_{\mathrm{u}}^0 - E_{\mathrm{p}}^0)}|\varphi_p^i\rangle + \sum_{p\neq n}\sum_{i} \frac{|\langle\varphi_p^i|H_1|\varphi_n\rangle|^2}{(E_{\mathrm{u}}^0 - E_p^0)^2}|\varphi_p^i\rangle + O(\lambda^3).
\tag{B.26}
$$

For additional details on the perturbation of the degenerate states, see Cohen-Tannoudi *et al*. and Merzbacher.

BIBLIOGRAPHY

Cohen-Tannoudji C, Diu B, Laloë F. Quantum mechanics. New York: Wiley; 1977.

Merzbacher E. Quantum mechanics. New York: Wiley; 1970.

APPENDIX C

ANGULAR MOMENTUM

Angular momentum is a very important problem in many fields, including semiconductor materials. One may encounter angular momentum when dealing with doping in semiconductors, solving Schrödinger equation of semiconductor energy bands, and many other cases. In this appendix, we present the most important properties of the angular momentum without going through derivations. In dealing with angular momentum, one needs to distinguish between the spin, the orbital angular momentum, and total angular momentum. For the orbital angular momentum of a spinless particle, we have three observables L_x, L_y, and L_z, which are the components of the orbital angular momentum operator L. The three components can be written as

$$\begin{aligned}\mathrm{L}_x &= YP_z - ZP_y\\ \mathrm{L}_y &= ZP_x - XP_z\\ \mathrm{L}_z &= XP_y - YP_x,\end{aligned} \tag{C.1}$$

where X, Y, and Z are the position observables and P_x, P_y, and P_z are the momentum observables. It was shown in Chapter 1 that $[X, P_x] = [Y, P_y] = [Z, P_z] = i\hbar$. Using this relation, we can write

$$\begin{aligned}\left[\mathrm{L}_x,\ \mathrm{L}_y\right] &= i\hbar\mathrm{L}_z\\ \left[\mathrm{L}_y,\ \mathrm{L}_z\right] &= i\hbar\mathrm{L}_x\\ \left[\mathrm{L}_z,\ \mathrm{L}_x\right] &= i\hbar\mathrm{L}_y.\end{aligned} \tag{C.2}$$

Introduction to Nanomaterials and Devices, First Edition. Omar Manasreh.

Similarly, the components of the total angular momentum J can be expressed as

$$\begin{aligned}[J_x, J_y] &= i\hbar J_z \\ [J_y, J_z] &= i\hbar J_x \\ [J_z, J_x] &= i\hbar J_y.\end{aligned} \tag{C.3}$$

It can be shown that $[J^2, J] = 0$. It is customary to use the following raising and lowering operators

$$J_{\pm} = J_x \pm J_y \tag{C.4}$$

The eigenvalues of J^2 usually takes the following form $j(j+1)\hbar$ where $j \geq 0$. It can be shown that for orthogonal wave vectors$|k, j, m\rangle$, the corresponding eigenvalues are

$$\begin{aligned}J_z|k, j, m\rangle &= m\hbar|k, j, m\rangle \\ J_{pm}|k, j, m\rangle &= \hbar\sqrt{j(j+1) - m(m \pm 1)}|k, j, m \pm 1\rangle,\end{aligned} \tag{C.5}$$

where m is the quantum number that indicates the projection of the angular momentum on the z-axis. Addition of two angular momenta or spin–orbit coupling is usually discussed thoroughly in quantum mechanics textbooks. We may revisit the spin–orbit coupling when dealing with quantization of energy levels in nanostructure and quantum wells.

APPENDIX D

WENTZEL-KRAMERS-BRILLOUIN (WKB) APPROXIMATION

The potential barriers and wells considered thus far are geometrically simple. If the barrier height is an arbitrary function of the position, the solution of Schrödinger equation becomes very complicated. A simple example in which the barrier height is a function of the distance is the triangular potential well, which is usually encountered at the semiconductor–heterojunction interfaces. This problem is discussed briefly in Chapter 2, in which the solution is expressed in Airy functions. Another example is the simple harmonic oscillator in which the potential is parabolic in distance. The solution of this problem is expressed in terms of Hermite polynomials. For an arbitrary potential barrier as shown in Fig. D1, we follow Merzbacher treatment, where he considered the Wentzel-Kramers-Brillouin (WKB) approximation. Let us consider Fig. D1a, in which we show an arbitrary spatially varying potential. The position a is called the *turning point* at which the wave function changes from propagating to decaying. The propagation vectors of both waves are given by

$$k(x) = \sqrt{\frac{2m}{\hbar^2}[E - V(x)]} \qquad \text{for } E > V(x) \tag{D.1}$$

and

$$\rho(x) = \sqrt{\frac{2m}{\hbar^2}[V(x) - E]} \qquad \text{for } E < V(x). \tag{D.2}$$

The WKB approximation suggests that the wave functions, either decaying or propagating, are wave-type functions generally defined as

$$\psi(x) \sim e^{i\varphi(x)}. \tag{D.3}$$

Introduction to Nanomaterials and Devices, First Edition. Omar Manasreh.

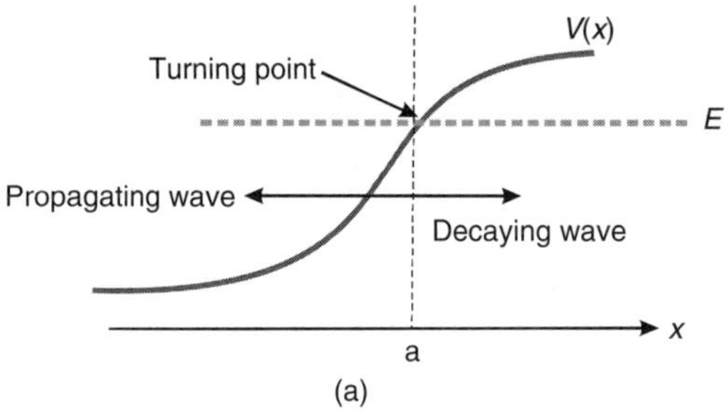

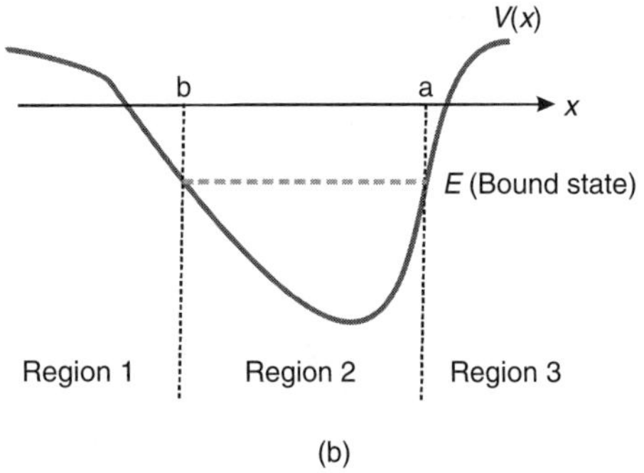

FIGURE D1 (a) Variation of potential barrier as a function of the distance showing the corresponding energy levels. (b) An arbitrary potential well used for WKB approximation.

By applying Schrödinger equation to the propagating wave function, we have

$$\frac{\partial^2 \psi(x)}{\partial x^2} + k^2(x)\psi(x) = 0. \tag{D.4}$$

Assuming that the proportionality constant of Equation D.4 is spatially invariant, the Schrödinger equation becomes

$$\frac{\partial^2 \mathrm{e}^{i\varphi(x)}}{\partial x^2} + k^2(x)\mathrm{e}^{i\varphi(x)} = 0, \tag{D.5}$$

which can be reduced to the differential equation of $\varphi(x)$ as

$$i\frac{\partial^2 \varphi(x)}{\partial x^2} - \left(\frac{\partial \varphi}{\partial x}\right)^2 + k^2(x) = 0. \tag{D.6}$$

This equation is equivalent to Schrödinger equation, except that it is nonlinear. One, however, can take advantage of the nonlinearity to solve Equation D.6. If we have a true free particle, the second derivative is very small assuming that the potential does not vary too much.

$$i\frac{\partial^2 \varphi(x)}{\partial x^2} = 0. \tag{D.7}$$

When Equation D.7 is omitted from Equation D.6, we obtain the following first crude approximation by replacing φ with φ_0:

$$\left(\frac{\partial\varphi_0}{\partial x}\right)^2 = k^2(x) \text{ or } \varphi_0 = \pm\int^x k(x)\mathrm{d}x + C. \tag{D.8}$$

The next approximation is to set Equation D.6 in the following form:

$$\left(\frac{\partial\varphi}{\partial x}\right)^2 = +k^2(x) + i\frac{\partial^2\varphi(x)}{\partial x^2}. \tag{D.9}$$

The nth approximation can be set for the right-hand side of this equation to obtain the $(n+1)^{\text{th}}$ approximation as follows:

$$\varphi_{n+1}(x) = \pm\int^x \sqrt{k^2(x) + i\varphi_n''(x)}\mathrm{d}x + C_{n+1}. \tag{D.10}$$

For $n = 0$, we have

$$\varphi_1(x) = \pm\int^x \sqrt{k^2(x) + i\varphi_0''(x)}\mathrm{d}x + C_1 = \pm\int^x \sqrt{k^2(x) \mp ik'(x)}\mathrm{d}x + C_1. \tag{D.11}$$

The correct $\varphi(x)$ is baseless unless $\varphi_1(x)$ is close to $\varphi_{(x)}$, which means

$$|k'(x)| \ll |k^2(x)|. \tag{D.12}$$

If this condition holds, the integrand can be expanded to obtain

$$\varphi_1(x) \approx \int^x \left[\pm k(x) + \frac{i}{2}\frac{k'(x)}{k(x)}\right]\mathrm{d}x + C_1 = \pm\int^x k(x)\,dx + \tfrac{i}{2}\ln(k(x)) + C_1. \tag{D.13}$$

All the above approximations are known as *WKB approximation*, which leads us to write the wave function as

$$\psi(x) \approx \frac{1}{\sqrt{k(x)}}\mathrm{e}^{\pm i\int^x k(x)\mathrm{d}x}. \tag{D.14}$$

The equivalent solution for the decaying wave is

$$\psi(x) \approx \frac{1}{\sqrt{k(x)}}\mathrm{e}^{\pm\int^x p(x)\mathrm{d}x}. \tag{D.15}$$

If $k(x)$ is regarded as the effective wave number, then for the propagating wave function we have $\lambda(x) = 2\pi/k(x)$. If the condition (Eq. D.12) holds, we have

$$\lambda(x)\left|\frac{\mathrm{d}p}{\mathrm{d}x}\right| \ll |p(x)|, \tag{D.16}$$

where $p(x) = \pm\hbar k(x)$ is the momentum that the particle would possess at point x. This condition (Eq. D.16) implies that the change of the momentum over a wavelength must be smaller compared to the momentum itself. This condition breaks down if $k(x)$ is zero or varies widely. The entire approach breaks down if the energy of the particle is close in value to the potential extremum because proceeding from left to right, the turning point a is reached before the particle gets sufficiently far away from the turning point b (Fig. D1b) for the WKB approximation to hold. Connecting the waves of one type to another (decaying and propagating) at the turning point requires mathematical details, which are not considered here, but the reader can find these details in other books (Merzbacher). The connecting formulas are written in terms of sine and cosine and are given as follows:
For $x = a$ we have

$$\begin{aligned} \frac{2}{\sqrt{k}} \cos\left(\int_x^a k\mathrm{d}x - \frac{\pi}{4}\right) &\Leftrightarrow \frac{2}{\sqrt{k}} \mathrm{e}^{-\int_a^x k\mathrm{d}x} \\ \frac{2}{\sqrt{k}} \sin\left(\int_x^a k\mathrm{d}x - \frac{\pi}{4}\right) &\Leftrightarrow -\frac{2}{\sqrt{k}} \mathrm{e}^{\int_a^x k\mathrm{d}x}, \end{aligned} \tag{D.17}$$

and for $x = b$, we have

$$\begin{aligned} \frac{2}{\sqrt{k}} \mathrm{e}^{-\int_x^b k\mathrm{d}x} &\Leftrightarrow \frac{2}{\sqrt{k}} \cos\left(\int_b^x k\mathrm{d}x - \frac{\pi}{4}\right) \\ -\frac{2}{\sqrt{k}} \mathrm{e}^{\int_x^b k\mathrm{d}x} &\Leftrightarrow \frac{2}{\sqrt{k}} \sin\left(\int_b^x k\mathrm{d}x - \frac{\pi}{4}\right). \end{aligned} \tag{D.18}$$

Let us now consider the WKB approximation to solve the bound states in an arbitrary potential well. Consider three regions in Fig. D1b where the potential is arbitrary. The WKB approximation is used in regions 1, 2, and 3 away from the turning points, and the connection Equations D.17 and D.18 are used near $x = a$ and $x = b$. The requirement is that $\psi(x)$ should be finite and the solution to Schrödinger equation must vanish as the particle moves outward from the turning points. The wave function can be written as

$$
\begin{aligned}
\psi_1(x) &\approx \frac{1}{\sqrt{\rho}} e^{-\int_x^b \rho(x)\mathrm{d}x} \qquad for\, x < b \\
\psi_2(x) &\approx \frac{2}{\sqrt{k}} \cos\left(\int_b^x k\mathrm{d}x - \frac{\pi}{4}\right) \qquad for\, b < x < a \\
&\approx \frac{2}{\sqrt{k}} \cos\left(\int_b^a k\mathrm{d}x - \int_x^a k\mathrm{d}x - \frac{\pi}{4}\right) \\
&\approx -\frac{2}{\sqrt{k}} \cos\left(\int_b^a k\mathrm{d}x\right) \sin\left(\int_x^a k\mathrm{d}x - \frac{\pi}{4}\right) \\
&+ \frac{2}{\sqrt{k}} \sin\left(\int_b^a k\mathrm{d}x\right) \cos\left(\int_x^a k\mathrm{d}x - \frac{\pi}{4}\right) \\
\psi_3(x) &\approx \frac{1}{\sqrt{\rho}} e^{\int_a^x \rho\mathrm{d}x} \qquad for\, x > a.
\end{aligned}
\tag{D.19}
$$

From the boundary condition at the turning point a from Equation (D.17), only the second term of $\psi_2(x)$ gives rise to a decreasing exponential satisfying the boundary conditions at infinity. Thus, the first term must be zero, which leads to the following relation:

$$
\cos\left(\int_b^a k\mathrm{d}x\right) = \left(n + \frac{1}{2}\right)\pi, \text{ where } n = 0, 1, 2, 3, \ldots \tag{D.20}
$$

This equation determined the possible discrete values of E. The energy E appears in the integrand as well as in the limits of integration, since the turning points a and b are determined such that $V(a) = V(b) = E$. Let us consider, for example, the triangular potential well shown in Fig. 2.16. Since the potential is sharp at $x = 0$, WKB approximation cannot be used at this point. The energy E_n can be related to the turning points $x_n.E_n(x_n)V(x_n) = eE_s x_n$, where E_s is the electric field. For the turning point of the decaying function, we have

$$
\psi_1(x) \simeq \frac{1}{\sqrt{\rho}} e^{-\int_{x_n}^{b} \rho(x)\mathrm{d}x} \qquad \text{for } x > x_n. \tag{D.21}
$$

This wave function must be connected to the cosine function according to Equation (D.17)

$$
\psi(x) \approx \frac{2}{\sqrt{k}} \cos\left(\int_x^{x_n} k\mathrm{d}x - \frac{\pi}{4}\right). \tag{D.22}
$$

Equation (D.22) must vanish at $x = 0$, so the bound states are found from

$$\cos\left(\int_{x}^{x_n} k\mathrm{d}x - \frac{\pi}{4}\right) = 0. \tag{D.23}$$

Using the relation

$$k(x) = \sqrt{\frac{2m}{\hbar^2}(E_n - V(x))} = \sqrt{\frac{2m}{\hbar^2}(eE_\mathrm{s}x_n - eE_\mathrm{s}x)} = \sqrt{\frac{2meE_\mathrm{n}}{\hbar^2}(x_n - x)}, \tag{D.24}$$

we can write:

$$\int_{x}^{x_n} k\mathrm{d}x - \frac{\pi}{4} = (2n+1)\frac{\pi}{2}, \tag{D.25}$$

which leads to

$$\int_{0}^{x_n} k\mathrm{d}x = (2n+1)\frac{\pi}{2} + \frac{\pi}{4}$$

$$\sqrt{\frac{2meE_\mathrm{s}}{\hbar^2}}\int_{0}^{x_n}\sqrt{x_n - x}\,\mathrm{d}x = \frac{\pi}{2}\left(2n + \frac{3}{2}\right) \text{ and} \tag{D.26}$$

$$\sqrt{\frac{2meE_\mathrm{s}}{\hbar^2}}\frac{2}{3}x_n^{3/2} = \frac{\pi}{2}\left(2n + \frac{3}{2}\right)$$

or (D.27)

$$x_n = \left(\frac{3\pi}{4}\left(2n + \frac{3}{2}\right)\right)^{2/3} \cdot \left(\frac{\hbar^2}{2meE_\mathrm{s}}\right)^{1/3}.$$

Finally, substituting $E_n = eE_\mathrm{s}x_n$ in the above equation, one finds that

$$E_n = \left(\frac{3\pi}{4}\left(2n + \frac{3}{2}\right)\right)^{2/3} \cdot \left(\frac{e^2E_\mathrm{s}^2\hbar^2}{2m}\right)^{1/3} \quad \text{for } n = 0, 2, 3, \ldots \tag{D.28}$$

The results obtained from Equation D.28 are exactly the same as those obtained using the Airy function (Eq. 2.139) for the triangular potential well.

BIBLIOGRAPHY

Merzbacher, E., "*Quantum Mechanics*", (Wiley, New York, 1970)

APPENDIX E

PARABOLIC POTENTIAL WELL

An example of an extremely important class of one-dimensional bound state in quantum mechanics is the simple harmonic oscillator whose potential can be written as

$$V(x) = \frac{1}{2}Kx^2, \tag{E.1}$$

where K is the force constant of the oscillator.

The Hamiltonian operator is given by

$$H = -\frac{\hbar^2}{2m}\frac{\partial^2}{\partial x^2} + \frac{1}{2}Kx^2. \tag{E.2}$$

The Schrödinger equation that gives the possible energies of the oscillator is

$$-\frac{\hbar^2}{2m}\frac{\partial^2\varphi(x)}{\partial x^2} + \frac{1}{2}Kx^2\varphi(x) = E_n\varphi(x). \tag{E.3}$$

This equation can be simplified by choosing a new measure length and a new measure of energy, each of which is dimensionless. $\zeta \equiv (\frac{mK}{\hbar^2})^{1/4}x$ and $\eta = \frac{2E_n}{\hbar\omega}$, where $\omega = \sqrt{\frac{K}{m}}$. With these substitutions, Equation E.3 becomes

$$\frac{\mathrm{d}^2\varphi(\zeta)}{\mathrm{d}\zeta^2} + (\eta - \zeta^2)\varphi(\zeta) = 0. \tag{E.4}$$

Introduction to Nanomaterials and Devices, First Edition. Omar Manasreh.

In looking for bounded solutions, one can notice that as η approaches infinity and becomes too small compared to ζ^2, the resulting differential equation can be easily solved to yield

$$\varphi(\zeta) \sim e^{\pm\frac{1}{2}\zeta^2}. \tag{E.5}$$

This expression for the asymptotic dependence is suitable only for the negative sign in the exponent. It is clear that because of the very rapid decay of the resulting Gaussian function as ζ goes to infinity, the function will still have the same asymptotic dependence; it is multiplied by any finite polynomial in ζ (Dicke and Wittke)

$$\varphi(\zeta) = H(\zeta)e^{-\frac{1}{2}\zeta^2}. \tag{E.6}$$

where $H(\zeta$ is a finite polynomial. By substituting Equation E.5 into Equation E.4, one can obtain

$$\frac{d^2H(\zeta)}{d\zeta^2} - 2\zeta\frac{dH(\zeta)}{d\zeta} + (\eta - 1)H(\zeta) = 0. \tag{E.7}$$

If we assume a solution to this equation in the form of a finite polynomial

$$H(\zeta) = A_0 + A_1\zeta + A_2\zeta^2 + \cdots + A_n\zeta^n, \tag{E.8}$$

a recursion formula connecting the coefficients can be obtained in the following form:

$$A_{n+2} = \frac{(2n + 1 - \eta)}{(n + 2)(n + 1)}A_n, \quad \text{for } n \geq 0. \tag{E.9}$$

For an upper cutoff to the coefficients such that the polynomial $H(\eta)$ remains finite, the condition

$$\eta = 2n + 1 \tag{E.10}$$

must be satisfied. Substituting $\eta = \frac{2E_n}{\hbar\omega}$ into Equation E.10, we obtain

$$E_n = \left(n + \frac{1}{2}\right)\hbar\omega. \tag{E.11}$$

The energy levels and the parabolic potential are shown in Fig. E1.

The polynomial solutions lead to wave functions that approach zero at $x = \pm\infty$ and allow normalization. These polynomials are called *Hermite polynomials*, and they are the acceptable solutions as wave functions. The Hermite polynomial is defined as follows:

$$H_n(\zeta) = (-1)^n e^{\zeta^2}\frac{d^n}{dx^n}(e^{-\zeta^2}). \tag{E.12}$$

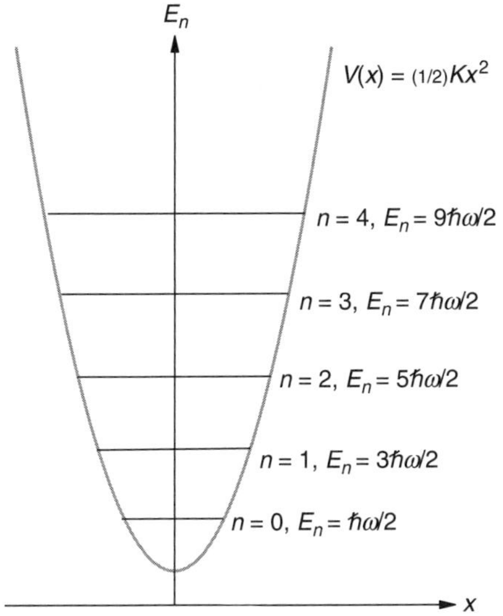

FIGURE E1 A parabolic one-dimensional potential well with a few of the allowed energy levels are shown.

Finally, the wave function can be written as

$$\varphi(\zeta) = N_n H_n(\zeta) e^{-\frac{1}{2}\zeta^2} = N_n(-1)^n e^{\frac{1}{2}\zeta^2} \frac{d^n}{dx^n}(e^{-\zeta^2}). \tag{E.13}$$

The normalization factor N_n can be found to be

$$N_n = \frac{1}{2^n!}\sqrt{\frac{\alpha}{\pi}}, \quad \text{where } \alpha = \left(\frac{mK}{\hbar^2}\right)^{1/2}. \tag{E.14}$$

The first lowest four wave functions are illustrated in Fig. E2a. The probability amplitude for the $n = 5$ eigenstate is shown in Fig. E2b along with the classical probability density. The first few Hermite polynomial functions are shown below.

Figure E2 shows several oscillations in the $\varphi^*(\zeta)\varphi(\zeta))$ curve, with their amplitudes fairly small near the origin and considerably larger near the end of the curve. As n is increased, the probability density becomes larger and larger near the end of the curves and smaller and smaller near the center of the curve, approaching the classical limit. According to classical mechanics (McKelvey), the probability of finding a particle in an interval dx is proportional to the time dt it spends in that interval. This is directly related to the velocity by $dx = \upsilon_x dt$. More precisely, if $T/2$ is the half period of oscillation, the fraction of time spent in dx is $dt/(T/2)$ or $2dx/(T\upsilon_x)$. This fraction is the classical analog of the probability density $\varphi^*(\zeta)\varphi(\zeta)$ For a classical harmonic oscillator, conservation of energy

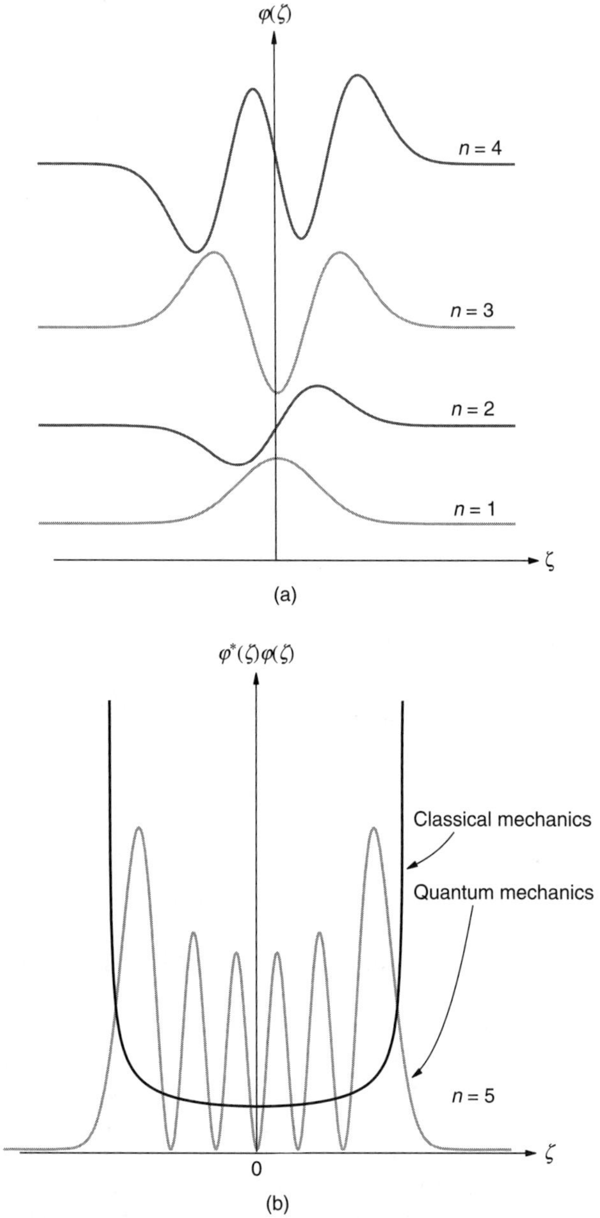

FIGURE E2 (a) Wave functions plotted for the lowest eigenstates. (b) The probability amplitude for $n = 5$ is shown along the classical probability density.

requires that the total energy E be

$$E = \frac{1}{2}mv_x^2 + \frac{1}{2}m\omega^2 x^2 = \frac{1}{2}m\omega^2 A^2. \tag{E.15}$$

Solving for v_x, we have $v_x = \omega(A^2 - x^2)^{1/2}$. The classical analog to $\varphi^*(\zeta)\varphi(\zeta)$ can then be written as $p(x)\mathrm{d}x$, where

$$p(x)\mathrm{d}x = \frac{2\mathrm{d}x}{Tv_x} = \frac{\mathrm{d}x}{\pi\sqrt{A^2 - x^2}}. \tag{E.16}$$

If the quantum energy is given by Equation E.11, with the help of Equation E.15, one can write the classical probability as

$$p(x) = \frac{2\mathrm{d}x}{Tv_x} = \frac{\mathrm{d}x}{\pi\sqrt{\frac{2n+1}{\alpha} - x^2}}, \tag{E.17}$$

where α is defined in Equation E.14.

N	$H_a(\zeta)$
0	1
1	2ζ
2	$4\zeta^2 - 2$
3	$8\zeta^3 - 12\zeta$
4	$16\zeta^4 - 48\zeta^2 - 12$
5	$32\zeta^5 - 160\zeta^3 + 120\zeta$

BIBLIOGRAPHY

Dicke RH, Witske JP. Introduction to quantum mechanics. Sydney: Addison Wesley; 1960.

McMelvey JP. Solid state physics for engineering and materials science. Malbar, Florida: Krieger Publishing Co.; 1993.

APPENDIX F

TRANSMISSION COEFFICIENT IN SUPERLATTICES

The following is a Matlab program for the calculation of the transmission coefficient in short-period superlattices, which is based on the propagation matrix method. This program is adapted from Levi's "*Applied quantum mechanics*," (Cambridge University Press, Cambridge, England, 2003).

```
%Superlattices.m
%resonant tunneling through 4 barriers of short period
   superlattices

clear
clf;
nbarrier=4; %number of potential barriers
bx=2.5e-9; %barrier width (m)
wx=2.e-9; %well width (m)
V0=0.68; %barrier energy (eV)

N=(2*nbarrier)+1; %number of samples of potential
for j=1:2:(2*nbarrier) %set up distance and potential array
 dL(j)=wx; V(j)=0;
 dL(j+1)=bx; V(j+1)=V0;
end
 dL(N)=wx; V(N)=0;
```

Introduction to Nanomaterials and Devices, First Edition. Omar Manasreh.

```
Emin=pi*1e-5; %add (pi*1.0e-5) to energy to avoid divide
   by zero
Emax=1.0; %maximum particle energy (eV)
npoints=2000; %number of points in energy plot
dE=Emax/npoints; %energy increment (eV)
hbar=1.05457159e-34; %Planck's constant (J s)
eye=complex(0.,1.); %square root of -1
m0=9.109382e-31; %bare electron mass (kg)
meff=0.22; %effective electron mass/ m0
m=meff*m0; %effective electron mass (kg)
echarge=1.6021764e-19; %electron charge (C)

for j=1:npoints
 E(j)=dE*j+Emin;
 bigP=[1,0;0,1]; %default value of matrix bigP
  for i=1:N
   k(i)=sqrt(2*echarge*m*(E(j)-V(i)))/hbar; %wave vec-
tor at energy E
 end
for n=1:(N-1)%multiply out propagation matrix
 p(1,1)=0.5*(1+k(n+1)/k(n))*exp(-eye*k(n)*dL(n));
 p(1,2)=0.5*(1-k(n+1)/k(n))*exp(-eye*k(n)*dL(n));
 p(2,1)=0.5*(1-k(n+1)/k(n))*exp(eye*k(n)*dL(n));
 p(2,2)=0.5*(1+k(n+1)/k(n))*exp(eye*k(n)*dL(n));
 bigP=bigP*p;
end
 Trans(j)=(abs(1/bigP(1,1)))^2; %transmission coefficient
end

figure(1); %plot potential and transmission coefficient
Vp=[V;V];Vp=Vp(:);
xcell=bx+wx;deltax=xcell/1000;x0=[wx-deltax,wx,xcell-
deltax,xcell];x=x0;for j=1:(N-3)/2,x=[x,x0+j*xcell];end;
   x=[0,x,(nbarrier+1)*xcell];
subplot(1,2,1),plot(x,Vp),axis([0,(nbarrier+1)*xcell,0,
   Emax]),xlabel('Distance (m)'),ylabel('Potential energy,
   V (eV)');
ttl=['Superlattices (',num2str(nbarrier),'-barriers),
   m_{eff}=',num2str(meff),' x m_0'];
title(ttl);
subplot(1,2,2),plot(Trans,E),axis([0,1,0,Emax]),xlabel
   ('Transmission coefficient'),ylabel('Energy,
    E (eV)');

figure(2);
```

```
subplot(1,2,1),plot(x,Vp);axis([0,(nbarrier+1)*xcell,0,
   Emax]),xlabel('Distance (m)');ylabel('Potential energy,
   V (eV)');
title(ttl);
subplot(1,2,2);plot(-log(Trans),E);axis([0,70,0,Emax]);
   xlabel('-ln trans. coeff.');ylabel('Energy, E (eV)');
dlmwrite('Superlattice4by3-Energy.xls',
   [E' -(-log(Trans'))],'');
```

APPENDIX G

LATTICE VIBRATIONS AND PHONONS

In semiconductor crystals, the atoms are tightly coupled to one another, and the binding energy is called *cohesive energy*, which is defined as the energy needed to separate the crystal into independent ions at a large distance from each other. The thermal kinetic energy of the atoms in the crystal is simply the vibrational energy of motion, which propagates in the crystal as waves. These waves are called *acoustical* or *sonic waves*. The quanta of these waves is called *phonon*. Phonons in semiconductors can absorb or scatter light in the infrared spectral region. To understand how the acoustical waves propagate in a solid, let us first consider a one-dimensional monatomic lattice, as shown in Fig. G1. By including only the nearest-neighbor interaction and assuming that the vibrational amplitudes are smaller than the lattice spacing, one can write the force on the nth atom as

$$F_n = m\frac{\partial^2 u_n}{\partial t^2} = \gamma(u_{n+1} - u_n) - \gamma(u_n - u_{n-1}) = \gamma(u_{n+1} + u_{n-1} - 2u_n), \tag{G.1}$$

where m is the mass of the atom and γ is the force constant. For a solution having the character of a traveling wave, we have for the nth atom the following solution (Eq. G.1):

$$u_n = Ae^{i(kx_n - \omega t)} = Ae^{i(kna - \omega t)}, \tag{G.2}$$

where k is the propagation constant, ω is the angular frequency, and $x_n = na$ (where a is the lattice constant). Similarly, the solutions for the nearest-neighbor

Introduction to Nanomaterials and Devices, First Edition. Omar Manasreh.

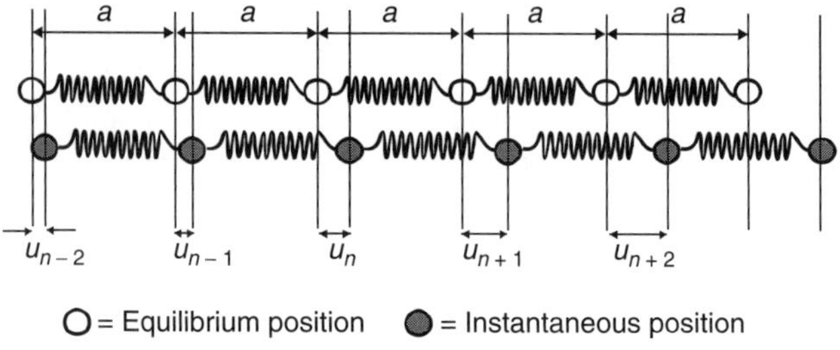

FIGURE G1 A one-dimensional illustration of a crystal with a lattice constant "a" showing the longitudinal displacement of a few atoms.

atoms are

$$u_{n+1} = Ae^{i(kx_{n+1}-\omega t)} = Ae^{i\{k(n+1)a-\omega t\}} = e^{ika}u_n$$
$$u_{n-1} = Ae^{i(kx_{n-1}-\omega t)} = Ae^{i\{k(n-1)a-\omega t\}} = e^{-ika}u_n. \tag{G.3}$$

The dispersion relation can be obtained by substituting Equations G.2 and G.3 into Equation G.1 and canceling u_n as follows:

$$\omega^2 = \frac{\gamma}{m}(2 - e^{ika} - e^{-ika}) = \frac{2\gamma}{m}(1 - \cos ka)$$
$$= \frac{4\gamma}{m}\sin^2\left(\frac{ka}{2}\right). \tag{G.4}$$

This equation can be rewritten as

$$\omega = \omega_m \left|\sin\left(\frac{ka}{2}\right)\right|, \tag{G.5}$$

where $\omega_m = \sqrt{4\gamma/m}$ and the absolute value sign is given to indicate that ω is a positive quantity. Displacement of the atoms shown in Fig. G1 produces a longitudinal acoustic wave with a frequency described by Equation G.5.

For transverse acoustical waves, the atoms are displaced as shown in Fig. G2. In addition to the atomic displacement shown in this figure, it is also possible to simultaneously displace the atoms perpendicular to the plane of the page. Thus, one may obtain two transverse modes. The equation of motion of the transverse modes of the nth atom can be obtained in a manner similar to that in the longitudinal mode

$$F_n = m\frac{\partial^2 u_n}{\partial t^2} = \gamma_t(u_{n+1} + u_{n-1} - 2u_n), \tag{G.6}$$

where γ_t is transverse force constant. The dispersion relation of the transverse mode is obtained as

$$\omega = \omega_m^t \left|\sin\left(\frac{ka}{2}\right)\right|, \tag{G.7}$$

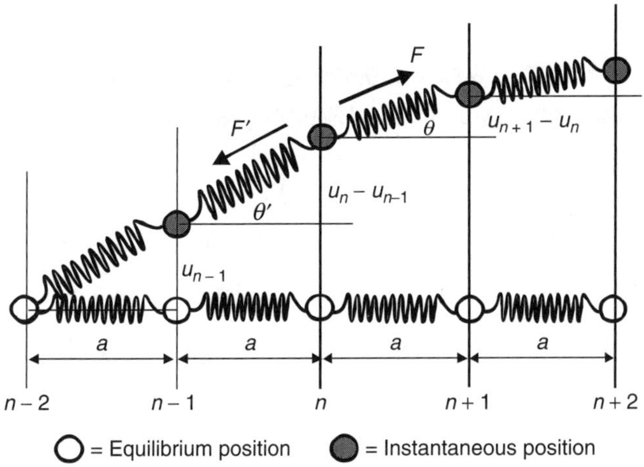

FIGURE G2 A one-dimensional model of a crystal with a lattice constant "a" showing the transverse displacement of a few atoms.

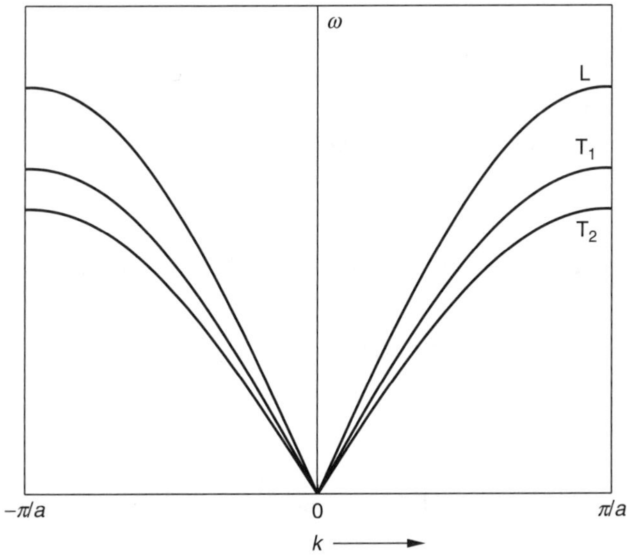

FIGURE G3 The angular frequency (ω) of the longitudinal (L) and transverse (T_1 and T_2) waves in a one-dimensional monatomic lattice is plotted as a function of the propagation constant (k).

where $\omega_m^t = \sqrt{4\gamma_t/m}$. A plot of the dispersion relations for both the longitudinal (L) and transverse modes (T_1 and T_2) is shown in Fig. G3. When the force constants are the same for both the transverse modes, the dispersion relation becomes degenerate with $T_1 = T_2$.

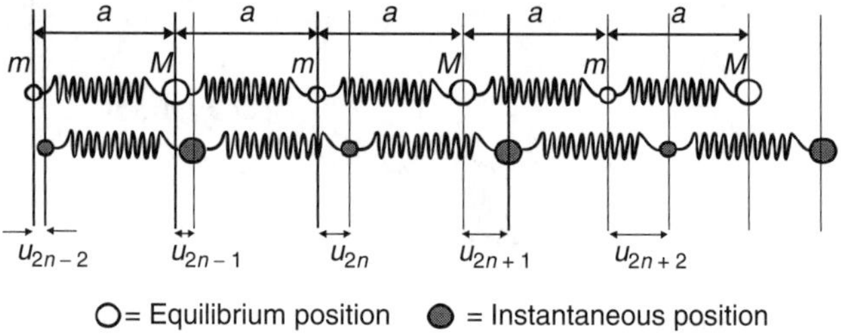

FIGURE G4 A one-dimensional chain of diatomic crystal with atomic masses M and m.

For a more complicated case, let us consider a linear one-dimensional diatomic lattice model, as shown in Fig. G4, in which the chain is composed of two alternating atoms of masses M and m (M is assumed to be larger than m). The equation of motions of atoms $2n$ (mass m) and $2n+1$ (mass M) can be written as

$$F_{2n} = m\frac{\partial^2 u_{2n}}{\partial t^2} = \gamma(u_{2n+1} + u_{2n-1} - 2u_{2n}) \quad \text{(a)} \tag{G.8a}$$

$$F_{2n+1} = M\frac{\partial^2 u_{2n+1}}{\partial t^2} = \gamma(u_{2n+2} + u_{2n} - 2u_{2n+1}), \quad \text{(b)} \tag{G.8b}$$

where γ is the force constant and u's are the atomic displacements from the equilibrium. The solutions to Equation G.8 can be expressed as

$$u_{2n} = A\mathrm{e}^{i(2kna-\omega t)} \quad \text{and} \quad u_{2n+1} = B\mathrm{e}^{i\{k(2n+1)a-\omega t\}}. \tag{G.9}$$

Similarly, the displacement of the atoms labeled $2n+2$ and $2n-1$ can be written as

$$\begin{aligned} u_{2n+2} &= A\mathrm{e}^{i\{k(2n+2)a-\omega t\}} = u_{2n}\mathrm{e}^{i2ka} \\ u_{2n-1} &= B\mathrm{e}^{i\{k(2n-1)a-\omega t\}} = u_{2n+1}\mathrm{e}^{-i2ka}. \end{aligned} \tag{G.10}$$

Substituting Equations G.9 and G.10 into Equation G.8 and evaluating the time derivative of u_{2n} and u_{2n+1}, the equations of motion can be rewritten as

$$\begin{aligned} (m\omega^2 - 2\gamma)u_{2n} + \gamma(1 + \mathrm{e}^{-2ika})\mathrm{u}_{2n+1} &= 0 \\ \gamma(1 + \mathrm{e}^{2ika})u_{2n} + (M\omega^2 - 2\gamma)\mathrm{u}_{2n+1} &= 0. \end{aligned} \tag{G.11}$$

These two homogeneous equations can be solved by equating the determinant to zero such that

$$\begin{vmatrix} (m\omega^2 - 2\gamma) & \gamma(1 + \mathrm{e}^{-2ika}) \\ \gamma(1 + \mathrm{e}^{2ika}) & (M\omega^2 - 2\gamma) \end{vmatrix} = (m\omega^2 - 2\gamma)(M\omega^2 - 2\gamma) - 4\gamma^2\cos^2(ka) = 0, . \tag{G.12}$$

where $1 + \cos(2ka) = 2\cos^2(ka)$ is used in the above expression. The solution of Equation G.12 can be expressed as

$$\omega^2 = \frac{\gamma(m+M)}{mM}\left[1 \pm \sqrt{1 - \frac{4\,mM\sin^2(ka)}{(m+M)^2}}\right]. \qquad \text{(G.13)}$$

The above dispersion relation has two roots (ω_+ and ω_-), which are plotted in Fig. G5. The two branches are called *optical* (ω_+) and *acoustical* (ω_-) *modes*. The maximum value of the optical branch is $\sqrt{\gamma(m+M)/(mM)}$, and there is a forbidden frequency gap at the Brillouin zone boundaries extended between $\sqrt{2\gamma/m}$ and $\sqrt{2\gamma/M}$. This gap is reduced to zero for $m = M$.

Phonon energy or frequency in semiconductors can be measured by infrared spectroscopy and Raman scattering techniques, provided that the phonon density of states is large. Phonon modes can be infrared active and/or Raman active. Selection rules that govern phonon absorption are outside the scope of this book, but detailed discussions and analyses are reported by many authors (Birman). To calculate the phonon density of states, let us consider the simplest case of a monatomic one-dimensional crystal. To determine the number of phonon modes with different values of k that fall in the frequency interval from ω to $\omega + d\omega$, one

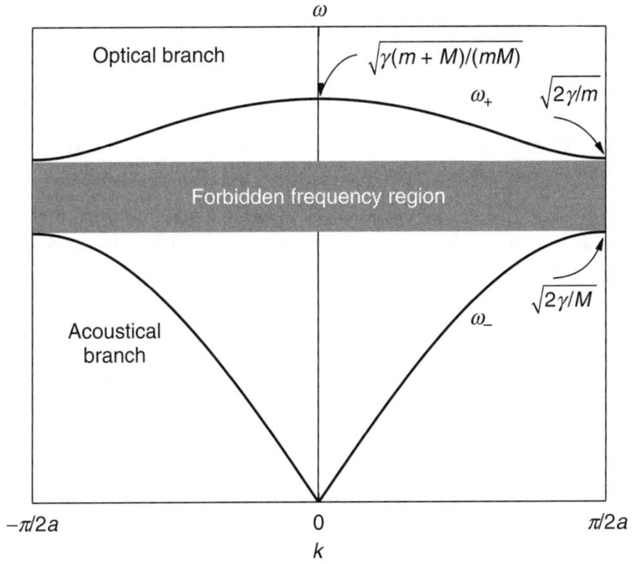

FIGURE G5 The dispersion curve for a diatomic one-dimensional lattice. The first Brillouin zone is extended between $-\pi/(2a)$ and $+\pi/(2a)$ since the unit cell is $2a$.

can obtain from the dispersion relation, expressed in Equation G.7, the following:

$$d\omega = \sqrt{\frac{4\gamma}{m}} d\left[\left|\sin\left(\frac{ka}{2}\right)\right|\right] = \sqrt{\frac{4\gamma}{m}} \frac{a}{2}\left|\cos\left(\frac{ka}{2}\right)\right| dk = a\sqrt{\frac{\gamma}{m}}\left|\cos\left(\frac{ka}{2}\right)\right| dk. \tag{G.14}$$

If one assumes that the one-dimensional crystal is a large circle that contains N atoms, where $N \gg 1$, the atomic displacements satisfy the following conditions: $u_{N+n} = u_n$ and $e^{ikNa} = 1$. From the second condition, one can obtain the following relation:

$$k = \frac{2\pi p}{Na}, \tag{G.15}$$

where p is an integer that satisfies the following relation: $-N/2<p<N/2$. These relations indicate that k is discrete with N possible values corresponding to N different standing waves. The number of modes, dn, in the interval dω is

$$dn = 2dp = \frac{N}{\pi}\sqrt{m/\gamma}\frac{d\omega}{|\cos(ka/2)|}. \tag{G.16}$$

With the help of Equation G.5, cos $(ka/2)$ can be expressed as

$$\cos(ka/2) = \sqrt{1 - \sin^2(ka/2)} = \sqrt{1 - \frac{\omega^2 m}{4\gamma}}. \tag{G.17}$$

The phonon density of states, g_{ph}, can now be defined as

$$g_{ph}(\omega) = \frac{1}{Na}\frac{dn}{d\omega} = \frac{2}{\pi a}\frac{1}{\sqrt{\omega_m^2 - \omega^2}}, \tag{G.18}$$

where $\omega_m = \sqrt{4\gamma/m}$. The density of states expressed in Equation G.18 approaches infinity as ω approaches ω_m, and it has a constant value when ω approaches zero. The same analysis can be applied for a diatomic chain using the dispersion relation described in Equation G.13. The analysis, however, is a little bit more complicated.

For three-dimensional crystals, such as Si and GaAs, the calculations of the phonon modes and their density of states are more extensive and require computer analysis. Table G1 summarizes the energy of the phonon modes in several semiconductor materials. A typical example of phonon Raman scattering is shown in Fig. G6 for semi-insulating GaAs sample. The Raman scattering spectrum in this figure was obtained using a Fourier-transform infrared spectrometer in conjunction with 1.06-μm Yttrium Aluminum Garnet (YAG) laser. The spectrum shows two sets of peaks *called Stokes* and *anti-Stokes phonon scattering*, which is known as the *Raman effect*. To understand this effect, consider an incident laser beam of an energy $\hbar\omega_i$, which is scattered by a semiconductor sample. The radiation consists of the laser beam ($\hbar\omega_i$) and weaker beams of energies

TABLE G1 The Energy of the Phonon Modes in Several Semiconductor Materials Reported in meV for Three Different Symmetry Points in the First Brillouin Zone

	Γ(000)		X(100)				L(111)			
Material	LO	TO	LO	TO	LA	TA	LO	TO	LA	TA
GaAs	35.8	33.0	35.4	30.9	27.8	9.7	29.2	32.1	25.6	7.6
InAs	29.9	26.9	20.0	26.2	18.0	13.5	23.7	26.6	18.0	9.0
InSb 24.5	22.0	16.0	21.7	14.7	4.9	19.6	20.9	12.3	4.1	—
AlSb	24.2	22.7	—	—	36.4	9.8	28.6	37.6	31.4	7.6
AlAs	49.5	44.2	—	—	12.7	12.7	—	—	—	9.9
GaSb	29.9	28.2	—	—	—	—	—	—	—	—
Si	63.8	63.8	50.3	56.9	50.3	18.4	51.5	60.1	46.2	14.0
Ge	36.9	36.9	28.2	33.7	28.2	10.1	30.3	34.4	26.4	8.0

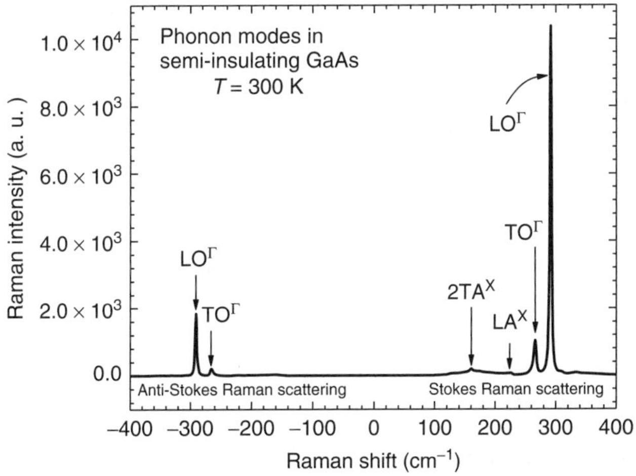

FIGURE G6 The Raman scattering spectrum obtained at 300 K for a semi-insulating GaAs sample. The spectrum shows Stokes and anti-Stokes phonon modes.

$\hbar\omega_i \pm \hbar\omega$. The beam with the energy $\hbar\omega_i - \hbar\omega$ is called the *Stokes Raman scattering line*, and the beam with the energy $\hbar\omega_i + \hbar\omega$ is called the *anti-Stokes Raman scattering line*. These lines are shown in Fig. G6. The most important aspect of Raman scattering is that ω is independent of ω_i. The effect was predicted by Smekal and experimentally measured by Raman. Raman scattering is considered as an inelastic scattering of light in which an internal form of motion (vibrational modes) of the scattering system is either excited or absorbed during the process.

Phonon modes in wurtzite structures such as GaN are more complicated than the phonon modes in diamond or zinc blende structures. For more discussion on the subject, see Manasreh *et al*. and Pattada *et al*. Raman spectroscopy is a very useful contactless technique in probing the charge carrier concentration in semiconductors. Charge carriers can be detected in Raman spectra through the coupling of the longitudinal optical phonon mode with plasma oscillations. The collective oscillation (plasmon) of an electron or hole gas in a solid is a longitudinal excitation, and its frequency, ω_p, can be written as

$$\omega_\text{p} = \sqrt{\frac{ne^2}{m^*\epsilon_0\epsilon_\infty}}, \tag{G.19}$$

where n is the charge carrier concentration, m^* is the charge carrier effective mass, ϵ_0 is the permittivity of space, and ϵ_∞ is the high frequency dielectric constant of the material, which is related to the refractive index, n_r, such that $\epsilon_\infty \approx n_r^2$.

A typical example of Raman scattering from the longitudinal phonon–plasmon coupled mode in a doped semiconductor quantum well is shown Fig. G7 for an InGaAsN/GaAs single quantum well grown by metalorganic chemical vapor deposition technique on semi-insulating GaAs substrate. The macroscopic electric field of the plasma wave interacts with the polarization field associated with the longitudinal optical (LO) phonons in polar semiconductors, such as GaAs (zinc blende) and GaN (wurtzite) materials. This coupling splits the LO phonons into

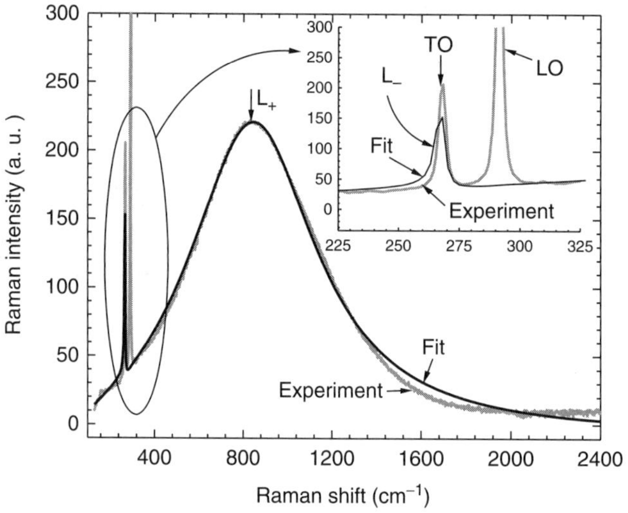

FIGURE G7 The Raman scattering spectrum obtained for an InGaAsN/GaAs single quantum well sample (gray line). The spectrum shows the LO, TO, and the L_+ branch of the LOPC mode. The solid black line is the result of the fitting analysis using Equations G.20 and G.21, which shows both the L_+ and L_- branches of the LOPC mode. The inset is the expansion of the spectral region in the vicinity of LO and TO phonon modes.

two longitudinal optical-plasmon coupled (LOPC) modes, known as L_+ and L_-. The low frequency mode, L_-, shifts from $0\,\text{cm}^{-1}$ to the transverse optical (TO) frequency, while the high frequency mode, L_+, shifts from LO frequency to the plasma frequency, ω_p, for increasing carrier concentration (Mooradian *et al*. and Abstreiter *et al*.).

The Raman intensity (I_s) is proportional to the imaginary part of the inverse of the total dielectric function (Manasreh 2002)

$$I_s \propto \text{Im}\left(-\frac{1}{\epsilon(\omega)}\right), \tag{G.20}$$

where the dielectric function contains the contribution from lattice vibration and the conduction electrons, which is given by

$$\epsilon(\omega) = \epsilon_\infty\left[1 + \frac{\omega_{\text{LO}}^2 - \omega_{\text{TO}}^2}{\omega_{\text{TO}}^2 - \omega^2 - i\omega\Gamma} - \frac{\omega_p^2}{\omega(\omega - i\gamma)}\right]. \tag{G.21}$$

The parameters Γ and γ are the damping constants of the phonon and plasmon, respectively. The plasmon frequency, ω_p, is obtained by fitting the LOPC Raman spectrum using Equations G.20 and G.21, with ω_p, Γ, and γ as fitting parameters. An example is shown in Fig. G7, in which the gray line spectrum is the experimental result and the thin black line spectrum is the theoretical fit. The plasmon frequency in this case is obtained as $909\,\text{cm}^{-1}$. Using this value in Equation G.19, the carrier concentration is obtained as $7.56 \times 10^{18}\ \text{cm}^{-3}$ for $\epsilon_\infty = 12.25$ and $m^* = 0.067m_0$.

As mentioned above, the LOPC mode splits into two modes known as L_+ and L_- *branches*. These two branches are approximately obtained by setting $\Gamma = \gamma = 0$ and solving Equation G.21 for $\epsilon(\omega) = 0$, which yields

$$L_\pm = \frac{1}{2}\left[\left(\omega_L^2 + \omega_p^2\right) \pm \sqrt{\left(\omega_L^2 + \omega_p^2\right)^2 - 4\,\omega_T^2\,\omega_p^2}\right]^{1/2}. \tag{G.22}$$

The fitting analysis of the experimental spectrum in Fig. G7 reveals the presence of both L_+ and L_-. The L_- region along with the LO and TO phonon modes are replotted in the figure inset for clarity. The same fitting procedure was repeated for several InGaAsN/GaAs single quantum well samples with different nitrogen contents. The plasmon frequency, ω_p, was obtained for each sample by fitting the experimental spectra as described above. In addition, the frequency maximum of the L_+ branch was obtained directly from the experimental LOPC mode spectra. To compare the experimental results to the theoretical predictions, L_+ and L_- modes are plotted as a function of ω_p using Equation G.22, as shown in Fig. G8. The experimental data were plotted in this figure as solid squares. The dashed lines represent the LO and TO phonon modes in the quantum well.

The plasmon frequency is used to calculate the carrier concentration in a series of samples with different nitrogen content. The results are shown in Fig. G9 in

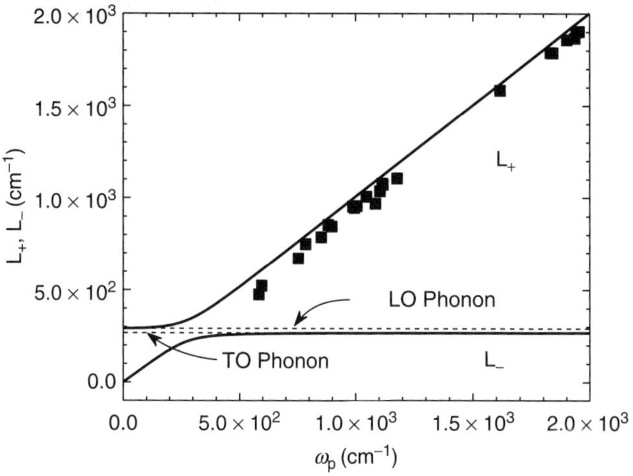

FIGURE G8 A plot of the L_+ mode as a function of the plasmon frequency for a series of InGaAsN/GaAs single quantum well samples (solid squares). The solid lines are plots of L_+ and L_- given by Equation G.22. The dashed lines represent the LO and TO phonon frequencies in InGaAsN quantum well.

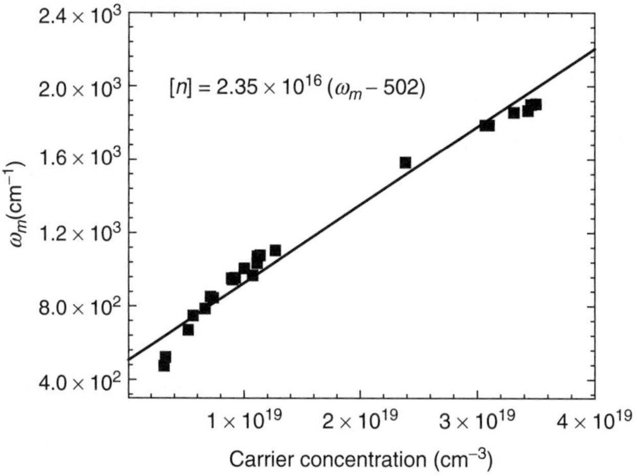

FIGURE G9 The frequency maximum, ω_m, of the L_+ branch as a function of the carrier concentration obtained from the data in Fig. G7. The solid line is a first-order linear fit of the data.

which the frequency maximum of the L_+ branch is plotted as a function of the calculated carrier concentration. The solid line in this figure is the result of the linear fit of the data from which the following empirical expression is obtained: $[n] = 2.35 \times 10^{16}(\omega_m - 502)$ cm^{-3}, where $[n]$ is the carrier concentration. This expression can be used to obtain the carrier concentration directly from the peak of L_+ mode, which is measured directly by Raman scattering in unit of cm^{-1}.

In addition to the determination of the carrier concentration using Raman scattering, the plasmon damping rate, γ, which is used in the fitting analysis, can be used to calculate the carrier drift mobility, μ, through the following relation: $\mu = e/(m * \gamma)$. The drift mobility values estimated from the plasmon damping rate are of the order of 100–200 $cm^2/V/s$ which are in good agreement with those reported by Young *et al*. Even though the carrier concentration and drift mobilities are estimated from fitting a simple model based on the Drude theory to Raman scattering spectra, the results provide a good indication of the material quality and its feasibility for device application.

BIBLIOGRAPHY

Birman JL. Theory of crystal space groups and lattice dynamics. Berlin: Springer; 1984.

Smekal A. Naturwissensch 1923;11:873.

Raman CV. Ind J Phys 1928;2:387.

Manasreh MO, Jiang HH, editors. Volume 13, III-nitride semiconductors: optical properties I. New York: Taylor and Francis; 2002.

Pattada B, Chen J, Manasreh MO, Guo S, Gotthold D, Pophristic M, Peres B. J Appl Phys 2003;93:5824.

Mooradian A, Wright GB. Phys Prev Lett 1966;16:999.

Absteiter G, Bauser E, Fischer A, Ploog K. Appl Phys 1978;A 16:345.

APPENDIX H

TUNNELING THROUGH POTENTIAL BARRIERS

The tunneling phenomenon was briefly discussed in Chapter 2 for a single potential barrier, a single potential well, and a double barrier structure. This effect was first reported by Esaki in a narrow germanium p–n junction (Esaki).At present, there are many devices based on the tunneling effect such as resonant tunneling diodes, point contact diodes, Schottky diodes, bipolar transistors, and field-effect transistors. One important aspect of the tunneling effect is that the tunneling time of carriers is proportional to the function $\exp(-2\rho L)$, where ρ is the decaying wave vector inside the barrier and L is the width of the potential barrier. The wave functions of the tunneling carriers are characterized as propagating waves in the wells and evanescent waves inside the barriers.

In Chapter 2, we have shown the form of the transmission coefficient for a particle tunneling through a rectangular barrier. When the product of the decaying wave vector and the width of the barrier is much larger than unity, the transmission coefficient is approximated by Equation 2.94. For a potential barrier with an arbitrary shape depicted in Fig. H1 (solid line), the exact derivation of the transmission coefficient becomes more complicated. It can be obtained with the help of approximation methods. For example, the Wentzel-Kramers-Brillouin (WKB) method becomes very handy in obtaining an approximate form of the tunneling probability.

The method of obtaining an approximate expression for the transmission coefficient in case of an arbitrary potential barrier is shown in Fig. H1. As opposed to the rectangular potential barriers shown as the dashed line in the figure, one can consider the barrier as being composed of small rectangular segments as shown in the figure. The form of the transmission coefficient for each segment is identical to the expression presented in Equation 2.94. The total transmission coefficient

Introduction to Nanomaterials and Devices, First Edition. Omar Manasreh.

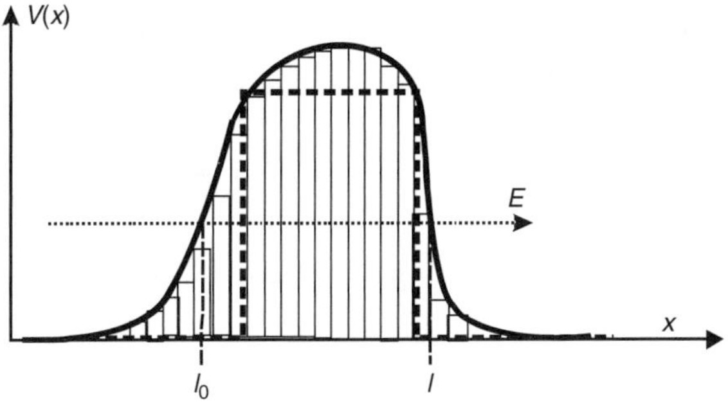

FIGURE H1 An arbitrary potential barrier divided into many small rectangular segments and overlapped on a rectangular potential barrier (dashed line) for comparison.

can now be approximated by the product of the transmission coefficients of the segments as

$$T(E) = T_1(E)T_2(E)T_3(E)\ldots T_n(E) = T_0 \mathrm{e}^{\sum_n -2\rho d_n}, \tag{H.1}$$

where the transmission through the nth segment of width d_n is $T_n(E) \propto \mathrm{e}^{-2\rho d_n}$. The number of segments can be made large enough so that the integration can be used instead of summation as follows:

$$T(E) \simeq T_0 \exp\left(-2\int_{l_0}^{l} \rho(x)\mathrm{d}x\right), \tag{H.2}$$

where the integration limits (l and l_0) correspond to the turning points, as shown in the figure, and $\rho(x)$ is the wave vector inside the barrier. We encountered the form of this transmission coefficient when we introduced the WKB approximation method discussed in Appendix D.

Let us first illustrate the transmission coefficient for an electron traveling through a δ-function barrier. This simple example is very important since the atoms in semiconductors or even small quantum dots can be seen by a traveling particle, such as an electron, as the δ-function potential. This approximation is very useful in understanding the basic idea of the tunneling through potentials. Let us assume that the δ-function potential barrier is represented by Fig. 2.2 and it is defined as $V(x) = \lambda\delta(x)$, where λ is the strength of the δ-function. The wave functions can be written as

$$\begin{aligned} \psi_{\mathrm{I}}(x) &= \mathrm{e}^{ikx} + A\mathrm{e}^{-ikx} \quad &\text{for } x < 0 \\ \psi_{\mathrm{II}}(x) &= B\mathrm{e}^{ikx} \quad &\text{for } x > 0, \end{aligned} \tag{H.3}$$

where $k = \sqrt{2mE/\hbar^2}$. The boundary conditions for δ-function potentials were discussed in Chapter 2, Section 2.11, and are given by

$$\psi_{\mathrm{I}}(0) = \psi_{\mathrm{II}}(0); \text{ and } \psi'_{\mathrm{II}}(0) - \psi'_{\mathrm{I}}(0) = \frac{2m\lambda}{\hbar^2}\psi_{\mathrm{II}}(0), \tag{H.4}$$

where m is the mass of the traveling particle. These boundary conditions yield a transmission coefficient of

$$T(k) = |B|^2 = \frac{k^2\hbar^4}{k^2\hbar^4 + m^2\lambda^2}. \tag{H.5}$$

This simple result shows that the transmission probability is unity as $\lambda \to 0$ and decreases drastically as the strength of the δ-function, depicted in λ, is increased. The transmission coefficient behavior as a function of λ^2 is shown in Fig. H2.

Another example of potential barriers is that when two semiconductor materials with different band gaps are grown to form a heterojunction, a band gap offset in both the conduction and valence bands is formed. In the absence of band bending, the conduction band discontinuity can be presented as in Fig. H3. The potential form in this figure can be obtained if the material with larger band gap is graded. For example, the potential profile can be formed by growing $Al_xGa_{1-x}As$ on $GaAs$, where x is incrementally varied from 0.3 to 0 during growth. Band bending usually occurs when there is a separation of charges at the modulation-doped heterojunction interface, as shown in Fig. 2.16. According to the WKB method (Appendix D), the transmission coefficient $[T(E)]$ is given by

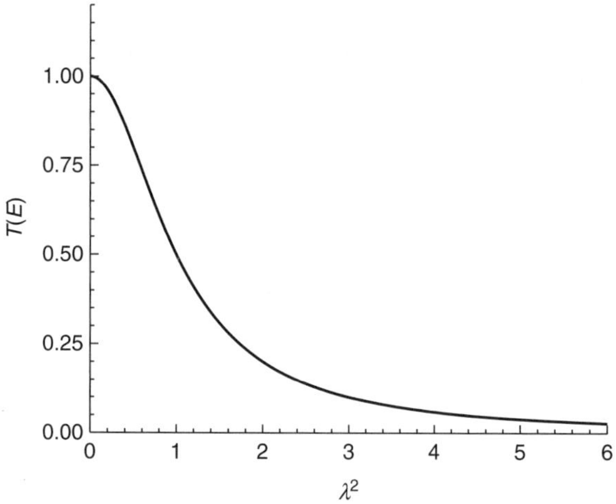

FIGURE H2 The transmission coefficient plotted as a function of λ for a particle traveling through a potential barrier with a strength λ.

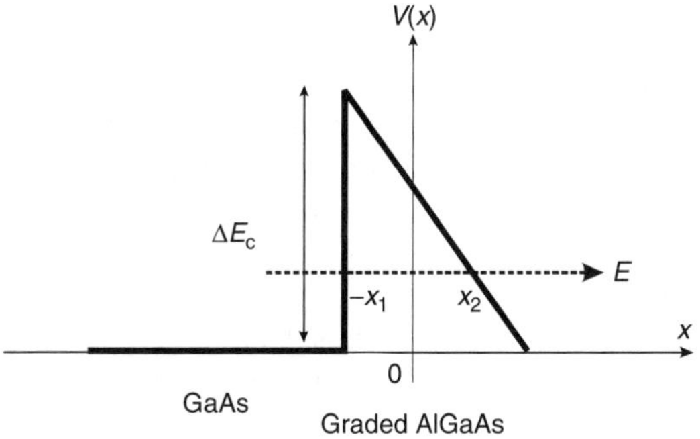

FIGURE H3 GaAs/AlGaAs heterojunction is plotted in the absence of band bending. The turning points are labeled $-x_1$ and x_2.

the following form:

$$T(E) \simeq \exp\left[-2\int_{-x_1}^{x_2} |k(x)|\mathrm{d}x\right], \tag{H.6}$$

where $-x_1$ and x_2 are the turning point and

$$k(x) = \sqrt{\frac{2m^*}{\hbar^2}(\Delta E_\mathrm{c} - e\mathcal{E}x - E)} \tag{H.7}$$

is the wave vector of an electron traveling from left to right with a kinetic energy E. Here ΔE_c is the conduction band offset and $V(x) = (\Delta E_c - e\mathcal{E}x)$ is the form of the graded potential. Notice that $e\mathcal{E}$ is the slope of the potential, which can be thought of as band bending due to an applied electric field $\mathcal{E}$. As a matter of fact, all the band structures in the real space exhibit a band bending similar to the potential profile shown in Fig. H3. The kinetic energy, E, of the particle is determined according to the WKB approximation from the following relation: $V(x_2 - x_1) = E$. Substitute Equation H.7 into Equation H.6 and perform the integration to obtain the following expression:

$$\begin{aligned} T(E) &\simeq \exp\left[-2\int_{-x_1}^{x_2}\left|\sqrt{\frac{2m^*}{\hbar^2}(\Delta E_\mathrm{c} - E - e\mathcal{E}x)}\right|\mathrm{d}x\right] \\ &\simeq \exp\left(\left(\frac{2m^*}{\hbar^2}\right)^{1/2}\left(\frac{4}{3e\mathcal{E}}(\Delta E_\mathrm{c} - E - e\mathcal{E}x)^{3/2}\right)\Bigg|_{-x_1}^{x_2}\right) \\ &\simeq \exp\left(-\frac{4}{3}\frac{\sqrt{2\mathrm{m}^*}}{\hbar e\mathcal{E}}(\Delta E_\mathrm{c} - E)^{3/2}\right). \end{aligned} \tag{H.8}$$

Notice that at $x = x_2$, we can approximate $(\Delta E_\mathrm{c} - E - e\mathcal{E}x) = 0$, and at $x = -x_1$, we have $(\Delta E_\mathrm{c} - E - e\mathcal{E}x) = (E_\mathrm{c} - E)$. The transmission coefficient is plotted in Fig. H4 in units of $\frac{4}{3}\frac{\sqrt{2m^*}}{\hbar e\mathcal{E}}$, and the band offset is taken as $\Delta E_c = 0.3$ eV. The transmission coefficient approaches unity when the energy of the electron reaches the value of the band offset as shown in the figure.

Another well-known barrier is the Schottky barrier formed between a metal and a semiconductor in contact, as shown in Fig. H5. The depletion region is formed near the interfaces where a build-in electric field exists because of the separation of charges. The Fermi energy (E_F) is pinned at an energy ($E_\varphi = eV_d$) below the conduction band of the semiconductor. The depletion potential (V_d) can be obtained as follows, assuming that the semiconductor is uniformly doped:

$$V_d = \int_0^d \mathcal{E}\mathrm{d}x = \int_0^d \frac{eN_d x}{\epsilon_0\epsilon_r}\mathrm{d}x = \frac{eN_d d^2}{2\epsilon_0\epsilon_r}, \tag{H.9}$$

where $\mathcal{E}$ is the build-in electric field given by Gauss's law as $\mathcal{E} = eN_d/\epsilon_0\epsilon_r$, N_d is the number of electrons transferred from the ionized donor atoms, ϵ_0 is the permittivity of space, and ϵ_r is the dielectric constant of the semiconductor. For n-type GaAs, E_ϕ is about 0.7 eV, with little variation for different metals.

If one assumes that the barrier has the following form: $V(x) = E_\phi[1 - (x/d)^2]$, the transmission coefficient of an electron traveling with an energy equal to the Fermi energy can be written as

$$\begin{aligned} T &\approx \exp\left(-\frac{2}{\hbar}\int_0^d \sqrt{2m^* E_\varphi\left(1-\left(\frac{x}{d}\right)^2\right)}\mathrm{d}x\right) \\ &= \exp\left(-\frac{\pi d}{2\hbar}\sqrt{2m^* E_\varphi}\right) = \exp\left(-\frac{d}{l}\right), \end{aligned} \tag{H.10}$$

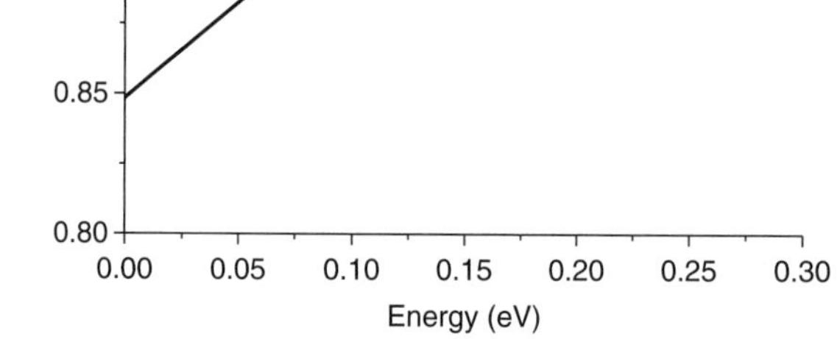

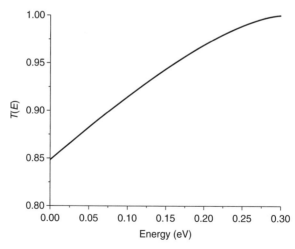

FIGURE H4 The transmission coefficient plotted as a function of the particle (electron) energy, E.

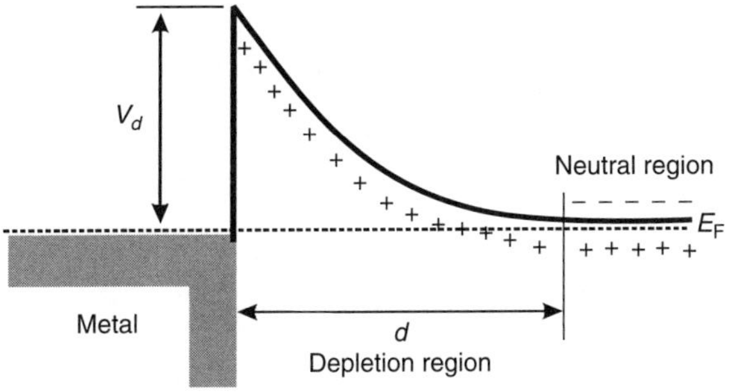

FIGURE H5 Schottky barrier formed between metal and n-type semiconductor, such as GaAs.

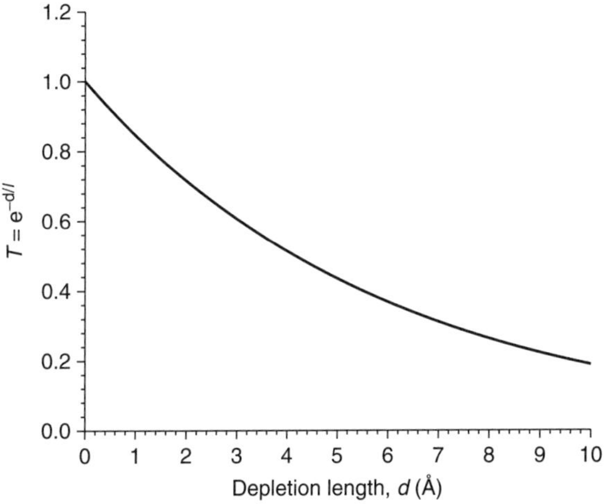

FIGURE H6 The transmission coefficient obtained for a Schottky potential barrier profile is plotted as a function of the depletion length inside the semiconductor, such as GaAs.

where l is defined as the decaying length and is given as $l = \sqrt{2\hbar^2/(\pi^2 m^* E_\varphi)}$, which is approximately 6.0 Å for GaAs. The transmission coefficient in Equation H.5 is plotted as a function of the depletion length, d, for $l = 6.0$ Å as shown in Fig. H6. It is clear from this figure that the transmission coefficient approaches unity when the depletion length is very small. This requires that the semiconductor material be heavily doped.

BIBLIOGRAPHY

Esaki L. New phenomenon in narrow germanium p-n junction. Phys Rev 1958;109:603.

INDEX

Introduction to Nanomaterials and Devices, First Edition. Omar Manasreh.